U.S. Department of the Interior
U.S. Geological Survey

Field Manual of Wildlife Diseases

General Field Procedures and Diseases of Birds

Biological Resources Division
Information and Technology Report 1999–001

Milton Friend and J. Christian Franson, Technical Editors

Elizabeth A. Ciganovich, Editor
Phillip J. Redman, Design and layout
Rosemary S. Stenback, Illustrator

U.S. Department of the Interior
Bruce Babbitt, Secretary

U.S. Geological Survey
Charles G. Groat, Director

Major funding support was provided by the U.S. Fish and Wildlife Service, Division of Federal Aid, Administrative Grant No. AP95-017.

Washington, D.C.

Any use of trade, product, or firm names in this publication is for descriptive purposes only and does not imply endorsement by the U.S. Government.

To order copies of this book telephone the Superintendent of Documents Telephone Order Desk at 202-512-1800 Monday through Friday from 7:30 a.m. to 4:30 p.m. Eastern time. Or visit http://www.gpo.gov/sales and enter the title in the Sales Product Catalog search box.

For additional information about this book:

USGS
Biological Resources Division
National Wildlife Health Center
6006 Schroeder Road
Madison, WI 53711
U.S.A.
World Wide Web: http://www.emtc.usgs.gov/nwhchome.html

For more information about the USGS and its products:

Telephone: 1-888-ASK-USGS
World Wide Web: http://www.usgs.gov

Library of Congress Cataloging-in-Publication Data
Field manual of wildlife diseases : general field procedures and
    diseases of birds / Biological Resources Division.
        p. cm. — (Information and technology report ; 1999-001)
    Includes bibliographic references

    1. Birds–Diseases Handbooks, manuals. 2. Wildlife diseases
Handbooks, manuals, etc. I. Geological Survey (U.S.). Biological
Resources Division. II. Series.
SF994.F54 1999
639.9'78—dc21   99-25869
                                                     CIP

Dedication

We dedicate this Manual to the countless field biologists within the U.S. Fish and Wildlife Service and the State wildlife agencies with whom we have had the privilege of working for nearly a quarter-century. Their endless assistance and devotion to the conservation of our Nation's wildlife resources has stimulated our own efforts to address wildlife health issues and made those efforts more rewarding than we originally believed was possible. We thank you for your efforts and hope that the material provided within these pages will be useful to you in the days ahead.

Milton Friend and J. Christian Franson

Acknowledgements

This publication was made possible through the generous contributions of time and effort by many individuals, and we offer our sincere thanks to all. The following people are particularly deserving of recognition. Ms. Debra Ackers endured much from us in typing draft manuscripts, making countless adjustments to the drafts, organizing and maintaining project files, tracking the progress of the many components of the Manual, arranging meetings among individuals involved with this project, and assisting in many other ways — always with a smile. Her contributions were invaluable. Dr. Louis N. Locke deserves special recognition for his technical input and laborious review of the draft manuscript. We thank Dr. James R. March and Ms. Barbara C. Scudder for contributing their time to read and comment on the final draft. Mr. Scott Hansen assembled the lists of chemicals and birds for Appendices E and G, respectively, and assisted us in locating many of the photographs used to illustrate the Manual. Ms. Kathryn Cleary, Ms. Karen Cunningham, and Mr. Harold Rihn provided tabulations of information from the National Wildlife Health Center database that were used in many of the graphics and tables. We thank the following individuals for helpful and timely reviews of various parts of the Manual: Dr. Donald Anderson, Mr. Tom Augspurger, Dr. Val Beasley, Dr. Rebecca Cole, Dr. Guy Connolly, Mr. Terry Creekmore, Mr. Doug Docherty, Mr. Monte Garrett, Dr. Robert Hallock, Dr. Larry Hansen, Dr. Wallace Hansen, Dr. Tuula Hollmén, Dr. David Jessup, Dr. Ken Langelier, Dr. Linda Lyon, Dr. Pierre Mineau, Dr. Patrick Redig, Dr. Milton Smith, Mr. Stanley Wiemeyer, and Dr. Thierry Work. We apologize in the event that we have failed to acknowledge some individuals who have assisted with this project over the extended period of time required for its completion.

Foreword

DO WILDLIFE DISEASES REALLY MATTER? The waterfowl manager who wakes up one morning to find ten thousand dead and dying birds in the marsh would think so. Yet virtually every wild bird and mammal harbors at least a few parasites seemingly without obvious adverse consequences. Parasites, viruses, bacteria, and fungi are component parts of the ecosystems in which wildlife are found, but do not necessarily cause disease. Millennia of coevolution have engendered a modus vivendi that assures the survival of both host and parasite populations.

Then why the ten thousand sick and dying birds? Ecosystems are changing. Waterfowl are concentrated on shrinking wetlands and remain there for longer periods of time, facilitating bird-to-bird spread of the bacteria that cause avian cholera. Or permitting the buildup of parasites in their hosts from a small, relatively benign number to massive numbers that cause disease and death. Water quality of wetlands changes, favoring the production of deadly botulinum toxin by bacteria and its mobilization up the food chain to waterfowl. New, totally artificial habitats are created with unpredictable results. The extreme temperature, salinity, and other conditions of the Salton Sea have created an unusual ecosystem in which botulism occurs in fish and in birds through biological cycles that are not yet understood. Wetland loss in southern California leaves few alternative places for waterbirds to go, so they are attracted to the Salton Sea.

Behavior changes. Mallard ducks take up residence on the ponds and lakes of city parks and lose their migratory habits. They share these bodies of water with exotic species, such as Muscovy ducks that have also taken up residence there after introduction by people, setting the scene for outbreaks of duck plague, and creating the risk of spread to migratory waterfowl that also use these areas. Raccoons and skunks become well adapted to urban life, bringing rabies and canine distemper with them into the city.

The environment changes the physiology of wild animals. Human activity introduces into wildlife habitats chemical compounds that adversely affect physiological processes such as reproduction and immune responsiveness. These compounds become incorporated into the ecosystems, often becoming more concentrated as they move up food chains. Their effects can influence wildlife populations. Some of these endocrine-disrupting chemicals, such as chlorinated hydrocarbons (DDE, PCBs), interfere with normal endocrine function by mimicking natural hormones, with resulting eggshell thinning and breakage. Effects of these chemical compounds on immune-system responses to infectious and parasitic agents are less well understood.

What to do? Incorporating disease-prevention measures into wildlife management practices requires more information than is usually available. The information-gathering process must begin in the field. Field biologists must monitor disease occurrence. This Field Manual is a valuable aid in identifying the diseases that are likely to be present, and in giving guidance on the gathering and treatment of specimens needed to establish the diagnosis in the laboratory.

But the wildlife field biologist is in a position to provide valuable information that goes beyond the collection of samples from sick and dead individuals. Although diseased individuals are the basic unit of surveillance, the occurrence of disease must be put into ecological perspective. A careful description of the ecological setting in which the disease is occurring, and any changes that have occurred over time, are ultimately as important as a careful description of the lesions observed in the individual, if the epidemiology of that disease is to be understood, and the disease prevented through sound wildlife-management practices.

It is my hope that the awareness of diseases affecting wildlife and the good disease-surveillance practices promoted by this manual will spread throughout the range of the species we are trying to mange and protect. We must know more than we do currently about disease occurrence throughout the ranges that the wildlife occupy. Many migratory species know nothing of international boundaries. Neither do their diseases. Until we have a much more complete picture of the disease-environment relationships of the blue-winged teal from its nesting ground in Canada, its migration route through the United States and overwintering areas in Central America or the Cienaga Grande de Santa Marta in Columbia, sound disease-prevention management of that species will not be possible. Similar considerations exist for other species.

Thomas M. Yuill
Madison, Wisconsin
May, 1999

Photo by J.Christian Franson

"Ingenuity, knowledge, and organization alter but cannot cancel humanities vulnerability to invasion by parasitic forms of life. Infectious diseases which antedated the emergence of humankind will last as long as humanity itself, and will surely remain, as has been hitherto, one of the fundamental parameters and determinants of human history."

(McNeill)

Introduction

"When one comes into a city in which he is a stranger, he ought to consider its situation, how it lies as to the winds and the rising of the sun; for its influence is not the same whether it lies to the north or to the south, to the rising or to the setting sun. These things one ought to consider most attentively, and concerning the waters which the inhabitants use, whether they be marshy and soft, or hard and running from elevated and rocky situations, and then if saltish and unfit for cooking; and the ground, whether it be naked and deficient in water, or wooded and well-watered, and whether it lies in a hollow, confined situation, or is elevated and cold ...From these things he must proceed to investigate everything else. For if one knows all these things well, or at least the greater part of them, he cannot miss knowing, when he comes into a strange city, either the diseases peculiar to the place, or the particular nature of the common diseases, so that he will not be in doubt as to the treatment of the diseases, or commit mistakes, as is likely to be the case provided one had not previously considered these matters. And in particular, as the season and year advances, he can tell what epidemic disease will attack the city, either in the summer or the winter, and what each individual will be in danger of experiencing from the change of regimen."

—Hippocrates, On Airs, Water, and Places, c. 400 B.C.

I was first employed in the field of wildlife conservation in 1956 as an assistant waterfowl biologist. Had I decided then to join some of my colleagues in preparing a manual about the diseases of wild birds similar to this publication, the task would have been much simpler. The number of chapters needed would have been far less because some of the diseases described in this Manual were not yet known to exist in free-ranging North American birds or, if they were known, they were not considered to be of much importance. This is especially true for diseases caused by viruses; also, organophosphorus and carbamate pesticides had not come into wide use. These types of differences are evident between this *Field Manual of Wildlife Disease — General Field Procedures and Diseases of Birds* and the *Field Guide to Wildlife Diseases — General Field Procedures and Diseases of Migratory Birds* that was published little more than a decade ago. The current Manual reflects both expanded knowledge about avian diseases and an increase in both the occurrence of disease in wild birds and the variety of agents responsible for illness and death of wild birds.

Landscape changes and environmental conditions that are related to them are a major factor associated with disease occurrence in wild birds. The direct association between environment and human health has been recognized since ancient times and was aptly stated by Louis Pasteur, "The microbe is nothing; the terrain everything." Despite this well documented relationship, which serves as a basic foundation for addressing many human and domestic animal diseases, there has been little consideration of "the terrain" as a factor for diseases of wild birds. We must learn to "read the terrain" in a manner similar to the teaching of Hippocrates and apply that knowledge to disease prevention or else the next edition of this Manual a decade from now will likely include another major expansion in the number of diseases being addressed.

Although this Manual is much larger than the 1987 *General Field Procedures and Diseases of Migratory Birds* the basic format and "terrain" approach of the previous publication were retained because of the positive comments that were received from its users. The format, the photographs previously used, and most of Section 1, General Field Procedures, have been basically retained, but the text for chapters about individual diseases (Sections 2 through 8) has been extensively reworked. This Manual also has separate sections that address biotoxins and chemical toxins in addition to major expansion of the number of individual diseases within the sections on bacterial, fungal, viral, and parasitic diseases. The presentations in the various sections have

---

Facing page quote from:

McNeill, W.H., 1976, Plagues and peoples: Anchor Press/ Doubleday, Garden City, N.Y., p. 291

been supplemented with introductory comments regarding the subject area, and most sections have been highlighted with descriptions of miscellaneous disease conditions that may interest users and readers.

As with the 1987 publication, the focus of this Manual is on conveying practical information and insights about the diseases in a manner that will help National Wildlife Refuge managers and other field personnel address wildlife health issues at the field level. The information represents a composite of our understanding of the scientific literature, of our personal experiences with and investigations of the various diseases, and of information generously provided by our colleagues within the wildlife disease and related fields. In presenting this information, we have borrowed freely from all of those sources. Because this is a synoptic field manual and not a textbook, literature citations are not provided in support of statements. Only a small portion of the specific literature that is the basis for the statements has been listed, and the supplementary reading lists are intended to provide entry into the scientific literature for more precise evaluation of specific topics.

The need to generalize and, thus, provide a practical overview of complex biological situations often results in a loss of precision for some information. We have attempted to provide detail where it is of significant importance and have been more general elsewhere. In all cases, we have attempted to represent the information objectively and accurately. For example, Appendix E presents specific brain cholinesterase values that are supported by laboratory data for different bird species to provide a baseline against which others can make judgements about mortality due to organophosphorus and carbamate pesticides. In contrast, representation of the geographic distribution, frequency of occurrence, and species susceptibility associated with specific diseases is of a general nature and is intended only for gross comparison. The differences in these representations of general information between the 1987 publication and this Manual are both a positive and a negative outcome of the last decade. These differences reflect enhanced information about disease in wild birds as a result of expanded study (a positive outcome), changes in disease patterns (a negative outcome due to expansion of disease), and both, depending on the disease.

Current understanding about wild bird diseases is being provided by those with technical knowledge about disease processes to those with technical knowledge and stewardship and conservation of our wild bird resources. Common language has been used whenever possible to aid in this communication and to stimulate greater interest in wildlife disease among others who may wish to read this Manual but who may not be familiar with some of the terms. Technical terms have been translated in a manner that we hope will be useful for readers as they pursue additional subject matter detail in the scientific literature. Technical terms have also been inserted into the text and defined where they provide value-added precision for the statements. It is my personal hope that a decade from now, when consideration is being given to a revision of this Manual, that a great deal of the preparation of the revision will be done by wildlife biologists who have become practitioners in the art of disease prevention and control because of an enhanced understanding of disease ecology that we have all gained through our collective efforts. The transition hoped for is no greater than other changes that have taken place since the 1987 publication of the original Field Guide. At that time, the National Wildlife Health Center (NWHC) was part of the Department of the Interior, U.S. Fish and Wildlife Service. Since then, the Center has become part of the Department of the Interior, U.S. Geological Survey, Biological Resources Division.

My professional situation has also changed. Those familiar with the 1987 publication will note that I was Director of the NWHC when that publication became available. In December 1997, Secretary of the Interior Bruce Babbitt asked me to accept the challenge of coordinating the science efforts that will aid and guide decisions for management actions to improve the health of the Salton Sea, California's largest inland body of water. Recurring major disease events involving migratory birds at the Sea since 1994 have focused public attention on it. These disease events became a catalyst for the expansion of efforts to improve the environmental quality of the Sea, and in June 1998, a combined National Environmental Policy Act (NEPA)/California Environmental Quality Act (CEQA) process was initiated to pursue attainment of that goal. I officially became part of the multiagency effort to "Save the Salton Sea" with my reassignment in April 1998 from Director of the NWHC to Executive Director, Salton Sea Science Subcommittee.

Milton Friend

Table of Contents

Section 1 General Field Procedures

- Chapter 1 Recording and Submitting Specimen History Data 3
- Chapter 2 Specimen Collection and Preservation 7
- Chapter 3 Specimen Shipment 13
- Chapter 4 Disease Control Operations 19
- Chapter 5 Euthanasia 49
- Chapter 6 Guidelines for Proper Care and Use of Wildlife in Field Research 53

Section 2 Bacterial Diseases

- Chapter 7 Avian Cholera 75
- Chapter 8 Avian Tuberculosis 93
- Chapter 9 Salmonellosis 99
- Chapter 10 Chlamydiosis 111
- Chapter 11 Mycoplasmosis 115
- Chapter 12 Miscellaneous Bacterial Diseases 121

Section 3 Fungal Diseases

- Chapter 13 Aspergillosis 129
- Chapter 14 Candidiasis 135
- Chapter 15 Miscellaneous Fungal Diseases 137

Section 4 Viral Diseases

- Chapter 16 Duck Plague 141
- Chapter 17 Inclusion Body Disease of Cranes 153
- Chapter 18 Miscellaneous Herpesviruses of Birds 157
- Chapter 19 Avian Pox 163
- Chapter 20 Eastern Equine Encephalomyelitis 171
- Chapter 21 Newcastle Disease 175

Chapter 22 Avian Influenza 181

Chapter 23 Woodcock Reovirus 185

Section 5 Parasitic Diseases

Chapter 24 Hemosporidiosis 193

Chapter 25 Trichomoniasis 201

Chapter 26 Intestinal Coccidiosis 207

Chapter 27 Renal Coccidiosis 215

Chapter 28 Sarcocystis 219

Chapter 29 Eustrongylidosis 223

Chapter 30 Tracheal Worms 229

Chapter 31 Heartworm of Swans and Geese 233

Chapter 32 Gizzard Worms 235

Chapter 33 Acanthocephaliasis 241

Chapter 34 Nasal Leeches 245

Chapter 35 Miscellaneous Parasitic Diseases 249

Section 6 Biotoxins

Chapter 36 Algal Toxins 263

Chapter 37 Mycotoxins 267

Chapter 38 Avian Botulism 271

Section 7 Chemical Toxins

Chapter 39 Organophosphorus and Carbamate Pesticides 287

Chapter 40 Chlorinated Hydrocarbon Insecticides 295

Chapter 41 Polychlorinated Biphenyls 303

Chapter 42 Oil 309

Chapter 43 Lead 317

Chapter 44 Selenium 335

Chapter 45 Mercury 337

Chapter 46 Cyanide 341

Chapter 47 Salt 347

Chapter 48 Barbiturates 349

Chapter 49 Miscellaneous Chemical Toxins 351

Section 8   Miscellaneous Diseases

Chapter 50 Electrocution 357

Chapter 51 Miscellaneous Diseases 361

Appendices

A Sample specimen history form 369

B Sources of wildlife disease diagnostic assistance 370

C Sources of supplies used for collecting, preserving, and shipping specimens 373

D Normal brain cholinesterase values for different bird species 375

E Common and scientific names of birds in text 377

F Common and scientific names other than birds 383

G Chemical names 384

H Conversion table 387

Glossary of technical terms 389

Index 400

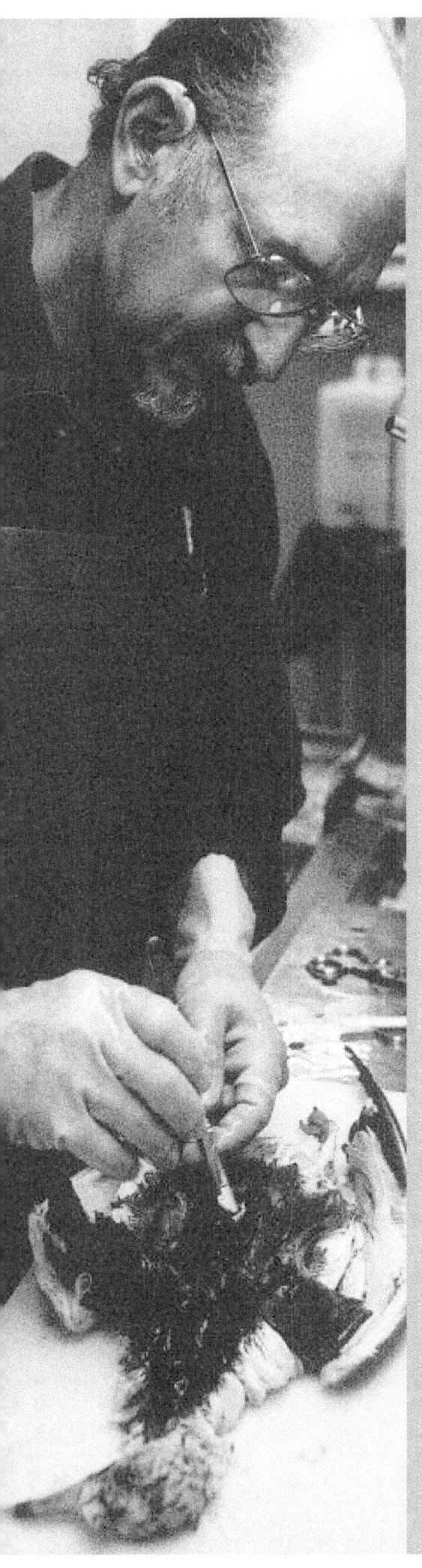

Section 1
Introduction to General Field Procedures

Recording and Submitting Specimen History Data

Specimen Collection and Preservation

Specimen Shipment

Disease Control Operations

Euthanasia

Guidelines for Proper Care and Use of Wildlife in Field Research

Dissecting a bird at the National Wildlife Health Center
Photo by Philip J. Redman

Introduction to General Field Procedures

"Given the conspicuous role that diseases have played, and in many parts of the world continue to play, in human demography, it is surprising that ecologists have given so little attention to the way diseases may affect the distribution and abundance of other animals and plants. Until recently, for example, ecology textbooks had chapters discussing how vertebrate and invertebrate predators may influence prey abundance, but in most cases you will search the index in vain for mention of infectious diseases." (May)

A basic premise for the preparation of this Manual is that disease in free-ranging wildlife is of concern and that disease prevention and control are desirable actions. However, these are not universally held perspectives. There are those who when confronted with disease outbreaks in free-ranging wildlife ask — "Why bother?" Also, the same individuals who may reject the need for response to one situation may demand a response to another situation. We acknowledge in this Manual the existence of this question by making reference to it, but we do not offer a direct response. To do so would require this Manual to address the full spectrum of individually held values, perspectives, interests, and beliefs within human society that form the basis for the underlying issues which create the question of "why bother?" Those factors would also need to be addressed within a context of the different roles and responsibilities of public agencies, and would need to include some additional considerations. Such an undertaking is outside the scope and purpose of this Manual. Although no direct response is offered, readers will gain considerable information regarding disease occurrence and impacts in the chapters that follow. This information should be of value in assisting readers to address the questions of "why bother?" from their own set of values and interests.

Section 1 of the Manual provides basic information regarding general field procedures for responding to wildlife disease events. Field biologists provide a critical linkage in disease diagnostic work and greatly affect the outcome of the laboratory efforts by the quality of the materials and information that they provide. The chapters in this section are oriented towards providing guidance that will assist field biologists in gathering the quality of information and specimens that are needed. Readers will find information regarding what to record and how; guidance for specimen collection, preservation, and shipment; and how to apply euthanasia when such actions are warranted. Disease operations are managed at the field level and they can be aided by general preplanning that can be utilized when disease emergencies arise; therefore, contingency planning is included within the Disease Control Operations chapter. Disease control techniques, including equipment that is used, are the main focus for this highly illustrated chapter. Section 1 is concluded with a chapter about the proper care and use of wildlife in field research. The guidelines provided address the continual need to consider animal welfare in all aspects of wildlife management.

Quote from:

May, R.M., 1988, Conservation and disease: Conservation Biology, v. 2, no. 1, p. 28–30.

Chapter 1
Recording and Submitting Specimen History Data

History can be defined as a chronological record of significant events. In wildlife disease investigations, determining the history or background of a problem is the first significant step toward establishing a diagnosis. The diagnostic process is often greatly expedited by a thorough history accompanying specimens submitted for laboratory evaluation. This information is also important for understanding the natural history or epizootiology of disease outbreaks, and it is difficult, if not impossible, to obtain the history after the outbreak has occurred. Detailed field observations during the course of a die-off and an investigation of significant events preceding it also provide valuable information on which to base corrective actions. The most helpful information is that which is obtained at the time of the die-off event by a perceptive observer.

What Information Should Be Collected

What seems irrelevant in the field may be the key to a diagnosis; therefore, be as thorough as possible. Avoid preconceptions that limit the information collected and that may imperceptibly bias the investigation. A sample specimen history form, which lists some categories of information that are helpful, is in Appendix A. A good description of unusual behavior or appearance, if any, an accurate list of what species were affected, and the number of animals that died are critical pieces of information. Send specimens and the written history to the laboratory as soon as possible. Photographs can be helpful if they convey specific information, such as environmental conditions during a die-off and the appearance of sick wildlife or gross lesions (Figs. 1.1, 1.2).

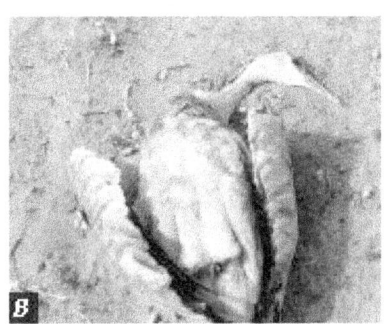

Figure 1.1 Examples of poor and good photography to record environmental conditions associated with wildlife disease problems. *(A)* Landscape photo displays topography and presence of a power line that may or may not be involved with the mortality event. Neither of the major factors involved with this event can be clearly seen. *(B)* Closeup photograph clearly shows both the species involved and the peanuts that proved to be contaminated with the mycotoxins that were the source of the problem. *(C)* Closeup photograph of sick bird clearly illustrates clinical signs of wing and neck droop; and the snow indicates the season.

Figure 1.2 The observer may use photography to illustrate field observations associated with wildlife morbidity and mortality. (A) For example, when sick birds are left undisturbed or approached quietly, they often remain motionless along the water's edge with their heads hanging down. When startled, these birds may attempt to escape by propelling themselves with their wings across water (B) or land (C) but are unable to fly. (D) This bird has lost the use of its legs, a common occurrence with avian botulism and certain toxins such as organophosphorus or carbamate compounds.

The following basic information is helpful for diagnosing the cause and assessing the severity of a wildlife health problem. Waterfowl are used as an illustrative example.

Environmental Factors

Determine if the start of mortality coincided with any unusual event. Environmental changes such as storms, precipitation, and abrupt temperature changes are potential sources of stress that can contribute to disease outbreaks. A food shortage may degrade the condition of birds and increase their susceptibility to disease. Water-level changes in an area may concentrate or disperse birds, alter the accessibility of toxins in food or water, or cause an invertebrate die-off that could lead to an avian botulism outbreak. Attempt to determine whether or not biting insect populations have increased or if such insects are present, because some insects are carriers of blood-borne infections in waterfowl.

The quality of the water used as a source for an impoundment may contribute to disease or mortality, for example, poor water quality may contribute to avian botulism or may be a primary cause of mortality if water contamination by toxic materials and substances such as oil, which can affect the integrity of feathers, is severe. Record recent pesticide applications and other habitat or crop management practices as well as previous disease problems in the area.

Estimating Disease Onset

When estimating the onset of disease, consider: (1) the earliest date when on-site activities could have resulted in the detection of sick or dead birds, if they were present, and the actual date when diseased birds were first seen, and (2) the proportion of fresh carcasses compared with the number of scavenged and decomposed carcasses. The abundance and types of scavengers and predators can be used to predict how long carcasses remain in the area. Other useful information about the onset of mortality can be gained from noting any differences in plumage, including stage of molt, if present, between live and dead birds. Size differences between live and dead nestlings and fledglings may also provide useful information for comparison with known growth rates. Also, air, water, and soil temperatures will affect the speed of decomposition and they should be considered in assessing how long birds have been dead. Include these observations in the history.

Species Affected

Much can be learned by knowing what species are dying. Those species present but unaffected are especially important to note, because some diseases infect a narrow host range and others infect a wide variety of species. For example, duck plague affects only ducks, geese, and swans, but avian cholera affects many additional species of water birds as well. Species with similar feeding habits may be dying as a result of exposure to toxins, while birds with different food requirements remain unaffected.

Age

Some disease agents may kill young birds but leave adults unaffected because of age-related disease resistance; other diseases kill birds of all ages, although young or old birds may be more susceptible because of additional stress placed on these age groups. When toxins are involved, differences in food habits may result in exposure of young birds, but not of adult birds, or vice versa.

Sex

Sex differences in mortality may be apparent in colonial nesters where females are incubating eggs, or in other situations where the sexes are segregated.

Number Sick/Number Dead

The longer a disease takes to kill, the more likely it is that significant numbers of sick birds will be found. For example, more sick birds will probably be observed during an avian botulism die-off than during an outbreak of a more acute disease such as avian cholera.

Clinical Signs

When observing sick birds, describe the clinical signs in as much detail as possible. Include any abnormal physical features and describe unusual behaviors, such as a sick bird's response to being approached. Photographs (Fig. 1.2) of various behaviors or conditions associated with a disease can be especially useful and should be included with the history.

Population at Risk

Try to determine what species, and in what numbers, are in the vicinity of the die-off. This information can provide clues about the transmissibility of disease, and it may be useful during control efforts.

Population Movement

Record recent changes in the number of birds in the area, as well as the species present. In particular note the presence of endangered species. If bird numbers have increased, try to determine where they came from; if bird numbers have decreased, attempt to determine where they have gone. This can often be accomplished when population movements are being monitored for census, hunting forecasts, and other purposes. State, Federal, and private refuge personnel and other natural resources managers are good primary sources of information.

Specific Features of Problem Areas

Describe the location of a die-off so that a relatively specific area can be identified on a road map. Also include any available precise location data, such as global positioning information or data that will facilitate entering of specific locations into geographical information system databases. Describe the problem area in terms that are sufficiently graphic so that someone with no knowledge of it can visualize its major characteristics, such as topography, soil, vegetation, climate, water conditions, and animal and human use.

Example description of die-off location

The problem area is a 10-acre freshwater pond located in Teno County, North Carolina, 1/2 mile east of County KV, 5 miles north of Highway 43. The pond has an average water depth of 6–12 feet and a sandy substrate. Vegetation around the pond border is bullbrush and reed canary grass. The surrounding uplands are essentially flat for one-half mile in all directions and lie fallow, covered with grasses and some shrubs. The area is coastal with enough relief to prevent saltwater intrusion into the pond even during major storms. Weather for the past 2 weeks has been pleasant and there has been no precipitation. Daytime temperatures are currently in the mid-80s (°F) and evening temperatures in the 70s. This is an isolated body of freshwater with good clarity, and sustains several hundred waterfowl, gulls, and small numbers of wading birds and shorebirds, and healthy warm water fish and amphibian populations. Cattle graze the adjacent area. There are no residential or industrial buildings within 1 mile of the site. Human visitation is frequent for bird watching, fishing, and hiking. Companion animals such as dogs are allowed on the area.

Identify where sick and dead birds are found. Especially note the locations of groups of dead birds and any differences of habitat where dead and sick birds are found. Birds found in agricultural fields may be dying of pesticide exposure, birds with more chronic toxicoses usually seek dense cover, and birds dying of acute diseases may be found in a variety of situations. Check any relation between specific bird use of the area and the location of affected birds, such as roost sites, loafing areas, and feeding sites.

If followup investigations are conducted after specimens have been submitted, summarize the findings and observations of those investigations in a supplemental report to the original history. Maintain a copy of the new report in station files, and provide a copy to the diagnostic laboratory where the specimens were sent. Both reports should contain the dates of the investigations, whether air or ground searches were performed, the number of investigators and the time spent on the investigation, the weather conditions, and the time of day when the site was investigated.

The insight provided by good specimen history data and by field observations is invaluable to disease specialists. This information enhances understanding of the ecology of disease, thereby serving as a basis for developing ways to prevent future die-offs or to reduce the magnitude of losses that might otherwise occur.

J. Christian Franson

Supplementary Reading

Wobeser, G.A., 1994, Investigation and management of disease in wild animals: New York, N.Y., Plenum Press, 265 p.

Chapter 2
Specimen Collection and Preservation

Specimens are used to provide supporting information leading to the diagnosis of a cause of disease or death. A specimen may be an intact carcass, tissues removed from carcasses, parasites, ingested food, feces, or environmental samples. The specimen should be as fresh and undamaged as possible.

Choosing a Specimen

An entire, fresh carcass is the best specimen to submit to the laboratory for diagnosis. This allows the diagnostician to assess all of the organ systems and to use appropriate organs for different diagnostic tests. Obtain the best specimens possible for necropsy; decomposed or scavenged carcasses are usually of limited diagnostic value. A combination of sick animals, animals that were euthanized after clinical signs were observed and recorded, and some of the freshest available carcasses compose an ideal specimen collection. The method of euthanasia should not compromise the diagnostic value of the specimen (see Chapter 5, Euthanasia). More than one disease may be affecting the population simultaneously, and the chances of detecting multiple diseases will be maximized if both sick and dead animals are collected. Specimens submitted should be representative of the species involved. If more than one species is affected, collect several specimens of each species; try to obtain a minimum of five specimens per species.

Tissue Collection

The primary consideration when collecting carcasses or tissues for diagnosis should be personal safety. Some wildlife diseases are transmissible to humans, and every carcass should be treated as a potential health hazard. Wear disposable rubber or plastic gloves, coveralls, and rubber boots. If gloves are not available, inverted plastic bags may be used (Fig. 2.1). Before leaving an area where carcasses are being collected, double-bag used gloves and coveralls, and disinfect boots and the outside of plastic bags with a commercial disinfectant or a 5 percent solution of household chlorine bleach. Also, double-bag specimens in plastic before removing them from the area. These precautions will help protect the people in the field and minimize transmission of disease to unaffected wildlife populations.

If it is impossible to submit an entire carcass for diagnosis, appropriate organs must be removed from specimens. If possible, do not dissect carcasses in the field without first consulting disease specialists about methods of dissecting and preserving tissues or parasites or both. Assistance can be obtained from a variety of sources (Appendix B). It is

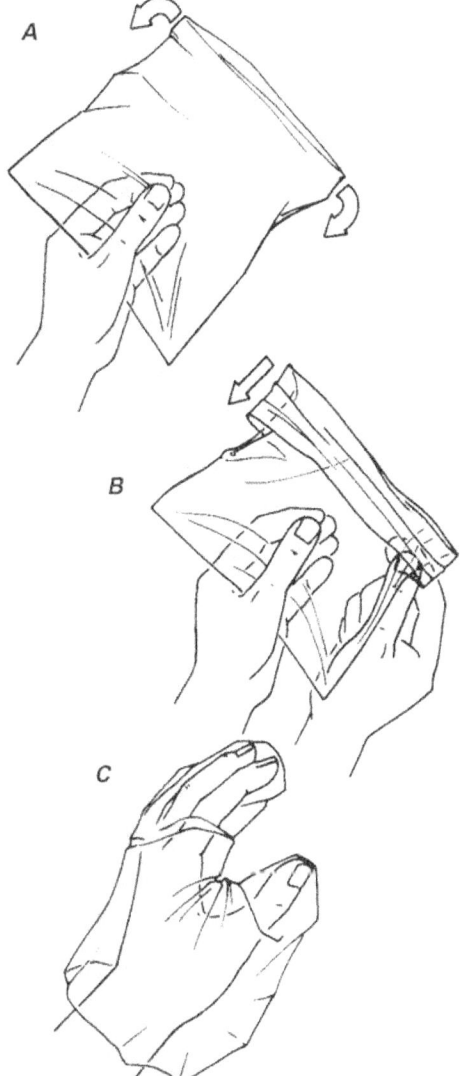

Figure 2.1 Use a plastic bag to protect hands from direct contact with animal tissues during the collection of specimens if plastic or other waterproof gloves are not available. **(A)** Grasp bag at the bottom and **(B)** with other hand pull open end down over hand holding bag **(C)**. Repeat for the "unbagged" hand. Reversing this process when handling small specimens will automatically place specimens in the bag, which then need only be sealed and put into a second bag for packaging and shipment.

best to become familiar with these sources and their ability to provide specific types of assistance before an emergency arises. The basic supplies and equipment that should be included in a field kit for specimen collection will vary with the species being sampled and the types of analyses that will be conducted. Keep a small kit packed in a day pack for ready use (Fig. 2.2). Sources of supplies used for collecting, preserving, labeling, and shipping specimens are listed in Appendix C.

Whirl-Pak® bags are very effective containers for tissue specimens. These bags have a sterile interior, are easy to carry in the field, and can be used to hold a variety of samples (Fig. 2.3). Specimen identification should be written directly on the bag with an indelible marker.

If lesions are noted, collect separate tissue samples for microscopic examination, microbiology, toxicology, and other analyses. With a sharp knife or scalpel cut a thin (1/8–1/4 inch, 3–6 millimeter) section of tissue that includes all or portions of the lesion and adjacent apparently healthy tissue (Fig. 2.4). Take care not to crush tissue in or around the lesion. Place the tissue sample in a volume of 10 percent buffered formalin solution equal to at least 10 times the tissue volume to ensure adequate preservation. Formalin is classified as hazardous; take appropriate measures to prevent skin contact or vapor inhalation. Jars, such as pint or quart canning jars, are convenient containers for preservation of tissues, but wide-mouth plastic bottles (Fig. 2.5) eliminate the potential breakage problems. After 2 or 3 days in 10 percent formalin, tissues can be transferred to Whirl-Pak® bags that contain enough formalin to keep the tissues wet. Write the specimen identification with indelible marker or pencil on a piece of index card, place the card inside the bag, and write the information directly on the bag with indelible marker. Pack the bags for shipping so as to prevent tissues from being crushed. Check with the courier regarding current requirements or restrictions for shipment of formalin.

If it is necessary to collect a blood sample from a live bird (if, for example, botulism is suspected), and syringes and needles are not available, sever the bird's head from its neck and collect the blood in a wide-mouth plastic jar.

Photographing external and internal lesions provides a record of the color, location, and appearance of lesions when appropriate camera equipment is available. Use a macro lens, high speed film, and a fast shutter speed to achieve maximum depth of field and sharply focused photographs with a hand-held camera. Include in the photograph for scale a coin or another readily recognized indicator of actual size. Explain on the history form submitted with the specimens what photographs were taken.

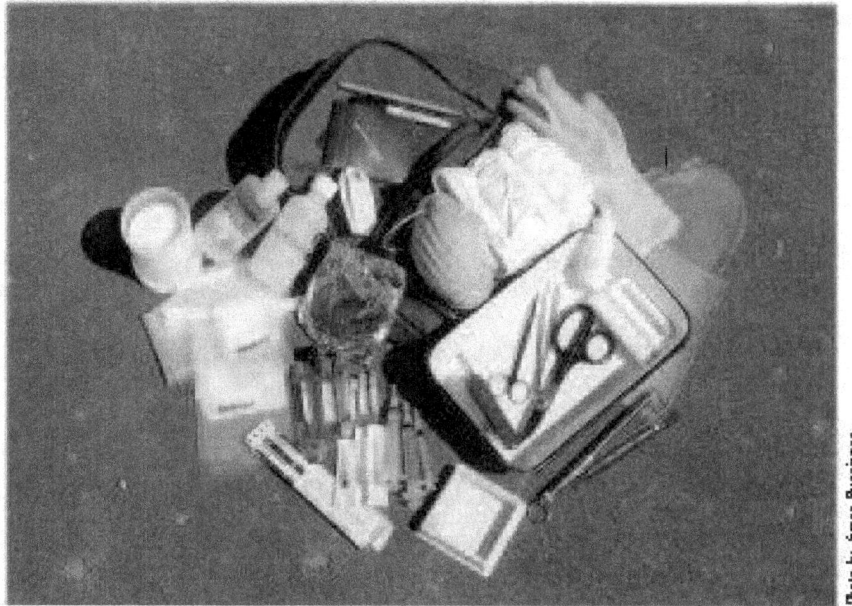

Figure 2.2 A basic necropsy kit that can be packed into a small day pack. Clockwise, from top of photo: Data recording: field notebook, tags, pencils, markers. Protective apparel: rubber gloves, disposable shoe covers and coveralls, mask. Necropsy equipment: disinfectant for cleaning instruments, scrub brush, heavy shears, forceps, scissors, scalpel handle and blades. Measuring equipment: hanging scale and ruler. Sampling materials: microscope slides, syringes and needles, swabs, blood tubes, aluminum foil, Whirl Pak® bags, plastic bags, wide mouth plastic jars. Preservatives: ethanol for parasites, formalin for tissue samples.

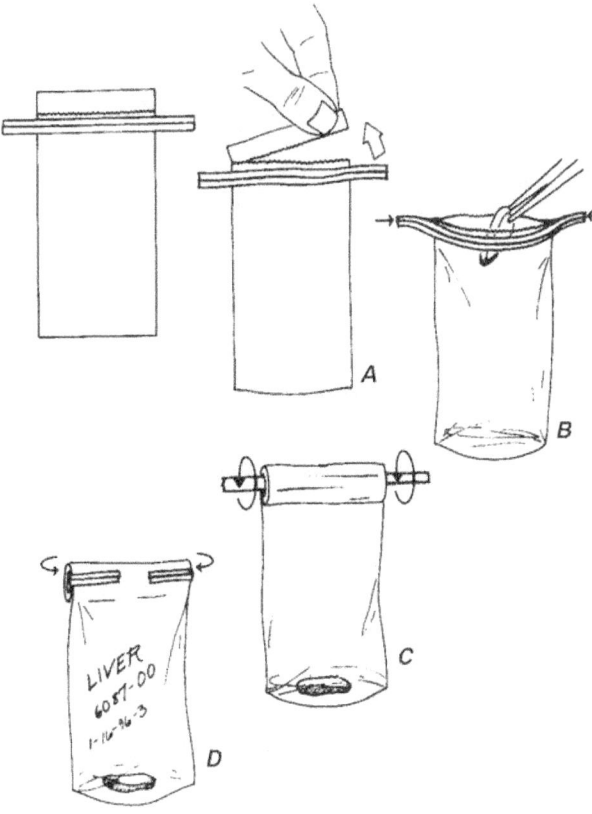

Figure 2.3 Using Whirl-Pak® bag for specimen collection. *(A)* Remove top at perforation. *(B)* Open bag by simultaneously pushing the protruding wire-reinforced tabs toward the center to insert the specimen and any appropriate preservative. *(C)* Close bag by pulling on tabs and then twirling bag while holding tabs. *(D)* Secure the closure by folding tabs around bags and label bag with type of specimen, date, and any identifying numbers.

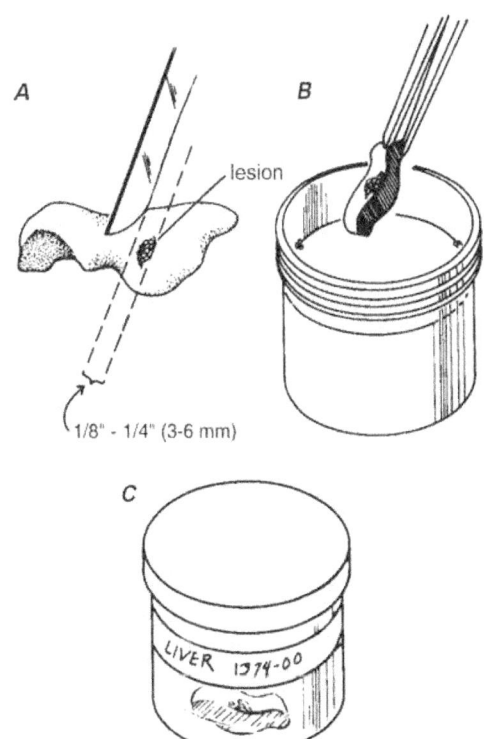

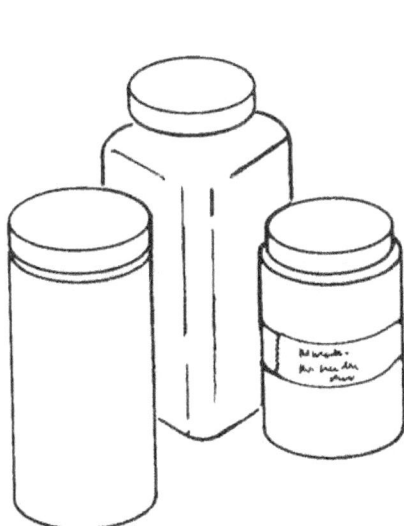

Figure 2.5 Plastic bottles used for tissue specimens. Regardless of size or shape, specimen bottles should have a wide mouth and threaded caps for secure closure.

Figure 2.4 Tissue sample collection for microscopic examination. *(A)* Tissue sample should include lesion, such as spots in liver, plus some apparently healthy tissue. The sample must be no thicker than 1/4 inch to ensure adequate chemical fixation by preservative. Use as sharp an instrument as possible (scalpel, knife, razor) for a clean cut. *(B)* Place tissue sample into container of 10 percent buffered formalin or other suitable fixative or preservative. The volume of formalin in the container should be about 10 times the amount of tissue sample. *(C)* Complete the process by securing the lid and properly labeling the container.

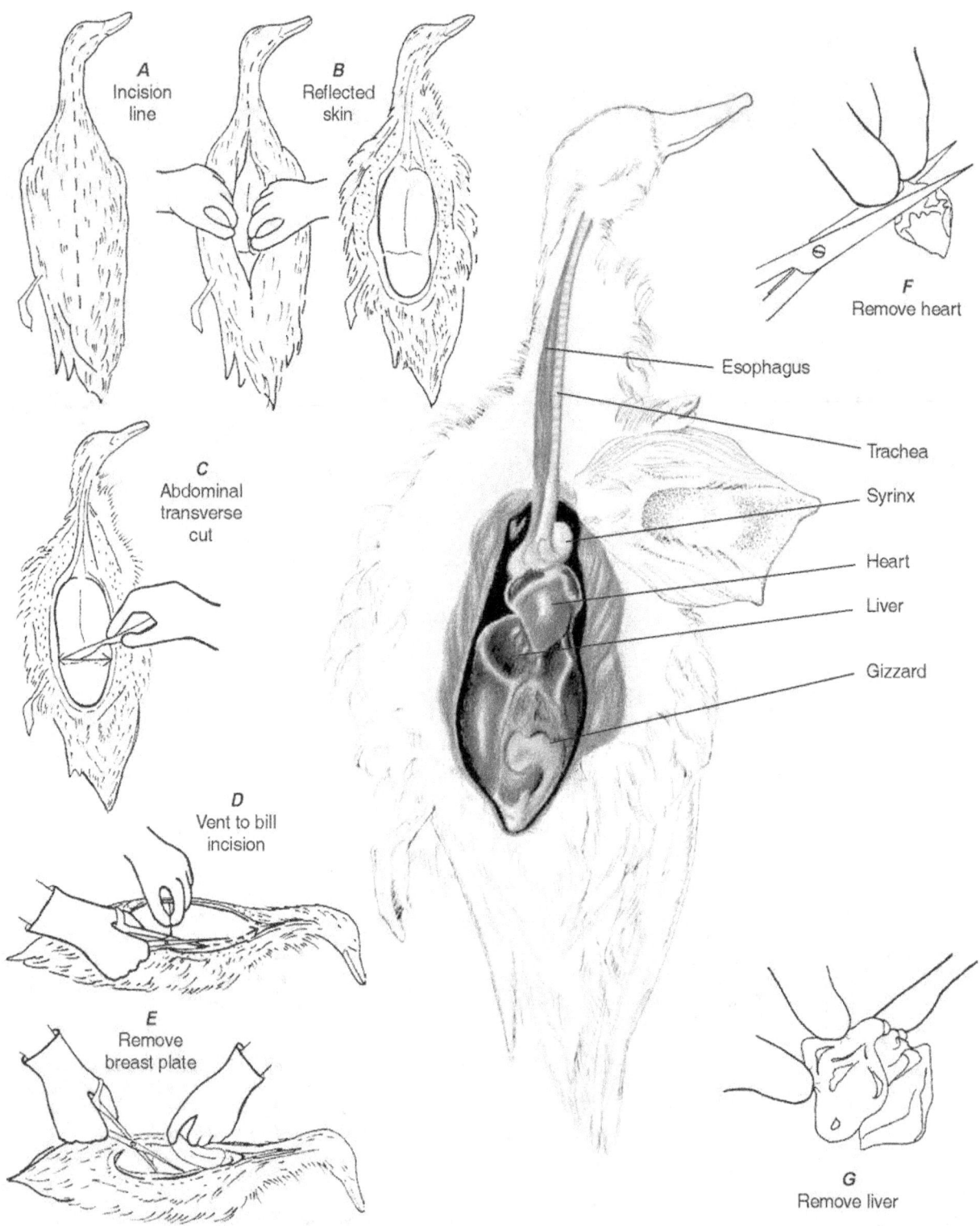

Figure 2.6 Dissecting a duck carcass: **(A)** incision line; **(B)** reflect the skin to expose the underlying anatomy; **(C)** make a transverse abdominal cut below the breast muscle; **(D)** extend cut through the ribs and wishbone; **(E)** remove breast plate; **(F)** dissect out heart; **(G)** remove liver; and **(H)** tie off and remove the gastrointestinal tract.

Avian Dissection

When dissecting a bird, it is always advisable to wear protective clothing, particularly disposable gloves. To begin, insert a scalpel or a knife to make a midline incision through the skin of the breast (Fig 2.6 A). Take care not to penetrate the body cavity, particularly in the abdominal region. Continue the skin incision to the vent and to the base of the bill. Reflect the skin away from the neck, breast, and abdominal areas. (B) Use the thumb and the first finger of each hand to reflect the skin to expose the underlying tissues. It is easiest to place the thumb and the first finger of each hand along the incision line in the breast area and then push and gently pull the skin to the side. When an opening in the skin has been established, work towards the bill and then the vent. (C) With a sharp blade, make a shallow transverse incision just below the breast muscles and sternum. (D) Insert the thumb of one gloved hand into the incision along the midpoint of the sternum and apply a slight pressure upwards. With a scissors in the other gloved hand, carefully cut through the ribs extending the cut on each side of the breast through the area of the wishbone. (E) Gently separate the breastplate from the carcass; use a scissors or other instrument to sever any connections and push aside the air sacs. (F) Dissect out the heart without cutting into other tissues. (G) Gently remove the liver and carefully cut away its area of connection with other tissues. (H) Tie off the gastrointestinal tract near the throat area, cut the esophagus above the tied-off area, and gently remove the entire gastrointestinal area.

Avian Anatomy

Figure 2.6 illustrates organs and tissues that may exhibit various lesions and that may be sampled for the diagnosis of disease agents described in this Manual. Species variation may result in some differences in the appearance and relative size of particular organs and tissues, but their location will be similar among species. Notable differences between the types of species illustrated are the small flat spleen in normal ducks and the larger oval spleen in pheasants. Also, pheasants have a crop and ducks do not; instead, the area just forward of the gizzard (the proventriculus) is more prominant in waterfowl.

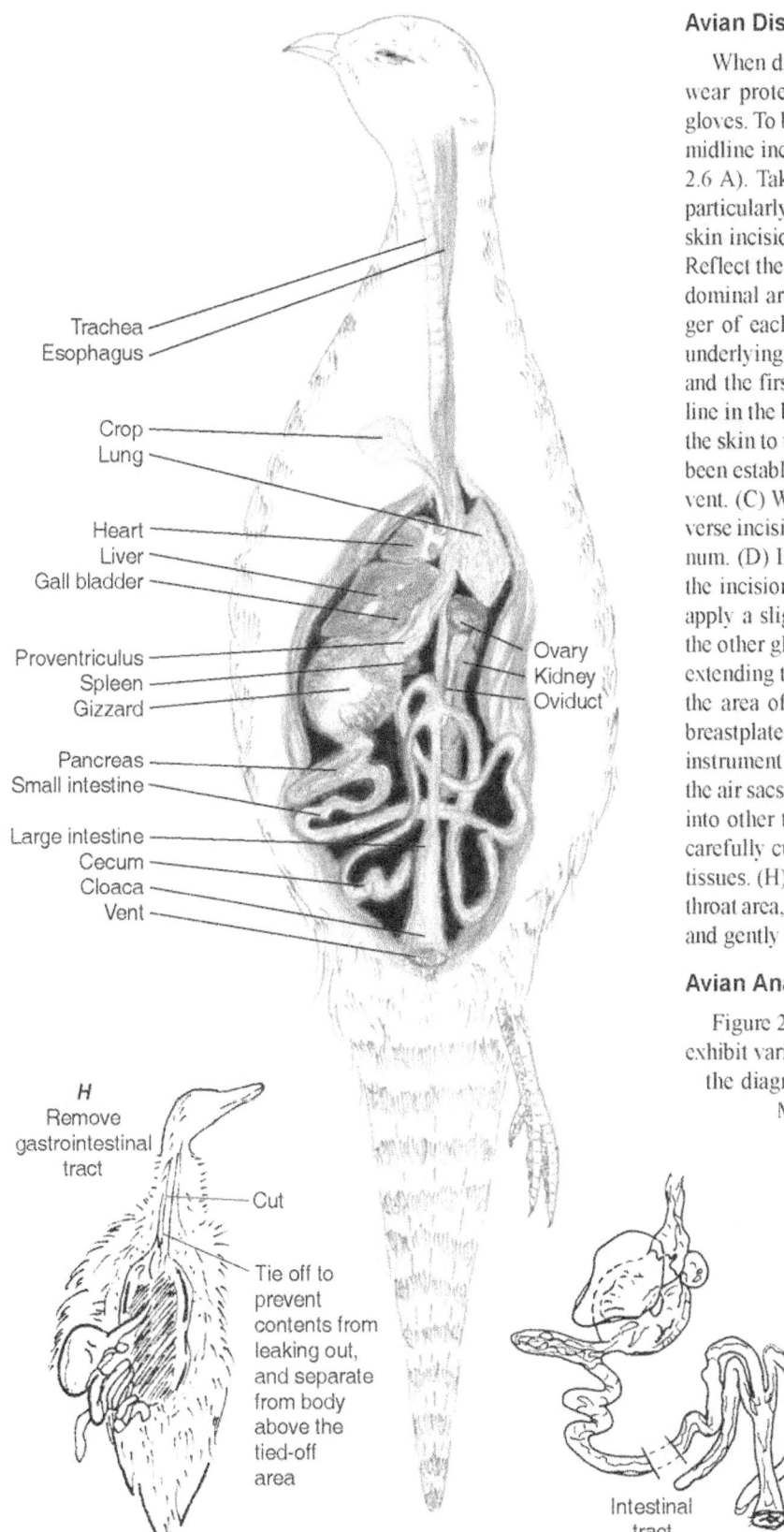

Specimen Collection and Preservation 11

Labeling Specimens

Proper labeling, maintaining label readability, and preventing label separation from specimens are as critical as proper specimen selection and preservation. The label should be as close to the specimen as possible; for example, a label should be attached to a carcass, attached to a tube of blood, or placed within the vial of preservative with a parasite. Double labeling, or placing a label on the outside of a plastic bag holding the specimen whenever practical, is worth the effort. The double labeling prevents confusion and potential

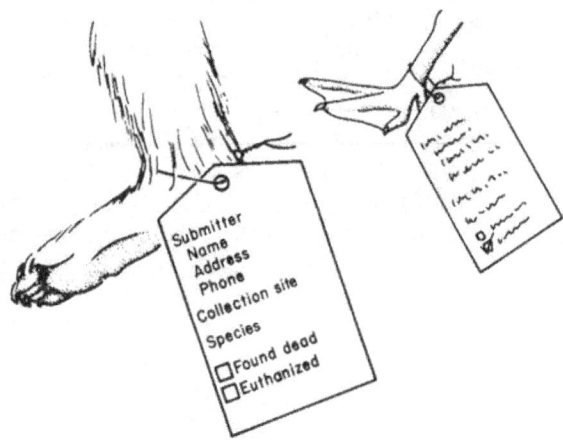

Figure 2.7 *Proper tagging of specimen. History of the specimen (see text for details) should be placed on back of tag.*

errors in specimen records at the diagnostic laboratory when specimens are received from multiple carcasses. Manila tags can be used, but take care to prevent their exposure to large amounts of fluids that may destroy the tag; tag destruction can be reduced by using tags with high rag content or even linen tags. Use soft lead pencil or waterproof ink on these tags; do not use ballpoint pen, nonpermanent ink, or hard lead pencil. The most durable tag is made of soft metal, such as copper or aluminum, and can be inscribed with ballpoint pen, pencil, or another instrument that leaves an impression on the tag.

Carcass

Identify each carcass with a tag fastened with wire to a leg (Fig. 2.7). If tags are not available, use a 3- by 5-inch card placed inside a plastic bag within the bag holding the carcass. Information on the tag should include the name, address, and telephone number of the submitter, collection site, species; whether the animal was found dead or was euthanized (indicate method); and a brief summary of any clinical signs. Place each tagged carcass in a separate plastic bag and seal the bag.

Tissues and Organs

When a specimen is in a plastic bottle, jar, or tube, wrap a piece of adhesive or masking tape entirely around the container and use an indelible marker to write on the tape. List the type of animal from which the sample was taken, the kind of tissue, and the date the sample was taken. When plastic bags are used as the first containers for tissues, they should be labeled with the same information directly on the bag. Do not insert tags inside containers with tissues and organs collected for microbiological or chemical analyses because the tag or the ink on it may contaminate the specimen. When chemically resistant tags are available, insert the tags into containers with preservatives such as formalin or alcohol.

Specimen Preservation

Chill or freeze all specimens, depending on how long it will take to ship to a diagnostic laboratory. Freezing reduces the diagnostic usefulness of carcasses and tissues, but if specimens must be held for 2 or more days, freezing the specimens as soon as possible after collecting them minimizes their decomposition. Formalin-fixed tissues should not be frozen. See Chapter 3, Specimen Shipment, for detailed instructions for packing and shipping specimens.

J. Christian Franson

(All illustrations in this chapter are by Randy Stothard Kampen, with the exception of Figure 2.6)

Supplementary Reading

Roffe, T.J., Friend, M., and Locke, L.N., 1994, Evaluation of causes of wildlife mortality, *in* Bookhout, T.A., ed., Research and Management Techniques for Wildlife and Habitats (5): Bethesda, Md., The Wildlife Society, p. 324–348.

Wobeser, G.A., 1997, Necropsy and sample preservation techniques, *in* Diseases of wild waterfowl (2nd ed): New York, N.Y., Plenum Press, p. 237–248.

Chapter 3
Specimen Shipment

Procedures for shipping specimens vary with different disease diagnostic laboratories. Therefore, it is important to contact the receiving laboratory and obtain specific shipping instructions. This will facilitate processing of specimens when they reach the laboratory and assure that the quality of specimens is not compromised. Time spent on field investigation, specimen collection, and obtaining an adequate history will be of little value if specimens become contaminated, decomposed, or otherwise spoiled during shipping to the diagnostic laboratory.

There are five important considerations for proper specimen shipment: (1) prevent cross-contamination from specimen to specimen, (2) prevent decomposition of the specimen, (3) prevent leakage of fluids, (4) preserve individual specimen identity, and (5) properly label the package. Basic supplies needed for specimen shipment are shown in Fig. 3.1.

Preventing Breakage and Leakage

Isolate individual specimens from one another by enclosing them in separate packages such as plastic bags. Protect specimens from direct contact with any coolant used (e.g., wet ice or dry ice), and contain all materials within the package so that leakage to the outside of the shipment container is prevented if breakage occurs (e.g., blood tubes) or materials thaw (wet ice and frozen carcasses) due to transit delays.

Containing Specimens

Plastic bags should be strong enough to resist being punctured by materials contained within them and from contact with other containers within the package.

Styrofoam® coolers, shipped in cardboard boxes, are useful for their insulating and shock absorbing qualities. Styrofoam® at least 1-inch thick is preferred. When possible, select Styrofoam® coolers that have straight sides. Coolers that are wider at the top than at the bottom are more likely to break during transit than those with straight sides. Fill the space between the outside of the Styrofoam® cooler and the cardboard box with newspaper or other packing material to avoid cooler breakage (Fig. 3.2). If coolers are not available, cut sheets of Styrofoam® insulation to fit the inside of cardboard boxes.

The cardboard box protects the Styrofoam® cooler from being crushed during transit and serves as containment for the entire package (Fig. 3.3). The strength of the box should

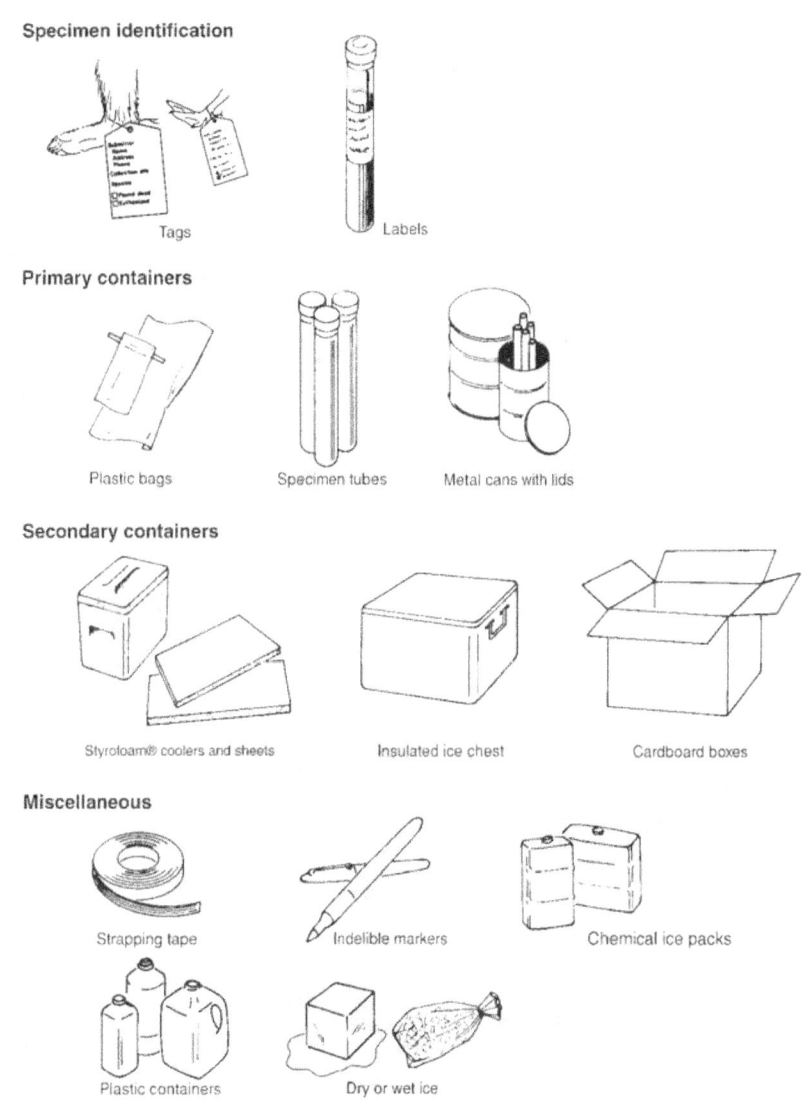

Figure 3.1 Basic specimen shipment supplies.

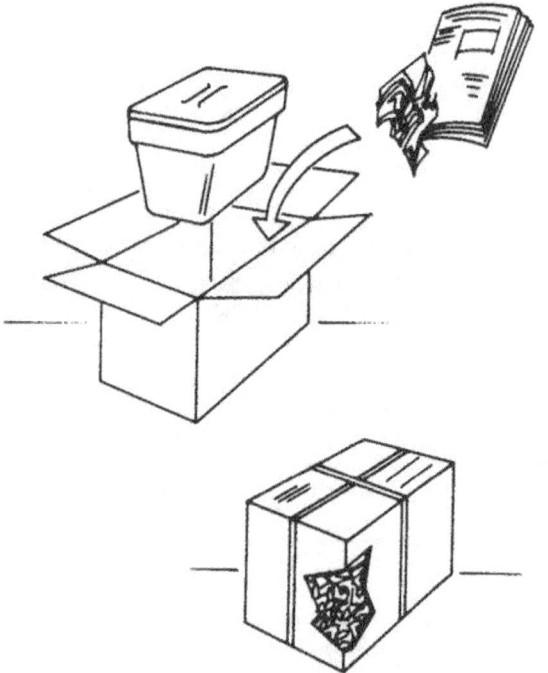

Figure 3.2 Proper packing to prevent Styrofoam® coolers from becoming crushed during transit. Place the sealed Styrofoam® cooler in a sturdy cardboard box. Use crumpled newspaper or other packing material to fill all space between the cooler and the box.

Figure 3.3 This Styrofoam® cooler was not packaged in a cardboard box for shipping.

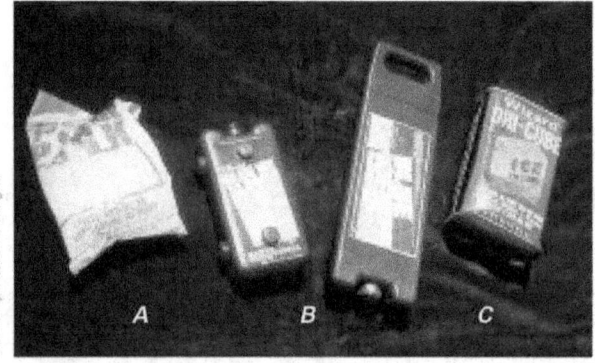

Figure 3.4 Chemical coolants are available in (A) soft plastic, (B) hard plastic, and (C) metal containers.

be consistent with the weight of the package. Cardboard boxes are not needed when hard plastic or metal insulated chests are used for specimen shipment, but boxes can be used to protect those containers from damage and to provide a surface for attaching labels and addresses to the shipment.

Cooling and Refrigeration

Chemical ice packs (Fig. 3.4) are preferable to wet ice because their packaging prevents them from leaking when they thaw. Ice cubes or block ice may be used if leakage can be prevented. This can be accomplished most easily by filling plastic jugs such as milk, juice, and soda containers with water and freezing them. The lids of these containers should be taped closed to prevent them from being jarred open during transit.

Use dry ice to keep materials frozen, but do not use it to ship specimens that should remain chilled because it will freeze them. Also, the carbon dioxide given off by dry ice can destroy some disease agents; this is of concern when tissues, rather than whole carcasses, are being shipped. Shipment of dry ice, formalin, and alcohol is regulated and should be cleared with the carrier before shipping.

Preparing Specimens for Shipment to the National Wildlife Health Center (NWHC)

Other disease diagnostic laboratories may require minor variations in shipping procedures.

1. Call the NWHC (608-270-2400) to determine the optimal type and number of specimens for diagnostic procedures, how these specimens are best preserved during transit (whether they should be chilled or frozen), and when they should be shipped. In most cases, the NWHC requests that specimens be shipped the same day or within 24 hours.

2. Double-bag carcasses (Fig. 3.5) and place them in a Styrofoam® cooler lined with a plastic bag. When both frozen and fresh whole carcasses are shipped in the same container, the frozen carcasses can be used as a refrigerant to keep the fresh carcasses chilled. This can be accomplished by interspersing individually bagged frozen carcasses among the individually bagged fresh carcasses or by placing the fresh carcasses between two layers of frozen carcasses (Fig. 3.6). Blood tubes and other breakable containers of uniform size can be protected by packing them in a common plastic bag that is sealed within a metal can or a hard plastic container with a lid (Fig. 3.7). Pack any space around the specimen containers within the can (side and top) with paper or some other absorbent material to prevent jarring that could cause breakage and to collect fluids if tubes do break. Seal the can within a plastic bag before placing it in the Styrofoam® cooler.

3. When using chemical ice packs, intersperse them among specimens; place within the Styrofoam® container other types of coolants in locations that will provide maximum cooling for all contents or, if dry ice is used, will keep everything frozen (Fig. 3.8). Fill all empty space within the

Styrofoam® cooler with newspaper to prevent materials from moving during transit. The insulating properties of newspaper will also help maintain cool temperatures within the package, and its absorbent qualities will help prevent fluid leakage outside of the box or container.

4. Close the plastic bag lining the cooler and seal the lid with strapping tape (Fig. 3.9). Tape the specimen data sheet and history, contained in an envelope within a waterproof plastic bag, to the top of the cooler (Fig. 3.10A).

5. Enclose the Styrofoam® cooler in a cardboard box and secure the contents with strapping tape (Fig. 3.10B).

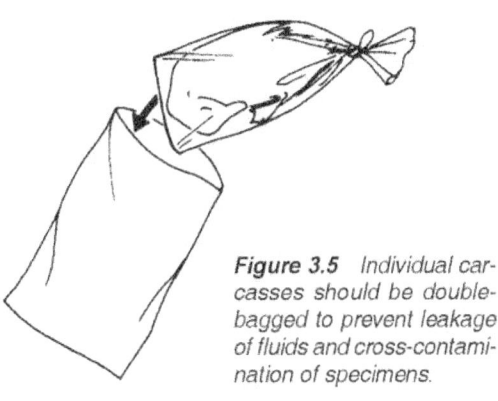

Figure 3.5 Individual carcasses should be double-bagged to prevent leakage of fluids and cross-contamination of specimens.

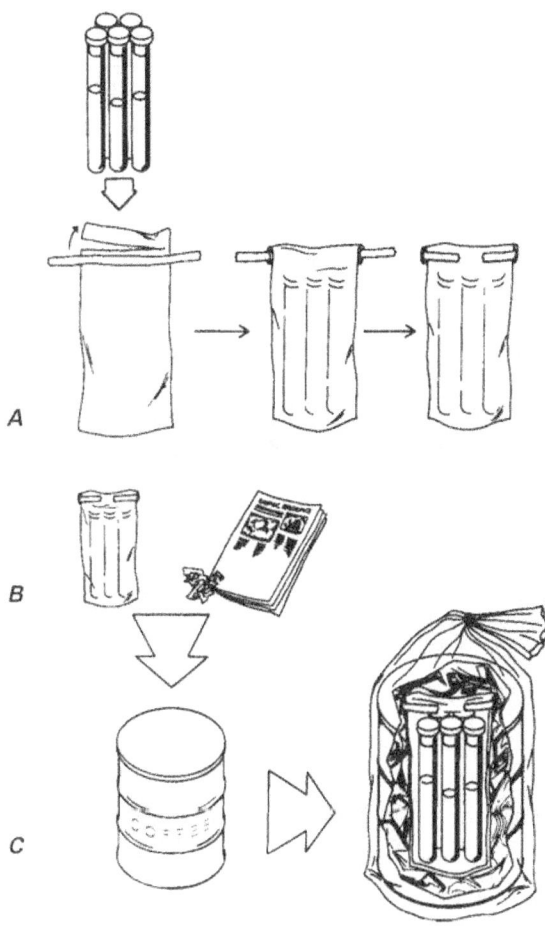

Figure 3.7 Packing sequence for blood tubes. **(A)** Pack blood tubes within Whirl-Pak® or other plastic bag; **(B)** place bag in metal can or hard plastic container and pack with crumpled newspaper or other absorbent, soft, space-filling material; and **(C)** enclose the can in a plastic bag, then seal the bag.

Figure 3.6 Frozen carcasses (white bags) can be used to keep fresh specimens (dark bags) chilled during short transit times of 24 hours or less. Fill the space between the carcasses and the top of the container with newspaper to provide additional insulation to maintain the cold temperature.

Specimen Shipment

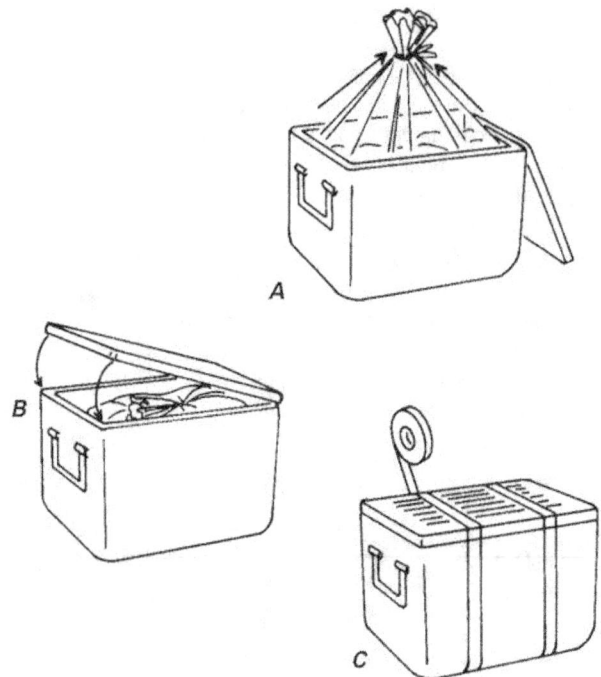

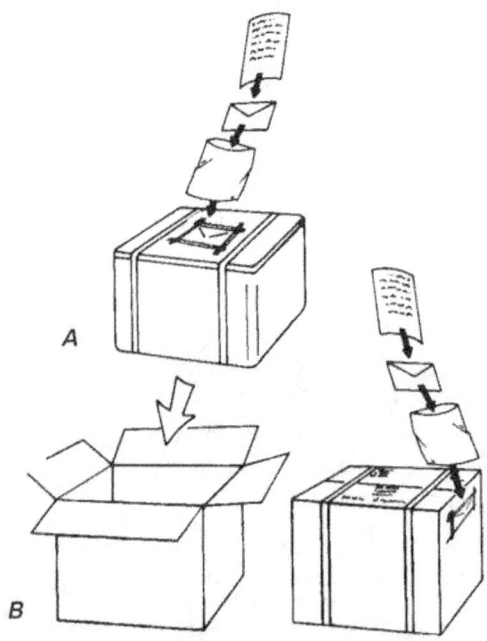

Figure 3.9 Closing a specimen container. **(A)** Secure the large plastic bag containing the specimens by tying the top; **(B)** close the container lid and **(C)** secure the container with several bands of strapping tape.

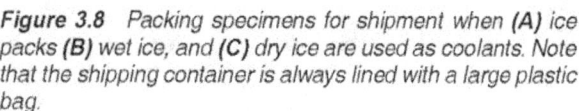

Figure 3.8 Packing specimens for shipment when **(A)** ice packs **(B)** wet ice, and **(C)** dry ice are used as coolants. Note that the shipping container is always lined with a large plastic bag.

Figure 3.10 Completing the packaging process. **(A)** Tape specimen data sheet and history, contained in an envelope within a waterproof plastic bag, to top of cooler. **(B)** Place cooler in cardboard box, secure box with several bands of strapping tape, and secure another copy of the specimen data sheet to the outside of the box. If the specimens were placed inside a Styrofoam® cooler, then use crumpled newspaper or other packing material to fill all spaces between the cooler and the box.

Federal Shipping Regulations for Packaging and Labeling

Your packaging and labeling of specimens must conform to the following regulations.

The Code of Federal Regulations (CFR) states under 50 CFR Part 14 of Fish and Wildlife Regulations that containers with wildlife specimens must bear the name and address of the shipper and consignee, and a list of the species and numbers of each species must be conspicuously marked on the outside of the container. You may instead conspicuously mark the outside of each package or container with the word "wildlife" or the common names of the species contained within the package. Secure an invoice or packing list that includes the name and address of the consignee and shipper and that accurately states the number of each species contained in the shipment to the outside of one container in the shipment.

In addition to Fish and Wildlife Service regulations, the interstate shipment of diagnostic specimens is subject to applicable packaging, labeling, and shipping requirements for disease-causing etiologic agents (42 CFR Part 72). These regulations do not require you to identify diagnostic specimens as etiologic agents when the disease agent is not known or is only suspected. However, all specimen packages sent to the NWHC should be prominently labeled with the words "DIAGNOSTIC SPECIMENS." You can meet packaging requirements under 42 CFR Part 72 by following recommendations 2 through 5 above for enclosing specimens within two containers before enclosing them within the package.

Hazardous Materials Regulations of the Department of Transportation apply whenever dry ice is contained within the shipping container (49 CFR Part 172, 173, 175). Always call the carrier ahead of time for the current shipping and package labeling requirements. At the time of this writing, the following must be clearly visible on containers with dry ice: DRY ICE 9, UN1845, weight of dry ice (kilograms), a hazardous materials miscellaneous 9 sticker, and the complete addresses of the shipper and recipient. The dry ice labeling should go on the side of the container, so it is visible if something is stacked on top of it. Always include the words "DIAGNOSTIC SPECIMENS (WILDLIFE)" on the container. A properly labeled container is illustrated in Fig. 3.11. Label containers with permanent markers, if possible.

Commercial Carriers

Specimens should be shipped by carriers that can guarantee 24-hour delivery to the location of the diagnostic laboratory. For many locations, commercial delivery services will pick up packages at the point of origin. When shipping arrangements have been made, contact the NWHC again and provide the airbill number and estimated time of arrival. This information is needed to allow prompt tracing of shipments that may not arrive on schedule and to schedule work at the laboratory.

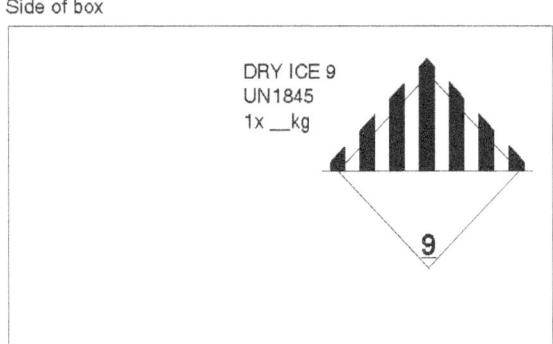

Figure 3.11 Proper package labeling.

J. Christian Franson

(All illustrations in this chapter are by Randy Stothard Kampen, with the exception of Figure 3.11)

Supplementary Reading

Code of Federal Regulations. Title 42; Part 72
Code of Federal Regulations. Title 49; Parts 172, 173, 175.
Code of Federal Regulations. Title 50; Part 14.

Chapter 4
Disease Control Operations

Individual disease outbreaks have killed many thousands of animals on numerous occasions. Tens of thousands of migratory birds have died in single die-offs with as many as 1,000 birds succumbing in 1 day. The ability to successfully combat such explosive situations is highly dependent on the readiness of field personnel to deal with them. Because many disease agents can spread through wildlife populations very quickly, advance preparation is essential for preventing infected animals from spreading disease to additional species and locations. Carefully thought-out disease contingency plans should be developed as practical working documents for field personnel and updated as necessary. Well-designed plans can prove invaluable in minimizing wildlife losses and the costs associated with disease control activities.

Although requirements for disease control operations vary and must be tailored to each situation, all disease contingency planning involves general concepts and basic biological information. This chapter, which is intended to be a practical guide, identifies the major activities and needs of disease control operations, and relates them to disease contingency planning.

Planning Activities

Identification of Needs

Effective planning for combating wildlife disease outbreaks requires an understanding of disease control operations and the basic needs such as personnel, equipment and supplies, permits, etc., that are associated with them (Tables 4.1 and 4.2). This information is the basis of disease contingency planning (Table 4.3; Figs. 4.1 and 4.2).

Biological Data Records

All disease outbreaks consist of three main components: a susceptible host population, a disease agent interface, and the environment in which the host and agent interact in a manner that results in disease. Disease control involves breaking the connections between these factors. Disease contingency plans expedite these efforts by providing basic information about the distribution and types of animal populations in the area, animal movement patterns, any history of disease problems on the area, and general environmental features. This information, along with facts gathered at the time of a disease outbreak, provides a profile for biological assessment and a basis for specific disease control actions.

Knowledge of the types of disease problems that have occurred in the area, their general locations, the month and year when they occurred, the species affected, and the general magnitude of losses is also of considerable value for planning a response to a disease outbreak. Incorporate a historical summary in tabular form in the contingency plan (Table 4.4). Animal population data are best represented by simple graphs and charts that convey general characteristics (Fig. 4.3); precise data are not needed. Generalized outline maps are useful for depicting concentration and feeding areas used by wildlife (Fig. 4.4) and major movement patterns (Fig. 4.5).

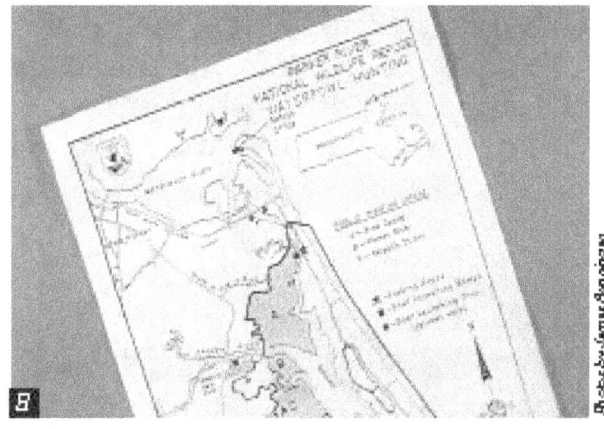

Figure 4.1 (A) Station brochures, animal lists, and other public-use documents provide a wide variety of site-specific background information and should be included as part of the station's disease contingency plan. (B) Documents containing maps of the area indicating access points provide essential information.

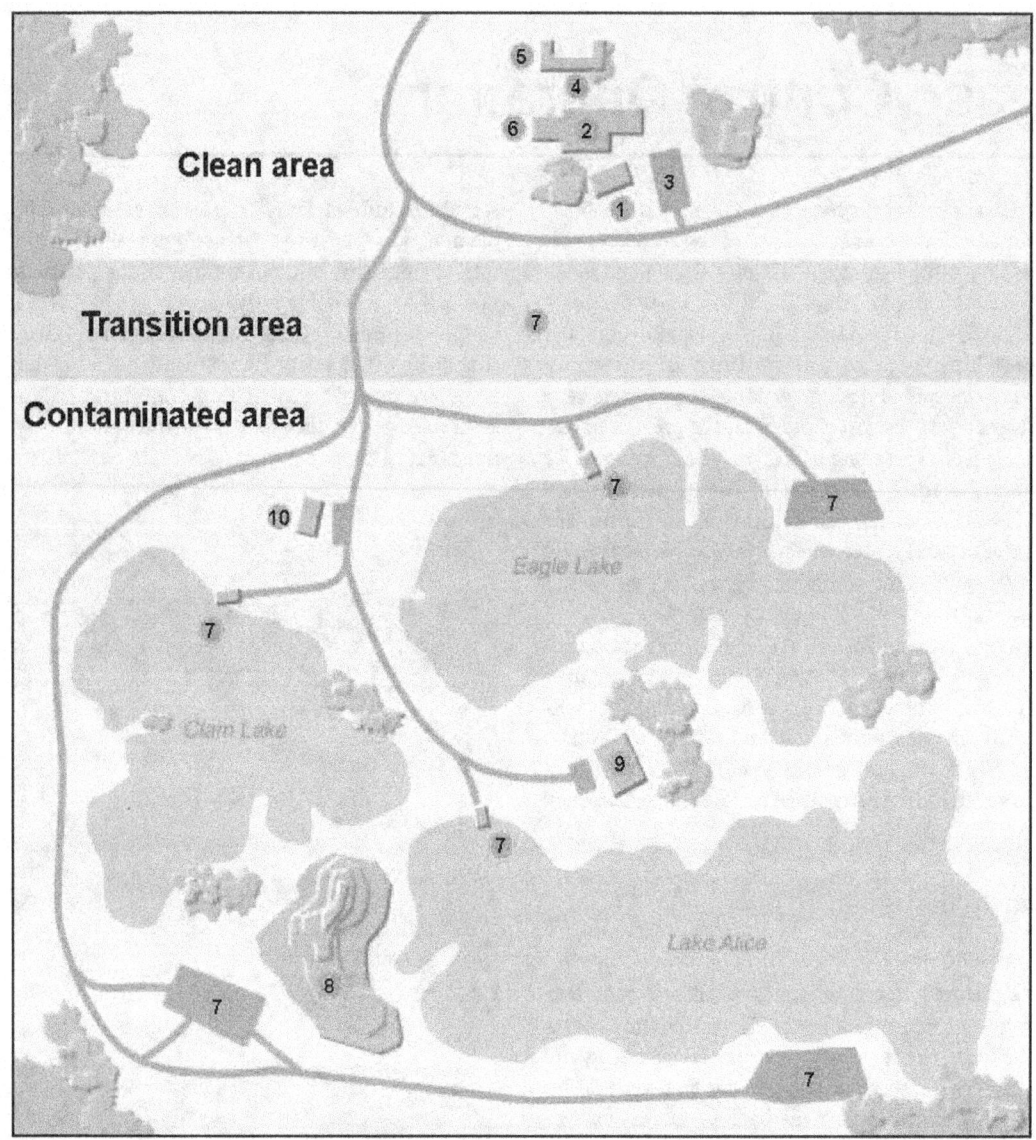

EXPLANATION

1. Command post and headquarters administrative area
2. Staff and press briefing room
3. Parking
4. Eating area and conference room
5. Staff rest area and visitors' center
6. Equipment and supply receipt—garage
7. Decontamination areas—boathouses, transition areas, parking lots
8. Carcass disposal site and observation hill
9. Animal holding—pole barn (has cement slab and electricity)
10. Laboratory investigations—shed (has cement slab, water, electricity)

Figure 4.2 Existing work areas used for disease control operations on a wildlife management area.

Response Activities

Response to wildlife die-offs will vary somewhat with the species but will always involve a set of common factors. Waterfowl die-offs are used to illustrate specific approaches to addressing these common factors. For large mammals, their size and weight pose additional needs regarding carcass transport and disposal.

Problem Identification

Early detection and rapid and accurate assessment of the causes of disease problems are essential to effective disease control operations. This is accomplished through surveillance of animal populations to detect sick and dead wildlife, and the prompt submission of specimens to qualified disease diagnostic facilities. The speed with which large numbers of animals can become exposed to disease agents and the differences in control activities required for different types of disease problems place a premium on both the speed and accuracy of diagnostic assessments. Once a disease problem has been identified, the following basic activities are carried out.

Carcass Removal: Protective Clothing and Supplies

Wildlife that have died from disease are often a primary source of the disease agent, and for most situations their carcasses need to be removed from the environment to prevent disease transmission to other animals through contact with or consumption of the carcass. Disease organisms released from tissues and body fluids as carcasses decompose also contaminate the environment. Some disease-causing viruses and bacteria can survive for several weeks or longer in pond water, mud, and soil.

Because carcass collection concentrates diseased material in a small area, it is essential that carcasses be handled so that they do not release infectious agents into the environment or jeopardize the health of personnel. Great care also needs to be taken to prevent mechanical movement of the disease agent from the problem area to other areas.

Personnel assigned to this task need to wear outer garments that provide a protective barrier against direct contact with disease organisms and that can be disinfected and removed before personnel leave the area. Typically, these include boots, coveralls or raingear, gloves, and a head covering (Fig. 4.6).

Use disposable coveralls and outer gloves when possible; the durability and cost of garments are considerations in decisions about whether or not disposable garments will be used. Personnel should remove coveralls and outer gloves before they leave the area, and the garments should be destroyed if they are disposable or they should be double-bagged before they are transported to a location where they can be thoroughly washed before they are reused. Dishwashing gloves, work gloves, and other types of rubber gloves are readily available at hardware and other retail stores, as are scrub brushes for cleaning (Fig. 4.7).

Carcass removal requires heavy-duty plastic bags or containers. Plastic body bags used by the military are excellent for containing wildlife carcasses. Plastic garbage cans lined with commercially available heavy-gauge leaf and litter plastic bags are also excellent containers for transporting carcasses. These containers are especially useful when personnel collect bird carcasses by boat (Fig. 4.8A), and for transporting carcasses in truck beds. Tie the bags shut and secure garbage can lids when transporting these containers to carcass disposal sites (Fig. 4.8B).

Depending on conditions, a variety of watercraft (Fig. 4.9) and all-terrain vehicles (Fig. 4.10) are useful for searching for carcasses and for transporting carcasses to collection and disposal sites. In some instances, the expense of helicopters may be warranted. Pickup trucks and other four-wheel vehicles are also indispensable under some field conditions.

Dogs have been used extensively in wildlife management, and they are a valuable search tool when they are appropriately chosen and handled. Use dogs whenever possible to locate carcasses if there is no disease risk to them. Infectious diseases of wild North American birds do not pose a significant health threat to dogs. Determine disease risk on a case-by-case basis by consulting with wildlife disease specialists. Local retriever clubs or kennels may provide dogs.

The contingency plan should identify sources of various equipment, whether equipment can be borrowed or rented, and contact persons and their telephone numbers. Commonly used supplies and equipment needed to support disease control operations are summarized in Table 4.2.

Carcass Disposal

The primary goal of carcass disposal is to prevent spread of the disease agent to other animals through environmental contamination. Because personnel will handle concentrated amounts of infectious or highly toxic agents, this activity requires proper training and supervision. Incineration, burying, rendering, and composting are the four basic disposal methods.

Incineration is generally the preferred method for disposing of carcasses and contaminated materials associated with wildlife disease outbreaks. However, air-quality standards often preclude open burning, even for disease emergencies. Consider purchasing or constructing portable incinerators (Fig. 4.11) for areas with recurring disease problems if local regulations allow using such equipment. Portable garbage incinerators can sometimes be borrowed from State parks and other sources. If portable incinerators are not available, open burning with tires or other fuel or both can be used, depending on local air pollution standards. Carcasses may be burned either above or below ground (Fig. 4.12). It is important to keep the fire contained and to get sufficient air movement under the carcasses to maintain a hot fire and completely burn the carcasses. Wood, coal, fuel oil, napalm, and

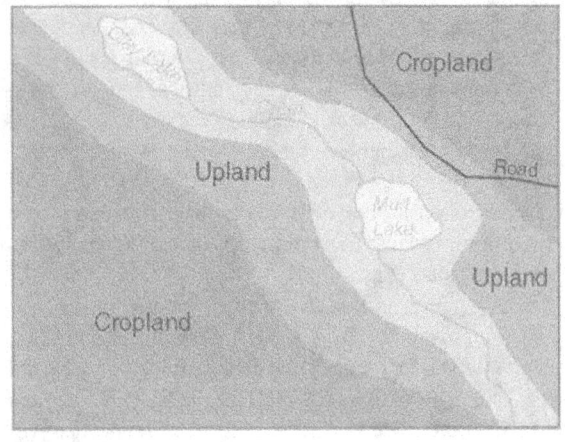

April–August: Nesting birds and broods on Mud and Clay Lakes and adjacent uplands.

September–mid-October: Fall migrants using Mud and Clay Lakes.

Mid-October–January: During hunting season, birds concentrated on Clay Lake and adjacent marsh.

February–March: After hunting season, birds distributed between Mud and Clay Lakes.

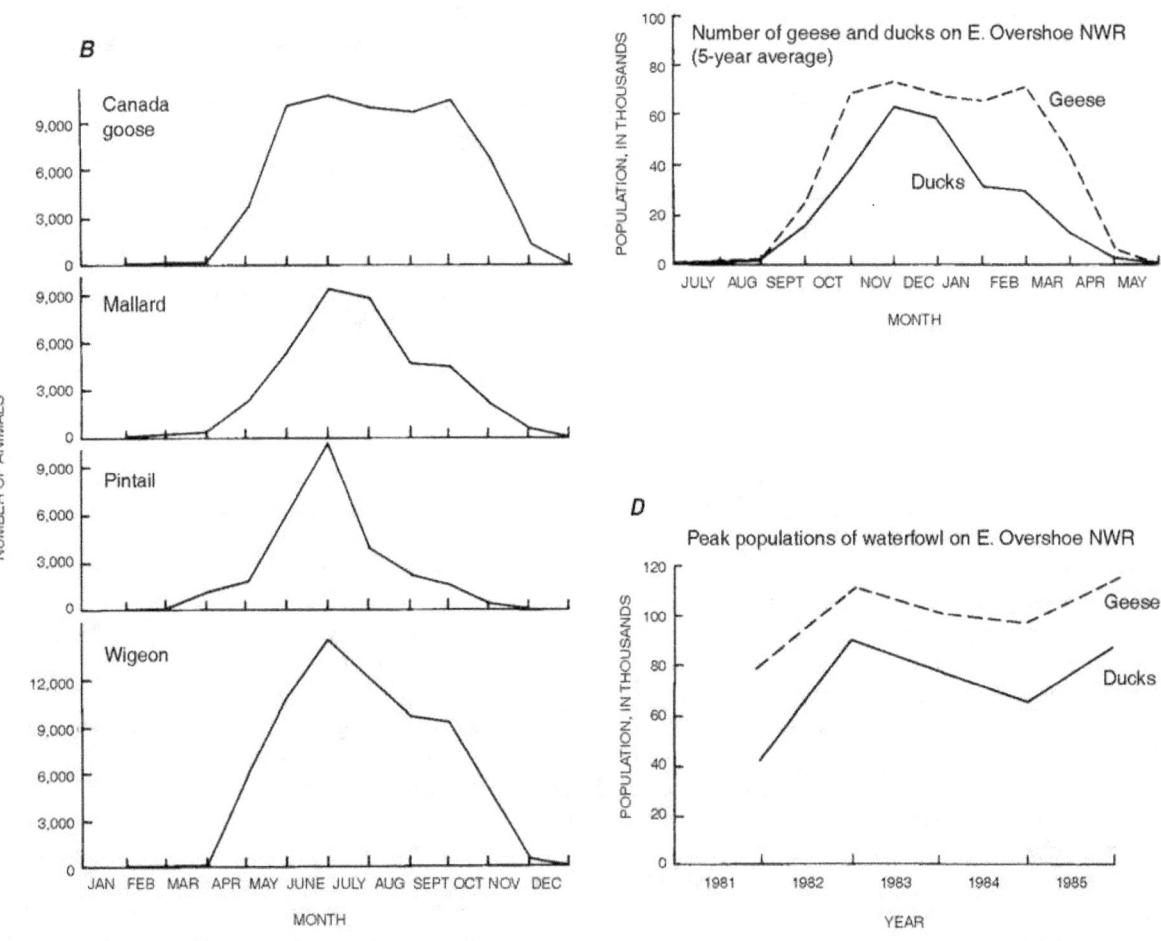

Figure 4.3 *Examples of how to present data on seasonal and annual wildlife use of a specific area. **(A)** General narrative format with map; **(B)** seasonal waterfowl populations by species, and total duck and goose use by **(C)** month and **(D)** year.*

22   Field Manual of Wildlife Diseases: Birds

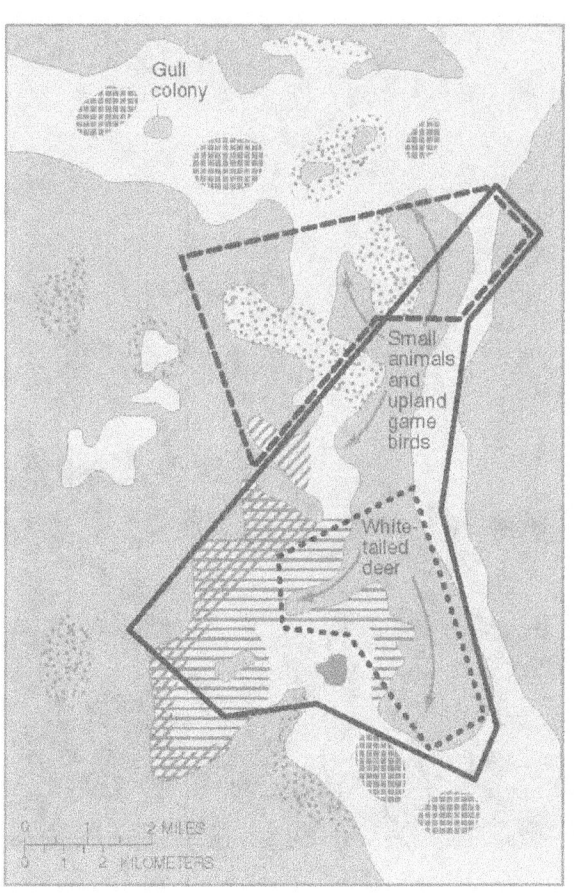

EXPLANATION
Major use areas

- ▨ Shorebirds and wading birds
- ▧ Bald eagle wintering roost site

Waterfowl

- ▦ Loafing areas
- ▤ Roosting areas
- ⋮ Feeding areas
- — National Wildlife Refuge boundary
- ••• National Wildlife Refuge Hunting Area
- – – State Game Management Area

Figure 4.4 Example of an outline map showing concentration and feeding areas used by wildlife.

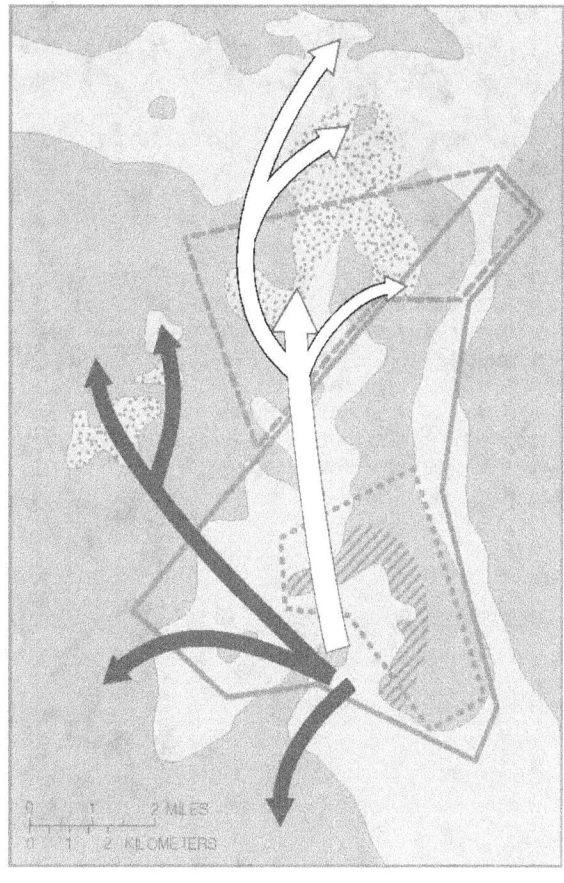

EXPLANATION
Major movement patterns

- ➡ Puddle duck and bay diving duck feeding patterns
- ⇨ Canada goose daily feeding flights

Major use areas

- ▨ White-tailed deer wintering area
- ⋮ Spring migration diving duck staging areas

Figure 4.5 Example of an outline map showing major movement patterns of species.

Figure 4.6 (A) Protective clothing such as coveralls, boots, head coverings, and gloves should be worn during carcass cleanup activities. (B) Before leaving the area, boots should be decontaminated and outer clothing removed and bagged for transportation to a location where they can be washed before being reused.

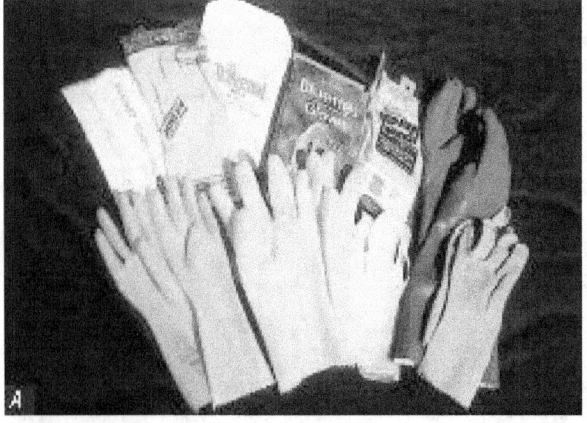

Figure 4.7 (A) Examples of readily available disposable and reusable gloves for disease control operations. Dishwashing gloves, surgical gloves, rubber work gloves, and other types can be purchased at drug and hardware stores and medical and laboratory supply houses. (B) A wide variety of scrub brushes needed for decontaminating boots, equipment, and other surfaces are also readily available from local merchants.

Figure 4.3. (A) Plastic barrel being used to transport carcasses from collection sites by airboat to disposal site. Note use of plastic bag to line barrel. The plastic bag containing carcasses can be secured, removed, and placed in a second plastic bag for further transportation if disposal site is not at the boat docking location, thereby allowing immediate reuse of the barrel. If the barrel containing carcasses is to be transported to some other location, the plastic bag should be tied closed and a cover placed on the barrel and secured. (B) Examples of improper transportation of carcasses to disposal site. Note untied bags, unbagged carcasses, wooden truck bed, and lack of tailgate. Carcasses and fluids contaminated with disease organisms could easily be released from the bags during transit. Fluids could contaminate the truck bed and leak to the ground through the cracks between the wooden boards. Wood absorbs fluids and is much more difficult to decontaminate than a nonporous surface. Also, carcasses could fall out of the truck because there is no tailgate.

Figure 4.9 Different types of (A) motorized and (B) nonmotorized watercraft are useful for carcass collection. Note the use of plastic bags for containment of carcasses and further transportation to disposal sites.

Figure 4.10 Selection of all-terrain vehicles should be matched to local conditions. All-terrain vehicles such as these three-wheel machines can (A) be equipped with small baskets to hold carcasses or live birds and (B) be used in water no more than 2-feet deep. Because of safety concerns, three-wheeled vehicles are not recommended. (C) Large tracked vehicles such as this equipment negotiate marshy terrain but are not amphibious. The major advantages of this equipment are the large capacity for carrying personnel, supplies, and equipment and excellent visibility afforded by the height of the vehicle. (D) Small amphibious vehicles such as this six-wheel machine are capable of transporting two persons and are more stable and versatile than three-wheel vehicles but are much slower on land surfaces.

Figure 4.11 Examples of portable incinerators used for disease control operations. (A) Garbage incinerator borrowed from State park to dispose of carcasses during Lake Andes duck plague die-off. (B and C) Locally designed and constructed incinerators in use during disease control operations. All of these are fueled with propane gas.

Figure 4.12 Examples of above-ground and in-trench methods for incineration of carcasses. **(A)** Portable grate the width of a pickup truckbed fashioned from metal pipes. **(B)** Simple grate suspended over pit into which carcass remains are placed for burial. **(C)** Major burning pit for large-scale operation—note surrounding area cleaned of vegetation for fire protection, the size and depth of pit, burning platform, rubber tires for fuel.—Figure 4.12 is continued on p. 30.

Disease Control Operations 29

Figure 4.12—continued (D) Intensity of heat generated by fire resulting in the bending of support pipes of the burning platform and metal grate. (E) Simple but sturdy above-ground structure of cinder blocks and steel grates elevated enough for fuel to be placed under the carcasses and for air to circulate upward. (F and G) Highly efficient above-ground burning platform constructed of a frame of used grader blades, wire mat platform, and (H) sheet metal heat deflector positioned at the rear of the platform. (I) Proper application of fuel oil (never use gasoline) for carcass incineration. Note that length of applicator prevents flashback or wind shift from endangering person applying fuel.

Disease Control Operations 31

other fuels have been successfully used. Never use gasoline because of the hazards involved. Incineration is facilitated by stacking or piling carcasses on the burning platform, soaking them with used oil or some other fuel, and waiting about 10 to 15 minutes before igniting them. The heat generated by large-scale carcass burning operations is intense enough to cause metal pipes to bend (Fig. 4.12D). Therefore, construct a sturdy carcass support surface so that it does not collapse into the fire.

During dry weather, burning carcasses in a pit surrounded by a vegetation-free area is more desirable than above-ground burning. In either situation, piling too many carcasses on the fire at once is a common mistake; burn carcasses one layer at a time (Fig. 4.13). When cinder blocks are used to support burning platforms, the length of the platform should be extended to keep the blocks out of direct heat or they will soon crumble.

When burning is not feasible or needed, burial is often a suitable alternative. Select burial sites carefully with consideration given to ground-water circulation and drainage, and any potential for later carcass exposure. Sprinkle lime or fuel oil on carcasses to discourage uncovering by scavengers and cover the carcasses with at least 3 to 4 feet of soil.

Composting is commonly used for the disposal of some domestic animal carcasses, and it is a technique that can be adapted to wildlife situations. The requirements for composting carcasses include an impermeable surface on which to place composting piles, a roof or other means of controlling moisture in the piles, and raw materials to mix with carcasses to achieve the correct carbon to nitrogen ratio for optimal decomposition of carcasses (Fig. 4.14).

When the combination of animal species, cause of mortality, and local situation allow, carcasses may also be disposed of by an animal rendering plant, and in rare instances infected wildlife may be killed and processed for food. Both of these methods are sometimes used for domestic species and captive-reared wildlife, but conservation laws generally prohibit the processing of free-living wildlife (with the exception of fish) as a commercial food source within the United States. Judgments on the use of rendering and food processing as animal disposal methods should be made only by qualified disease control specialists.

To the extent possible, dispose of carcasses on-site to reduce the risk associated with transporting contaminated material. Regardless of whether burning, burial, or large-scale composting is used, earth-moving equipment is needed. The disease contingency plan should identify how and where bulldozers, backhoes, and similar equipment can be obtained.

Animal Relocation

It is often as necessary to deal with the live, apparently healthy population during disease control activities as it is to remove and dispose of animals dying from disease. Depending on individual circumstances, consider denying animal use of specific sites by dispersing animals from the problem area, concentrating and holding wildlife within a specific area, or trapping animals for sampling.

Scare devices such as propane exploders (Fig. 4.15A) and cracker shells (Fig. 4.15B) may be useful for keeping wildlife away from a toxin or infectious agent within a specific area. Hazing wildlife with airplanes, helicopters, airboats, snowmobiles, and other motorized equipment has also been successful for moving them away from disease problem areas. Conversely, wildlife can be concentrated in an area for euthanasia, and they can be lured to other areas by broadcasting and dumping large amounts of grain and other feed to prevent their movement to problem areas, by knocking down standing grain to make it more available to them, by providing water through pumping operations and diverting water flow, and by providing refuge by closing the area to hunting and other interactions between wildlife and humans (Fig. 4.16). Take care to assure that grain used for attracting wildlife is not moldy and does not contain dangerous concentrations of mycotoxins.

Figure 4.13 Examples of **(A)** correct and **(B)** incorrect layering of carcasses for burning. Carcasses must be burned one layer at a time to prevent charred outer carcasses from insulating inner carcasses from incineration. (Illustration by Randy Stothard Kampen)

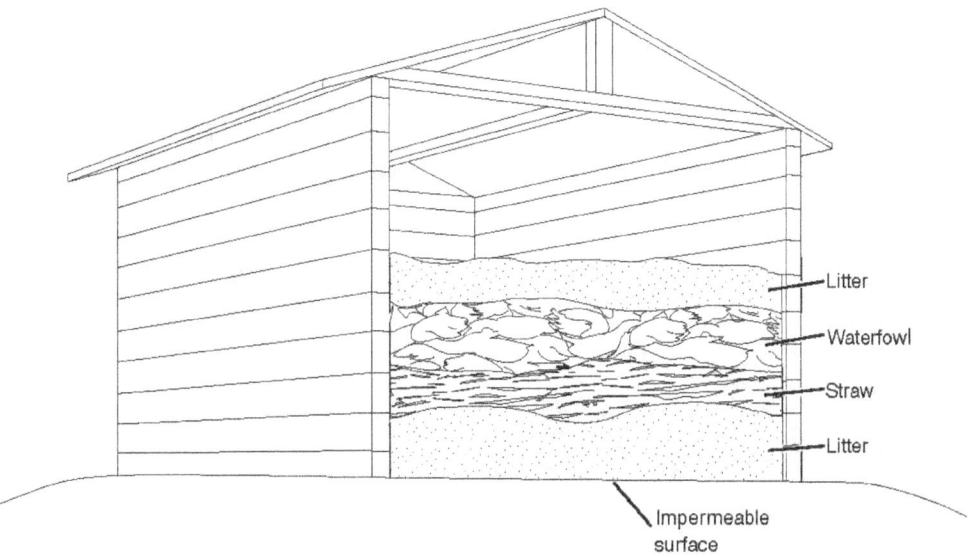

Figure 4.14 Example of a simple composting bin for waterfowl carcasses. Litter (bedded manure from poultry houses is a good source), straw, and carcasses are added proportionally to achieve the appropriate moisture content and carbon to nitrogen ratio. (Modified from Rynk, 1992.)

Food and water are also helpful in trapping wildlife for assessing disease control activities. When birds have been lured to a site, they may be captured by such means as drugs incorporated within feed, rocket nets, drop nets, walk-in and swim-in traps, or other means of preventing escape (Fig. 4.17).

A timely response to disease outbreaks can be facilitated if such factors as need for special permits, area closures, possible involvement of endangered species, and water purchase can be anticipated and addressed before an urgent situation arises.

Because of the potential complexity of biological interactions in animal relocation, field managers should seek the advice of disease control specialists whenever possible before taking independent action. As a general rule, animal dispersal is not recommended when infectious disease is involved unless it can be assured that the population being dispersed will not infect other wildlife. Also, it is important that water manipulation not produce conditions favorable to development of botulism or other disease problems.

Disinfection

The purpose of disinfection is to prevent the mechanical transmission of disease agents from one location to another by people, equipment, and supplies. Some viruses, bacteria, and other infectious agents have considerable environmental persistence. Disinfection of the local environment involved in a disease outbreak may be required to prevent recurrence of the disease when the site is used by other animals. Disinfection of a disease outbreak site should always be done under the direct guidance of disease control specialists.

Wash thoroughly the clothing worn during disease control (coveralls and clothes worn under protective raingear) before it is used again. Personnel should shower and shampoo their hair before leaving the site, if possible, but always before they go to other wildlife areas. Disinfect boots before entering vehicles when in contaminated areas, and disinfect all equipment to the extent possible before it is moved from the area (Fig. 4.18A and B). Give special attention to the underside of vehicles (Fig. 4.18C and D). Put motor vehicles through a car wash before moving them to other areas, and wash and clean boats and all-terrain vehicles before they leave the area. Large volume tanks and pumps that can be operated from mobile units such as trucks (Fig. 4.19) and boats are especially useful for holding and dispensing disinfectant.

Disinfection procedures require a suitable disinfectant, containers for that disinfectant once it has been diluted to appropriate strength, and a way of applying the disinfectant. Commercial disinfectants are available from farm supply stores and veterinarians. Refuge managers and other field managers should consider keeping a supply of disinfectant for general use. Chlorine bleach is a highly suitable disinfectant and it is available at most grocery stores. For general

Figure 4.15 Wildlife can be discouraged from use of areas by (A) propane exploders that function by the ignition of propane gas within the "cannon" due to the striking of a flint at a timed interval. With the exception of placing the cannon and maintaining a fuel supply, this activity does not require the presence of personnel. (B) Manual firing of cracker shells has also been used successfully to discourage wildlife use of areas. These fireworks-like shells should only be fired through a break-open type shotgun so that the barrel can be checked between shots to assure that there are no obstructions remaining in the barrels. These shells should not be used where they can fall into dry vegetation because of fire hazard.

Figure 4.16 Closure of areas is often needed to assist disease control operations. (A) Sign used to close Lake Andes National Wildlife Refuge during the duck plague die-off. (B) Sign used to delineate refuge area so that bird disturbance and movement was minimized during another South Dakota disease control operation.

use, dilute one part chlorine bleach with 10 parts water. Use stronger concentrations of one part bleach to five parts water for disinfecting heavily contaminated areas.

Stiff bristle brushes, buckets, and containers that can be used for foot baths and pressure or hand sprayers that can be used to dispense the disinfectant are also needed. The station contingency disease plan should identify readily available sources of these supplies and equipment.

When the disease problem involves an infectious agent, personnel handling contaminated materials should refrain from working with similar species or those susceptible to the disease for at least 7 days following completion of their disease control activities. For example, a field manager involved in an intensive avian cholera disease control operation on Monday should not band waterfowl in that refuge or elsewhere until Tuesday of the next week.

Personnel

Labor-intensive operations such as carcass removal and disposal sometimes require more personnel than are usually employed on an area. In some instances, specialized help such as low level aircraft flights for surveillance may be needed. The use of nonstation personnel for routine operations has a potential educational value. For example, the use of local sportsmen clubs to help with carcass collections during a major lead poisoning die-off has been highly effective in changing negative attitudes towards nontoxic shotuse.

Sportsmen clubs; retriever clubs; biology and wildlife classes at local universities and colleges; local chapters of conservation organizations such as the Audubon Society; the active military and National Guard, who also may provide valuable technical assistance; and similar groups have all provided volunteer assistance in combating disease problems

Figure 4.17 Various types of capture devices are useful for disease operations. (**A**) Rocket net being fired over Canada geese. (**B**) Snow geese captured by cannon nets. (**C**) Constructing a funnel trap to capture birds in a zoological park.—Figure 4.17 is continued on p. 36.

Photos by Milton Friend

Figure 4.17—continued (D) Capturing birds within a funnel trap. **(E)** Capturing flightless Canada geese in a drive trap. **(F)** Capturing waterfowl in a large, baited funnel trap. **(G)** Using drugged grain to capture birds in residential situations. When drugs are used, maintain close surveillance of the situation so that animals that become drugged, such as the bird **(H)** lying on its back, can be promptly collected before they are seized by other animals or drown if they venture into the water before the drug takes effect.

36 Field Manual of Wildlife Diseases: Birds

Photos by Milton Friend

Figure 4.18 Equipment and personnel should be disinfected to the extent possible before leaving disease operation areas. **(A)** Initial disinfection procedures should take place well within the contaminated area. **(B)** Boots and other items in contact with the ground should receive a second application of disinfectant at the point where entry is made into the "clean area," as is being done at the location where the specimen chest is being transferred. **(C and D)** Various types of spray units can be used to apply disinfectant to the underside of vehicles. Tires and wheel wells are the primary areas of concern as they may contain contaminated soil or animal fecal material from the disease area.

Figure 4.19 (A) Portable tank and pump mounted on a truck bed for dispersing disinfectant during duck plague control operation and (B) application of that disinfectant to a structure used to house birds. The long length of hose on this unit allowed all areas of major bird use to be reached from service and perimeter roads.

at various times and places. Sound judgment must be exercised in the selection and utilization of volunteers because of legal liability in case of an accident. Contingency plans should list groups and organizations and contact persons for each group, their telephone numbers, and an approximation of the work force and times of its availability (e.g., weekends only or Wednesday only). For technical assistance, list the specific type of personnel needed, such as bulldozer operator or helicopter pilot.

In addition to preparing a station contingency plan, wildlife personnel should become familiar with the other phases of disease control operations. Table 4.1 provides a descriptive outline of these phases. Especially relevant to field managers are the equipment and supply needs identified under the Disease Response Section of Table 4.3.

Response Modifications

Disease control operations can be seriously undermined without current assessment of wildlife morbidity and mortality and the cause of disease problems. When infectious or highly toxic agents are involved, early detection of disease problems is critical to preventing the problem from becoming widespread. Also, failure to accurately assess the cause of the die-off can result in control actions actually contributing to the magnitude of losses and spread of the problem. Different types of disease problems require different types of response. Do not assume that the current die-off is due to the same cause as previous die-offs that have occurred on the area or that only one disease agent is responsible. It is not uncommon for two or more causes of wildlife mortality to occur simultaneously in an area. Control of these different diseases may require opposite types of actions, thereby requiring that a more comprehensive strategy be developed for the disease control operation.

Refuge managers and other field biologists greatly influence the effectiveness of disease control operations by their responsiveness, knowledge of the local situation, how well they are prepared, the flexibility they maintain, their resourcefulness, and when possible, their ability to obtain appropriate technical assistance and training for combating disease problems. Timely and properly carried-out disease control activities can significantly reduce the magnitude of wildlife losses that might otherwise occur. When carrying out control activities, always consider the safety of the personnel involved.

Milton Friend and J. Christian Franson

Supplementary Reading

Friend, M., 1995, Disease considerations for waterfowl managers *in* Whitman, W. R., and others, eds., Waterfowl Habitat Restoration Enhancement and Management in the Atlantic Flyway (3): Dover, Del., Delaware Department of Natural Resources and Environmental Control, p. J24–J117.

Roffe, T.J., Friend, M., Locke, L.N., Evaluation of causes of wildlife mortality, *in* Bookout, T. A., editor, 1994, Research and management techniques for wildlife and habitats, Fifth ed.,: Bethesda, Md., The Wildlife Society, p. 324–348

Rynk, R., ed., 1992, On-farm composting handbook: Ithaca, N.Y., Northeast Regional Agricultural Engineering Service, 186 p.

Wobeser, G. A., 1994, Investigation and management of disease in wild animals: New York, N.Y., Plenum Press, 265 p.

Table 4.1 Outline of disease control operations.

I. Planning

   A. *Identify needs*

      1. Sources of additional personnel to help during disease emergencies. Potentially, these include
         a. State and Federal agencies
         b. Active military and National Guard
         c. Private conservation agencies
         d. Local sporting clubs
         e. Local universities

      2. Sources and availability of equipment and supplies for disease control operations (Appendix C)

      3. Special needs
         a. Burning permits
         b. Endangered species consultations
         c. Lodging and meal facilities for work crews
         d. Ability to attract and hold wildlife in site-specific areas by providing food, water, refuge, or other means
         e. Ability to deny wildlife use of specific areas by scaring devices and other means
         f. Ability to capture wildlife for sampling, immunization, or other needs

   B. *Record biological information*
      1. Daily and seasonal wildlife movement patterns within the general area
      2. Migration patterns and population peaks for major and endangered species
      3. Past history of diseases

   C. *Prepare contingency plan (See Tables 4.2 and 4.3.)*

II. Initial Response

   A. *Identify problems*

      1. Obtain diagnosis by submitting carcasses to a qualified diagnostic laboratory as soon as mortality or morbidity is evident. (See Chapters 3 and 4 for shipping procedures.)

      2. Conduct field investigation to determine extent of problem (i.e., species, number of wildlife, and geographic area involved).

      3. Identify special biological, political, or physical considerations associated with problem. Before proceeding further with II. B and C., seek the advice of a specialist.

   B. *Establish control of area*
      1. Close affected area, when warranted, to all but authorized personnel.

      2. Identify special work areas for disease control activities.
         a. Carcass disposal sites
         b. Laboratory investigations area
         c. Briefing area for news media and staff
         d. Vehicle parking
         e. Assembly areas for arriving workers
         f. Command post

      3. Initiate carcass cleanup, but do not dispose of carcasses without guidance from disease control specialists.

Table 4.1 Outline of disease control operations (continued).

   C. *Communications*

      Notify appropriate agency and nonagency personnel of die-off.

III. Disease Control

   A. *Response*

      1. Disease control actions are dictated by the type of disease, environmental factors, species involved, and other circumstances. Typically, actions associated with major die-offs require:
         a. Bringing personnel, equipment, and supplies on-site
         b. Organizing workforce, briefing workers about the problem, and assigning duties
         c. Carcass pickup and disposal
         d. Monitoring cause of mortality to detect changes in the cause of the problem (die-offs often involve more than a single cause and different control actions may be required for these different causes)
         e. Decontamination of personnel and equipment
         f. News media briefing sessions and "show-me" trips[1]

   B. *Management*

      1. Disease management activities often involve:
         a. Population manipulation such as removal, controlled movement including relocation and local concentration of wildlife populations, and population dispersal
         b. Habitat manipulation to prevent, attract, or maintain wildlife use of an area

      2. Decontamination of the infected environment, such as:
         a. Chemical treatment of land, water, and structures
         b. Vegetation and water removal (desiccation) to allow air and sunlight (ultraviolet) to destroy microorganisms

   C. *Controlled burning to remove vegetation and dispose of mechanical structures*

IV. Surveillance

   A. *Monitoring*

      After disease control operations have ended, the area should be kept under surveillance for 10 to 30 days to watch for additional flareups.

   B. *Investigations*

      This stage is also an appropriate time to conduct followup investigations of factors that helped cause and sustain the problem, and to carry out wildlife and environmental sampling to discern disease exposure patterns and environmental reservoirs of disease agents.

V. Analyses

Each disease control operation provides a learning experience. It is important to the success of future operations to evaluate what was done, the degree of success achieved, problems encountered, and what should have been done differently.

---

[1] Media briefing sessions and "show-me" trips should be conducted by personnel with comprehensive knowledge of the situation.

Table 4.2 Equipment and supplies used in disease control operations.

Activity	Equipment and supplies
A. Carcass Collection	
1. Transportation of personnel	a. All-terrain and four-wheel vehicles, snowmobiles b. Airboats, canoes, other boats c. Helicopter d. Waders, snowshoes
2. Transportation of carcasses	a. Large, heavy-duty plastic bags b. Plastic trash cans with lids c. Sleighs and trailers d. Trucks, boats e. Strapping tape and other means of securing closure of containers
B. Carcass Disposal	
1. Burial	a. Earth-moving equipment for digging trenches or pits (bulldozer, backhoe) b. Shovels c. Lime or fuel oil to spread on carcasses d. Any applicable permits
2. Incineration	a. Portable incinerators and fuel b. Local permanent incinerator c. Earth-moving equipment for digging trenches or pits (bulldozer, backhoe) d. Burning permits e. Shovels f. Metal grates and cinder blocks for building burning platforms g. Sheet metal or metal roofing for heat reflectors h. Fuel for burning carcasses (wood, coal, rubber tires, fuel oil, napalm) i. Fire suppression equipment
3. Composting	a. Composting bin made of pressure-treated lumber b. Straw and manure to alternate with layers of dead birds c. Trucks to transport carcasses, straw, and manure
C. Sanitation Procedures	
1. Decontamination of environment	a. Chemical disinfectants and structures b. Pumps and suction apparatus for drainage of water areas c. Buckets, brushes d. Spray application by aircraft, power systems mounted in trucks and boats, and hand-carried spray units
2. Protection of personnel and prevention of mechanical movement of disease agents to secondary locations by people and equipment	a. Raingear, coveralls, rubber gloves, rubber foot gear, hats b. Spray units and chemical disinfectants c. Plastic bags for transportation of field clothes to laundry d. Brushes, buckets e. Disposable gloves, hats, coveralls, and foot coverings

Table 4.2 Equipment and supplies used in disease control operations (continued).

Activity	Equipment and supplies
D. Field Communications	
1. Field activities	a. Portable radios or cellular telephones for communication between field personnel b. Radios in vehicles for communication between field units and between units and command post
2. Information activities	a. Word processor or typewriter for preparing briefing documents b. Maps, acetate, and other supplies for overlays depicting die-off and control activity information c. Telephone lines for communication with others d. Transportation for news media "show-me" trips
E. Surveillance and Observation	
1. Field activities	a. Aircraft and pilots certified for low-level flights (500 feet and below) for monitoring wildlife populations and environmental conditions b. Binoculars and spotting scopes
2. Office activities	a. Maps, acetate, and other supplies for tracking the progress of events and wildlife populations associated with die-off b. Telephone for contacting others to trace movement of migrant bird populations that might enter problem area or that have departed problem area
F. Wildlife Population and Habitat Manipulation	
1. Denying wildlife use of an area	a. Aircraft, boats, snowmobiles, and other motorized means of hazing wildlife populations b. Propane exploders c. Cracker shells, break-open shotguns, and protective face shield d. Audio systems and other scare devices e. Pumps for draining water or adding water to areas
2. Concentration and maintenance of wildlife in a specific area	a. Grain and other sources of food b. Pumps and water to provide habitat c. "No Hunting" and "Area Closed" signs to provide temporary refuge area
G. Wildlife Sampling and Monitoring	
1. Wildlife capture	a. Cannon nets and other capture equipment b. Grain and other baits to lure wildlife to capture site
2. Wildlife marking	a. Visible marking devices such as paint, neck collars, and other devices b. Permanent marking devices such as leg bands and ear tags (see Bookhout, 1994) c. Temporary marking devices such as radio transmitters

Table 4.3 Station disease contingency plan.

I. Introduction

A. Size, configuration, and other important characteristics of station area conveyed with help of tables, maps, photographs, station brochures, public use maps, and similar documents

B. Record of previous disease outbreaks, including nature of disease, species involved, magnitude of die-off, and season and year (Table 4.4)

II. Disease Surveillance

A. Brief outline of current surveillance activities on station and adjacent areas — State, Federal, and private

B. Identify disease reporting and notification procedures (names, titles, organization, and telephone numbers of persons to be contacted)

III. Disease Response

A. Logistical considerations

1. Personnel sources (telephone numbers, addresses, names of contact persons)
   a. Local, State, and Federal agencies (military, university)
   b. Sporting clubs and volunteers

2. Equipment (types and numbers on-site, and sources off-site)
   a. Vehicles (conventional and all-terrain)
   b. Aircraft (fixed-wing and rotary)
   c. Earth-moving equipment (backhoe, bulldozer)
   d. Pumps (for flooding or draining marshes)
   e. Boats (motor, self-propelled, air boats)
   f. Radios (portable and fixed); during nonfire seasons the
      National Interagency Fire Center
      3905 Vista Avenue
      Boise, Idaho 83704
      (208) 389-2458
      is a potential source for obtaining assistance for very large communication needs
   g. Incinerators
   h. Composting bins
   i. Decontamination units (sprayers)
   j. Scaring devices (propane exploders, sirens)
   k. Freezers
   l. Portable toilets (construction-site type)

3. Supply sources (Identify sources, addresses, and telephone numbers of local or closest sources.)
   a. Disinfectants and chemicals
   b. Plastic bags
   c. Fuel for carcass burning
   d. Field clothes (gloves, rainwear, coveralls, boots)
   e. Plastic trash barrels, tubs, scrub brushes
   f. Scaring devices (cracker shells, fireworks); provide contact telephone number and address for local animal damage control office
   g. Dry ice and liquid nitrogen
   h. Grain and other wildlife foods
   i. Nearest shipping address for air and ground receipt of goods and supplies

Table 4.3   Station disease contingency plan (continued).

---

    4. Lodging for temporary personnel assigned to disease control operation

    5. Food
       a. On-site capabilities
       b. Off-site capabilities (Give consideration to early and late hours.)

    6. Identify working areas (Diagrams are sufficient; limited narrative may also be required; Fig. 4.2.)
       a. Clean areas
          1. Command post (must have adequate telephones)
          2. News media briefing room
          3. Parking
          4. Eating areas
          5. Staff assembly and rest areas
          6. Equipment and supply receipt
          7. Other
       b. Transition areas
          1. Decontamination of personnel
          2. Decontamination of equipment
       c. Contaminated areas
          1. Carcass disposal
          2. Laboratory investigations
          3. Animal holding

B. *Biological considerations (Provide data in charts, figures, photographs, maps, tables.)*

    1. Species and population data
       a. Major species (Identify by season of presence, relative abundance, and peak population periods.)

    2. Wildlife movement patterns (Figs. 4.3 through 4.5)
       a. Daily
       b. Seasonal
       c. Production and dispersal patterns

    3. Weather patterns
       a. Freeze-up and ice-out periods
       b. Major periods of precipitation and drought
       c. Other (temperature profiles, major periods of haze, fog, and high winds)

    4. Habitat and population manipulation potential
       a. Methods (water manipulation capability, feeding)
       b. Anticipated population response to habitat (movement, concentration, dispersal)

C. *Communications (Provide lists of principal local and regional contact personnel and telephone numbers.)*
    1. State agencies
       a. Conservation
       b. Agriculture
       c. Health department
       d. University diagnostic laboratories

Table 4.3 Station disease contingency plan (continued).

    2. Federal agencies
       a. Environmental Protection Agency
       b. U.S. Department of Agriculture
       c. U.S. Public Health Service

    3. Other organizations
       a. Cooperating organizations (e.g., area representatives of Audubon Society, National Wildlife Federation, Ducks Unlimited)
       b. Local sporting clubs
       c. Private wildlife area managers
       d. Local game breeder organizations
       e. Local domestic animal husbandry and production operations

    4. Media
       a. Television
       b. Radio
       c. Newspapers

IV. Supplemental Information

    A. Location of nearby laboratories (hospitals, universities, county and State facilities)

    B. Federal and State permit status for biological collections

    C. Burning permits

    D. Regulatory requirements

    E. Background information (e.g., water sources, water-quality data, potential sources of disease transmission between wildlife and domestic animal concentrations)

    F. Identification and location of adjacent or nearby wildlife refuges, management areas, and private reserves

    G. Identification of unusual or politically sensitive aspects of area

Table 4.4 Example of a disease outbreak summary for a wildlife management area[1]. [—, no data available]

Disease	Date	Location	Principal species involved	Estimated population at risk[2]	Carcass count	Estimated total	Control efforts	Diagnostic laboratory
Unknown	Aug.–Oct. 1972	Unit 6 and Yellowleg Flat on adjacent State land	Northern pintail, teal	5,000	1,500	3,000	None	None
Lead poisoning	April 1980	Mud Lake	Redhead, tundra swan	—	75	200	None	State diagnostic laboratory
Lead poisoning	Dec. 1983	Mud Lake	Mallard, Canada goose	10,000	—	150	Blinds were relocated in 1984	Veterinary Science Dept., State college
Unknown	Jan. 1983	Mud Lake	Muskrat	3,000	100	500	None	Veterinary Science Dept., State college
Avian botulism	July–Sept. 1985	Units 3, 4, 5	Shorebirds, northern pintail, teal	25,000	2,200	10,000	Drained Unit 4, flooded 3 and 5	National Wildlife Health Center

[1] The material in this table is fictitious and for illustration only.
[2] Number of animals using the area involved in the die-off.

Chapter 5
Euthanasia

Background

Euthanasia means to cause humane death. Some current euthanasia techniques may become unacceptable over time and be replaced by new techniques as more data are gathered and evaluated. The following information and recommendations are based largely on the 1993 report of the American Veterinary Medical Association (AVMA) Panel on Euthanasia. The recommendations in the panel report were intended to serve as guidelines, and they require the use of professional judgement for specific situations. Ultimately, it is the responsibility of those persons carrying out euthanasia to assure that it is done in the most humane manner possible.

There is no perfect euthanasia technique appropriate to all situations. What is sought in each instance is immediate insensitivity of the animal to pain as a result of depression of the central nervous system (brain and spinal cord). The AVMA panel in its evaluation considered the following to be important factors to consider when selecting a euthanasia method:

Considerations for selecting a euthanasia method

- Does the method cause the animal to lose consciousness and die without causing the animal pain, distress, anxiety, or apprehension?
- How much time does the method require to induce unconsciousness?
- Is the method reliable?
- Does the method put personnel at risk of injury or health problems?
- Is the method irreversible?
- Is the method compatible with the purpose of euthanasia?
- Will the method cause distress and anxiety among observers and personnel?
- Does the method interfere with or detract from the subsequent evaluation, examination, and use of tissue?
- Are drugs required by the method available? Can the drugs be abused by humans?
- Is the method appropriate for the animal age and species?
- Is the equipment required by the method in proper working order?
- Is the method cost-effective?

Methods of euthanasia are physical or chemical. Physical methods of euthanasia include cervical dislocation, decapitation, stunning and removal of blood, and gunshot. Chemical methods of euthanasia involve introducing a toxic agent into the body by injection or inhalation. After completing euthanasia, be certain that specimens being collected are properly identified, preserved, and packaged for transportation to the diagnostic laboratory (see Chapter 2, Specimen Collection and Preservation, and Chapter 3, Specimen Shipment). Be sure to indicate the euthanasia technique used.

Physical Euthanasia

Cervical Dislocation

Cervical dislocation can be used without any special equipment to euthanize small birds and ducks. The dislocation must take place at the base of the brain, or within the upper one-third of the neck (the cervical spine). Grasp the base of the bird's skull in one hand and its body, usually at the base of the neck, in your other hand. Pulling rapidly and firmly in opposite directions will separate the spinal cord (Fig. 5.1). Cervical dislocation can be used for larger birds, like geese, by separating the upper cervical spine with an emasculatome, which is available from veterinary supply

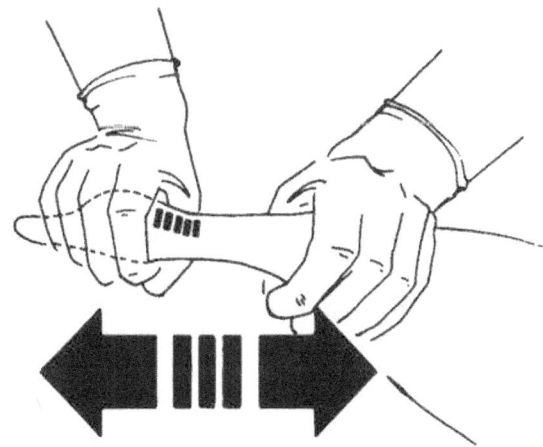

Figure 5.1 Cervical dislocation procedures. The brain can be separated from the spine in small- to medium-sized animals by grasping the animal at the base of the skull with one hand, at the base of the neck with the other, and pulling rapidly and firmly in opposite directions with a strong snapping action.

stores. As with all methods, learn how to properly use this instrument before applying it to a live animal.

Cervical dislocation may upset the casual observer because animals, especially birds, convulse for several seconds to minutes after death. These movements are due to spinal reflexes and the animals do not feel pain. This technique is effective, rapid, inexpensive, and only minimally affects diagnostic testing.

Decapitation

Severing the head from the neck is an effective method of euthanasia for small mammals and any size bird, but it is often used for larger waterfowl. Use a knife, machete, hatchet, or bolt cutters to ensure that the spinal cord, encased in the cervical spine, is severed. The same convulsions seen after cervical dislocation will follow decapitation. This technique has similar attributes as cervical dislocation. However, take care to prevent injuries to personnel resulting from the use of the sharp implements, and to prevent exposing personnel to toxic or infectious agents that may be in the blood.

Stunning and Exsanguination (Removal of Blood)

This method requires striking the center of the skull to render the animal unconscious, followed by severing the major blood vessels in the neck, and allowing the animal to bleed out. Do not use this technique if the brain is required for diagnostic tests.

Gunshot

Shooting animals in the head, or the neck if the brain is needed for diagnostic purposes, with a small caliber rifle can be used as a method of euthanasia. Training and experience are required to assure a humane death, and also to reduce the human safety hazards.

Chemical Euthanasia

Extreme caution is required for the use of chemical euthanasia, because of the potential hazards for humans. These procedures should be carried out only by trained individuals who are properly authorized to use the appropriate chemicals.

Inhalant Anesthetics

Several inhalant anesthetics have been used for wildlife euthanasia. Halothane is often the inhalant selected because it rapidly induces unconsciousness. Enflurane also rapidly induces unconsciousness, but seizures under deep anesthesia from enflurane are more common than from halothane. Methoxyflurane vaporizes slowly and, therefore, has a longer anesthetic induction time, which can cause the bird to become agitated. Isoflurane has a rapid induction time, but its odor can cause the animal to hold its breath, thereby delaying unconsciousness. Nitrous oxide has a low potency and is available only in gas form; other anesthetics are purchased as a liquid, and they vaporize at room temperature and normal air pressure. Nitrous oxide can be used in combination with other inhalants to speed anesthesia, but it should not be used alone because animals often become agitated and distressed before they lose unconsciousness.

To administer an inhalant anesthetic for euthanasia of an individual bird, prepare a cone (from a syringe case or other plastic material) that will fit snugly when it is placed over the beak and nares (Fig. 5.2). Pour a small amount of the anesthetic agent on a piece of cotton, tissue, or cloth, and place it in the narrow part of the cone. Restrain the bird; put the open end of the cone over the beak and nares, and continue restraining the bird until it becomes unconscious. Restraint can then be discontinued, but keep the cone in place for several minutes before checking to assure that the bird is

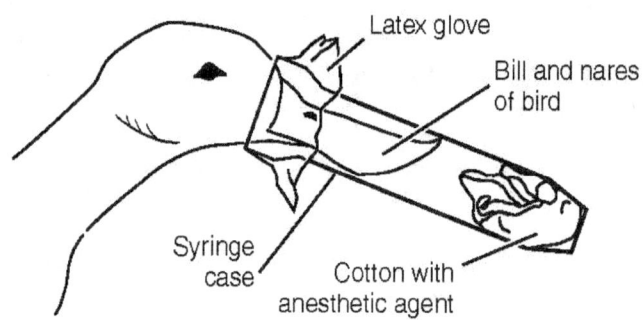

Figure 5.2 A cone prepared from an empty syringe case can be used for euthanasia. Tape a piece of latex glove over the open end of the cone, and cut a slit in the latex so that the bill and nares fit through it. Place the anesthetic agent on a piece of cotton in the end of the cone.

dead. Alternatively, place an individual bird, or several small birds, in a cage or crate; cover it with plastic or place the cage in a covered plastic barrel. Place the cotton, tissue, or cloth soaked with anesthetic agent inside the chamber with the birds and tie or otherwise seal the plastic to prevent the vaporized agent from escaping (Fig. 5.3). Cold temperatures will decrease the rate at which the liquid becomes gas. Small mammals can be euthanized by similar procedures.

A animal exposed to anesthetic gas may pass through an "excitation phase" before it becomes unconscious; it may vocalize and appear to struggle for a short time. This behavior may be distressing to the casual observer and it can be dangerous for the handler, depending on the species. It is important to assure that the animal is dead, and not just unconscious, before shipment, necropsy, or disposition. After removing the animal from the gas environment, it may wake up quickly, with little warning. Remember this when working with raptors, carnivores, and other biting animals.

Because all of these gases constitute a human health hazard, including the potential to cause spontaneous abortion and congenital abnormalities, the workplace must be well-ventilated.

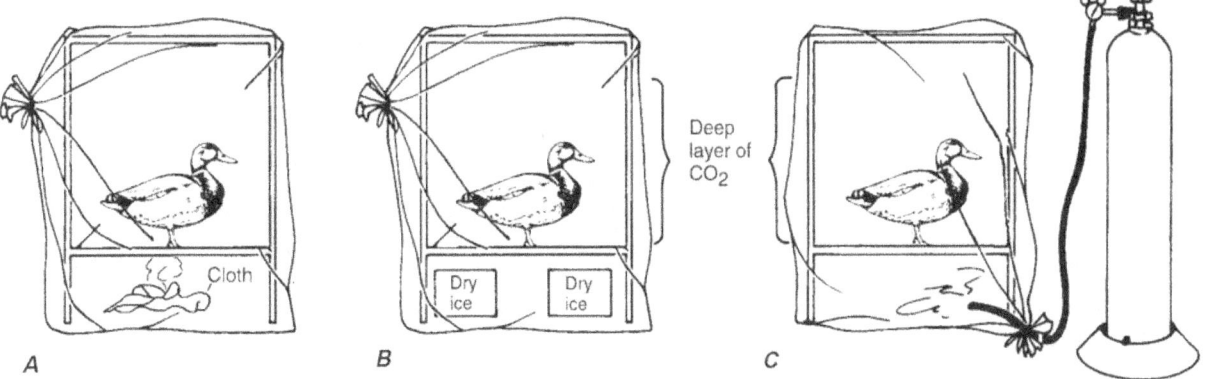

Figure 5.3 Use of cage enclosed with plastic for euthanasia of birds. **(A)** Anesthetic agent placed on a piece of cloth under the cage. **(B)** Evaporation of dry ice. **(C)** Direct application of carbon dioxide gas. Because this gas is heavier than air, a deep layer of gas must be built up so that the animals being euthanized cannot get above the gas. The chamber containing the animals must not be airtight or gas buildup may result in an explosion. Openings should be at the top of the chamber.

Toxic Gas

Toxic gases such as carbon monoxide (CO) or carbon dioxide (CO_2) may be useful when many small birds or animals must be killed. Keep in mind that, even at concentrations of less than 1 percent, carbon monoxide is lethal and represents a substantial human safety hazard because it is highly toxic and difficult to detect. In concentrations exceeding 10 percent, carbon dioxide can be flammable and explosive. Work with this gas, as with anesthetic gases, must be conducted in an open area away from electrical equipment.

Carbon monoxide and carbon dioxide may be purchased as compressed gases in cylinders. Dry ice can also be used as a source of carbon dioxide. If dry ice is used, protect animals from contact with it. Cages covered with plastic bags (Fig. 5.3) or plastic garbage cans can be used as killing chambers, but the cages must be vented to allow displacement of air within the chamber by the toxic gas. Leave the animals in the chamber until breathing and heartbeat have ceased.

Lethal Injection

To administer lethal injections, personnel must be trained in injection techniques and proper doses as well as in the safe handling and disposal of needles, syringes, and drugs. Federal drug regulations make the use of these agents, except by licensed veterinarians, largely impractical. Lethal injections can be used for any animal that can be given an intravenous injection, but they are probably most useful for mammals and large birds, such as geese.

Sleepaway® (made by Ft. Dodge Laboratories, Inc., Ft. Dodge, Iowa) and Beuthanasia – D Special® (made by Burns-Biotic Laboratories, Inc. Omaha, Neb.) are concentrated barbiturate solutions plus additives. The solutions are inexpensive, but, due to the potential for human abuse, require licensing by the Federal Drug Enforcement Administration (DEA) for purchase, use, and storage. Considerable record-keeping of use of the drug is required by the DEA.

Lethal injections may not be appropriate in certain instances because drug residues interfere with some tests. Check first with the diagnostic laboratory to see if the proposed euthanasia technique is compatible with the testing to be performed.

The need for individual handling and injection of each animal generally precludes using this technique for euthanasia of more than a few birds or animals per event. Proper disposal of carcasses is needed to prevent secondary poisoning of scavenger species in situations where more birds or animals are euthanized than are needed for diagnostic testing.

J. Christian Franson
(Modified from an earlier chapter by Patricia A. Gullet)

Supplementary Reading

Andrews, E.J., Chairman, 1993 Report of the American Veterinary Medical Association panel on euthanasia, 15 January 1993: Journal of American Veterinary Medical Association 202:229–249.

Chapter 6
Guidelines for Proper Care and Use of Wildlife in Field Research

Prologue

Public attitudes towards animals continue to change over time. These changes apply to wildlife along with other species, and in recent years, attitudes have been increasingly oriented toward assuring that all species receive proper care whenever human interactions are involved. Guidance regarding the application of euthanasia is provided in the previous chapter. This chapter provides basic guidelines for the proper use of wildlife in field investigations. We believe this previously published information from The Wildlife Society is sufficiently important to include in this field manual. The Wildlife Society has been kind enough to grant permission for this reproduction. The scope of this chapter extends to all wildlife, and the application of this material extends beyond research to all wildlife investigations. This chapter is reproduced, with the addition of illustrations and minor modifications, as it appeared in Research and Management Techniques for Wildlife and Habitats (Bookhout, 1994), and, thus, it deviates from the format for the rest of Volume I.

Introduction

Philosophy

Scientists do not operate in a vacuum, but rather in an arena with responsibilities to the organisms they study and to society. Professional scientists must consider the effects of their activities on the organisms under study, on the validity of study results, and on the use of these organisms by other segments of society. The Wildlife Society recognizes these relationships and supports the sound application of responsible methods for the conduct of animal research in all field and laboratory investigations. This position reflects our ethical and moral concerns regarding human interactions with each other and with other species, and recognizes the scientific benefits of investigations that are not compromised by the manner in which animals are handled or maintained. These concerns are the foundation for our philosophy that responsible methods of animal investigations must include all animal species. Wildlife professionals are urged to apply high standards of animal care and maintenance, and responsible methods of experimental procedures, in conducting each animal investigation.

Purpose

These guidelines are intended for field research involving wild animals. The variety of wild vertebrates investigated and of conditions encountered precludes provision of specific information applicable to each situation. Lists of useful references for those seeking more specific information are provided in the Appendices.

Background

The Animal Welfare Act (7 U.S.C. 2131, and following) was enacted on 23 December 1985, with amendments including Parts 1, 2, and 3 (9CFR); Fed. Register 4(168) 3611236163, effective 30 October 1989. The Act established definitions of terms (Part 1) used in the regulations (Part 2) and standards (Part 3) for the humane handling, care, treatment, and transportation of regulated animals used for research or exhibition purposes, sold as pets, or transported in commerce. Excluded from the provisions of the Act are cold-blooded vertebrates, birds, rats *(Rattus)* and mice *(Mus)* bred for use in research, horses and other farm animals used or intended for use as food and fiber, and livestock and poultry used or intended for use in improving animal nutrition, breeding, management, or production efficiency, or for improving the quality of food or fiber. Also excluded are field studies as defined by the Act, i.e., "any study conducted on free-living wild animals in their natural habitat, which does not involve an invasive procedure, and which does not harm or materially alter the behavior of the animals under study." Collection of blood samples, ear-notching, branding, and collection of routine weight and measurement data are examples of exempted activities.

Exclusion of animal species under the Act removes reporting requirements and reduces oversight by the U.S. Department of Agriculture, but does not negate coverage of these species under guidelines established by other agencies. Thus, fish, amphibians, reptiles, birds, and mammals are covered by the National Science Foundation (NSF) and the National Institutes of Health (NIH) guidelines. This coverage is extended to research grants funded by these agencies and to Federal agencies, such as the U.S. Fish and Wildlife Service, that function under the guidelines of the Interagency Research Animal Care Committee.

Role of Institutional Animal Care and Use Committees

A major requirement of the Animal Welfare Act and NIH/NSF guidelines is establishment of institutional facility Animal Care and Use Committees (ACUCs). The function of ACUCs is critical to the conduct of scientific investigations. Each ACUC must consist of at least three members, one of whom is the attending veterinarian of the research facility (or another veterinarian with delegated program responsibility) and one of whom is not affiliated in any way with the facility other than as a committee member. The purpose of the ACUC is to evaluate the care, treatment, housing, and use of animals and to certify compliance with the Act. This process involves evaluation of experimental protocols to ensure that animal pain and distress are minimized. ACUC oversight includes laboratory and field studies. Consensus recommendations on effective ACUCs for laboratory animals were provided by Orlans and others (1987). Differences between laboratory and field studies (Orlans, 1988) do not negate the need for application of responsible methods for care and use of animals during field research activities. ACUCs and field investigators must work together in reaching agreement on appropriate protocols and methods for specific circumstances of the field research to be undertaken. "Standards for humane treatment of wild vertebrates must continue to be constantly developed, applied, and re-examined. Practices that are acceptable today may well prove unacceptable to tomorrow's scientific community, and/or to society in general" (Canadian Council on Animal Care, 1984, p. 192). Wildlife professionals are strongly encouraged to serve on ACUCs and contribute their specific knowledge about the needs of free-living wildlife to help guide Committee actions involving protocol reviews for field investigations. Wildlife professionals also are encouraged to publish manuscripts that document the proper care and maintenance of free-living wildlife species during field investigations. Development of this information by knowledgeable field biologists provides specific species information for guiding ACUC decisions involving protocol reviews.

Field research study conditions for wildlife

Irrespective of the species or circumstances involved, wildlife professionals should satisfy the following conditions for all field research studies. Written assurance that these conditions will be met is a prerequisite for project consideration and funding by many granting agencies. These conditions also are principal points for evaluation by the ACUC.

1. Procedures employed should avoid or minimize distress to animals consistent with sound research design.

2. Procedures that may cause more than momentary or slight distress to animals should be performed with appropriate sedation, analgesia, or anesthesia, except when justified for scientific reasons in writing by the investigator in advance.

3. Animals that otherwise would experience severe or chronic distress that cannot be relieved will be euthanized at the end of the procedure or, if appropriate, during the procedure.

4. Methods of euthanasia will be consistent with recommendations of the American Veterinary Medical Association (AVMA) Panel on Euthanasia (Andrews and others, 1993) unless deviation is justified for scientific reasons in writing by the investigator. However, species differences must be considered. As noted elsewhere, "The AVMA recommendations cannot be taken rigidly for ectotherms; the methods suggested for endotherms are often not applicable to ectotherms with significant anaerobic capacities" [American Society of Ichthyologists and Herpetologists (ASIH), the Herpetologists' League (HL), and the Society for the Study of Amphibians and Reptiles (SSAR), 1987, p. 2].

5. Living conditions of animals held in captivity at field sites should be appropriate for that species and contribute to their health and well-being (Fig. 6.1). Specific considerations include appropriate standards of hygiene, nutrition, group composition and numbers, provisions for refuge and seclusion, and protection from weather and other forms of environmental stress. The housing, feeding, and nonmedical care of these animals must be directed by a scientist trained and experienced in the proper care, handling, and use of the species being maintained or studied. Some experiments (e.g., competition studies) will require the housing of mixed species, possibly in the same enclosure. Mixed housing also is appropriate for holding or displaying certain species.

Figure 6.1 (A) Temporary "field hospital" for recovery of waterfowl with avian botulism and (B) a more permanent structure used for the same purpose. The permanent structure provides shade and has a cement floor for easy cleaning and disinfection and has a water trough the birds can swim in. For both situations, periodic inspection of the pens during each day is needed for the detection and prompt removal of dead birds. Prolonged use of the temporary hospital should be avoided because of fecal contamination that cannot be readily neutralized. By segmenting the temporary facility into separate pens, "pasture rotation" followed by treatment of vacated areas can help provide reasonably clean holding areas. An alternative would be to construct pens that can be easily moved. A tarpaulin or other covering placed over the top of the temporary structure or placement of such structures under the shade of trees will enhance bird survival by minimizing heat stress.

Wildlife Observations and Collections

General

Before initiating field research, investigators must be familiar with the target species and its response to disturbance, sensitivity to capture and restraint, and, if necessary, requirements for captive maintenance to the extent that these factors are known and applicable.

To the extent feasible, animals with dependent young should not be removed from the wild unless the young also are collected or removed alive and provided for in a manner that facilitates their survival beyond the period of dependency. Whenever possible, voucher specimens of animals, their tissues, and parasitic and microbial fauna collected during field investigations should be deposited in catalogued scientific collections available to others within the scientific community, to provide for maximum use of animals collected.

The number of animals required for investigations depends on questions being investigated, but provision of adequate sample size is essential to assure scientific validity of results and avoid unnecessary repetition of studies. Removal of animals from a population (either for translocation or by lethal means) should be restricted to the fewest animals necessary to achieve established goals, but should never jeopardize the population's well-being.

Investigator Disturbance and Impacts

Potential gains in knowledge from field investigations must be balanced against the potential adverse consequences associated with the conduct of the study (Animal Behavior Society/Animal Society for Animal Behavior, 1986). A high level of sensitivity to the potential, indirect effects of investigator presence and study procedures must be maintained, and appropriate steps must be taken to minimize these effects. Examples of secondary impacts associated with field investigations may include nest desertion, abandonment of young, increased vulnerability to predation, traumatic injuries and mortality resulting from panic escape response, cessation of breeding activities, increased energy use by disrupted species, altered feeding behavior, habitat abandonment, long-term marring of fragile habitats, increased vulnerability to hunting, introduction of disease, and spread of disease. These effects may impact either research (target) or other (nontarget) species. Investigators should use available information on secondary impacts as a basis for taking appropriate precautions to minimize known potential impacts.

Such factors as frequency and timing of investigator presence can influence greatly research effects on target and nontarget species. When applicable, remote methods of data collection can be used to minimize disturbance. Also, habitat conservation should be practiced rigorously during all field investigations, and every reasonable effort should be made to leave the study area and access to it as undisturbed as possible.

Museum Collections and Other Killed Specimens

Collection of animals often is an essential component of field investigations. These collections may involve systematic zoology, comparative anatomy, disease assessments, food preference studies, environmental contaminant evaluations, and numerous other justifiable causes and scientific needs.

Assessment of the need should involve appropriate evaluations to determine that the proposed collections will provide scientific data that are not duplicative of information already available in the scientific literature (unless confirmation of these data is needed), or that are presently available in accessible scientific collections and repositories. These evaluations also should assess whether suitable information can be obtained from alternative methods that do not require taking live animals. Methods of collection must be responsible, minimize the potential for the taking of non-target species, and not compromise the purpose of the study. In some instances it is possible and practical to capture animals and then apply approved euthanasia methods (see Andrews and others, 1993). However, for many field studies the only practical means of animal collection are those involving direct killing as the initial step in the collection process. Under these conditions, methods of vertebrate collection must be as species or age-class specific as possible. Methods must not be employed that compromise data evaluation. Appropriate provisions also must be made for proper collection and preservation of biological materials associated with the purpose of the study. Improperly collected or preserved specimens that fail as useful and valid sources of scientific information negate the purpose of collecting the animals.

When shooting is the collection method, the firearm and ammunition should be appropriate for the species and purpose of the study. The shooter should be sufficiently skilled to be able to kill the animal cleanly. If an animal is wounded, immediate attention must be given to appropriate follow-up actions to kill it quickly. Attention also must be given to the animal's location to assure it can be killed cleanly and that it will be readily accessible for retrieval and data collection.

Kill traps, with attendant baits and attractants, are acceptable and effective for animal collection when used in a manner that minimizes the potential for collecting nontarget species. All traps should be checked regularly, at least daily, to prevent specimen loss from scavengers and predators and should be rendered nonfunctional when not in use.

Live traps for nocturnal species should be set before dusk, checked as soon as possible after dawn, and closed during the day to prevent capture of nontarget species. Live traps for diurnal species should be shaded or positioned to avoid full exposure to the sun. Live traps for nonfossorial mammals should enclose a volume of space adequate for movement within the trap; for fossorial mammals, trap diameter should approximate that of the burrow. The live-trap mechanism should not cause serious injury to the animal, and trap doors should be effective in preventing the captive animal from becoming stuck or partially held in the door opening (Ad Hoc Committee on Acceptable Field Methods in Mammalogy, 1987). Pitfalls used as live traps should contain adequate food to last until the next trap check and should be covered to keep out rain or punctured to permit drainage.

Blood and Tissue Collections

Only properly trained individuals proficient in the required techniques should attempt to take tissue samples from live animals. Collection of tissue samples requires proper animal restraint to avoid traumatic injuries to the animal and to the investigator taking the samples. Use of anesthetics is required when the sample procedure will cause more than slight or momentary pain. The institution/facility ACUC is the proper source for evaluating collection methods and use of anesthetics for noninvasive and invasive procedures for tissue collections from live animals.

Blood is the most common tissue sampled from live animals. A conservative rule of thumb is that the amount of blood drawn at one time from a healthy animal that is to be kept alive should be no more than 1 percent of its body weight. However, the amount of blood taken should be limited to actual needs, rather than the maximum amount that can be safely taken, to reduce stress from handling. Appropriate equipment (e.g., needle size) and sample site should be selected to provide the amount of blood needed for the species involved.

The three most common sites for bleeding birds are the jugular vein of the neck, medial-metatarsal vein of the leg, and brachial vein of the wing (Fig. 6.2). The jugular is preferred for bleeding most birds because of its accessibility and size and the relative ease with which large samples can be taken. The medial-metatarsal vein is not recommended for use in raptors, nor is the brachial vein in large birds such as cranes. Feathers should not be plucked to locate these veins. Birds also can be bled from a variety of other sites including the heart and occipital venous sinus. However, there is seldom reason to assume the risk associated with these sites for nonlethal sampling, even though successful application of these techniques has been demonstrated.

Multiple sites also are available for drawing blood samples from mammals (Fig. 6.3A). Venipuncture of the cephalic,

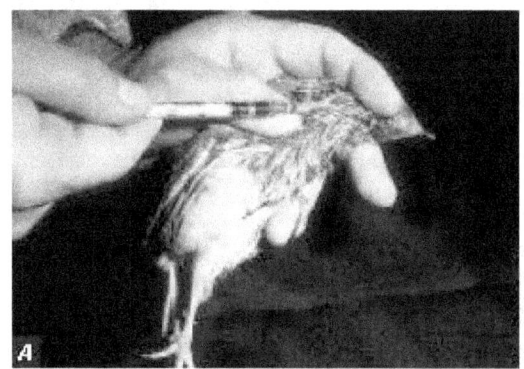

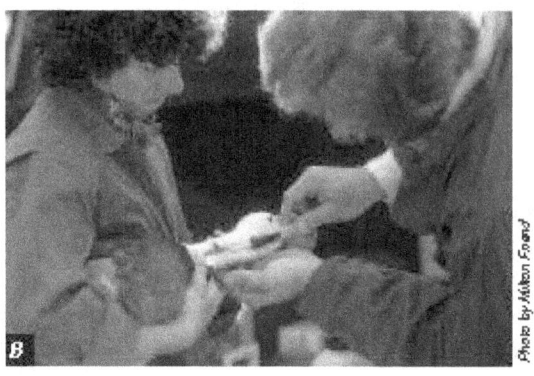

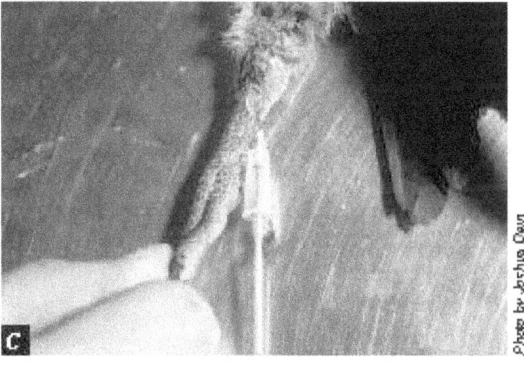

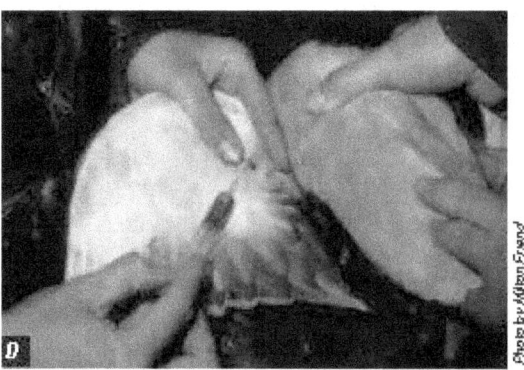

Figure 6.2 Blood can be drawn from a variety of sites and not jeopardize the well-being of birds when properly trained investigators utilize appropriate techniques and equipment for that task. *(A)* Proper restraint for jugular bleeding of small birds is shown and is best accomplished by the person doing the bleeding. *(B)* For larger birds such as this blue goose, the handler supports the body weight and restrains the wings by cradling the bird against her body while controlling the head with her other hand. *(C)* The bleeder normally controls the leg that blood is being drawn from when the medial-metatarsal vein is used. *(D)* Bleeding from the brachial vein. Care must be exercised so as not to apply excessive torque to the wing.

femoral, tarsal, or jugular vein, the orbital sinus, or various venous plexuses are common procedures. In some instances cardiac bleeding also is acceptable. Need for anesthesia with any of these procedures depends upon methods of restraint, species being bled, physical condition of the animal, and volume of blood needed. In reptiles, such as turtles, sites for blood collection are more limited (Fig. 6.3B).

Restraint and Handling

General

Safety of both wild animals and scientists who are studying them should be the primary consideration when physical contact between them is judged to be necessary and unavoidable. Nondomesticated animals almost without exception will try to elude capture, handling, and restraint. The means by which a particular animal may try to prevent capture will vary with the species, sex, physiologic condition, and temperment of the individual. In attempts to elude capture, wild animals are capable of inflicting severe damage to themselves and their potential captors.

Behavioral characteristics of wild animals often may be used to assist the potential captor. For instance, animals in a small pen or cage often voluntarily will enter a smaller container to hide and evade capture. If that container provides adequate restraint, the potentially dangerous work of securing the animal can be accomplished more easily. Every effort involving contact between wild animals and humans should be carefully conceived and skillfully executed. Personnel involved must know the habits and behaviors of the animal to be handled; the plan must have suitable alternatives; and a genuine regard for the physical, physiological, and psychological welfare of the animal must be of deep concern to those actually handling the animals. If the planned and alternate procedures do not appear to be satisfactory, the responsible thing to do is cease immediately and return to the planning stage. Trying to enforce unworkable procedures in a particular situation is a virtual guarantee of injury to either the animals or the humans involved.

Physical Restraint

For many situations physical restraint is the most appropriate method of animal handling, because of risks from chemical immobilization to the animal and humans when potentially toxic drugs are used. When physical restraint is selected, an adequate number of sufficiently trained and equipped personnel must be available to complete the task safely. Location and type of capture, as well as procedures to be performed and time required to accomplish them, will influence the particular type of physical restraint. Gloves, catch poles, ropes, nets, body bags, holding boxes, corrals, squeeze chutes, or more sophisticated mechanical holding devices may be required for specific situations (Fig. 6.4).

For some highly excitable or anatomically fragile species, prolonged physical restraint without some chemical tranquilization may result in self-inflicted trauma, physiological disturbances, or, occasionally, death. Investigators have an obligation to make every effort to avoid physical restraint procedures that result in cardiogenic shock, capture myopathy, and other stress-induced causes of mortality in their animal subjects (Fig. 6.5). Stress-related damage may not be immediately apparent but may lead to debility or death after release.

Chemical Restraint

Use of chemicals or drugs to render a wild and potentially dangerous animal safe to handle has many applications in wildlife research and management (Pond and O'Gara, 1994). Use of anesthetics, analgesics, and sedatives is mandatory for the control of pain and distress before potentially painful procedures such as surgery are performed on animals. Use of drugs and "tranquilizer guns," however, is not the panacea to wild-animal restraint. Chemicals used for tranquilization and immobilization, if not correctly handled and delivered, may be dangerous to the target animals and humans (Fig. 6.6). In addition, during the drug induction phase or during recovery, an unrestrained animal may be subject to increased potential for accidental injury or death including predation. While under the effects of the drug the animal may become hyper- or hypothermic, depending on chemicals used and ambient temperature, it may vomit and aspirate the vomitus, or pregnant females may abort. A darted animal may be able to elude its captors and hide before being completely anesthetized, a particularly acute hazard when chemicals are employed that require administration of an antidote. All of these circumstances and possibilities must be understood and evaluated by the researcher before a chemical is selected as the best method of restraint in a given instance.

If chemical restraint is selected, it is imperative for all members of the capture team to have a working knowledge of the chemical or drugs being used, even if they are to be handled and delivered by a veterinarian. It also is the responsibility of researchers to know the effects, side effects, advantages, and disadvantages of the drugs being used, and to have knowledge of such factors as the minimum and maximum induction times and potential for adverse drug reactions. This type of information is necessary to evaluate the danger to target animals, and to humans that might be exposed to the drugs. Searchers should be capable of monitoring the condition of anesthetized animals and be able to apply resuscitative routines in a life-threatening emergency. Specific recommendations for drug use and their dosage, drug delivery systems, and physical restraint techniques applicable to the specific species are available in the published literature (Pond and O'Gara, 1994). Information on use of these methods exists in guidelines on acceptable field techniques by various professional societies (See "Professional society guidelines" at the end of this chapter).

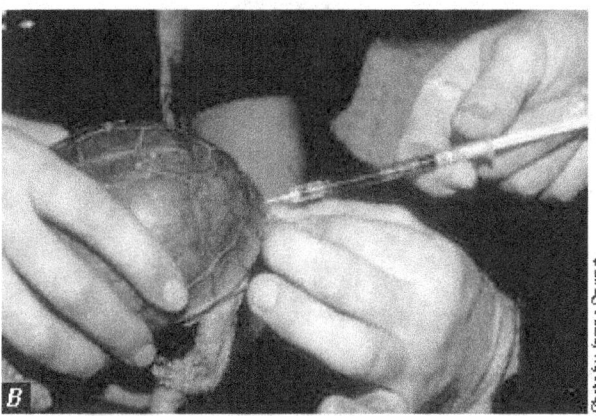

Figure 6.3 (A) Blood collection from the tarsal vein of a deer and (B) from the tail vein of a tortoise.

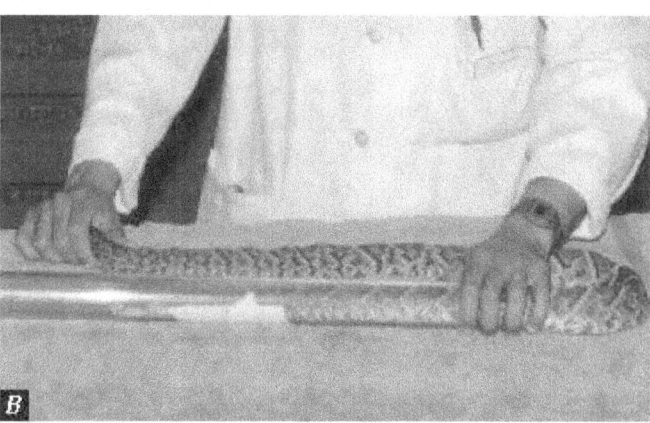

Figure 6.4 (A) Squeeze chutes and head restraints can allow a blood sample to be safely taken from the jugular of large ungulates. (B) Poisonous animals such as this rattlesnake should only be handled by well-trained personnel that have experience with these types of species.

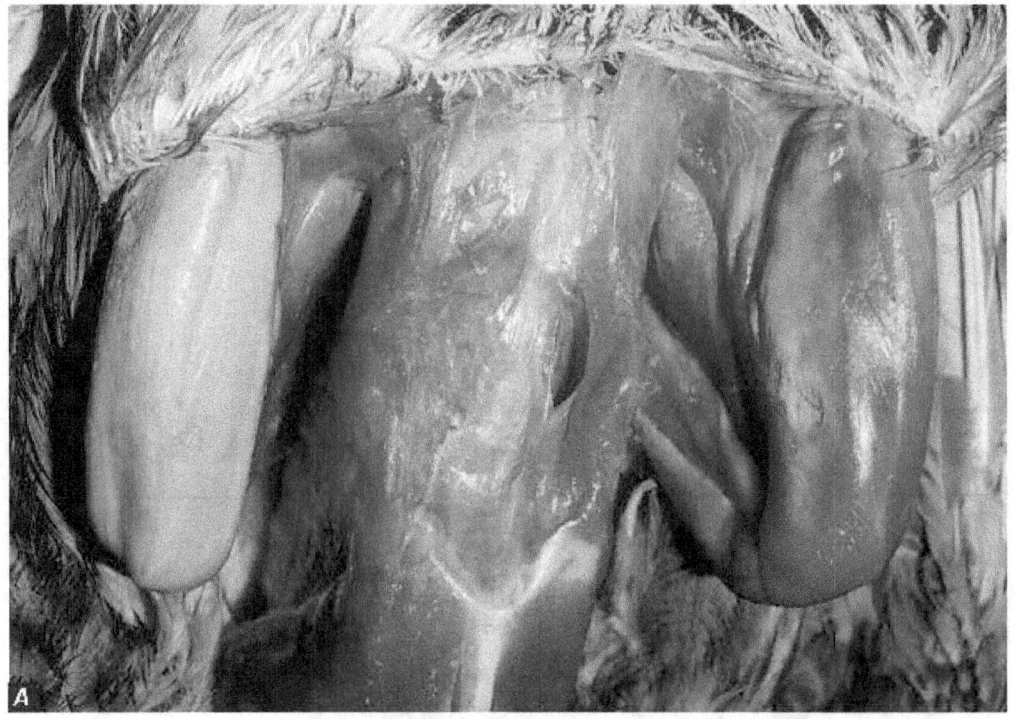

Figure 6.5 (A) The pale coloration of the muscle tissue of the right leg and discolored areas of muscle tissue in the left leg of this whooping crane are lesions of capture myopathy due to stress associated with improper/extended restraint. (B) The light area in this piece of leg muscle from an antelope is also due to capture myopathy.

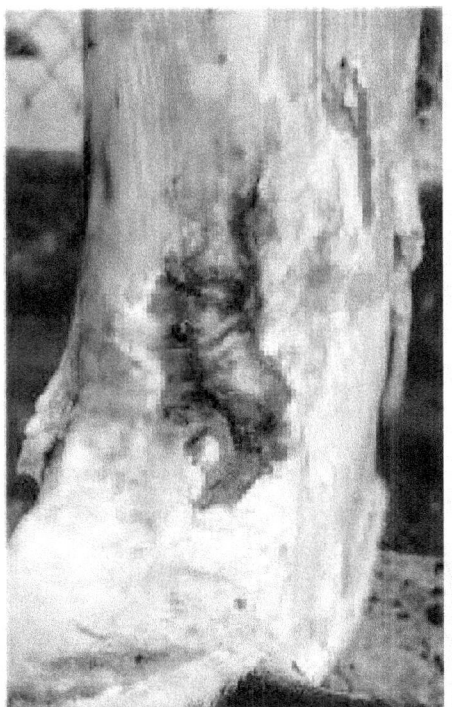

Figure 6.6 Extensive tissue damage and hemorrhage, such as seen in the tissues of this black bear, can occur from immobilization with a CO_2 projected dart.

Animal Marking

Developing means of reliably identifying individual animals to achieve field research objectives often is necessary. In addition to requiring individual identification, researchers may need information on nonconspicuous aspects of physiology or movements, or other aspects of animal ecology that can be determined directly or indirectly through specially designed markers.

Consideration for animal marking

Before initiating any marking procedure for wild animals, researchers must resolve the following questions to determine whether marking is required and appropriate for the particular situation.

1. Do naturally occurring differences in the morphology of the animals under consideration provide sufficient identification to achieve research objectives?

2. How many animals must be individually identifiable?

3. If animals must be physically marked, can a sufficient number of animals be marked in the time available?

4. Are the risks (to both the animal and researcher) associated with capture, handling, and marking, and subsequent well-being, minimal and acceptable in both responsible and scientific contexts?

If the marking process causes pain or distress, as defined by the Animal Welfare Act, appropriate analgesics or anesthetics should be used.

Criteria for Marking

When answers to the four initial questions lead to a decision to initiate an animal-marking program, researchers must search among a wide array of potential techniques with varying strengths and weaknesses to select the method(s) most suited to their particular project (Nietfeld and others, 1994). Technological and methodological constraints and available resources can vary widely from project to project and will require each researcher to examine each potential marking technique in terms of a standard set of criteria. Specific criteria relate to impacts of marking on the organism, validity of the study, and other constraints such as legal requirements.

Evaluation criteria for marking techniques

The following are essential criteria for evaluation of marking techniques:

1. Marks should have minimal effect on the anatomy and physiology of the organism, i.e., no immediate or long-term physical hindrance.

2. Marks should not influence the organism's behavior, i.e., they should not reduce an organism's ability to secure food or inhibit breeding activity (unless the marks are intended as a reproductive inhibitor).

3. Marks that make an organism more conspicuous must be evaluated carefully to ensure that they neither cause others of the same species to react differently to it than to other conspecifics nor subject it to increased selection by potential predators (unless this is a purpose of the study) (Fig. 6.7).

4. Marks should be retained for the minimal period required to achieve project goals.

5. Unambiguous marks that are quick and easy to apply should be selected to avoid extensive handling or error potential.

6. Marks must comply with Federal, State, and other agency rules and regulations.

The first three criteria focus on the well-being of the organism being studied and the potential for marks to influence research results by affecting the fitness or behavior of the organisms. Criteria 4 and 5 may affect the validity of the research design, and criterion 6 reflects other constraints placed upon the researcher. Violation of any of the first five criteria may result in biased research results, so researchers should specifically address these criteria in any evaluation of research resulting from a sample of marked organisms.

Although marks that may be applied to organisms are commonly perceived as passive and visual, markers also

Figure 6.7 (A) Color marking waterfowl should be done with rapidly drying paints and (B and C) the painted feathers held separated until the paint dries to prevent the feathers from sticking to one another and hindering normal flight.

exist that are active and visual (lights), that are auditory, that feature radiotelemetry, or that rely on chemical detection. A vast literature exists of techniques and potential concerns regarding the marking of organisms from insects to whales, and it has been summarized in detail elsewhere (see "Professional society guidelines"; Day and others, 1980; Orlans, 1988).

Other Professional and Ethical Considerations

Many organisms of interest to wildlife professionals are free-ranging and may be enjoyed by other segments of society in many ways, from observation or photography to harvest as meat or trophies. Professional ethics dictate that those other potential uses of organisms be considered and accommodated insofar as possible. Wild animals and birds are valued in part because they are wild, and the presence of human-caused marks may detract from that value. Accordingly, short-lived and inconspicuous marks should be selected whenever they can meet the objectives of proposed research. Scientists have an ethical responsibility to attempt to remove collars or other external markers at the conclusion of their research if possible and feasible. Furthermore, professional and ethical considerations dictate that permanent markers that injure or change the appearance of an animal (e.g., toe-clipping, branding, and tattooing) be employed only under the most humane conditions and when alternate methods are not available to achieve desired research objectives.

Housing and Maintenance of Field Sites

General

Proper care and responsible treatment of incarcerated animals must depend on scientific and professional judgement, on concern for the animal, on knowledge of animal behavior and animal husbandry, and on familiarity with the species. Investigators working with species unfamiliar to them should obtain all pertinent information before confining those animals. It also may be necessary to test and compare several methods of housing to determine the most appropriate one for the well-being of the animal and the purpose of the study. Findings should be part of a permanent record system and animal logbook associated with the study and the maintenance facility.

Housing

Housing for wild vertebrates should approximate natural conditions as closely as possible. Housing should provide safety and comfort for the animal as well as meet the study objectives. Methods of housing should provide for behavioral needs, safety, adequate exercise and rest, and conditions for the general well-being of the animal. Considerations depend on the animal involved and include isolation or refuge areas, natural materials, dust and water baths, natural foods, sunlight, and fresh air. Housing should incorporate as many aspects of natural living as possible, such as brushy areas for escape, resting cover, shade and protection from environmental elements, a natural stream traversing the pen, rocky areas for hoofed animals that need to wear down their hooves, and social groups of animals kept together. Housing of compatible species in a common pen also will provide for social interaction. Frequency of cleaning should be a compromise between level of cleanliness necessary to prevent disease and amount of stress imposed by cleaning (Fig. 6.8).

In general, housing must be of adequate size to allow for the physical and behavioral needs of the animals, while allowing scientists to collect appropriate data. For many housing situations, the pen can be large and natural, with a smaller internal or attached catch pen to restrain animals for experimental techniques. Pen construction materials must provide for the safety of the animals, as well as prevent the animals from escaping. Materials should be of sufficient durability to last for the intended period of confinement. When long-term confinement (weeks or longer) is necessary, or pens are to be reused, materials with impervious surfaces should be used to facilitate sanitation and minimize the potential for survival of animal pathogens. All animals that are inherently dangerous, are environmentally injurious, or have a propensity for escape require special attention. Double walls or

Figure 6.8 A high quality enclosure for New Zealand black stilt that approximates several aspects of the natural habitat and provides safety and comfort for the birds.

double enclosures, covered tops of enclosures, and construction with metal bars or chain link may be required, depending on the species. Mesh size and spacing between fencing materials must be small enough to prevent the head of an animal from extending through the fence. Smaller fencing mesh also is more visible to animals. Colored flagging material may be necessary for animals to visualize fencing until they become accustomed to it. Animals should be released into the housing in a calm and unstressed manner so that initial mortality and morbidity from fence encounters are minimal. A small dose of tranquilizer often will reduce the immediate flight response when an animal is released into the housing and may help prevent initial injuries. Once animals have investigated the limits of the housing, injury occurrence is minimized if investigators do not cause undo flight reactions.

Adequacy of housing often can be judged on normal behavior patterns, weight gains and growth, survival rates, reproductive success, and physical appearance of the animals involved in the research project. Established guidelines for housing laboratory and farm animals were provided by the Canadian Council on Animal Care (1980, 1984). Additional guidelines for housing requirements of fish, amphibians reptiles, wild birds, and small mammals were reported by the appropriate professional societies and appear in the Animal Welfare Act (see also "Professional society guidelines" at the end of this chapter).

Nutrition

Nutrition must meet the needs of the animal unless deviations are an approved purpose of the investigation. Researchers are responsible for determining the appropriate nutritional needs of study animals prior to placing them in confinement and for obtaining adequate food supplies to sustain the animals during the period of confinement. Feeding and watering should be under the direct supervision of an individual trained and experienced in animal care for the species being maintained. Animal care personnel must be familiar with the animals being studied so abnormalities in appearance and behavior that may be indicative of nutritional deficiencies can be recognized quickly.

Transportation

General Considerations

A variety of vehicles such as conventional motor vehicles, all-terrain vehicles, snow machines, rotary and fixed-wing aircraft, and boats are used to transport wild animals. The species involved, method of transportation selected, and length of time an animal is to be transported are important factors regarding the type of care and conditions of containment required to maintain the animal in a state of well-being (Fig. 6.9). To the extent possible, selection of transportation vehicles should take into account maintenance of the animal in a comfortable environment. Veterinary assistance may be required to prescribe and administer appropriate tranquilizers or other drugs when conditions of transportation are likely to result in a high level of stress to the animal due to its behavioral and physiological characteristics, restrictions of confinement, engine noise, and rigors of the trip. The transportation process should be as brief as possible. This can be expedited by proper and adequate planning to assure that transportation vehicles and housing units in appropriate numbers and size are available and ready for use as needed; that food, water, bedding, and other needs to provide for the animals also are available; that individuals involved in the transportation process are trained in the procedures to be used in containment and transportation of the

Figure 6.9 (A) Restraint of bighorn sheep being translocated via helicopter. The legs have been immobilized to prevent injury to the animal and holders. (B) Blinders on this caribou reduces stress from the presence of humans. Legs are restrained similar to the procedure shown for the bighorn sheep.

animals; and that all permits, health certificates, and other paperwork have been completed to the extent possible.

When interstate movement of animals or shipment by commercial carriers is involved, scheduling of transportation segments to minimize the number of transfers and delays between transfers, having someone involved with the project meet the shipment at each transfer point, and, when appropriate, arranging for prompt clearance of animals by veterinary and customs inspectors can result in major reductions in transit time. The receiving party should be on-site when the animals reach their destination.

For some species, periodic rest periods are required to allow the animals to feed undisturbed. Other species are best transported when they are normally inactive and do not feed. Ventilation within the housing unit and transportation vehicle should provide for adequate air movement to keep animals comfortable and avoid buildup of exhaust gases. Subdued lighting and visual barriers between animals and humans and between animals and their transportation environment should be provided to help keep the animals calm. The U.S. Fish and Wildlife Service has published rules for the Humane and Healthful Transport of Wild Animals and Birds to the United States (see Fed Reg. 50 CFR Part 14).

Confinement During Shipping

Animal containers should be inspected to assure they have no sharp edges, protrusions, or rough surfaces that could cause injury during transport. When appropriate, containers also should be padded to help prevent injury. The floor of shipping containers should allow reasonable footing to prevent falling due to a slippery surface. Also, containers should not have coatings or be constructed of materials that are toxic and could be consumed by the animal through licking or chewing during transportation. In general, housing units of porous materials, such as cardboard boxes, should not be reused; all other containers used to house animals should be suitably disinfected between uses (Fig. 6.10). That portion of the transportation vehicle used to contain the housing units also should be disinfected.

Grouping or separation of animals being transported at the same time should take into consideration the species, age, and other appropriate factors. Direct contact generally should be maintained between females and their dependent young, particularly if abandonment may result (unless the young are to be maintained by some other means). Birds should be isolated in separate cells within the shipping container; if this cannot be done, each individual should have sufficient space to assume normal postures and engage in comfort and maintenance activities unimpeded by other birds (Ad Hoc Committee on the Use of Wild Birds in Research, 1988).

Health Aspects

For short-term transportation (less than 30 min), basic considerations are to prevent pain, injury, and undue stress. Thermoregulation capabilities of the species must be considered when an animal is removed from its existing environment and placed in the transportation environment. Transported animals should be protected from exposure to inclement weather, harsh environmental conditions, and major temperature fluctuations and extremes.

Bedding, feed, and water should be provided, as appropriate, and the animals should be observed periodically to determine their state of well-being during transportation. On-site veterinary assistance may be warranted to monitor animals and to provide life-support assistance should a medical emergency occur during transportation or at the release or field study site. Selection of veterinary assistance should focus on the individual's knowledge and experience with the wildlife species involved. Any animals that die during transit should be removed as soon as practical from the sight and olfactory detection of other animals being transported. These carcasses should be retained for pathological examinations regarding cause of death. Similarly, animals that become severely injured or clinically ill should be removed and responsibly euthanized. Euthanasia should not take place in the presence of other live animals. Sick animals disposed of in this manner also should be retained for pathological assessments. Determinations of cause of death are needed to assess whether the remaining animals are at risk from pathogens associated with the dead animals.

Surgical and Medical Procedures

Guidelines for wildlife medical procedures

Wildlife field research can involve surgical and medical procedures such as implanting radio transmitters and surgical sex determination in birds. Incorporation of such techniques into a research protocol should follow these guidelines:

1. Surgical and medical techniques used should be based on accepted protocols for the studied species or for the most closely related domesticated species. The Canadian Council on Animal Care's (1984) Guide to the Care and Use of Experimental Animals, Volume 2, is a good source of such information.

2. Protocols should be developed and, if possible, implemented in collaboration with a qualified veterinarian. Only properly trained personnel, conversant in all techniques necessary, should conduct the procedures.

3. Protocols must be reviewed carefully by the ACUC with special attention paid to limiting pain during the actual procedure and post-procedure period.

4. Adequate anesthesia and/or analgesia must be provided.

Figure 6.10 (A) Canada geese restrained within burlap bags with openings for the head and neck for short-distance transportation by vehicle. (B) Porous materials such as these bags and the cardboard boxes these Hungarian partridge are being released from should not be reused for animal transport. More permanent holding containers such as (C) plastic poultry crates and (D) large animal crates should be thoroughly washed and disinfected between uses.

Minor Procedures

Minor medical procedures such as collection of blood, administration of drugs intravenously or intramuscularly, biopsies of superficial structures such as skin, and sutured attachment of radio transmitters usually can be performed safely and responsibly in the field without complicated equipment. However, it is the researcher's responsibility to choose the least invasive and least painful technique, minimize the duration of the procedure, use the most appropriate equipment and aseptic technique, and provide analgesia or sedation when indicated.

Major Procedures

As defined by the Animal Welfare Act, major operative procedures are (p. 36,121) "any surgical intervention that penetrates and exposes a body cavity or any procedure which produces permanent impairment of physical or physiological functions." Major surgical procedures, when survival of the animal is intended, should be performed only under proper anesthesia and with sterile technique. Examples of major procedures used in wildlife research include laparotomy, surgical flight restraint, and sterilization. These procedures should be performed only in a clean space set aside for sterile surgery, with surgical instruments and drapes of the proper type, and with anesthesia protocols judged to be safe and responsible for the species involved. Necessary equipment and trained personnel to deal with surgery or anesthesia-related emergencies (i.e., severe blood loss, cessation of breathing or cardiac function, severe hypo- or hyperthermia, acid-base imbalances) should be available at all times. This will maximize the success and subsequent scientific return from those often costly procedures and, therefore, minimize the number of animals needed and amount of animal distress (Fig. 6.11).

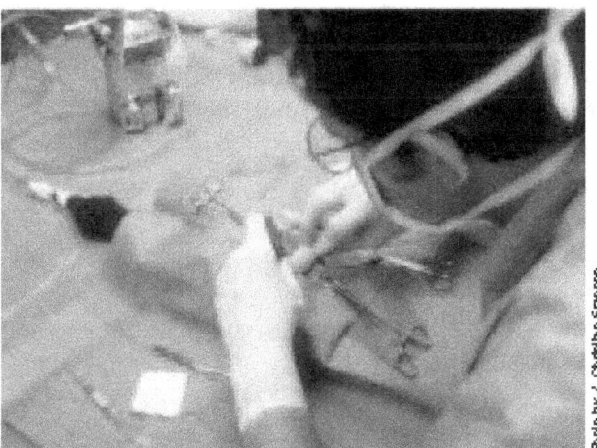

Figure 6.11 Invasive surgical procedures should be done only by properly trained personnel knowledgeable of techniques necessary to successfully carry out the procedure and appropriately respond to medical emergencies that might arise.

Medical Considerations

Wildlife field researchers should have access to veterinary consultation and take responsibility to prepare themselves to deal with any health problems that might arise in their study population. Sometimes intervention and control of a natural disease process may not be advisable and may interfere with the study's goals. However, if the health problem arises due to the researcher's work, or if it will interfere with the study, the researcher must be ready to respond. Preparations should include gaining familiarity with the common diseases and health problems of the species under study, establishing a contact with a veterinary consultant, and having appropriate treatment or control equipment and drugs on hand or easily accessible. The researcher also is responsible for evaluating the possible impact of disease in the study animals on the larger population or ecosystem as a whole, and for making the maintenance of their welfare a priority as decisions are made. This is especially true when release or translocation of animals is part of a study; disease must be considered in evaluating the advisability of the program.

Euthanasia

Euthanasia is defined under the Animal Welfare Act as (p. 36,121) "the humane destruction of an animal accomplished by a method that produces rapid unconsciousness and subsequent death without evidence of pain or distress, or a method that utilizes anesthesia produced by an agent that causes painless loss of consciousness and subsequent death." Euthanasia may not be an approved component of a field study, but it may become a necessary health care option in a study involving capture, restraint, or surgical procedures. Therefore, all wildlife researchers involved in invasive studies must be familiar with the approved euthanasia methods for their study species (Andrews and others, 1993) and have the appropriate equipment/drugs on hand so euthanasia can be performed quickly.

Disease Considerations

Field investigators need to be fully aware of disease concepts so they may avoid introduction of new disease problems into animal populations or the spread of disease to other populations and locations as a result of their studies. Disease introductions and spread occur as a result of animals brought to the field research site to serve as biological sentinels, as decoys to lure and capture other animals, for species introductions or releases to supplement existing populations, for behavioral studies, for assistance in tracking or retrieving animals, and for other purposes. All of these uses of animals involve acceptable methods for scientific research and wildlife management. However, under no circumstances should the well-being of free-ranging wildlife populations be unduly jeopardized by disease risks associated with animal use in field research. Field investigators have ethical and professional obligations to take appropriate actions for minimizing the introduction of the following: (a) new disease agents, (b) vectors (e.g., ticks and internal parasites) capable of efficiently transmitting indigenous, dormant diseases or those not currently being effectively transmitted, and (c) species that can serve as amplification hosts for transmitting indigenous diseases to other species (Fig. 6.12).

In addition, animals that are highly susceptible to diseases indigenous to the study location should not be released into the wild without using applicable prophylactic measures, unless these animals are to serve as biological sentinels for disease investigations. Biological sentinels should be monitored closely and euthanized by approved, responsible methods as soon as is practical after study objectives have been met.

Disease introduction and spread can result from mechanical means such as contaminated personnel, supplies, and equipment in addition to the biological processes identified above. Steps taken to address disease prevention are far more cost effective than disease control activities initiated after a problem has developed.

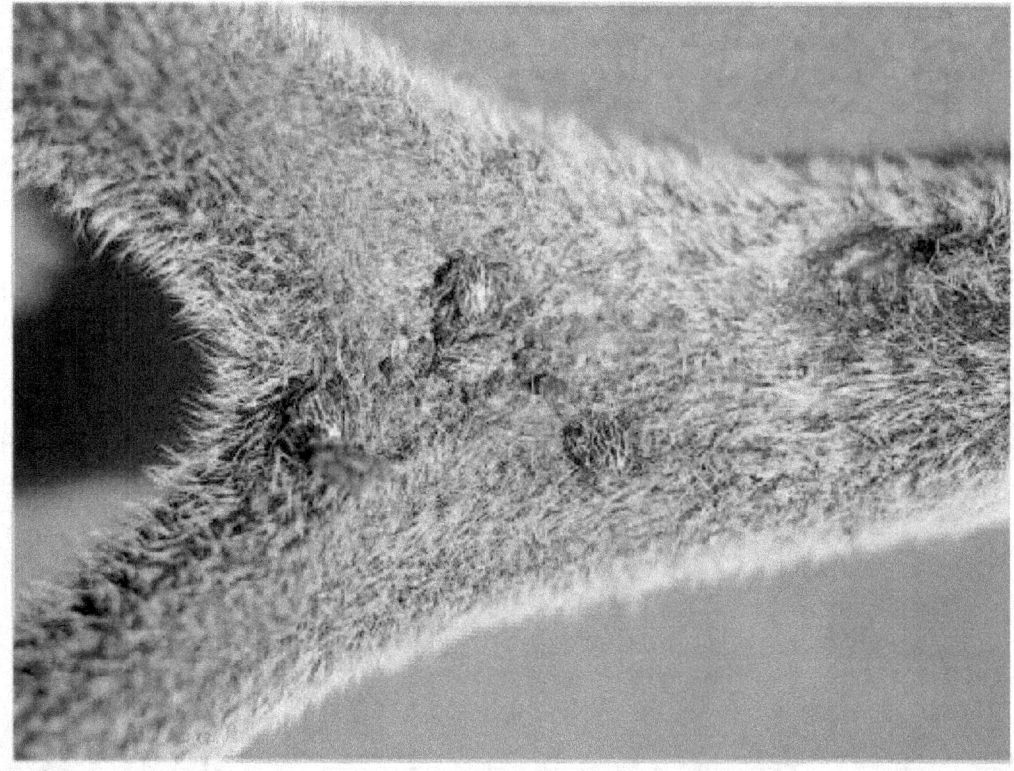

Figure 6.12 Wildlife are often referred to as a "biological package" as the relocation of animals may involve life forms other than the animals themselves. The ticks feeding on this velvet covered antler could be disease carriers. Once introduced into a new area, the ticks may become an important vector for transmission of an indigenous disease. Disease potential is an important consideration that should be adequately addressed when translocating wildlife.

Wildlife disease prevention during field research

Protection of free-ranging wildlife from disease is aided by the following actions:

1. Appropriate health certification should be required for all animals being brought to the site of field investigations. State veterinary officials should be contacted to determine what specific testing must be done when animals are moved into their jurisdiction.

2. Appropriate disinfection procedures should be used for investigators and their equipment when disease risks are present.

3. Prior knowledge of disease activity at the study site should be obtained to guide actions involving the research study.

4. Source for any animals being brought to a field investigation site (captive-reared and relocated wild stock) should be evaluated for inherent disease problems, and appropriate steps should be taken to avoid disease introductions.

5. To the extent possible, animals should be held under surveillance for 15–30 days prior to their release into the wild, and only healthy animals should be released. These animals should not be mixed with other species during transportation and should be isolated from other animals during the surveillance period.

6. Any animals that die should be examined by a disease diagnostic laboratory having competency for determining cause of death in the species involved; these findings should be used to guide appropriate actions (Fig. 6.13).

7. Animals that become clinically ill should be examined by disease specialists, and their counsel should be used to protect the well-being of other animals within the study area.

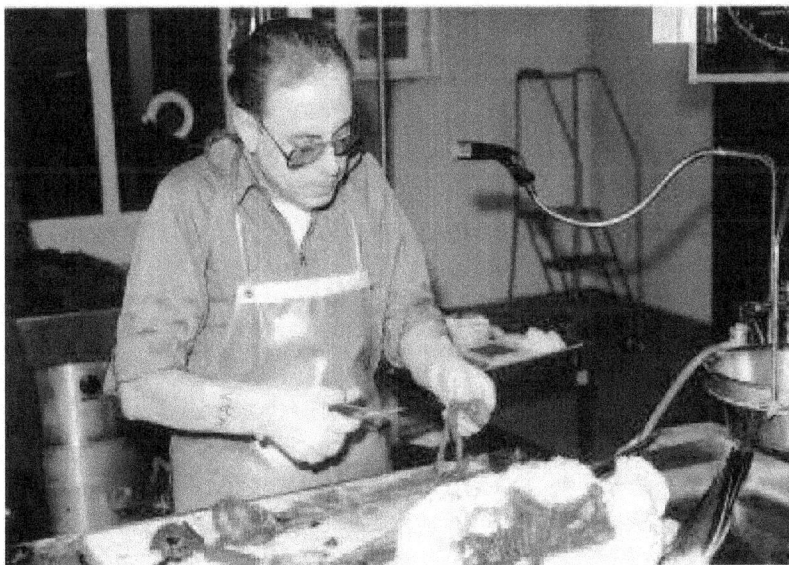

Figure 6.13 Timely diagnosis of causes of wildlife morbidity and mortality is invaluable for the detection of emerging hazards that can jeopardize the well-being of the population being studied and may be of great potential consequences. Submission of animals that die to competent laboratories provides information useful for intervention.

Animal Disposition at Completion of Study

When live animals are in the possession of investigators or under their control at the time of study completion, an evaluation must be made as to whether these animals can be released to a free-ranging existence, should be maintained under controlled conditions, or should be euthanized.

Animal release guidelines

As a general rule, field-captured animals should be released only:

1. At the site of the original capture, unless conservation efforts or safety considerations dictate otherwise. Prior approval for releases at noncapture sites should be obtained from appropriate State/Federal agencies. Relocation release sites should be within the native range of the species, or established range for introduced species, and be in habitat suitable for species survival;

2. When the released animal can be reasonably expected to function normally within the population;

3. When local and seasonal conditions are conducive to survival;

4. When the ability to survive in nature has not been irreversibly impaired; and

5. When release is not likely to spread pathogens or contribute to disease processes in other ways.

The decision of whether to release captive-reared animals into the wild after completion of a field research project demands more rigorous evaluation than for field-captured animals. In addition to evaluating the future well-being of the animal being released, impacts on other animals of the same species and competition and risks for other species sharing that environment also must be considered. Rarely, if ever, will releases of captive-reared animals at the completion of research studies be justified on the basis of animal welfare considerations.

When animals are to be released, efforts should be made to enhance their chances of survival. Animals should be in good physical condition and released when weather conditions are favorable, at a time of day when they are able to locate food and cover that meet survival needs.

Animals that cannot be released should be considered for distribution to other scientists for further study. However, if the animal was subject to a major invasive procedure, it may not be appropriate for additional experimentation. Animals not suitable for research may be suitable display animals that can be donated to a zoo or other type of educational institution.

When animals must be euthanized, responsible methods appropriate for the species and circumstances must be used. Care must be taken to assure that the animal is dead before disposal of the carcass. Also, disposal procedures must prevent carcasses containing toxic substances or drugs from the research investigations or euthanasia procedures to enter the food web of other animals. To the extent feasible, euthanized animals should be properly preserved and used as voucher specimens or for teaching purposes.

Safety Considerations

Researchers working with free-ranging wildlife are subject to enhanced levels of exposure to wildlife diseases transmissible to humans. Disease transmission may involve direct contact with infected animals such as those with rabies, contact with disease vectors such as ticks transmitting Lyme disease, or contact with contaminated environments such as bird roosts harboring histoplasmosis. Field investigators should become familiar with the common diseases of wildlife species they are working with and the relative prevalence of those diseases in the populations they are studying. Consultation with a physician regarding immunization or other preventative treatment is advised when serious diseases for humans commonly occur in the populations being studied. Investigators who become ill should seek medical assistance and advise their physicians of their exposure to potentially hazardous animals, diseases, and environmental conditions.

Acknowledgments

These guidelines were prepared by a committee of The Wildlife Society appointed by J. G. Teer during his Presidency. The committee acknowledges the contributions of F. J. Dein for his review of these guidelines and valuable input provided in enhancing the final content.

Milton Friend, Dale E. Toweill, Robert L. Brownell, Jr., Victor F. Nettles, Donald S. Davis, and William J. Foreyt

Literature Cited

Ad Hoc Committee on Acceptable Field Methods in Mammalogy, 1987, Acceptable field methods in mammalogy: Preliminary guidelines approved by the American Society of Mammalogists: Journal of Mammalogy, v. 68, (supplement 4), 18 p.

Ad Hoc Committee on the Use of Wild Birds in Research, 1988, Guidelines for use of wild birds in research: Auk, v. 105, (supplement 1), 41 p.

American Society of Ichthyologists and Herpetologists (ASIH), The Herpetologists League (HL), and The Society for the Study of Amphibians and Reptiles (SSAR), 1987, Guidelines for the use of live amphibians and reptiles in field research: Journal of Herpetology, v. 4, p. 1–14.

Andrews, E.J., and others, 1993, Report of the AVMA panel on euthanasia: Journal of American Veterinary Medical Association, v. 202, p. 229–249.

Animal Behavior Society/Animal Society for Animal Behavior, 1986, ABS/ASAB guidelines for the use of animals in research: Animal Behavior Society Newsletter, v. 31, p. 7–8.

Bookhout, T.A., editor, 1994, Research and management techniques for wildlife and habitats, Fifth ed.,: The Wildlife Society, Bethesda, Md., 740 p.

Canadian Council on Animal Care, 1980, Guide to the care and use of experimental animals, v. 1: Canadian Council on Animal Care, Ottawa, Ont., 120 p.

_____ 1984, Guide to the care and use of experimental animals. vol. 2: Canadian Council on Animal Care, Ottawa, Ont., 208 p.

Day, G.I., Schemnitz, S.D., and Taber, R.D., 1980, Capturing and marking wild animals, *in* S. D. Schemnitz, ed. Wildlife management techniques manual, Fourth ed., rev: The Wildlife Society, Washington, D.C., p. 61–88.

Nietfeld, M.T., Barrett, M.W., and Silvy, N., 1994, Wildlife marking techniques, *in* T. A. Bookhout, ed., Research and management techniques for wildlife and habitats, Fifth ed.: The Wildlife Society, Bethesda, Md., p. 140–168.

Orlans, F.B., editor, 1988, Field research guidelines: impact on animal care and use committees: Scientists Center for Animal Welfare, Bethesda, Md., 23 p.

Orlans, F.B., Simmonds, R.C., and Dodds, W.J., eds., 1987, Effective animal care and use committees: Scientists Center for Animal Welfare, Bethesda, Md., 178 p.

Pond, D.B. and O'Gara, B.W., 1994, Chemical immobilization of large mammals, *in* T. A. Bookhout, ed., Research and management techniques for wildlife and habitats, Fifth ed.: The Wildlife Society, Bethesda, Md., p. 140–168.

Professional society guidelines for use of live animals in field research

Ad Hoc Committee on Acceptable Field Methods in Mammalogy, 1987, Acceptable field methods in mammalogy: preliminary guidelines approved by the American Society of Mammalogists: Journal of Mammalogy, vol. 68, (supplement 4), 18 p.

Ad Hoc Committee on the Use of Wild Birds in Research, 1988, Guidelines for use of wild birds in research: Auk, vol. 105, (supplement 1), 41 p.

American Society of Ichthyologists and Herpetologists (ASIH), American Fisheries Society (AFS), and the American Institute of Fisheries Research Biologists (AIFRB), 1987, Guidelines for use of fishes in field research: Copeia 1987 (supplement), 12 p.

American Society of Ichthyologists and Herpetologists (ASIH), The Herpetologists League (HL), and The Society for the Study of Amphibians and Reptiles (SSAR), 1987, Guidelines for the use of live amphibians and reptiles in field research: Journal of Herpetology, vol. 4, p. 1–14.

Sources of assistance for technical information, implementation, and interpretation of the Animal Welfare Act

Animal Welfare Information Center
National Agricultural Library
10301 Baltimore Ave., 5th Floor
Beltsville, MD 20705–2351
(301) 504-6212
fax (301) 504-7125

National Library of Medicine
8600 Rockville Pike
Bethesda, MD 20894
(301) 594-5983

Scientists Center for Animal Welfare
7833 Walker Dr., Suite 340
Greenbelt, MD 20770–3229
(301) 345-3500

U.S. Department of Agriculture
Animal and Plant Health Inspection Service
4700 River Rd.
Riverdale, MD 20737
(301) 734-7833

Section 2
Bacterial Diseases

Avian Cholera

Tuberculosis

Salmonellosis

Chlamydiosis

Mycoplasmosis

Miscellaneous Bacterial Diseases

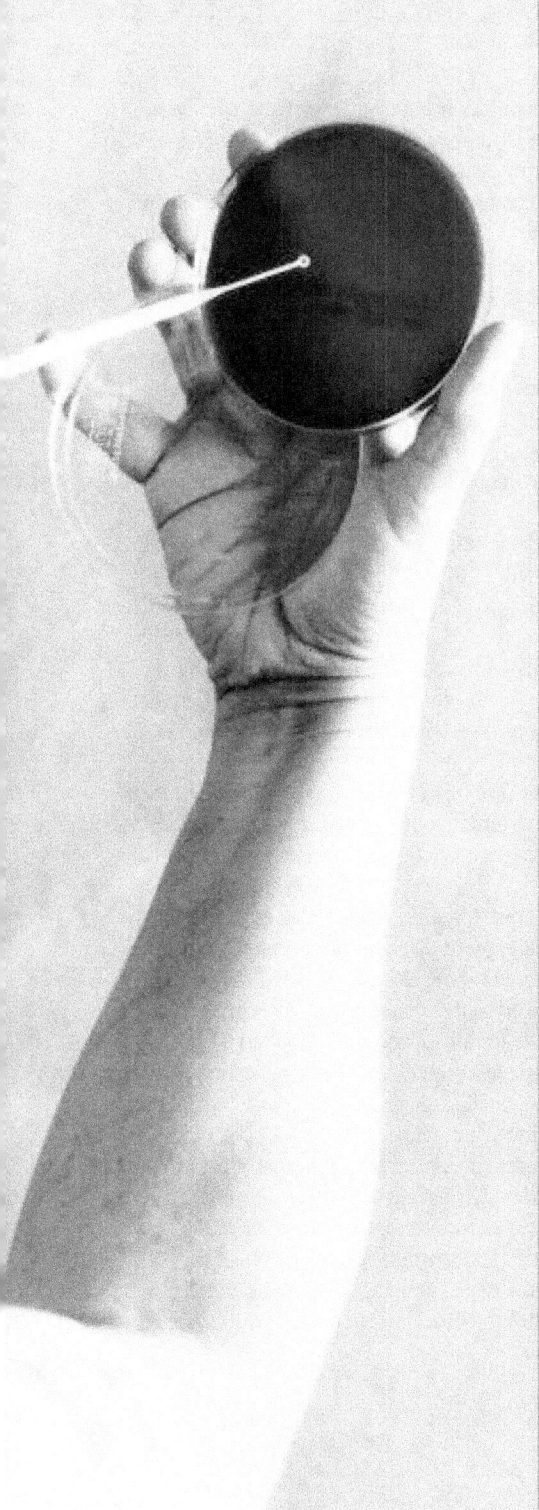

Inoculating media for culture of bacteria
Photo by Phillip J. Redman

Introduction to Bacterial Diseases

"Consider the difference in size between some of the very tiniest and the very largest creatures on Earth. A small bacterium weighs as little as 0.00000000001 gram. A blue whale weighs about 100,000,000 grams. Yet a bacterium can kill a whale...Such is the adaptability and versatility of microorganisms as compared with humans and other so-called 'higher' organisms, that they will doubtless continue to colonize and alter the face of the Earth long after we and the rest of our cohabitants have left the stage forever. Microbes, not macrobes, rule the world."
(Bernard Dixon)

Diseases caused by bacteria are a more common cause of mortality in wild birds than are those caused by viruses. In addition to infection, some bacteria cause disease as a result of potent toxins that they produce. Bacteria of the genus Clostridium are responsible for more wild bird deaths than are other disease agents. *Clostridium botulinum*, which causes avian botulism, is primarily a form of food poisoning and it is included within the section on biotoxins (see Chapter 38). Other *Clostridium* sp. that colonize intestinal tissues produce toxins that cause severe hemorrhaging of the intestine, thus leading to tissue death or necrosis and intoxication of the bird due to the exotoxins produced by the bacterial cell. The descriptive pathology is referred to as a necrotizing gastroenteritis or necrotic enteritis and the disease as clostridial enterotoxemia. The classic example in gallinaceous birds such as quail, turkey, pheasant, grouse, and partridge, is ulcerative enteritis or quail disease, which is caused by *Clostridium colinium*; quail are the species most susceptible to that disease. Necrotic enteritis of wild waterbirds, especially geese, has been reported with increasing frequency during recent years. *Clostridium perfringens* has been associated with these deaths.

The frequency of wild bird mortality events and the variety of infectious bacterial diseases causing that mortality has increased greatly during recent decades. Avian cholera has become the most important infectious disease of waterbirds, but it did not appear in North American waterfowl or other waterbirds until 1944. Most of the geographic expansion and increased frequency of outbreaks of avian cholera has occurred since 1970. Avian tuberculosis is a historic disease of captive birds, but it is relatively rare in North American wild birds. The high prevalence of avian tuberculosis infection that has occurred since 1982 in a free-living foster-parented whooping crane population has challenged the survival of that subpopulation of cranes. Salmonellosis has become a major source of mortality at birdfeeders throughout the Nation, and mycoplasmosis in house finches has become the most rapidly spreading infectious disease ever seen in wild birds. This disease reached the Mississippi River and beyond within 2 years of the 1994 index cases in the Washington, D.C. area.

Avian botulism has also expanded in geographic distribution and has gained increased prominence as a disease of waterbirds. It is undoubtedly the most important disease of waterbirds worldwide. Much of the geographic expansion of avian botulism has occurred during the past quarter-century.

As a group, bacterial diseases pose greater human health risks than viral diseases of wild birds. Of the diseases addressed in this section, chlamydiosis, or ornithosis, poses the greatest risk to humans. Avian tuberculosis can be a significant risk for humans who are immunocompromised. Salmonellosis is a common, but seldom fatal, human infection that can be acquired from infected wild birds. This section provides individual chapters about only the more common and significant bacterial diseases of wild birds. Numerous other diseases afflict wild birds, some of which are identified in the chapter on Miscellaneous Bacterial Diseases included at the end of this section.

Timely and accurate identification of causes of mortality is needed to properly guide disease control operations. The magnitude of losses and the rapidity with which those losses can occur, as reflected in the chapters of this section, should be a strong incentive for those who are interested in the conservation of wild species to seek disease diagnostic evaluations when sick and dead birds are encountered. In order to accurately determine what diseases are present, specimens need to be sent to diagnostic laboratories that are familiar with the wide variety of possible diseases that may afflict wild birds. Those laboratories must also have the capability to isolate and identify the causative agents involved. Several sources of wildlife disease expertise that might be called upon when wildlife mortality occurs are identified within Appendix B.

Quote from:

Garrett, L., 1994, The coming plague—Newly emerging diseases in a world out of balance: Farrar, Straus, and Giroux, New York, N.Y., p. 411.

Chapter 7
Avian Cholera

Synonyms

Fowl cholera, avian pasteurellosis, avian hemorrhagic septicemia

Cause

Avian cholera is a contagious disease resulting from infection by the bacterium *Pasteurella multocida*. Several subspecies of bacteria have been proposed for *P. multocida*, and at least 16 different *P. multocida* serotypes or characteristics of antigens in bacterial cells that differentiate bacterial variants from each other have been recognized. The serotypes are further differentiated by other methods, including DNA fingerprinting. These evaluations are useful for studying the ecology of avian cholera (Fig. 7.1), because different serotypes are generally found in poultry and free-ranging migratory birds. These evaluations also show that different *P. multocida* serotypes are found in wild birds in the eastern United States than those that are found in the birds in the rest of the Nation (Fig. 7.2).

Acute *P. multocida* infections are common and they can result in bird deaths 6–12 hours after exposure, although 24–48 hours is more common. Susceptibility to infection and the course of disease — whether or not it is acute or chronic — is dependent upon many factors including sex, age, genetic variation, immune status from previous exposure, concurrent infection, nutritional status, and other aspects of the host; strain virulence and other aspects of the bacterium; and dose and route of exposure. Infection in poultry generally results when *P. multocida* enters the tissues of birds through the mucous membranes of the pharynx or upper air passages. The bacterium can also enter through the membranes of the eye or through cuts and abrasions in the skin. It is assumed that transmission is similar in wild birds.

Environmental contamination from diseased birds is a primary source for infection. High concentrations of *P. multocida* can be found for several weeks in waters where waterfowl and other birds die from this disease. Wetlands and other areas can be contaminated by the body discharges of diseased birds. As much as 15 milliliters of nasal discharge containing massive numbers of *P. multocida* have been collected from a single snow goose. Even greater amounts of bacteria enter the environment when scavengers open the

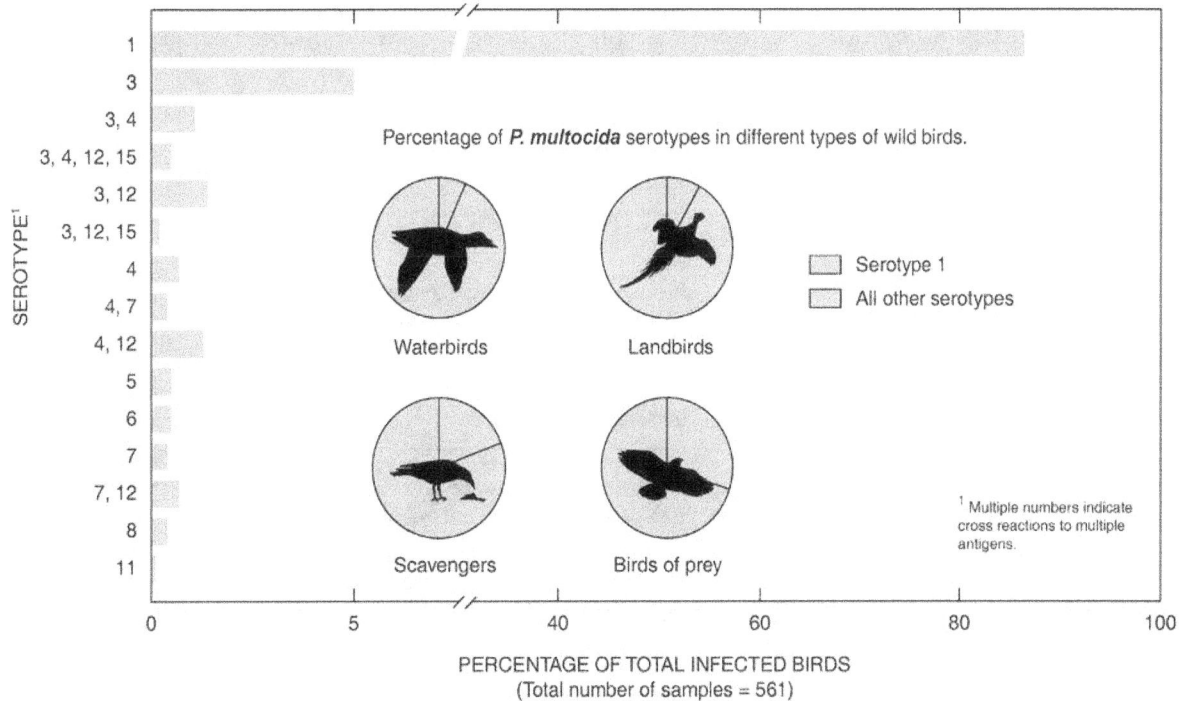

Figure 7.1 Distribution of **Pasteurella multocida** serotypes from 561 wild bird isolates from the United States.

Figure 7.2 *Distribution of Pasteurella multocida serotypes from 561 wild birds isolated by waterfowl flyway.*

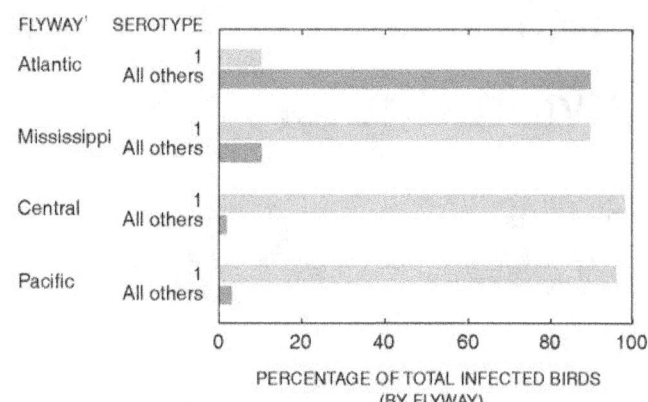

¹Flyways are administrative units for the management of waterfowl that are geographic representations of the primary migratory patterns of waterfowl.

Route of transmission and field situation	Comments
Bird-to-bird contact	Secretions from infected birds shedding *P. multocida*. Requires close contact, such as when individuals struggle over aquatic plants that they are feeding upon.
Ingestion	Probably most common route for transmission. Consumption of diseased carcasses by scavengers and predators. Ingestion of *P. multocida* in food and water from contaminated environments.
Aerosol	May be important in heavily contaminated environments, such as during major die-offs. Activities that result in splashing of surface waters result in bacteria-laden sprays when water becomes contaminated.
Insects	Biting insects that feed on birds after having fed upon contaminated carcasses or contaminated environments (ticks, mites, flies). Insects fed upon by birds (maggots, flies) following ingestion of *P. multocida* by the insect when feeding.
Animal bites	Not thought to be an important route for infection of wild birds. Nonfatal bites from small mammals, such as raccoon, can result in *P. multocida* infections that become systemic and possibly initiate disease outbreaks. Thought to occur in some domestic turkey flocks, not yet demonstrated in wild birds.
Fomites (inanimate objects)	Contaminated cages, equipment, and clothing used in field operations can serve as mechanical transport mechanisms for introducing *P. multocida*. Environmental persistence of *P. multocida* is sufficient for this to be a consideration when personnel and equipment are used to combat an avian cholera outbreak and then are to be redirected for other activities.

Figure 7.3 *Potential means for transmission of avian cholera to free-ranging wild birds.*

carcasses of diseased birds. Avian cholera can be transmitted within this contaminated environment in several ways. Ingestion of bacteria in contaminated food and water, including scavenging of diseased carcasses, is an important source of infection for wild birds. The disease can be transmitted by direct bird-to-bird contact, either between infected and noninfected live birds, or between infected carcasses that serve as "decoys" and noninfected live birds. Aerosol transmission is also thought to take place. In wetlands where avian cholera breaks out, the highest concentrations of *P. multocida* are found near the water surface rather than deep in the water column. Birds landing, taking flight, bathing, and otherwise causing disturbance of the water surface cause bacteria-laden aerosols, which can serve to infect those birds. Other means for transmission of avian cholera have also been reported, each of which may occur for specific situations, but none of which are primary means for disease transmission in wild birds (Fig. 7.3).

The role of disease carriers as a means for initiating avian cholera outbreaks in wild birds has long been postulated because chronically infected birds are considered to be a major source for infection of poultry. It has been reported that the only limit to the duration of the chronic carrier state is the lifespan of the infected bird. Disease carriers have been conclusively established for poultry, and *P. multocida* can commonly be isolated from the mouth area or tonsils of most farm animals, dogs, cats, rats, and other mammals (Fig. 7.4). However, types of *P. multocida* that are found in most mammals do not generally cause disease in birds (see Species Affected, this chapter). The role of disease carriers among migratory species of wild birds has long been suggested by the patterns of avian cholera outbreaks in wild waterfowl, but it has not been clearly established by scientific investigations. Recent findings by investigators at the National Wildlife Health Center (NWHC) have provided evidence that disease carriers exist in snow goose breeding colonies. Shedding of *P. multocida* by disease carriers is likely to be through excretions from the mouth, which is the area where the bacteria are sequestered in carriers and is the means for dissemination of *P. multocida* by poultry. Poultry feces very seldom contain viable *P. multocida*, and there is no evidence that *P. multocida* is transmitted through the egg.

Species Affected

It is likely that most species of birds and mammals can become infected with *P. multocida*; however, there are multiple strains of this bacterium and those different strains vary considerably in their ability to cause disease in different animals. These differences are most pronounced for cross-infections between birds and mammals. Strains isolated from birds will usually kill rabbits and mice but not other mammals. Strains isolated from cattle and sheep do not readily cause clinical disease in birds. However, some strains from pigs have been shown to be highly virulent (very few organ-

Domestic animals

Cattle, horses, swine, goats, sheep
Dogs, cats
Gerbils, rabbits

Big game

Elk and deer
Caribou and reindeer
Bighorn sheep
Bison
Pronghorn antelope

Carnivores

Bears
Lynx, bobcat, puma
Foxes
Weasels, mink
Raccoon

Rodents

Rats, mice, voles
Muskrats, nutria
Chipmunks

Pinnipeds

Sea lions, fur seals

Rabbits

Cottontail rabbits

Domestic poultry

Chickens, turkeys
Ducks, geese
Pigeons

*Figure 7.4 Partial list of domestic species and wild mammals from which **Pasteurella multocida** has been isolated.*

isms cause serious disease) for poultry. Also, cultures from the mouths of raccoons were pathogenic or caused disease in domestic turkeys. Bite-wound infections by raccoons have been postulated as a source of avian cholera outbreaks in poultry. An interspecies chain of avian cholera transmission has been described in free-ranging California wildlife that involved waterbirds, mice, and avian scavengers and predators (Fig. 7.5).

More than 100 species of free-ranging wild birds are known to have been naturally infected with *P. multocida* (Fig. 7.6) in addition to poultry and other avian species being maintained in captivity. Infection in free-ranging vultures has not been reported, although a king vulture is reported to have died from avian cholera at the London Zoo. As a group, waterfowl and several other types of waterbirds are most often the species involved in major avian cholera mortalities of wild birds. Scavenger species, such as crows and gulls, are also commonly diagnosed with avian cholera, but deaths of raptors, such as falcons and eagles, are far less frequent (Fig. 7.7). However, there have been several reports of avian cholera in birds kept by falconers, both from birds consuming infected prey when being flown and from being fed birds that died from avian cholera. Waterfowl and coots experience the greatest magnitude of wild bird losses from this disease (Fig. 7.8). In general, species losses during most major outbreaks are closely related to the kinds of species present and to the numbers of each of those species present during the acute period of the die-off. During smaller events, although several species may be present, mortality may strike only one or several species and the rest of the species that are present may be unaffected. Major outbreaks among wild birds other than waterbirds are uncommon.

Impacts on population levels for various species are unknown because of the difficulty of obtaining adequate assessments in free-ranging migratory birds. However, the magnitude of losses from individual events and the frequency of outbreaks in some subpopulations have raised concerns about the biological costs from avian cholera. Disease that is easily spread through susceptible hosts can be devastating when bird density is high, such as for poultry operations and wild waterfowl aggregations (Fig. 7.9). Mortality from avian cholera in poultry flocks may exceed 50 percent of the population. An outbreak in domestic geese killed 80 percent of a flock of 4,000 birds. Similar explosive outbreaks strike in free-ranging migratory birds. Peak mortality in wild waterfowl has exceeded more than 1,000 birds per day.

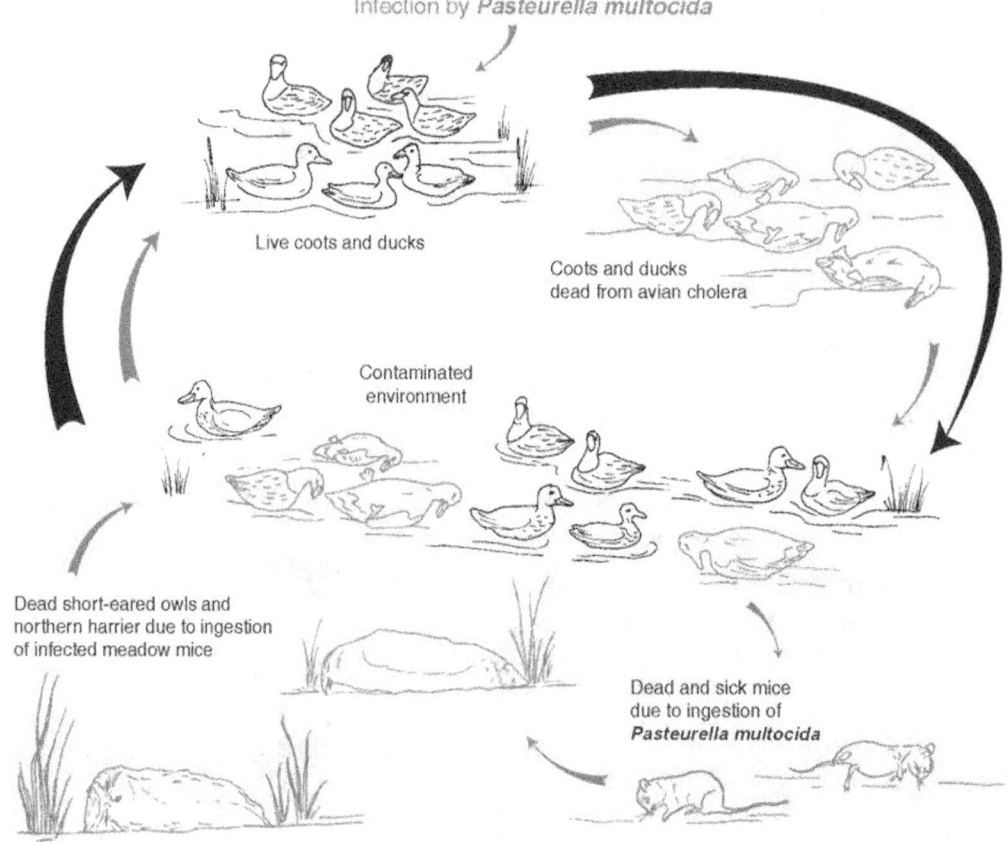

Figure 7.5 *Example of an interspecies chain for transmission of avian cholera that occurred in a California wetland.*

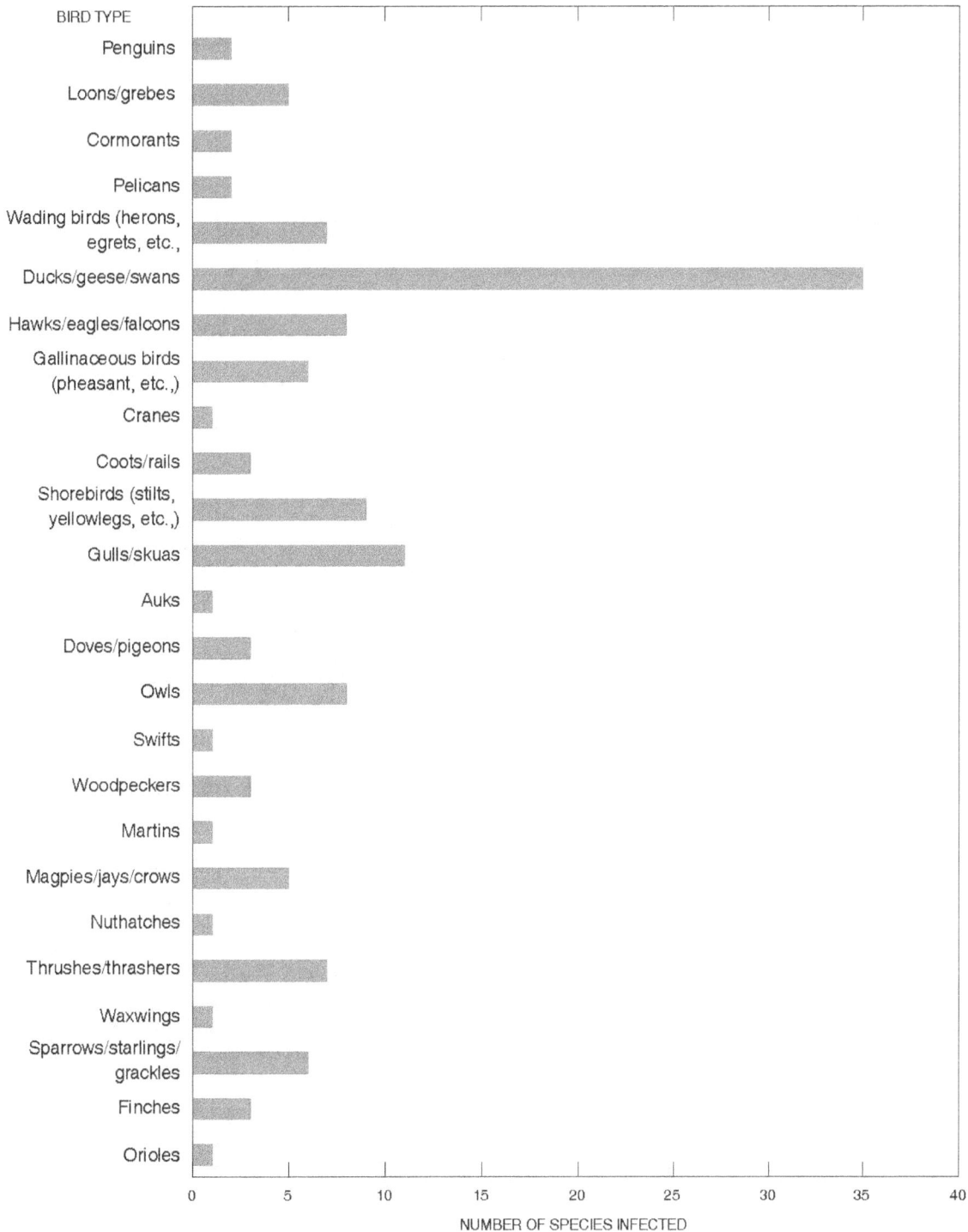

Figure 7.6 Free-ranging wild birds that have been diagnosed with avian cholera.

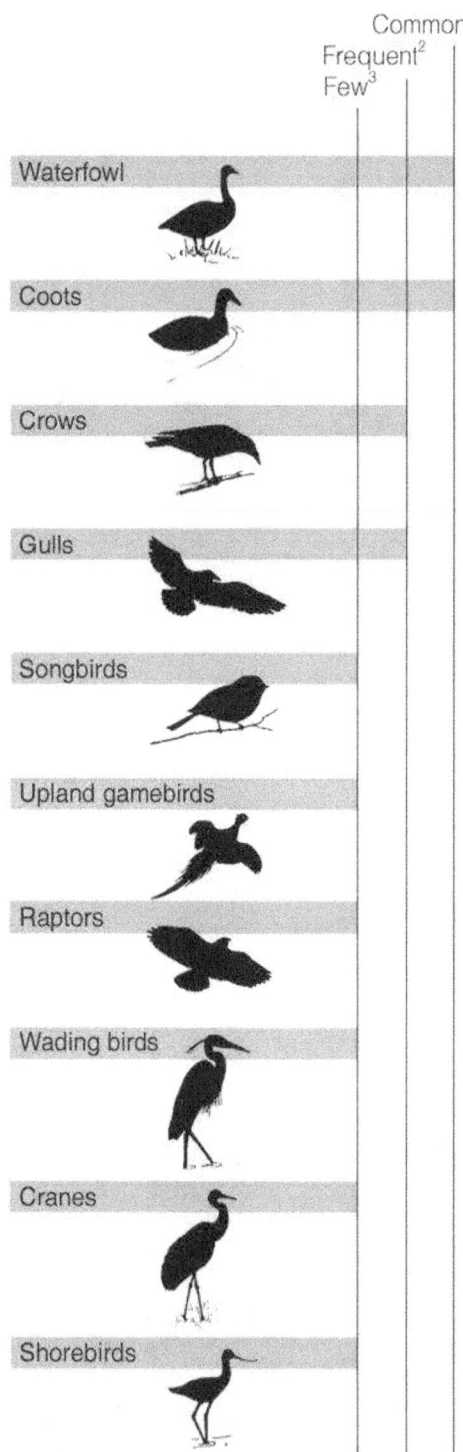

[1] Major die-offs occur almost yearly.
[2] Mortality in these species is common but generally involves small numbers of birds.
[3] Small number of reports, generally involving individual or small numbers of birds.

Figure 7.7 *Relative occurrence of avian cholera in wild birds.*

Studies by researchers at the NWHC indicate that some flocks of snow geese wintering in California have significantly reduced survival rates because of this disease. Evaluation of band returns from midcontinent white-fronted geese and field assessments of other waterfowl populations also suggest decreased survival rates due to avian cholera during some years. Avian cholera has periodically caused heavy losses of breeding eiders and these outbreaks devastate those colonies.

Avian cholera is clearly an important disease of North American waterfowl and it requires more intensive studies to adequately assess impacts on population dynamics. Avian cholera now rivals avian botulism for the dubious honor of being the most important disease of North American waterfowl. Its threat to endangered avian species is continually increasing because of increasing numbers of avian cholera outbreaks and the expanding geographic distribution of this disease.

Distribution

Avian cholera is believed to have first occurred in the United States during the middle to late 1880s, but it was unreported as a disease of free-ranging migratory birds prior to the winter of 1943–44 when many waterfowl died in the Texas Panhandle and near San Francisco, California. Avian cholera outbreaks involving free-ranging wild birds have now been reported coast-to-coast and border-to-border within the United States. Although avian cholera is found in many countries, there have been few reports in the scientific literature of die-offs from avian cholera affecting free-ranging wild birds in countries other than the United State and Canada. This disease undoubtedly causes more infections and deaths than are reported, and it is an emerging disease of North American free-ranging migratory birds.

Sporadic cases of avian cholera have been documented in the United States since the early 1940s, and perhaps before, in species such as crows, starlings, grackles, sparrows, and other birds that are closely associated with poultry operations. Most of these wild species are now seldom found to be infected, perhaps due to changes in poultry husbandry and waste disposal practices. Avian cholera also broke out in California in free-ranging quail during the early 1940s and in cedar waxwings in Ohio during 1968. However, waterfowl are the primary species that are affected by this disease.

The emergence of avian cholera as a significant disease for North American waterfowl began about 1970. The frequency and severity of avian cholera outbreaks vary greatly among years and geographic areas but the pattern of continual spread is of major concern (Fig. 7.10). The first outbreaks in eastern Canada involving wild waterfowl were reported during 1964 in eiders nesting on islands in the St. Lawrence Seaway. The first outbreaks in western Canada took place in snow geese during 1977; this disease has occurred annually in western Canada ever since. Several suspect diagnoses of avian chol-

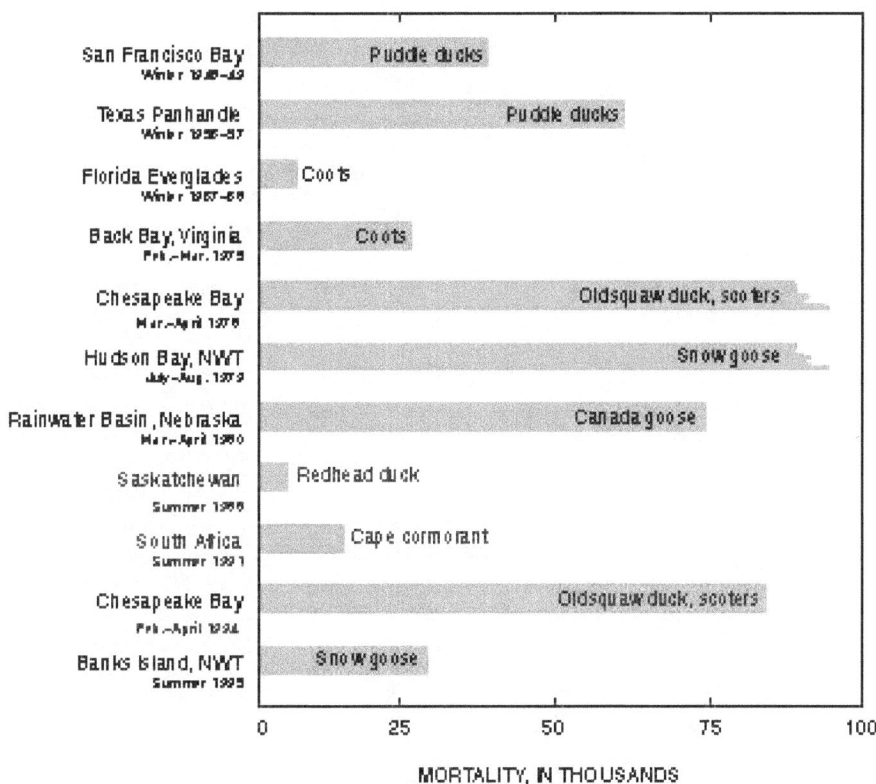

Figure 7.8 Examples of major avian cholera outbreaks in wild birds. (Broken bars indicate very high but indeterminate mortality.)

Figure 7.9 (A) Dense aggregations of waterfowl facilitate the rapid spread of avian cholera because of the highly infectious nature of this disease. (B) Large-scale mortality has occurred in such situations.

EXPLANATION

Frequency of occurrence of avian cholera, by State *(Map A)*

- ■ Annual to nearly annual occurrence, often resulting in deaths of thousands of birds during individual events
- ▨ Frequent occurrences, most resulting in death of moderate to small numbers of birds
- ▦ Occasional occurrences, many of which result in large-scale mortality
- ▢ Occasional occurrences, most resulting in death of moderate to small numbers of birds
- □ Not reported

Time period of first reported occurrence of avian cholera, by State *(Map B)*

- ■ 1944 — 53
- ▨ 1954 — 63
- ▦ 1964 — 73
- ▢ 1974 — 83
- ▨ 1984 — 93
- ▦ 1994 — 97
- □ Not reported

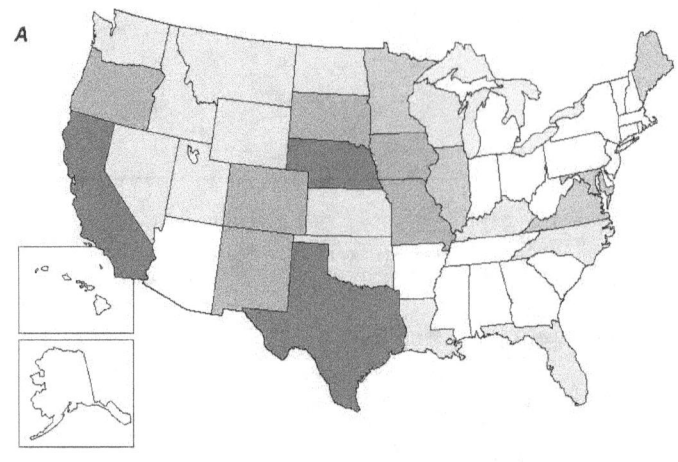

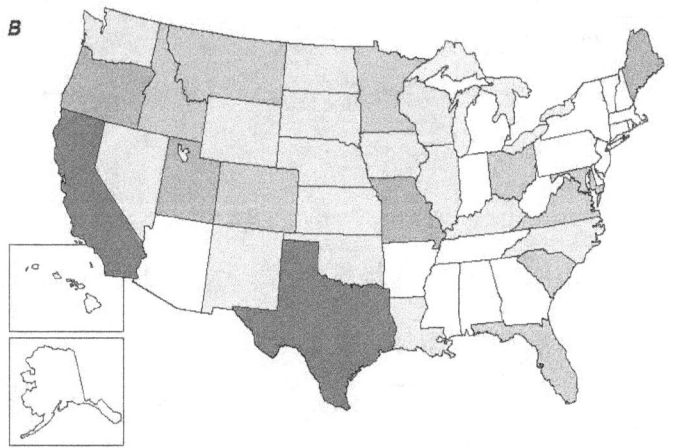

Figure 7.10 *(A) Reported frequency of avian cholera in free-ranging waterfowl in the United States. (B) Reported occurrence of avian cholera in free-ranging migratory birds in the United States.*

era have been reported for waterfowl mortality events in Mexico during recent years, but these events lack laboratory confirmation. The absence of confirmed reports of this disease in wild waterfowl in Mexico is likely due to lack of surveillance and reporting rather than to the absence of avian cholera.

In the United States, there are four major focal points for avian cholera in waterfowl: the Central Valley of California; the Tule Lake and Klamath Basins of northern California and southern Oregon; the Texas Panhandle; and Nebraska's Rainwater Basin below the Platte River in the south-central part of the State. The movement of avian cholera from these areas follows the well-defined pathways of waterfowl movement. The spread of this disease along the Missouri and Mississippi River drainages is also consistent with waterfowl movement. No consistent patterns of avian cholera outbreaks exist within the Atlantic Flyway. There are periodic outbreaks in eider ducks nesting off of the coast of Maine and occasional major die-offs of sea ducks, including eiders, within the Chesapeake Bay of Maryland and Virginia (Fig. 7.11).

Seasonality

Losses can occur at any time of the year. For poultry, outbreaks of avian cholera are more prevalent in late summer, fall, and winter. Those time periods have no special biological associations, except, possibly, with production schedules in response to holiday market demands that influence poultry age-classes within production facilities. Chickens become more susceptible as they reach maturity. Turkeys are much more susceptible than chickens, and turkeys die at all ages, but the disease usually occurs in young mature turkeys. Losses in domestic ducks are usually in birds older than 4 weeks of age.

For wild waterfowl, a predictably seasonal pattern exists in areas where avian cholera has become well established. This pattern is closely associated with seasonal migration patterns and it has resulted in avian cholera becoming a "disease for all seasons," killing waterfowl during all stages of their life cycle (Fig. 7.12). Some areas experience prolonged periods of avian cholera mortality. Outbreaks in California normally start during fall and continue into spring. Other areas have seasonal avian cholera outbreaks in the same geographic location. For example, Nebraska, which has had outbreaks most springs since 1975, now frequently also has outbreaks in the fall.

Field Signs

Few sick birds are seen during avian cholera outbreaks because of the acute nature of this disease. However, the number of sick birds increases when a die-off is prolonged over several weeks. Sick birds often appear lethargic or drowsy (Fig. 7.13), and they can be approached quite closely before they attempt to escape. When captured, these birds often die quickly, sometimes within a few seconds or minutes after being handled. Other birds have convulsions (Fig. 7.14), swim in circles, or throw their heads back between their wings and die (Fig. 7.15). These signs are similar to those seen in duck plague and in some types of pesticide poisoning. Other signs include erratic flight, such as flying upside down before plunging into the water or onto the ground, and attempting to land a foot or more above the surface of the water; mucous discharge from the mouth; soiling and matting of the feathers around the vent, eyes, and bill; pasty, fawn-colored, or yellow droppings; and blood-stained droppings or nasal discharges, which also are signs of duck plague (duck virus enteritis or DVE).

Always suspect avian cholera when large numbers of dead waterfowl are found in a short time, when few sick birds are seen, and when the dead birds appear to be in good flesh. Death can be so rapid that birds may literally fall out of the sky or die while feeding, with no signs of illness. When sick birds are captured and die within a few minutes, avian cholera should also be suspected. None of the signs described above are unique to this disease; these signs should be recorded as part of any history being submitted with specimens and are considered along with lesions seen at necropsy.

Gross Lesions

Under most conditions, birds that have died of avian cholera will have substantial amounts of subcutaneous and visceral fat, except for seasonal losses of fat. The most prominent lesions seen at necropsy are in the heart and liver and, sometimes, the gizzard. Hemorrhages of various sizes are frequently found on the surface of the heart muscle or the coronary band or both (Fig. 7.16). Hemorrhages are also sometimes visible on the surface of the gizzard. Areas of tissue death that appear as small white-to-yellow spots are commonly seen within the liver. Where the area of tissue death is greater, the spots are larger and, in some instances, the area of tissue death is quite extensive (Fig. 7.17).

The occurrence of the abnormalities described for the heart, liver, and gizzard are dependent upon how long the bird lived after it became infected. The longer the survival time, the more abundant and dramatic the lesions. In addition, there may be changes in the color, size, and texture of the liver. There is darkening or a copper tone to the liver, and it may appear swollen and rupture upon handling. Because birds infected with avian cholera often die so quickly, the upper portions of the digestive tract may contain recently ingested food. All of these findings are similar to what might also be seen with duck plague; therefore, laboratory diagnosis is needed.

Freshly dead ducks and geese that have succumbed to avian cholera may have a thick, mucous-like, ropy nasal discharge. The lower portions of the digestive tract (below the gizzard) commonly contain thickened yellowish fluid (Fig.

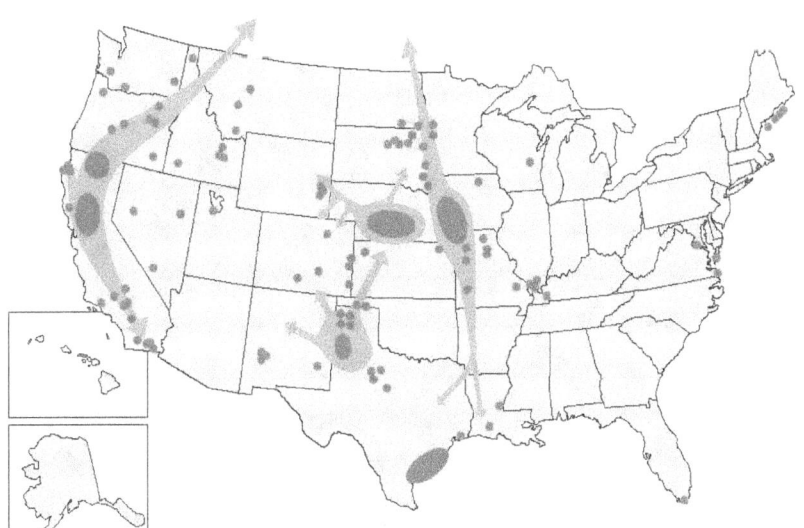

EXPLANATION
Avian cholera in waterfowl outbreak sites, 1944–97

• Outbreak site

↗ Migratory movements of waterfowl

Figure 7.11 The occurrence of avian cholera in waterfowl seems to be closely related to bird movements west of the Mississippi River. There is no apparent pattern for outbreaks along the Atlantic seaboard.

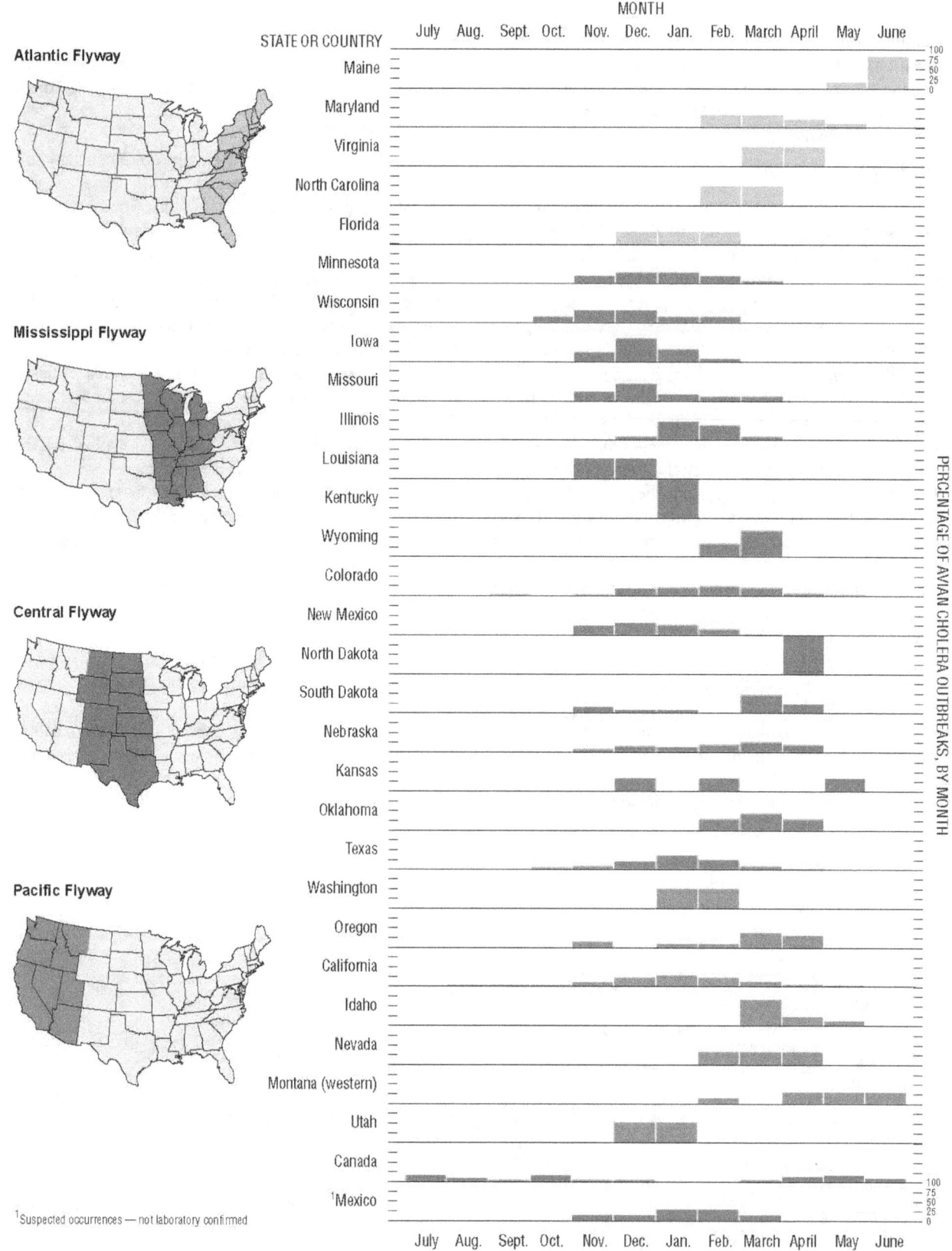

Figure 7.12 Relative monthly probability for the occurrence of avian cholera in migratory waterfowl, expressed as a percentage of outbreaks throughout the year. Information from the National Wildlife Health Center database.

84 Field Manual of Wildlife Diseases: Birds

Figure 7.13 Lethargic appearance of drake northern pintail with avian cholera.

Figure 7.14 Avian cholera-infected crow in convulsions.

Figure 7.15 Avian cholera-infected drake mallard. (A) Note tossing of head toward back and circular swimming as evidenced by ripples in water. (B) Bird at death with head resting on back.

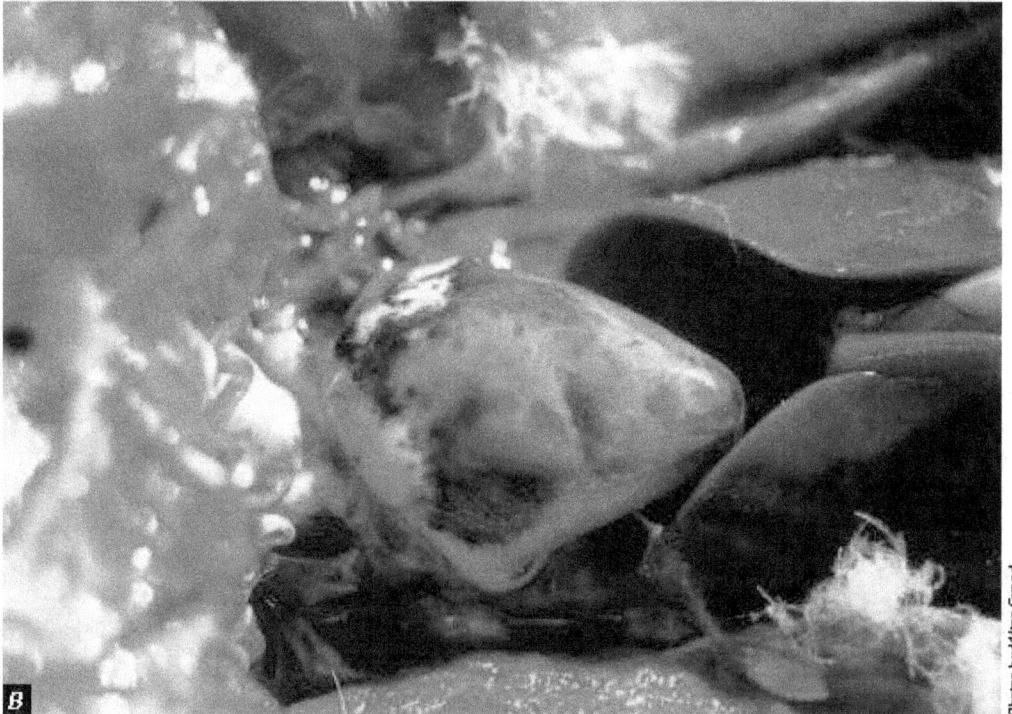

Figure 7.16 Hemorrhages of varying degrees of severity are often seen on the hearts of avian cholera-infected birds. **(A)** Pinhead-sized hemorrhages along fatty areas of the heart are readily evident in this bird. **(B)** Broad areas of hemorrhage also occur.

Figure 7.17 Lesions in the livers of avian cholera-infected birds generally appear as small, discrete, yellowish spots, which are dead tissue. Note the variation in size and appearance of these lesions. (A) Note also the absence of any apparent heart lesions in one bird, (B) only a few minor hemorrhages on the coronary band of another bird, (C) and more extensive hemorrhages on the heart muscle of the third bird. Also note the abundance of fat covering the gizzards of all these birds. This fat attests to the excellent condition these birds were in before exposure to the bacterium and to the rapidity with which each bird died.

Figure 7.18 The thickened, yellowish fluid present in the intestines of this avian cholera-infected bird contains millions of bacteria. These bacteria contaminate the environment when the carcass decomposes or is scavenged, serving as a source of infection for other wildlife.

7.18). Both of these fluids are heavily laden with *P. multocida* and care must be taken to not contaminate the environment, field equipment, or oneself with these fluids.

Diagnosis

As with all diseases, isolation of the causative agent is required for a definitive diagnosis. A whole carcass provides the diagnostician with the opportunity to evaluate gross lesions seen at necropsy and also provides all appropriate tissues for isolation of *P. multocida*.

When it is not possible to send whole carcasses, send tissues that can be collected in as sterile a manner as possible in the field. The most suitable tissues for culturing are heart blood, liver, and bone marrow. Remove the entire heart and place it in a Whirl-Pak® bag for shipment as identified in Chapter 2, Specimen Collection and Preservation; do not attempt to remove the blood from the heart. The liver should also be removed and placed in a separate bag. A major portion of this organ (at least half) should be submitted if it cannot be removed intact. These samples must be refrigerated as soon as possible after collection and kept cool during shipment. When shipment is to be delayed for more than 1 day or when transit time is expected to exceed 24 hours, freeze these specimens.

P. multocida persists several weeks to several months in bone marrow. The wings of badly scavenged or decomposed carcasses should be submitted whenever avian cholera is suspected as the cause of death, and when more suitable tissue samples are not available.

Control

Numerous factors must be considered in combating avian cholera (Fig. 7.19). Avian cholera is highly infectious and it spreads rapidly through waterfowl and other bird populations. This process is enhanced by the gregarious nature of most waterfowl species and by dense concentrations of migratory waterbirds resulting from habitat limitations. The prolonged environmental persistence of this bacterium further promotes new outbreaks (Table 7.1). Pond water remained infective for 3 weeks after dead birds were removed from one area in California; bacterial survival in soil for up to 4 months was reported in another study; and the organism can persist in decaying bird carcasses for at least 3 months.

Early detection of avian cholera outbreaks is a first line of defense for controlling this disease. Frequent surveillance of areas where migratory birds are concentrated and the timely submission of carcasses to disease diagnostic laboratories allows disease control activities to be initiated before

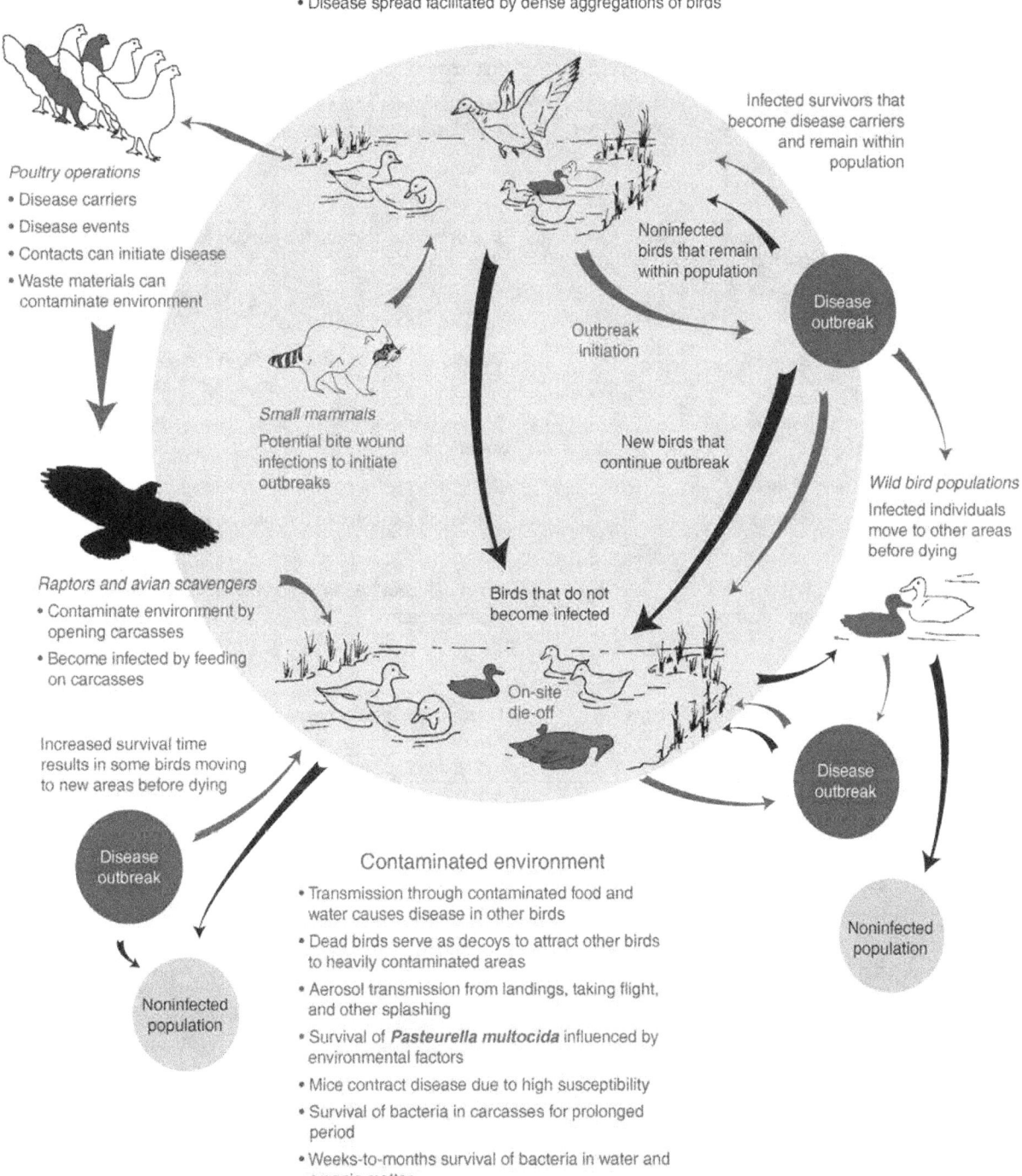

Figure 7.19 Some of the many interrelated factors associated with avian cholera outbreaks in free-ranging wild birds.

Avian Cholera 89

Table 7.1 Examples of reported environmental persistence for **Pasteurella multocida**.

Substrate	Survival time	Comments
General	Highly variable	Amount of moisture, temperature, and pH affect survival of *P. multocida*.
		Survival in soils enhanced when moisture content is 50 percent or greater.
		Survival in water is enhanced by high organic content and turbidity.
		Survival in wetland waters enhanced by presence of magnesium and chloride ions.
Garden soil	3 months	
Unspecified soil	113 days at 3 °C; 15–100 days at 20 °C; 21 days at 26 °C	Soil chemistry information needed to properly evaluate data.
Poultry yard	2 weeks	Infectious for birds after last death and removal of all birds.
Water	3 weeks	Following removal of 100 dead snow geese; no other waterfowl use.
	99 days	Water contaminated with turkey litter.
	30 days	In marsh near carcass that had been opened.
Infected tissues	120 days but not 240 days	American coot hearts buried in marsh after birds died from avian cholera.
Fomites (Inanimate objects)	8 days but not 30 days	Dried turkey blood on glass at room temperature.

the outbreaks reach advanced stages. The opportunity to prevent substantial losses is greatest during the early stages of outbreaks, and costs are minimal in comparison with handling a large-scale die-off. Control actions need to be focused on minimizing the exposure of migratory and scavenger bird species to *P. multocida* and minimizing environmental contamination by this organism.

The NWHC recommends carcass collection and incineration as standard procedures. Carcass collection contributes to avian cholera control in several ways. Several milliliters of fluids containing large concentrations of *P. multocida* are often discharged from the mouths of birds dying from this disease, resulting in heavy contamination of the surrounding area. Carcass decomposition results in additional contamination. These carcasses attract (decoy) other birds, thereby increasing the probability for infection. Scavenging of carcasses also transmits the disease through the direct consumption of diseased tissue (oral exposure).

Care must be exercised during carcass collection to minimize the amount of fluid discharged from the mouths of birds into the environment. Birds should be picked up head first, preferably by the bill, and immediately placed in plastic bags. Care must also be taken to avoid contaminating new areas while carcasses are transported to the disposal site. Double-bagging is recommended to prevent fluids leaking from punctures to the inner bag. Bags of carcasses should always be securely closed before they are removed from the area.

Prompt carcass removal also prevents scavenging by avian species that can mechanically transport infected material to other sites or by feeding or drinking at other locations following consumption of infected tissue. This situation is aggravated by apparently longer disease-incubation times in gulls, crows, and some other avian scavengers. Instead of dying within hours or 1–2 days after exposure to virulent strains of *P. multocida*, avian scavengers more typically die after several days to 1–2 weeks, and they may die far from the site of exposure. When these birds die, they may serve as new potential focal points for contamination.

In some instances, population reduction of gulls and crows has been used to limit the role of these species in spreading and transmitting avian cholera. This technique has limited application and it is not recommended as normal operating procedure. To be most effective, population reduction must be undertaken before there is a major influx of scavengers in response to carcass availability. Also, the techniques used must not result in dispersal of infected birds out of the area.

> Population reduction of infected American coots, gulls, terns, and eiders has also been used to directly combat avian cholera. Destruction of migratory birds infected with this disease can be justified only under special circumstances and conditions:
>
> 1. The outbreak must be discreet and localized rather than generalized and widespread.
>
> 2. Techniques must be available that will allow complete eradication without causing widespread dispersal of potentially infected birds.
>
> 3. The methods used must be specific for target species and pose no significant risk for nontarget species.
>
> 4. Eradication must be justified on the basis of risk to other populations if the outbreak is allowed to continue.
>
> 5. The outbreak represents a new geographic extension of avian cholera into an important migratory bird population.

Habitat management is another useful tool for combating avian cholera outbreaks. In some instances, it may be necessary to prevent further bird use of a specific wetland or impoundment because it is a focal point for infection of waterfowl migrating into the area. Drainage of the problem area in conjunction with creation or enhancement of other habitat within the area through water diversion from other sources or pumping operations denies waterfowl the use of the problem area and redistributes them into more desirable habitat. The addition of a large volume of water to a problem area can also help to dilute concentrations of *P. multocida* to less dangerous levels. These actions require careful evaluation of bird movement patterns and of the avian cholera disease cycle. Movement of birds infected with avian cholera from one geographic location to another site is seldom desirable.

Under extreme conditions, disinfection procedures to kill *P. multocida* may be warranted in wetlands where large numbers of birds have died during a short time period. The environmental impact of such measures must be evaluated and appropriate approvals must be obtained before these actions are undertaken. A more useful approach may be to enhance the quality of the wetland in a way that reduces the survival of *P. multocida*; the best means of accomplishing this is still being investigated.

Hazing with aircraft has been successfully used to move whooping cranes away from a major outbreak of avian cholera. This type of disease prevention action can also be accomplished by other methods for other species. Eagles can be attracted to other feeding sites using road-killed animals as a food source, and waterfowl can be held at sites during certain times of the year by providing them with refuge and food. During an avian cholera outbreak in South Dakota, a large refuge area was temporarily created to hold infected snow geese in an area by closing it to hunting. At the same time, a much larger population of snow geese about 10 miles away was moved out of the area to prevent transmission of the disease into that population. The area closed to hunting was reopened after the desired bird movement had occurred.

Vaccination and postexposure treatment of waterfowl have both been successfully used to combat avian cholera in Canada goose propagation flocks. The NWHC has developed and tested a bacterin or a killed vaccine that totally protected Canada geese from avian cholera for the entire 12 months of a laboratory study. This product has been used for several years with good results in a giant Canada goose propagation flock that has a great deal of contact with free-flying wild waterfowl and field outbreaks of avian cholera. Before use of the bacterin, this flock of Canada geese suffered an outbreak of avian cholera and was successfully treated with intramuscular infections of 50 milligrams of oxytetracycline followed by a 30-day regimen of 500 grams of tetracycline per ton of feed. A NWHC avian cholera bacterin has also been used to successfully vaccinate snow geese on Wrangle Island, Russia, and Banks Island, Canada. Vaccine use in these instances was in association with studies to evaluate avian cholera impacts on survival rates rather than to control disease in those subpopulations.

As yet, there is no practical method for immunizing large numbers of free-living migratory birds against avian cholera. However, captive propagation flocks can be protected by this method. Endangered species can be trapped and immunized if the degree of risk warrants this action. Live vaccines should not be used for migratory birds without adequate safety testing.

Human Health Considerations

Avian cholera is not considered a high risk disease for humans because of differences in species susceptibility to different strains of *P. multocida*. However, *P. multocida* infections in humans are not uncommon. Most of these infections result from an animal bite or scratch, primarily from dogs and cats. Regardless, the wisdom of wearing gloves and thoroughly washing skin surfaces is obvious when handling birds that have died from avian cholera.

Infections unrelated to wounds are also common, and in the majority of human cases, these involve respiratory tract exposure. This is most apt to happen in confined areas of air movement where a large amount of infected material is present. Processing of carcasses associated with avian cholera die-offs should be done outdoors or in other areas with adequate ventilation. When disposing of carcasses by open burning, personnel should avoid direct exposure to smoke from the fire.

Milton Friend

Supplementary Reading

Botzler, R.G., 1991, Epizootiology of avian cholera in wildfowl: Journal of Wildlife Diseases, 27 p. 367–395.

Brand, C.J., 1984, Avian cholera in the Central Mississippi Flyways during 1979–80: Journal of Wildlife Management, 48, p. 399–406.

Rimler, R.B., and Glisson, J.R. 1997, Fowl cholera, *in* Calnek, B.W., and others, eds., Diseases of poultry (10th ed.): Ames, Iowa, Iowa State University Press, p. 143–159.

Wobeser, G.A., 1997, Avian cholera, *in* Diseases of wild waterfowl (2nd ed): New York, N.Y., Plenum Press, p. 57–69.

Chapter 8
Tuberculosis

Synonyms
Mycobacteriosis, tuberculosis, TB

Cause

Avian tuberculosis is usually caused by the bacterium *Mycobacterium avium*. At least 20 different types of *M. avium* have been identified, only three of which are known to cause disease in birds. Other types of Mycobacterium rarely cause tuberculosis in most avian species; however, parrots, macaws, and other large perching birds are susceptible to human and bovine types of tuberculosis bacilli. Avian tuberculosis generally is transmitted by direct contact with infected birds, ingestion of contaminated feed and water, or contact with a contaminated environment. Inhalation of the bacterium can cause respiratory tract infections. Wild bird studies in the Netherlands disclosed tuberculosis-infected puncture-type injuries in birds of prey that fight at the nest site (kestrels) or on the ground (buteo-type buzzards), but tuberculosis-infected injuries were not found in accipiters (falcons), which fight in the air and seldom inflict such wounds.

Species Affected

All avian species are susceptible to infection by *M. avium*. Humans, most livestock species, and other mammals can also become infected. Recent molecular studies with a limited number of isolates from birds, humans, and other mammals clearly indicated that *M. avium* can be transmitted between birds and pigs, but the studies did not disclose a similar cross transmission between birds and humans for the isolates tested. It is generally accepted that pigs, rabbits, and mink are highly susceptible to *M. avium*; deer can also become infected. Dogs appear to be quite resistant to the avian type of tuberculosis (Fig. 8.1).

In captivity, turkeys, pheasants, quail, cranes, and certain birds of prey are more commonly infected than waterfowl. However, when avian tuberculosis becomes established, it can be a common and lethal disease in captive waterfowl flocks. Chronic infections exist in some captive nene goose flocks, making these flocks unsuitable donors to supplement the wild population of this endangered species. Pheasants are unusually susceptible to avian tuberculosis.

In free-ranging wild birds, avian tuberculosis is found most often in species that live in close association with domestic stock (sparrows and starlings) and in scavengers (crows and gulls). The prevalence of tuberculosis in free-ranging North American birds has not been determined, although generally less than 1 percent of birds examined at postmortem are affected. Sampling biases due to the limited numbers of speci-

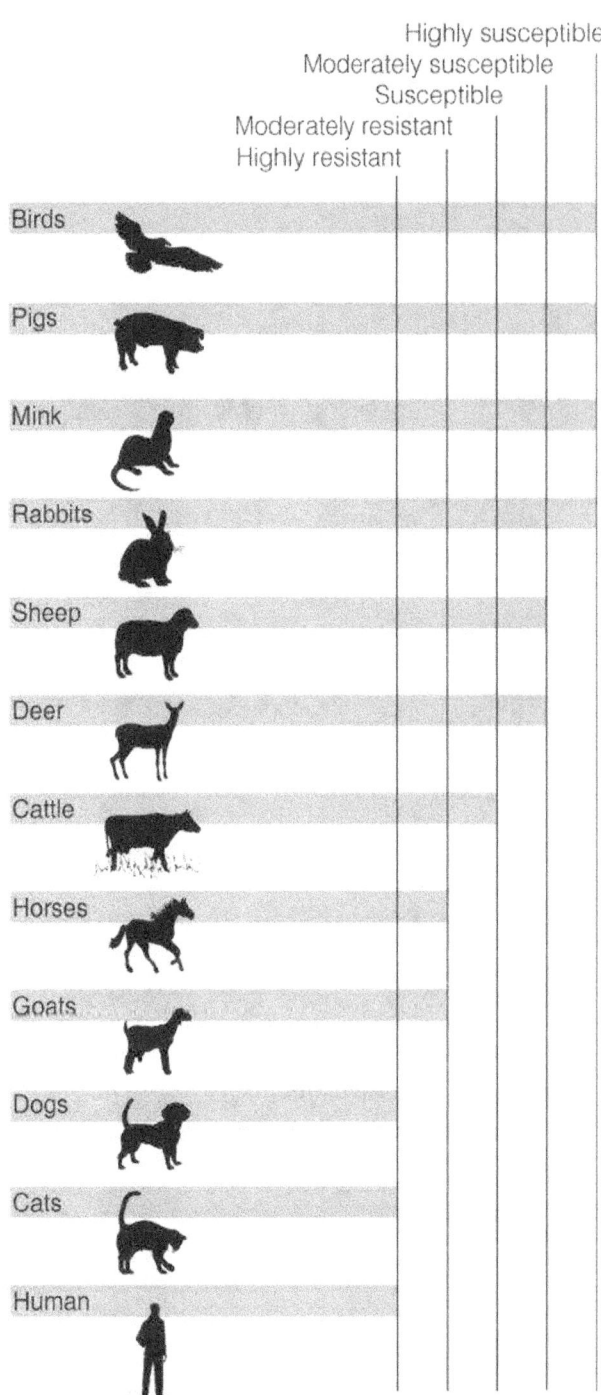

Figure 8.1 Relative susceptibility of various animal groups to **M. avium**.

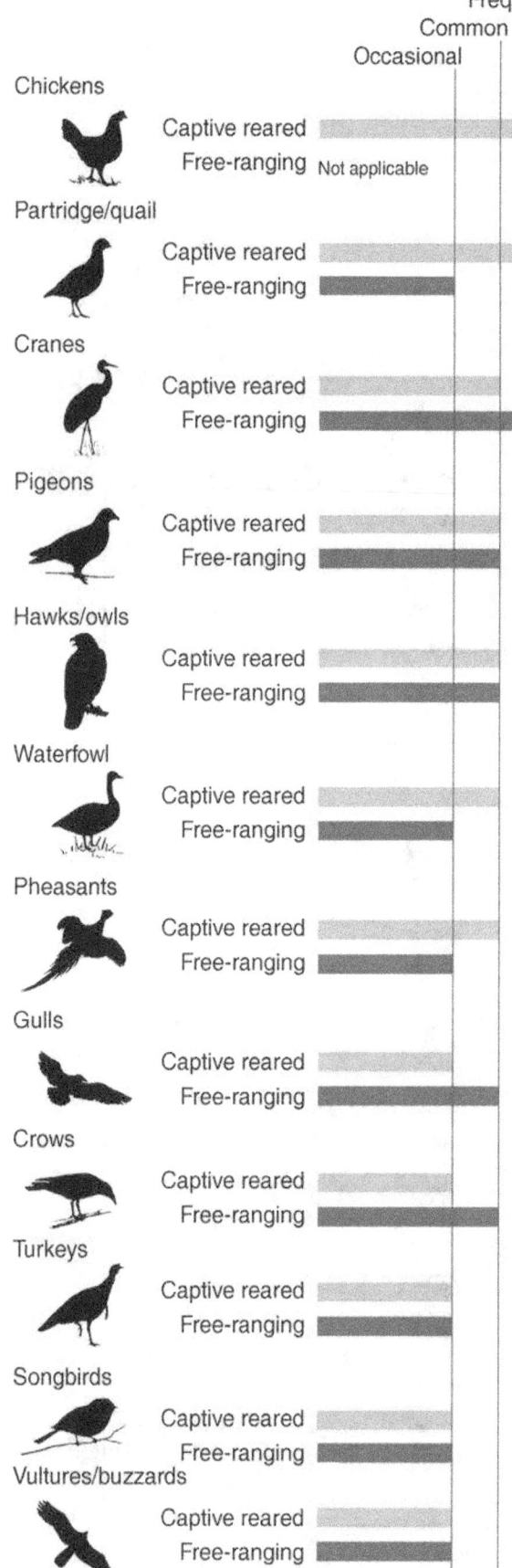

mens examined preclude extending findings to reflect actual prevalence (Fig. 8.2). A decade-long study of nearly 12,000 wild birds necropsied in the Netherlands disclosed that 0.7 percent of the birds had tuberculosis. The sample included waterbirds, birds of prey, songbirds, and pheasants. Studies in the United States disclosed that 0.3 percent of 3,000 waterfowl necropsied were infected with tuberculosis, and a study in British Columbia found tuberculosis in 0.6 percent of more than 600 wild birds. Tuberculosis in whooping cranes stands in marked contrast to other wild birds; approximately 39 percent of the western population's free-ranging whooping cranes necropsied at the National Wildlife Health Center have been infected with avian tuberculosis.

Distribution

Avian tuberculosis is a ubiquitous and cosmopolitan disease of free-ranging, captive, and domestic birds. The disease is most commonly found in the North Temperate Zone, and, within the United States, the highest infection rates in poultry are in the North Central States. Distribution of this disease in free-ranging wild birds is inferred from birds submitted for necropsy; however, the sampling underrepresents both the geographic distribution and the frequency of infection for individual species. Avian tuberculosis likely exists in small numbers of free-ranging wild birds wherever there are major bird concentrations.

Seasonality

Seasonal trends of tuberculosis in wild birds have not been documented. The chronic nature of this disease guarantees its presence yearround for both wild and captive birds.

Factors that may influence seasonal exposure to tuberculosis in migratory birds are changes in habitat used, food base during the year, and interspecies contacts. Contaminated sewage and wastewater environments containing tubercle bacilli are more likely to be used by waterfowl during fall and winter than during warmer months. Wastewater sites are often closed to hunting, thereby serving as refuge areas, and warm water discharges to these sites maintain open water in subfreezing temperatures, thus inviting ready use by waterfowl. Predatory and scavenger species such as raptors and crows often ingest many different food items during different periods of the year; scavengers, therefore, may be exposed to tuberculosis through contaminated food yearround. Contact between wild birds and poultry and livestock is often restricted to specific periods of the year owing to husbandry practices. Wild birds may be exposed to *M. avium* in manure that is spread on fields during early spring.

Environmental conditions can greatly affect the susceptibility of birds to tuberculosis and the prevalence of tuberculosis in captive birds. Captive birds that are on an inadequate

Figure 8.2 *Relative occurrence of avian tuberculosis in birds.*

diet and that are maintained in crowded, wet, cold, poorly ventilated, and unhygienic aviaries have increased susceptibility to tuberculosis.

Field Signs

No clinical signs specifically identify avian tuberculosis in birds. Advanced disease and clinical signs are seen most often in adult birds because of the chronic, insidious nature of the disease. Infected birds are often emaciated, weak, and lethargic, and they exhibit wasting of the muscles. These signs are similar to those of lead poisoning and other debilitating conditions. Other signs depend on which body system is affected and signs may include diarrhea, lameness, and unthrifty appearance. Darkening and dulling of plumage have been reported in the United Kingdom for wood pigeons infected with tuberculosis, but not for other species.

Gross Lesions

Typical cases of avian tuberculosis in wild birds involve emaciated carcasses with solid-to-soft or crumbly, yellow-to-white or grey nodules that are less than 1 millimeter to several centimeters in size and that are deeply embedded in infected organs and tissues. The liver (Fig. 8.3A) most often contains such nodules, but the spleen (Fig. 8.3B), lung, and intestines (Fig. 8.3C) may also contain similar nodules. Aggregations of these nodules may appear as firm, fleshy, grape-like clusters. Abscesses and nodular growths (Fig. 8.4) have been reported on the skin of birds in the same locations where pox lesions are commonly seen—around the eyes, at the wing joints, on the legs, side of the face, and base of the beak. Other birds have died of avian tuberculosis without any obvious clinical signs or external lesions.

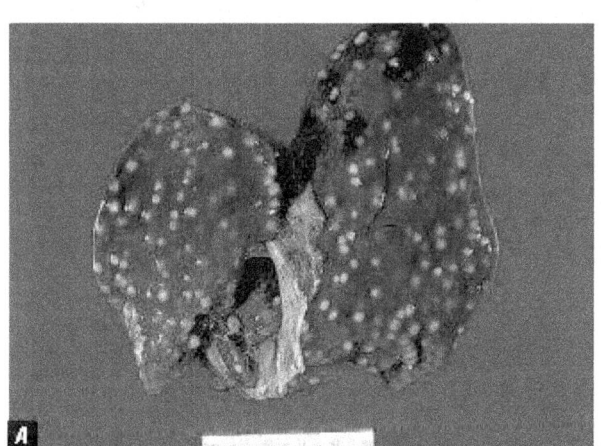

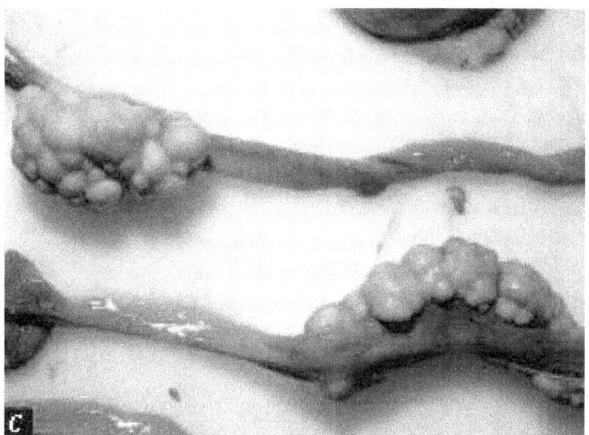

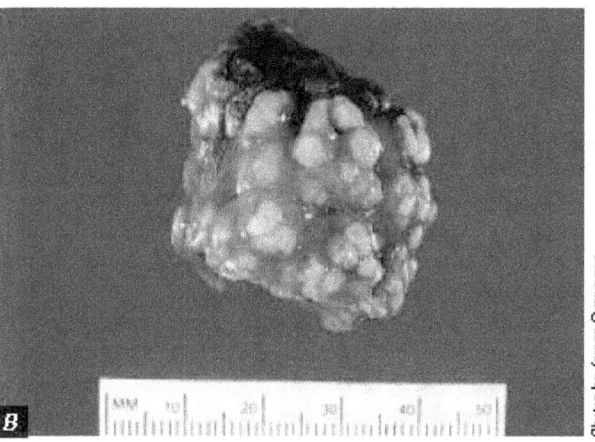

Figure 8.3 The raised, firm nodules in these organs are typical lesions of avian tuberculosis; *(A)* liver; *(B)* spleen; and *(C)* intestine.

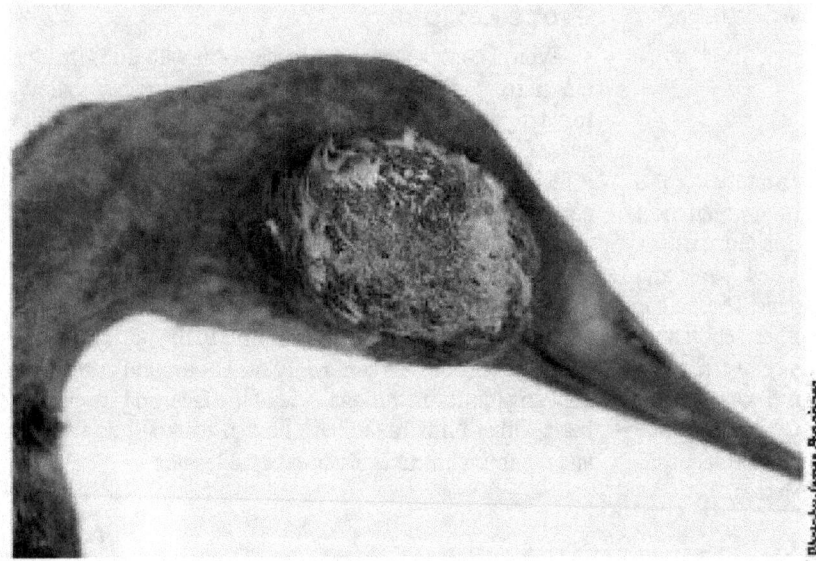

Figure 8.4 Nodular lesion, which was caused by avian tuberculosis, on the skin of a canvasback.

Nodular tuberculosis lesions in internal organs are often grossly similar to those of aspergillosis, and laboratory diagnosis is required to differentiate the two diseases as well as others that produce similar lesions. Less typical lesions resemble those of other diseases. Sometimes the primary lesions seen at necropsy are enlarged livers and spleens that are so fragile that they easily rupture upon being handled. Most of these cases have livers and spleens with a tan-to-green translucence due to amyloid deposits. Less commonly, in situations where nodules are not formed nor is amyloid deposited, the liver and spleen can be large, pale, and firm.

The location of primary lesions is an indication of route of exposure. Intestinal lesions suggest ingestion of *M. avium* in contaminated feed or water. Lesions in the lungs and other areas of the respiratory tract suggest inhalation as the route of exposure.

Diagnosis

Typically, tuberculosis is discovered in captive birds during routine investigation of mortality, and in wild birds during carcass examinations associated with die-offs due to other causes. The gross lesions described above (Fig. 8.3) are suggestive of tuberculosis, but a definitive diagnosis is based on bacteriological isolation and identification of the organism. Because *M. avium* is slow-growing and other bacteria can easily overgrow it, a noncontaminated sample is needed for examination. Whole carcasses are preferred, but when a whole carcass cannot be submitted, remove the leg at the hip joint, wrap it in clean aluminum foil, place it in a plastic bag, and freeze it for shipment to a qualified disease diagnostic laboratory. The marrow within the femur has the lowest potential for being contaminated and it provides a good sample for the bacteriologist. When carcass or tissue submissions to a laboratory are not possible within a short time, tissue preserved in 10 percent buffered formalin solution is useful for diagnostic purposes (see Chapter 2, Specimen Collection and Preservation).

The bacterium can also be isolated from infected tissues that show gross lesions. Microscopic studies can provide a diagnosis of tuberculosis, although such studies cannot determine the species of *Mycobacterium*. Because this disease is transmissible to humans, extra care must be taken when handling infected carcasses.

Control

Tuberculosis is difficult to detect in free-ranging birds despite its broad geographic distribution. Tuberculosis rarely causes a major die-off, and there are no practical nonlethal testing procedures for mobile wild birds. Therefore, there is no focal point and, hence, no method developed for disease control in wild bird populations. By contrast, tuberculosis can cause die-offs in captive flocks, and mortality has been reported in sea ducks and other birds, including chukar partridge and pheasants. Some captive flocks of wild birds have experienced losses of nearly 30 percent or more from tuberculosis.

Close monitoring of the health of bird populations — free-ranging or captive — is an essential first step toward detecting tuberculosis so that control efforts can be developed and initiated when feasible. Monitoring can best be accomplished by the timely submission of carcasses to disease diagnostic laboratories. Tuberculosis testing of birds maintained in captivity and laboratory analyses of fecal samples from captive and wild flocks also can be used to identify the presence of

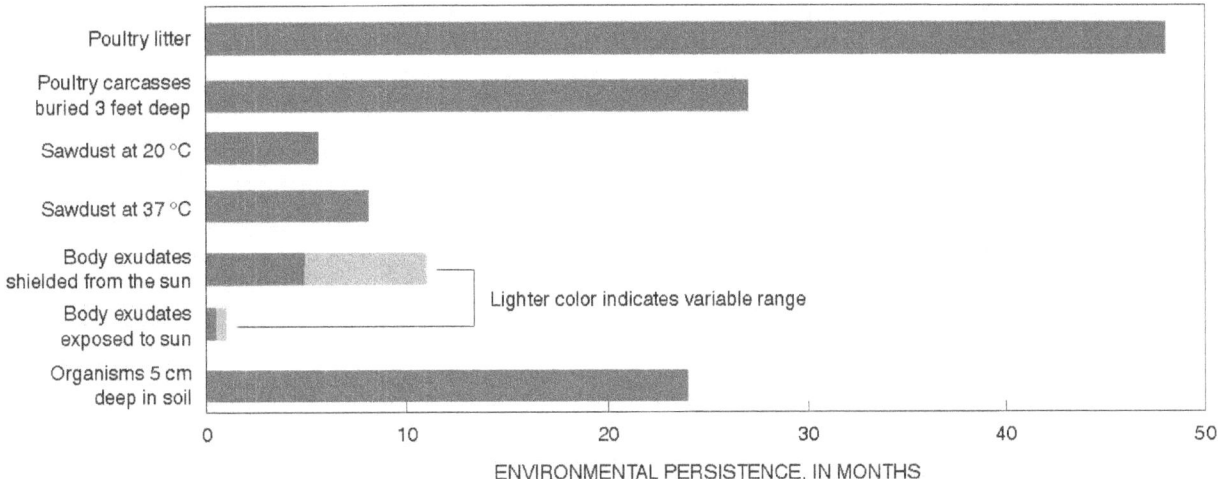

Figure 8.5 Examples of environmental persistence of **M. avium**.

this disease. These tests do not detect all infected birds, but the tests are useful for identifying infected flocks.

Fecal contamination of the environment is the major means of tuberculosis dissemination; ingestion of the bacterium in contaminated feed and water is the most common means of disease transmission. Because this bacterium can survive outside of the vertebrate host for long time periods in an organic substrate (Fig. 8.5), a few infected animals can contaminate an area that has prolonged bird-use patterns. The long-term environmental survival of *M. avium* that is shed by disease carriers when combined with repeated site use and, possibly, a high degree of susceptibility to avian tuberculosis may be the factors contributing to the high prevalence of this disease observed in whooping cranes. A site can also be contaminated by wastewater discharges containing *M. avium* and by the application of contaminated manure for fertilizer. Tuberculosis outbreaks in birds have been associated with sewage effluents and discharges from slaughter houses, meat processing plants, and dairies. In one instance, an outbreak occurred in a captive waterfowl flock when contaminated water was sprayed into the enclosure. These events illustrate the importance of disease prevention for addressing tuberculosis in free-ranging and captive wild birds.

The use of wastewater for maintaining captive waterfowl and other wild birds is questionable without adequate testing or treatment or both to assure that the wastewater does not contain tubercle bacilli. Also, the use of wetlands for wastewater discharges and the use of wastewater to create wetlands for migratory bird habitat should be carefully considered because of the possible presence of *M. avium* in the wastewater. Other actions that should be considered include preventing land use that could place tuberculosis-infected swine in close proximity to major wild bird concentrations and not using unexamined chicken and pigeon carcasses as food for raptors being reared in captivity for release into the wild.

Infected flocks of captive birds should be destroyed because treatment is ineffective and because not all infected birds will be detected by current testing procedures. Because of the long-term environmental persistence of the tubercle bacilli, additional bird use of the site should be avoided for approximately 2 years. Vegetation removal and turning of the soil several times during this period will facilitate sunlight-induced environmental decay of the bacilli. Eradication of free-ranging migratory flocks is rarely feasible. However, when a major outbreak of tuberculosis occurs in wild birds, the circumstances should be assessed, and limited population reduction should be considered if the remaining population-at-risk is well defined, limited in immediate distribution, and involves species that can withstand this action. Habitat manipulation, such as drainage, and scaring devices, such as propane exploders, can sometimes be used to deny birds use of areas where tuberculosis outbreaks occur.

The insidious nature of avian tuberculosis combined with the long environmental persistence of the causative bacterium strongly indicate a need to prevent the establishment of this disease in wild bird populations. When the disease becomes established in free-ranging populations, interspecies transmission and the mobility of free-ranging birds could serve to spread it widely. The continued persistence of avian tuberculosis as a major cause of avian mortality in zoological collections attests to the difficulty of disease control.

Human Health Considerations

There are many authenticated cases of *M. avium* infection in people, although humans are considered highly resistant to this organism. Avian tuberculosis is generally considered noncontagious from an infected person to an uninfected person. Infection is more likely to occur in persons with pre-existent diseases, especially those involving the lungs, and in persons whose immune systems are impaired by an illness, such as AIDS or steroid therapy.

Milton Friend
(Modified from an earlier chapter by Thomas J. Roffe)

Supplementary Reading

Karlson, A.G., 1978, Avian tuberculosis, *in* Montali, R.J., ed., Mycobacterial infections of zoo animals: Washington, D.C., Smithsonian Institution Press, p. 21–24.

Smit, T., Eger, A., Haagsma, J., and Bakhuizen, T., 1987, Avian tuberculosis in wild birds in the Netherlands: Journal of Wildlife Diseases 23, p. 485–487.

Thoen, C.O., 1997, Tuberculosis, *in* Calnek, B.W., and others, eds., Diseases of poultry (10th ed.): Ames, Iowa, Iowa State University Press, p. 167–178.

Wobeser, G.A., 1997, Tuberculosis, *in* Diseases of wild waterfowl (2nd ed): New York, N.Y., Plenum Press, p. 71–75.

Chapter 9
Salmonellosis

Synonyms

Salmonellosis; paratyphoid; bacillary white diarrhea (a synonym for pullorum disease); pullorum disease[1], fowl typhoid[1]

Cause

Avian salmonellosis is caused by a group of bacteria of the genus salmonella. Approximately 2,300 different strains of salmonellae have been identified, and these are placed into groupings called "serovars" on the basis of their antigens or substances that induce immune response by the host, such as the production of specific antibody to the antigen. Current taxonomic nomenclature considers the 2,300 different serovars to be variants of two species, *Salmonella enterica* and *S. bongori*. *S. enterica* is further subdivided into six subspecies on the basis of biochemical characteristics. This results in complex nomenclature for each serovar, such as, *S. enterica* subsp. *enterica* serovar *typhimurium*. Readers should be aware of this convention for naming salmonellae because they will find this nomenclature in the current scientific literature. In this chapter, different serovars of salmonellae will be referred to by their previous, less complex nomenclature, such as *S. typhimurium*.

Pullorum disease, (*S. pullorum*) and fowl typhoid (*S. gallinarum*) are two classic and distinctive diseases of poultry that have received considerable attention because of their economic impacts. Wild birds have been infected with pullorum disease and fowl typhoid, but wild birds are more commonly infected by the variants of salmonellae that are collectively referred to as paratyphoid forms, of which *S. typhimurium* is a prominent representative. The paratyphoid forms constitute the great majority of salmonellae, and they are becoming increasingly important as causes of illness and death in wild birds (Table 9.1).

Salmonella infections can be transmitted in many ways (Table 9.2), and the importance of different modes for transmission varies with the strain of salmonellae, behavioral and feeding patterns of the bird species, and husbandry practices when human intervention becomes part of the hatching and rearing processes. For example, ovarian transmission of *S. typhimurium* occasionally occurs in turkeys, but it is uncommon in chickens. Egg transmission and environmental contamination of rearing facilities are of more importance for infecting poultry than are contaminated feeds. For wild birds and humans, contaminated foods are the primary source for infection; food and water become contaminated by fecal discharges from various sources. Rats, mice, and other species, including reptiles and turtles, in addition to birds, are sources of fecal discharges of paratyphoid forms of salmonellae. Inhalation of the bacterium during close confinement in high humidity environments such as hatching and brooder operations, direct contact with infected birds and animals, and insects are other demonstrated transmission routes for salmonellosis.

Intestinal microflora are an important factor influencing infection and disease by salmonellae in poultry. Very small numbers of salmonellae can cause infection of poultry during the first few weeks of life. Thereafter, the infectious dose becomes progressively higher, apparently because poultry acquire intestinal microflora that protect them against infection even in the presence of a highly salmonella-contaminated environment. This may explain the high prevalence of salmonellosis occasionally found in chicks of some colonial nesting species, such as gulls and terns, and in heron and egret rookeries, but the lower-than-expected infection rates in adult birds from those same colonies. Experimental studies with full-grown herring gulls disclosed a rapid elimination of salmonella bacteria from the intestines of these birds, which suggests that adult herring gulls may be passively, rather than actively, infected and may simply serve as a mechanical transport mechanism for the movement of salmonellae ingested from contaminated environments.

Individual infected birds can excrete salmonella bacteria for prolonged periods of time ranging from weeks to months. Prolonged use of sites by birds and high density of individuals at those sites can result in cycles of salmonellosis within those populations. Persistently contaminated environments result from a small percentage of birds which remain as lifelong carriers that intermittently excrete salmonellae into the environment. The environmental persistence of these bacteria is another factor influencing the probability for infections of birds using that site (Table 9.3). The common practices of using sewage sludge and livestock feces and slurry as fertilizer provide another means for infecting wild birds. Tests of sewage sludge often disclose contamination with salmonellae. Survival periods for salmonellae in cattle slurry samples have been reported to range from 11 to 12 weeks and for months in fields where the slurry has been applied as fertilizer. There are numerous reports of the isolation of salmonellae from rivers and streams as a result of pollution by sewage effluent and slurry runoff from fields.

---

[1] Distinct forms of salmonellosis caused by specific variants of salmonellae.

Table 9.1 Characteristics of important salmonellae-causing disease in birds.

Characteristic	*Salmonella pullorum*	*Salmonella gallinarum*	*Salmonella typhimurium*
Common name	Pullorum disease	Fowl typhoid	Salmonellosis
Natural hosts	Chickens (primary), turkeys	Chickens, turkeys	Wide range of vertebrates; not restricted to birds.
Age susceptibility	Mortality usually confined to the first 2–3 weeks of age.	Generally infects growing and adult birds; disease also infects young due to egg transmission.	All ages affected; more common in young and often in association with concurrent disease agents.
Transmission	Infected hatching eggs followed by spread from infected chicks to uninfected chicks that hatch.	Infected carrier birds most important; egg transmission of secondary importance.	Contaminated environment resulting in ingestion through food and water; egg transmission can also occur.
Relative occurrence in wild birds	Rare in free-ranging species; not maintained within wild populations.	Uncommon in free-ranging species; not maintained within wild populations.	Prevalence varies with species; most common in those species associated with landfills, sewage lagoons, and other waste-disposal sites and those with close associations with livestock and poultry operations.
Other naturally infected avian species	Ducks, coots, pheasants, partridges, guinea fowl, sparrows, European bullfinch, magpies, canaries, hawk-headed parrot.	Ducks, swans, curlews, pheasants, quail, partridge grouse, guinea fowl, peafowl, wood pigeon, ring dove, rock dove, owls, rooks, jackdaws, sparrows, blackbirds, goldfinches, ostrich, parrots.	Wide range of species; commonly found in gulls and terns and passerine birds using birdfeeding stations. Also reported in herons, egrets, ducks, geese, cormorants, cranes, owls, eagles, falcons, hawks, and other species.
Current geographic occurrence	Rare in most advanced poultry-producing areas.	Essentially eliminated from commercial poultry within the United States. Low incidence in Canada, USA, and several European countries; significant disease in Mexico, Central and South America, Africa, and Middle East.	Worldwide due to wide range of species infected.
Relative human health significance	Occasional infections following massive exposure (contaminated food); prompt recovery without treatment.	Rare and of little public health significance.	One of the most common causes of food-borne disease in humans.

Table 9.2 Pathways for transmission of **Salmonella** sp. in birds.

Type of transmission	Means	Consequences/processes
Vertical (from parent to offspring)	Through contaminanted eggs from infected female; embryo may be infected or surface of egg becomes infected as it passes down oviduct.	Infection of hatchlings at age of greatest susceptibility. Infected hatchlings become source of infection for other hatchlings.
Horizontal	Bird-to-bird contact	Infected birds shed organism in feces. Birds in close contact inhale salmonellae that become airborne or ingest salmonellae when pecking at contaminated surfaces of infected birds.
	Contaminated environments	Multiple sources of fecal contamination from a wide variety of warm- and cold-blooded species results in ingestion of salmonellae when pecking at contaminated feathers, litter, and other materials. Infected birds and other animals that are fed upon by birds with predatory and scavenging food habits become exposed to salmonellosis. Birds that feed in landfills, dung piles, wastewater discharge areas, and sewage lagoons are at highest risk to acquire infections.
	Contaminated feeds	Salmonella-contaminated feed has been the source of salmonella outbreaks in poultry. Little is known about levels of salmonella contamination in commercial feed used at birdfeeding stations.
	Inapparent infections	Stress of translocation or conditions causing birds to be brought into rehabilitation can result in shedding of salmonellae by carrier birds or result in clinical disease in birds with subclinical infections. Disease can be transmitted to other birds in close proximity; contamination of the environment can result in further transmission, and release of actively shedding birds can serve to spread the disease and contaminate other environments.

Table 9.3 Examples of reported environmental persistence for **Salmonella** sp. in different substrates. [—, no data available.]

Substrate	Temperature			Ambient	Serovar
	11 °C	25 °C	38 °C		
Poultry feed	18 months	16 months	40 days	—	*S. typhimurium*
Poultry litter	18 months	18 months	13 days	—	*S. typhimurium*
Soil from vacated turkey pens	—	—	—	6–7 months	Unspecified paratyphoid form
Urban garden soil	—	—	—	280 days	*S. typhimurium*
Hatchery fluff	—	—	—	5 years	Unspecified paratyphoid form
Avian feces	—	—	—	28 months	Unspecified paratyphoid form
Reptilian feces	—	—	—	30 months	Unspecified paratyphoid form
Manure	—	—	—	36 months	Unspecified paratyphoid form

Species Affected

All species of birds should be considered susceptible to infection by salmonellae. The outcome of salmonella infections is reported to be highly dependent upon the age of the birds, concurrent stress, serovar and strain virulence, and susceptibility of the host species.

Salmonellosis has been studied as a disease of poultry since at least 1899. Wild bird surveys have often been concurrent with studies of this disease in poultry and as sources for human infections. These and other investigations have resulted in numerous strains of *Salmonella* sp. being isolated from free-ranging (Fig. 9.1) and captive wild birds. However, findings from these studies have also disclosed a much lower infection rate than anticipated and have caused numerous investigators to conclude that in general, salmonellosis is not an important disease of free-ranging wild birds.

The historic patterns of salmonellosis in wild birds are of isolated mortality events involving individual or very small numbers of birds and incidental findings associated with concurrent infections involving other disease agents. Before the 1980s, major mortality events from this disease were rare in free-ranging wild birds.

Prior to the 1980s most isolations of *Salmonella* sp. from free-ranging wild birds were made from apparently healthy birds, were incidental findings from birds with other disease conditions, or were from lethal cases of salmonellosis involving small number of birds. This is no longer the situation. Large-scale mortalities of birds using feeding stations have become common in the United States (Fig. 9.2), and such mortalities are also reported from Canada and Europe, including Scandinavia. Typically, these events are caused by *S. typhimurium* and usually involve passerine birds (Fig. 9.3). European starling, blackbirds, common grackle, and mourning dove are also among the species that have been found dead from *S. typhimurium* at birdfeeding stations.

Salmonellosis has also been the cause of die-offs of aquatic birds including several species of ducks, mute swan, various species of gulls and terns, American coot, double-crested cormorant, eared grebe, and several species of egrets and herons. However, large-scale mortality events in free-ranging populations, except for songbirds and colonial nesting birds, have rarely been reported.

Many species of captive-reared birds commonly become infected with salmonellae and die from salmonellosis. Aquatic species have died from salmonellosis in zoological gardens and other captive collections. Gamebirds, such as grouse and pheasants, being reared in captivity for sporting purposes and cranes being reared for species conservation efforts are often victims of salmonellosis. Mortality is generally confined to chicks.

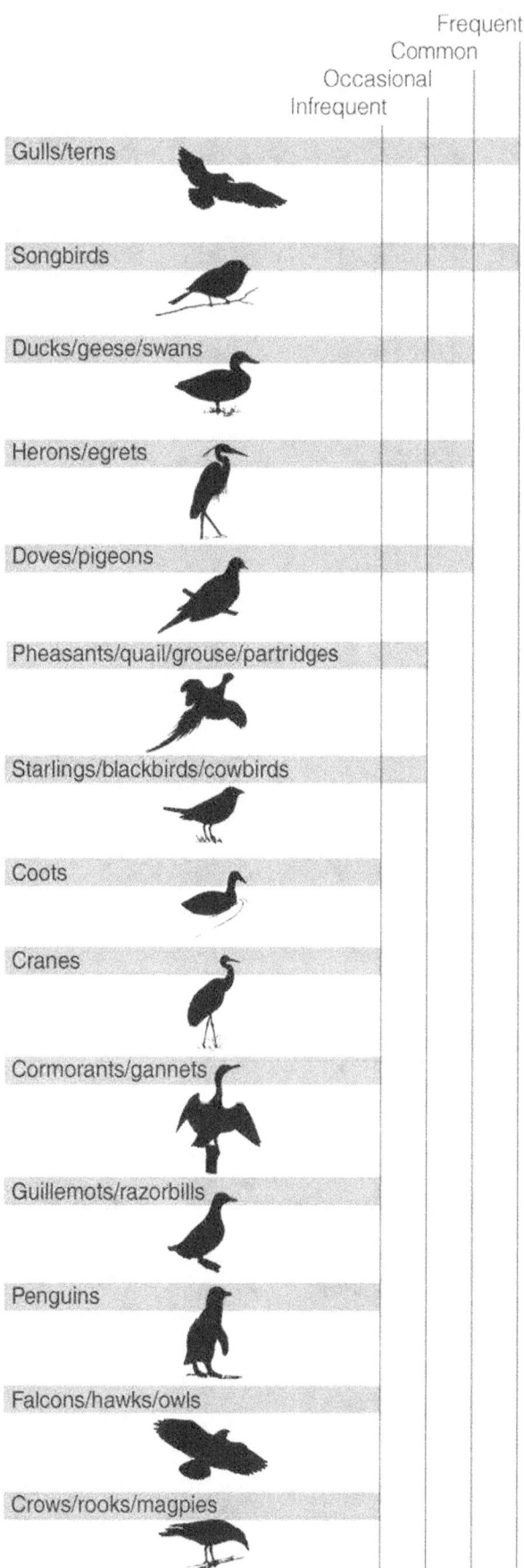

Figure 9.1 Relative rates of isolation of **Salmonella** sp. in free-ranging wild birds.

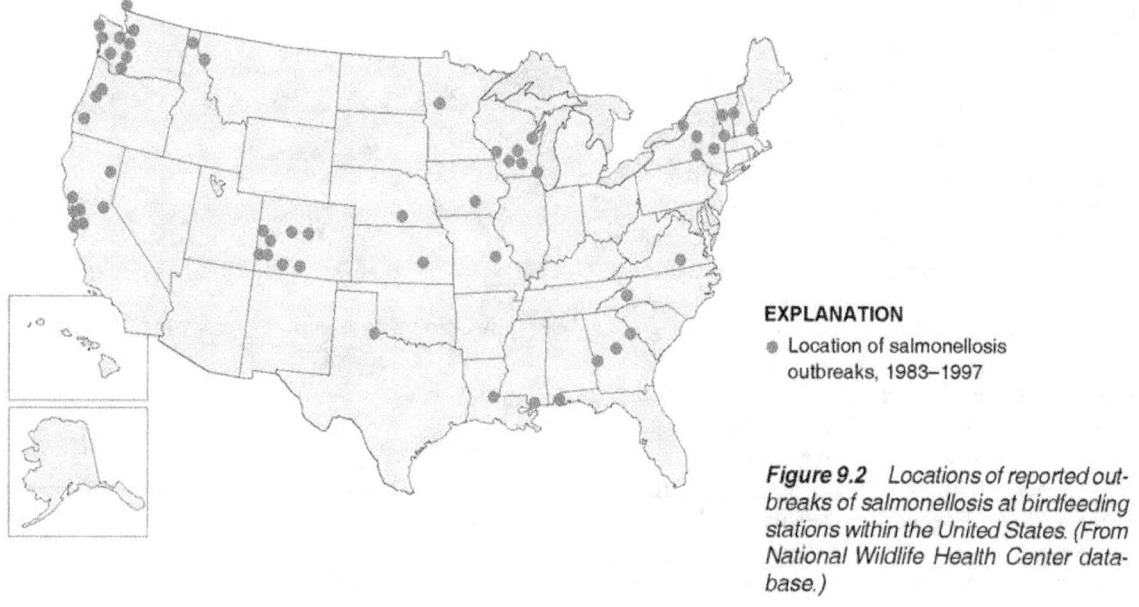

Figure 9.2 Locations of reported outbreaks of salmonellosis at birdfeeding stations within the United States. (From National Wildlife Health Center database.)

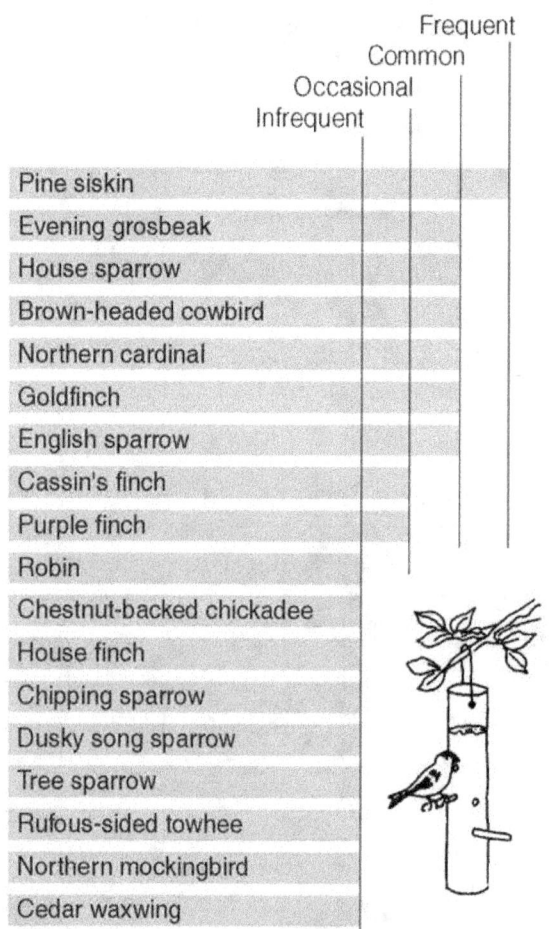

Figure 9.3 Relative occurrence of species found dead from salmonellosis outbreaks at birdfeeding stations within the United States.

Distribution

Extensive and prolonged control programs have essentially eliminated pullorum disease as a disease confronting commercial poultry production in most of the world and fowl typhoid from most Western countries. In contrast, salmonellosis due to paratyphoid infections occurs worldwide (Table 9.1) and is increasingly prevalent among wild birds in a wide variety of habitats. Salmonellosis in songbirds is clearly an emerging disease of urban and suburban environments and it has also been introduced into remote bird populations, such as Antarctic penguins and skua. The geographic distribution of salmonellosis in free-ranging wild birds is closely associated with sources of environmental contamination that enters the food web of birds and is passed to other species when infected individuals are fed upon by predators and scavengers.

Seasonality

Salmonellosis can present itself at any time of year. Outbreaks at birdfeeding stations are closely associated with the periods of greatest use of those stations (Fig. 9.4); fall and spring die-offs of songbirds from salmonellosis are common in England. Other outbreaks occur among the young of colonial nesting species, such as gulls and terns, shortly after the young are hatched during the summer (Fig. 9.5).

Field Signs

There are no distinctive signs associated with salmonellosis in wild birds. Different species and ages of birds may have different signs even if they are infected with the same serovar; young birds typically exhibit more pronounced signs of disease. Infection may result in acute disease with sudden onset of death, or it may result in a more prolonged course of infection that may become septicemic or be characterized by the presence and persistence of bacteria in the blood, or result in localized infection within the body. The disease in poultry has been described to result in gradual onset of depression over a few days and by unthrifty appearance. These birds huddle, are unsteady, shiver, and breathe more rapidly than normal; their eyes begin to close shortly before death; and they exhibit nervous signs including incoordination, staggering, tremors, and convulsions. Blindness has also been reported in some birds.

The rapid death of songbirds at feeding stations has often caused observers to believe the birds had been poisoned. Neurological signs, such as those described above for poultry, have also be reported in infected songbirds. In contrast, young domestic ducklings are reported to die slowly, exhibiting tremors and gasping for air. Their wings often droop and they sometimes stagger and fall over just before death. Like infected chickens, these birds often have pasted vents and eyelids that are swollen and stuck together by a fluid discharge. Commonly reported signs among all species include ruffled feathers, droopiness, diarrhea, and severe lethargy. Chronically infected birds often appear severely emaciated.

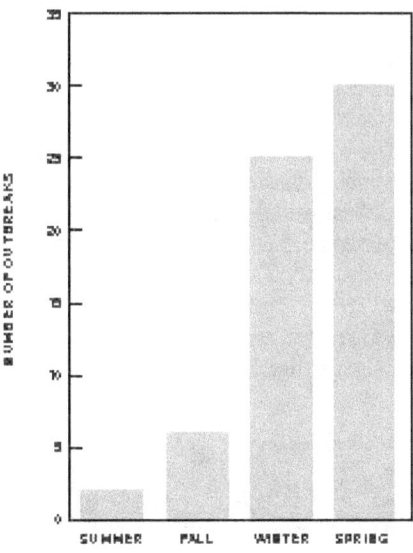

Figure 9.4 Seasonal occurrence of salmonellosis outbreaks at bird feeding stations within the United States.

Figure 9.5 (A) Salmonellosis can cause large-scale losses of colonial nesting birds. (B) Young birds are especially vulnerable.

Gross Lesions

The occurrence and types of gross lesions are highly variable depending on the course of the infection, the virulence of the organism, and the resistance of the host. In acute cases, obvious lesions can be completely absent. Livers often become swollen and crumbly with small reddened or pale spots if the course of the disease has been prolonged. In other infections, so-called paratyphoid nodules develop in the liver and extend into the body cavity. These are small tan-to-white granular nodules that are best seen under a microscope. In some birds, these nodules are more visible and appear as plaques or granular-abscess-like lesions seen within breast muscle and other tissues and organs. Infected songbirds often have yellow, cheesy nodules visible on the surface of the esophagus. When the esophagus is cut open, the nodules may be seen as large, diffuse plaque-like lesions or as discrete, nodular areas within the esophagus (Fig. 9.6).

An acute intestinal infection can be recognized by the reddening of the internal lining of the posterior two-thirds to one-half of the small intestine, the ceca, which are the blind pouches that extend from both sides of the beginning of the large intestine, and the colon. As the disease progresses, the intestinal lining becomes coated with a pale, tightly adher-

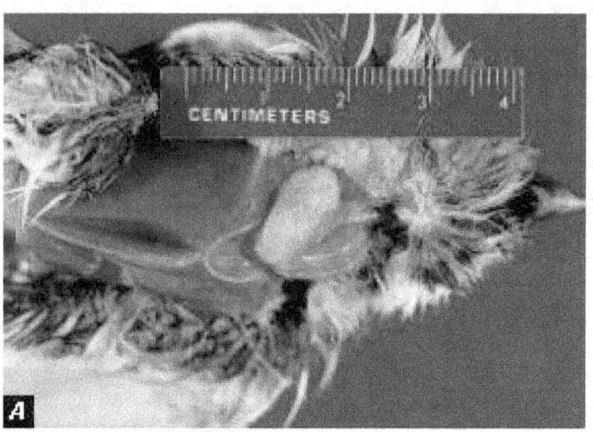
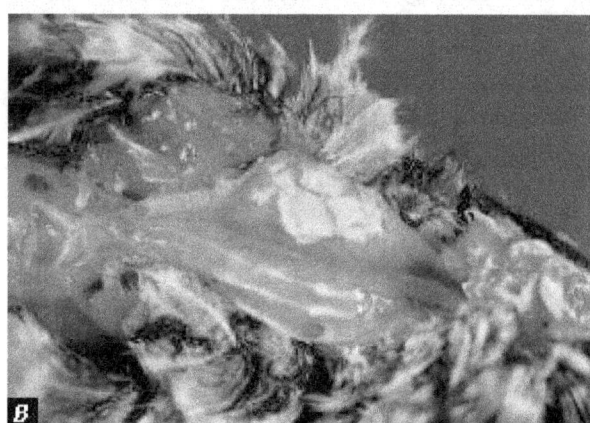
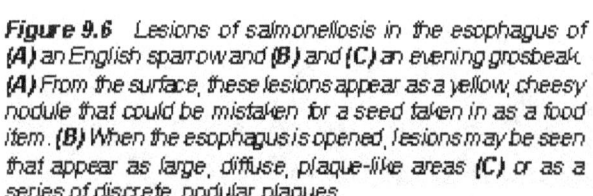

Figure 9.6 Lesions of salmonellosis in the esophagus of (A) an English sparrow and (B) and (C) an evening grosbeak. (A) From the surface, these lesions appear as a yellow, cheesy nodule that could be mistaken for a seed taken in as a food item. (B) When the esophagus is opened, lesions may be seen that appear as large, diffuse, plaque-like areas (C) or as a series of discrete, nodular plaques.

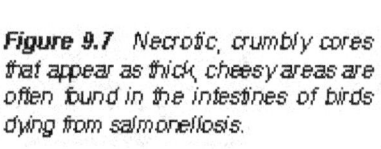
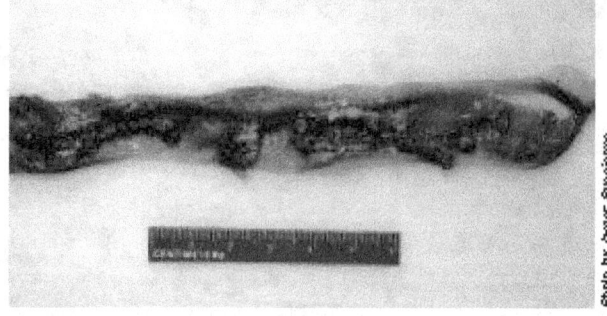

Figure 9.7 Necrotic, crumbly cores that appear as thick, cheesy areas are often found in the intestines of birds dying from salmonellosis.

ing, fibrinous material. In some infected birds, the intestinal ceca contain thick, crumbly necrotic cores (Fig. 9.7). Enlargement and impaction of the rectum are commonly reported in domestic ducklings.

Arthritis in the wings of pigeons is common. Domestic ducks with paratyphoid infections often have arthritis of the hips and knee joints. Small external abscesses about 1 millimeter in diameter have been described for infected pigeons and house sparrows. These abscesses appear in small bunches along the underside of the bird along the mid-to-posterior areas of the body.

Diagnosis

Gross lesions of salmonellosis can be similar to several other diseases, including avian cholera and colibacillosis. Diagnosis requires laboratory isolation and identification of *Salmonella* sp. from infected tissues in conjunction with pathological findings. Therefore, whole carcasses should be submitted for examination. Birds with markedly abnormal behavior patterns, such as convulsions and tumbling, often have lesions observable by microscopic examination of the brain. Isolation of salmonellae from the intestine without significant lesions and accompanying isolation of the bacteria from other tissues generally indicates that the bird was a carrier, rather than a victim, of salmonellosis.

Salmonellae are often confined to the gut. The ceca offer the greatest potential for obtaining positive cultures for most strains of salmonellae. Therefore, when whole carcasses cannot be submitted, submit the intestine as a minimum sample. The liver and heart should also be removed and submitted, if possible. Wrap each different tissue in a separate piece of aluminum foil. Place the foil-wrapped specimens in tightly sealed plastic bags, and ship them frozen to the diagnostic laboratory (Chapter 2, Specimen Collection and Preservation and Chapter 3, Specimen Shipment).

Fecal droppings can be checked for *Salmonella* sp., but these need special handling and they should not be submitted as diagnostic specimens without prior discussions with the diagnostic laboratory. Submission of whole eggs should be considered when low hatchability is encountered. Egg shells and shell membranes can also be cultured for salmonellae; this is an effective means of detecting salmonellae in eggs that have hatched, provided that the egg fragments have not been subjected to environmental conditions that would destroy the bacteria. Eggs, too, should only be submitted following consultation with disease specialists.

Control

Prevention of infection by pathogenic forms of *Salmonella* sp. and control of salmonellosis is warranted for wild bird populations despite the fact that *Salmonella* sp. have been isolated from a wide variety of wild bird species from many different types of habitats. Surveys have disclosed that the prevalence of salmonellae in most wild bird populations is generally low. Other studies have indicated a rapid elimination of salmonellae from the intestines of their avian host, suggesting passive, rather than active, infection in some instances. The relatively recent increase in the frequency of occurrence of large-scale salmonella outbreaks in wild birds, especially songbirds, is without precedent and it suggests that environmental contamination is an important source for infection of birds.

Landfills and waters where sewage effluent is discharged are common feeding areas for gulls, the wild bird species group with the highest prevalence of salmonella infections. Ducks and other waterbirds also feed heavily in areas of sewage effluent, and they generally have a higher prevalence of salmonellae than most land birds except for pigeons and sparrows, two species that feed in manure piles. Raptors are thought to become infected from the prey they feed upon (often small rodents such as mice).

Eliminating point sources of infection should be the focus for combating salmonellosis in wild bird populations (Fig. 9.8). Disease prevention should be practical at birdfeeding stations; the public should be educated to maintain clean feeders and to remove spilled and soiled feed from the area under the feeder. Feeders occasionally should be disinfected with a 1:10 ratio of household bleach and water as part of the disease-prevention program. In the event of a die-off from salmonellosis, more rigorous disinfection of feeding stations is necessary and station use should be discontinued temporarily.

Other potential point sources of infection include garbage, sewage wastewater, and wastewater discharges from livestock and poultry operations. The potential for contaminating migratory bird habitat with *Salmonella* sp. should be considered when wastewater is intentionally used to create wetland habitat; when existing wetlands are used to receive wastewater discharges; when agricultural fields on wildlife areas are to receive manure and slurries as fertilizer; and when development of landfill, livestock, and poultry operations are proposed in areas where contamination of environments used by migratory birds is likely. A 1995 outbreak of *S. enteriditis* in California poultry was traced to sewage treatment plant wastewater which entered a stream that bordered the poultry farm. Contamination of feral cats and wildlife by the waters of the stream was thought to be the source of entry of *S. enteriditis* in the poultry.

Control of salmonellosis in captive flocks of migratory birds is necessary to prevent major losses, especially in young birds. Control of this disease should be of continual concern whenever migratory birds and other wild birds are being propagated for release programs or are being maintained in captivity during rehabilitation. The conditions causing birds to be brought to rehabilitation and the stresses of confinement may result in inapparent infections developing into systemic clinical salmonellosis that may jeopardize the well-being of the infected bird and of other birds within the facil-

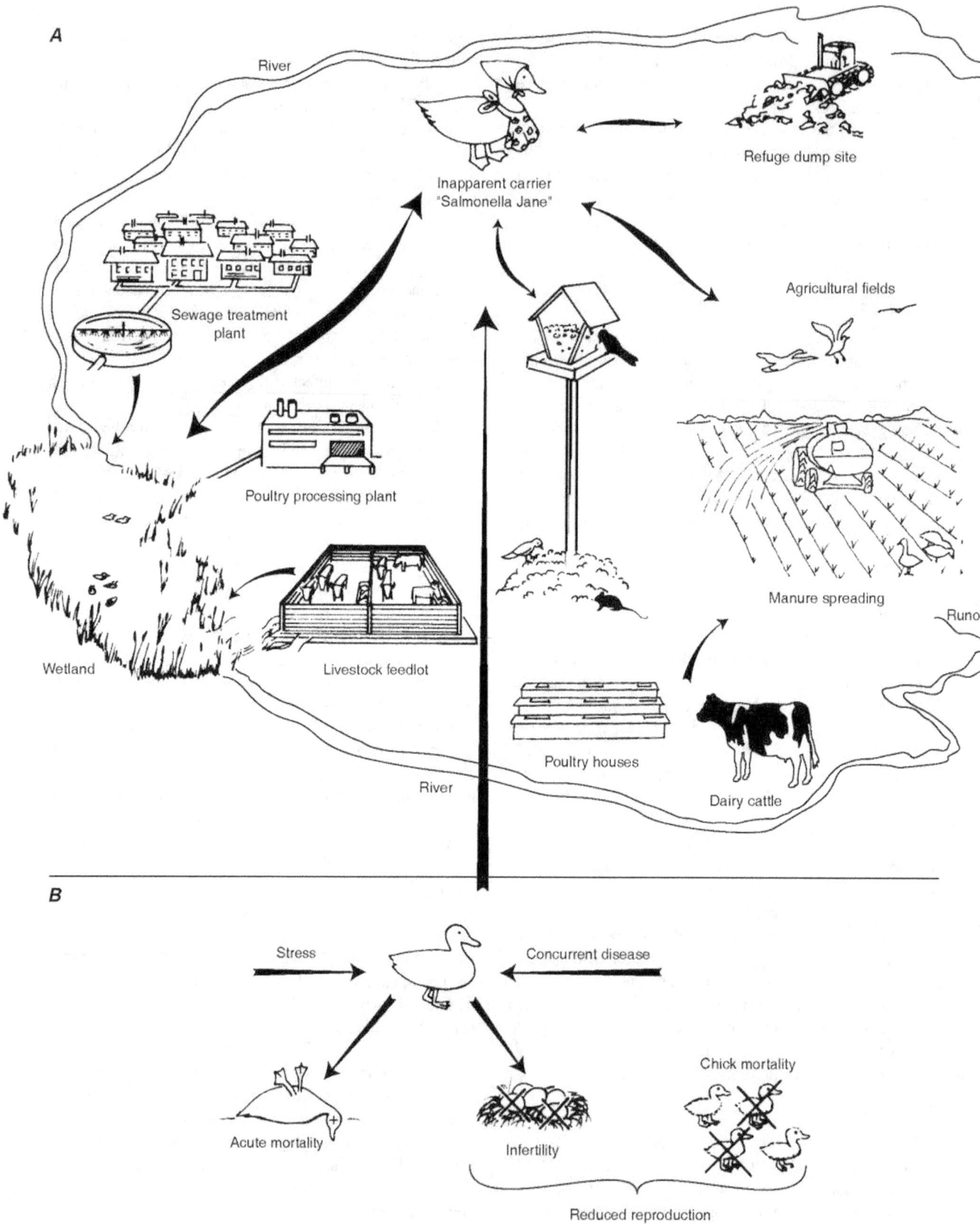

Figure 9.8 (A) Sources and (B) consequences of salmonellosis in wild birds.

ity. Strict sanitation measures need to be instituted and judiciously followed. Salmonella carriers can be identified by fecal culturing and should be destroyed. Multiple periodic fecal cultures are required to identify carrier birds because salmonellae are intermittently shed from the intestine. All birds that die should undergo necropsy and appropriate laboratory testing to determine the cause of mortality and any actions required to prevent further losses.

Infected adults should never be used for breeding. Antibiotic therapy may aid in overcoming an outbreak of salmonellosis, but antibiotic therapy will not eliminate carriers and vertical transmission via eggs could result in new outbreaks and disease spread. Storage of food in rodent- and insect-proof containers should be part of a disease prevention program. Many outbreaks in domestic poultry operations have been traced to food contaminated by rodent feces because rats and mice are common sources of salmonellae.

Human Health Considerations

Bacteria of the genus Salmonella are well-documented human pathogens. "Food poisoning" characterized by acute intestinal pain and diarrhea is the most common form of human infection. However, more serious forms of salmonellosis also affect humans. The general level of *Salmonella* sp. in most species of wild birds is low, but extra care with personal hygiene is warranted by people who handle these birds or materials soiled by bird feces. This consideration is not limited to situations where disease is apparent, and it extends to routine maintenance of birdfeeders, cleaning transport cages, and handling birds during banding and other field activities.

Milton Friend
(Modified from an earlier chapter by Richard K. Stroud and Milton Friend)

Supplementary Reading

Gast, R.K., 1997, Paratyphoid infections, *in* Calnek, B.W., and others., eds., Diseases of poultry (10th ed.): Ames, Iowa, Iowa State University Press, p. 97–121.

Snoeyenbos, G.H., 1994, Avian salmonellosis, *in* Beran, G.W., and Steele, J.H., eds., Handbook of zoonoses (2d ed., Section A): Bacterial, rickettsial, chlamydial, and mycotic: Boca Raton, Fla., CRC Press, p. 303–310.

Steele, J.H., and M.M. Galton, 1971, Salmonellosis, *in* Davis, J.W., and others, eds., Infectious and parasitic diseases of wild birds: Ames, Iowa, Iowa State University Press, p. 51–58.

Chapter 10
Chlamydiosis

Synonyms
Parrot fever, psittacosis, ornithosis, parrot disease, Louisiana pneumonitis

Cause
Chlamydiosis refers to an infection with organisms of the genus *Chlamydia* sp., which are bacteria that live within animal cells. *Chlamydia psittaci* is the species generally associated with this disease in birds. The severity of the disease differs by the strain of *C. psittaci* and the susceptibility of different species of birds. As a result, chlamydiosis may range from an inapparent infection to a severe disease with high mortality. The organism is excreted in the feces and nasal discharges of infected birds and can remain infective for several months. Infection commonly occurs from inhaling the bacteria in airborne particles from feces or respiratory exudates. Because of the organism's resistance to drying, infected bird feces at roosts are especially hazardous.

Species Affected
Chlamydiosis was first recognized as an infectious disease affecting parrots, parakeets, and humans involved in the international parrot trade in the late 1920s to 1930s. Chlamydiosis has since become known as a serious disease of domestic turkeys in the United States, of domestic ducks and geese in central Europe, and as a common infection of domestic and feral pigeons worldwide. The feral city pigeon is the most common carrier of *Chlamydia* sp. within the United States.

Chlamydial infections have been reported from at least 159 species of wild birds in 20 orders, but most isolations have been made from six groups of birds (Figure 10.1). Psittacine birds such as parakeets, parrots, macaws, and cockatiels are most commonly identified with this disease, while among other caged birds *Chlamydia* sp. occurs most frequently in pigeons, doves, and mynahs. Waterfowl, herons, and pigeons are the most commonly infected wild birds in North America (Figure 10.2). Chlamydiosis also occasionally infects gulls and terns, shorebirds, songbirds, and upland gamebirds.

Distribution
Among free-living birds, avian chlamydiosis has been found worldwide in the feral pigeon, in gulls and fulmars on islands of coastal Great Britain, in waterfowl and shorebirds in the Caspian Sea, and in herons, waterfowl, gulls, and doves in the United States. Infected parrots and parakeets have been found throughout the tropics and Australia.

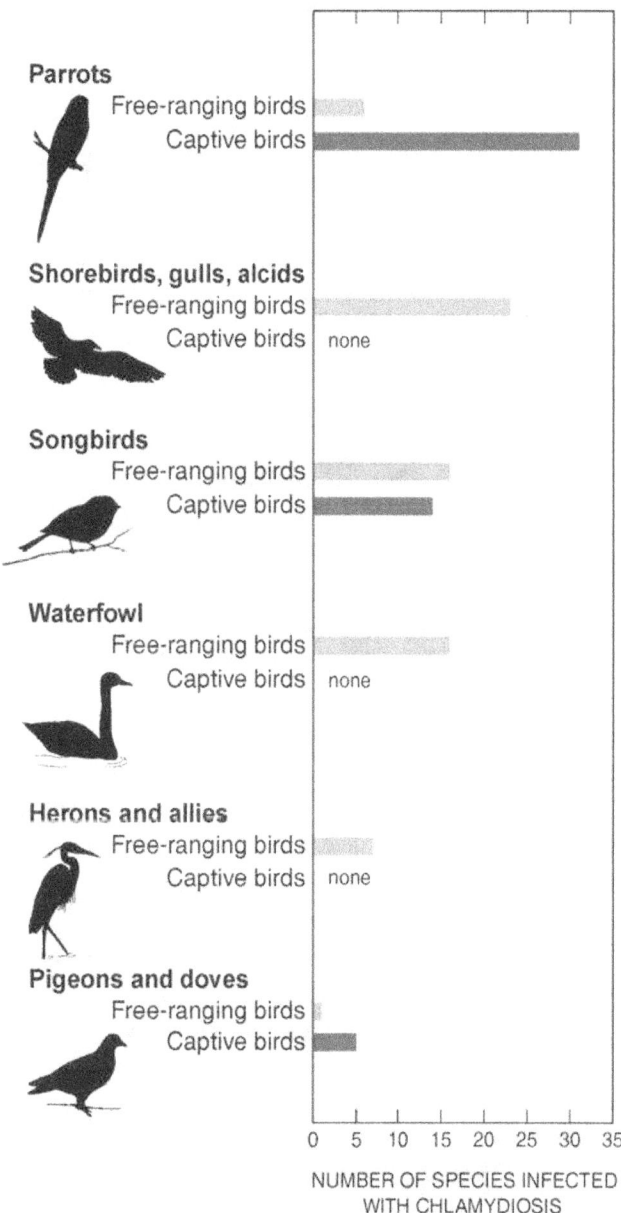

Figure 10.1 Relative occurrence of reported chlamydiosis in the most frequently infected groups of birds. (Adapted from Burkhart and Page, 1971).

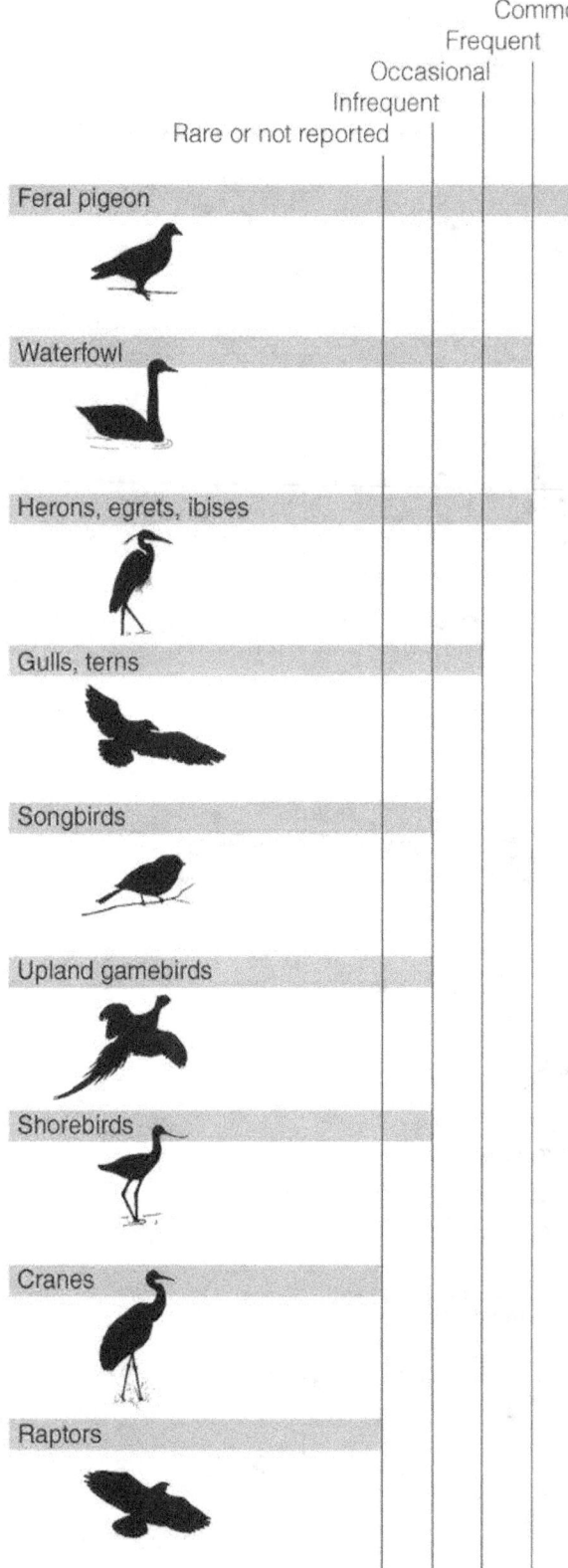

Figure 10.2 Relative occurrence of chlamydial infections in wild birds in North America.

Seasonality

Individual cases may occur at any time because of healthy carriers and latent infections within bird populations. Shipping, crowding, chilling, breeding, and other stressors have been attributed to active shedding of the infectious agent among captive birds with latent infections. Groupings of wild birds together in flocks, such as during spring and fall migrations, may facilitate the transmission of chlamydiosis. In caged birds, the onset of disease following exposure to *C. psittaci* occurs across a broad range of time from as quickly as 3 days to as long as several weeks. Young birds are more susceptible than adults, and the disease can spread rapidly among colonial nesting birds.

Field Signs

Signs of infection depend on the species of bird, virulence of the strain of *Chlamydia sp.*, the physiological condition of the bird as influenced by stressors, and route of exposure to the organism. Chlamydiosis in wild birds is often inapparent and infected birds can serve as asymptomatic carriers. Infection may also result in an acute, subacute, or chronic form of disease. *C. psittaci* can cause severe, acute disease that may be rapidly fatal in highly susceptible species. Birds often become weak, stop eating, and develop purulent (fluid containing pus) discharges of the eyes and nares. Birds tend to become motionless, remain in a fixed position, huddled up with ruffled feathers (Fig 10.3). Birds may have diarrhea, sometimes rust-colored because of the presence of blood, and respiratory distress is common. Feces from birds that stop eating are often dark green. In an outbreak of chlamydiosis in wild gulls, primarily fledglings died and the birds that were found dead were typically thin. Captive snowy and American egrets with chlamydiosis exhibited weakness, abnormal gait, ruffled feathers, diarrhea, and rapid weight loss; the birds generality died 1–2 days after the onset of signs. In other species of egrets, the infection may be inapparent even though the organism can be isolated from swabs of the cloaca or respiratory tract.

Feral pigeons exhibit many of the same signs; however, their diarrhea is likely to be more frequently tinged with blood. Mortality rates in young pigeons are often very high. Purulent discharges from the eyes of a very sick pigeon should cause the observer to think first of chlamydiosis. Sudden death without any signs of illness has been reported among captive cage birds (Java finch, parrots) and among wild parrots in Australia where king parrots were reported to have fallen out of trees and died within minutes.

Gross Lesions

The most common anatomical change in infected birds is an enlargement of the spleen or splenomegaly or of the liver or hepatomegaly or both, up to three-or-four times normal size (Figure 10.4). During an outbreak of chlamydiosis in gulls, splenomegaly was noted in each of nine birds exam-

Figure 10.3 Classic appearance of an immature little blue heron with severe chlamydia infection.

Human Health Considerations

Chlamydiosis can be a serious human health problem, infecting more frequently those who work with birds. The close association between parrots and this disease in humans prompted the United States and most nations of Western Europe to outlaw the importation of parrots and parakeets from 1930 to 1960. Individuals who work in areas in which there is a strong possibility of inhaling airborne avian fecal material should consider wearing a mask or respirator. Dry, dusty areas with bird droppings can be wetted down with a 5 percent solution of household bleach, or a commercial disinfectant. Working with large numbers of birds in dusty, closely confined areas should be avoided as much as possible.

ined and hepatomegaly was noted in four of the nine. Pericarditis, which is an inflammation and thickening of the pericardial sac that surrounds the heart (Figure 10.5), is a striking lesion sometimes seen with acute or subacute chlamydiosis. The air sacs may be thickened and the lungs are often congested, appearing darker than normal.

Diagnosis

Diagnosis is based upon the isolation of *Chlamydia* sp. from tissues of infected birds. Whole birds should be submitted. When this is not possible, selected tissues should be collected (Chapter 2, Specimen Collection and Preservation and Chapter 3, Specimen Shipment). The lungs, spleen, liver, and affected air sacs are the preferred tissues for microbial examination. Because *C. psittaci* is also a human pathogen, care must be taken in handling carcasses and tissues.

Diagnosis cannot be based on gross lesions alone because the lesions of some other diseases are similar. Chronic avian cholera infection can produce similar gross lesions in gulls, avian malaria can cause enlarged spleens, and early stages of aspergillosis can produce somewhat similar changes in the lungs and air sacs.

Control

Chlamydia sp. are present in the tissues, feces, discharges from the eyes and nares, and may also be present on plumage of infected birds. When the excreta and discharges dry, the resulting material can become airborne. Infection may be transmitted by direct contact with affected birds, or by inhaling dried bird fecal material or respiratory exudates that contain *Chlamydia* sp. organisms. Sick birds should be collected and euthanized and carcasses should be picked up. The removal and incineration of carcasses will help reduce the amount of infective material in the area. However, the level of human activity in the area should be carefully considered because it may cause redistribution of birds that could result in the spread of infection to new areas.

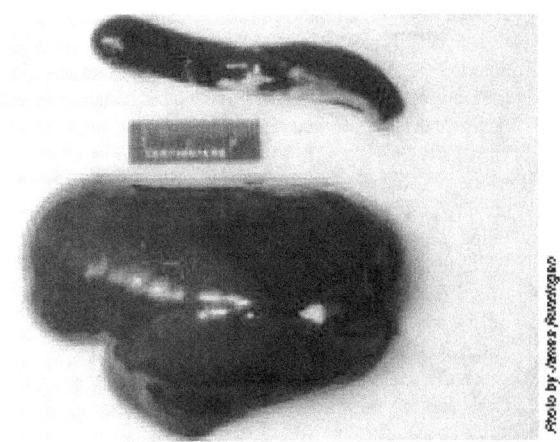

Figure 10.4 Enlarged spleen (top) and liver (bottom) of a ring-billed gull affected with chlamydiosis. (From Franson and Pearson, 1995. Reprinted with permission from the Journal of Wildlife Diseases).

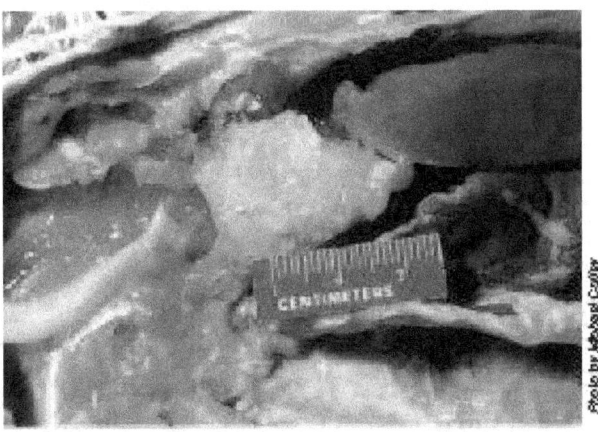

Figure 10.5 Pericarditis in a ring-billed gull that died of chlamydiosis (From Franson and Pearson, 1995. Reprinted with permission from the Journal of Wildlife Diseases.)

Outbreaks have occurred among poultry slaughterhouse workers and there have also been several severe cases among wildlife biologists. These biologists were thought to have become infected from handling snow geese, common egrets, snowy egrets, white-winged doves, and ducks.

Before the availability of antibiotics, chlamydiosis was fatal in about 20 percent of the human cases. Today, such fatalities are rare. However, persons working with birds should inform their physicians of that fact to help avoid potential situations where early signs of chlamydiosis could be overlooked or dismissed.

J. Christian Franson
(Modified from an earlier chapter by Louis N. Locke)

Supplementary Reading

Brand, C.J., 1989, Chlamydial infections in free-flying birds: Journal of the American Veterinary Medical Association, v. 195, no. 11, p. 1,531–1,535.

Burkhart, R. L., and Page, L.A., 1971, Chlamydiosis (ornithosis-psittacosis), *in* Davis, J.W., and others, eds., Infectious and parasitic diseases of wild birds: Ames, Iowa, Iowa State University Press, p. 118–140

Franson, J.C., and Pearson, J.E., 1995, Probable epizootic chlamydiosis in wild California (*Larus californicus*) and ring-billed (*Larus delawarensis*) gulls in North Dakota: Journal of Wildlife Diseases, v. 31, no. 3, p. 424–427.

Grimes, J.E., 1994, Avian chlamydiosis, *in* Beran, G. W., and others, eds., Handbook of zoonoses (2nd ed.): Boca Raton, Fla., CRC Press, p. 389–402.

Wobeser, G.A., 1997, Chlamydiosis, *in* Diseases of wild waterfowl (2nd ed): New York, N.Y., Plenum Press, p. 88–91.

Chapter 11
Mycoplasmosis

Synonyms
Chronic respiratory disease, infectious sinusitis, house finch conjunctivitis

Cause

Mycoplasmosis is caused by infection with a unique group of bacteria that lack cell walls but possess distinctive plasma membranes. Mycoplasma are also the smallest self-replicating life-forms, and they are responsible for a variety of diseases in humans, animals, insects, and plants. These bacteria can cause acute and chronic diseases in hosts that they infect, and they are also implicated with other microbes as causes of disease when the immune system of the host has become impaired through concurrent infection by other disease agents or through other processes. This chapter focuses on mycoplasmal infections of birds, the most significant of which are caused by *Mycoplasma gallisepticum* (MG), *M. meleagridis* (MM), and *M. synoviae* (MS). Only MG is of known importance for wild birds.

Species Affected

Until recently, mycoplasmosis has not been considered an important disease of wild birds. During late winter 1994, eye infections in house finches caused by MG were first observed in the Washington, D.C. area. Since then, mycoplasmosis has rapidly spread throughout much of the eastern range of the house finch. Mycoplasmosis has also appeared in wild populations of American goldfinch within the eastern United States. Clinical or observable disease caused by MG has not previously been found in wild passerine birds in the United States despite a long history and common occurrence of MG in poultry wherever poultry are raised. Molecular studies of isolates from the songbirds shows that those isolates are similar but that they are distinctly different from isolates obtained from poultry.

M. gallisepticum is a known pathogen of upland gamebirds raised in captivity, and it has been isolated from ducks and geese. Studies of mycoplasmosis in Spain have resulted in isolation of MG from free-ranging peregrine falcons, and isolation of MG from a yellow-naped Amazon parrot is further evidence of a diverse host range that can become infected by this organism (Table 11.1). Strain differences of MG exist and differ in their ability to cause clinical disease. Also, isolates of the same strain can vary widely in their ability to cause clinical disease in different species. This variance in the ability to cause clinical disease is, in part, shown by the greater numbers of birds that have antibody to MG than by the presence of mycoplasmosis in species and popu-

Table 11.1 Reported occurrence of selected avian mycoplasmas of poultry in selected wild avian species. [Frequency of occurrence: ● frequent, ● common, occasional, ○ infrequent or not reported. Square symbol indicates free-ranging species. All other reports are natural infections in captive-reared birds.]

	Mycoplasma sp.			
Type of bird	M. gallisepticum (MG)	M. meleagridis (MM)	M. synoviae (MS)	M. gallinarum
Chicken	●	○	common	common
Domestic turkey	●	common	common	occasional
Pigeons	○	○	○	occasional
Peafowl/guinea fowl	○	○	common	○
Pheasants/quail/partridge	occasional	○	○	○
Wild turkey	○	□	○	○
Ducks/geese	occasional	○	○	○
Birds of prey	□	○	○	□
Songbirds	■	○	□	○
Parrots	○	○	○	○

lations tested. The isolates of MG from wild songbirds do not cause significant disease in chickens.

Chickens and turkeys are commonly infected with MG, and direct contact of susceptible birds with infected carrier birds causes outbreaks in poultry flocks. Aerosol transmission via dust or droplets facilitates spread of MG throughout the flock. Transmission through the egg is also important for poultry, and MG is thought to spread by contact with contaminated equipment. The highly gregarious behavior of house finches and their use of birdfeeders likely facilitates contact between infected birds or with surfaces contaminated with the bacteria. Infected finches are thought to be responsible for spreading this disease because they move between local birdfeeders and to distant locations during migration.

M. meleagridis causes an egg-transmitted disease of domestic turkeys, and it appears to be restricted to turkeys. Clinical disease has not been documented in wild turkeys, and reports of infection in other upland gamebirds have not been confirmed. Airborne transmission and indirect transmission by contact with contaminated surfaces also happen. *M. synoviae* has a broader host range than MM. Chickens, turkeys, and guinea fowl are the natural hosts. Several other species have been naturally infected, and others have been infected by artificial inoculation. Transmission is similar to that for MG, except that MS spreads more rapidly.

Many other avian mycoplasmas have been designated distinct species, some of which are identified in Table 11.2. The number of mycoplasma species identified from birds has increased rapidly during recent years and it will continue to grow. For example, *M. sturni* was recently isolated from the inner eyelids (conjunctiva) of both eyes of a European starling that had the clinical appearance of MG infection in house finches. Enhanced technology is providing greater capabilities for studying and understanding the biological significance of this important group of microorganisms. Too little is known about mycoplasma infections in wild birds to cur-

Table 11.2 Primary hosts of some mycoplasma species isolated from birds. [—, no data available.]

Mycoplasma species	Primary host						
	Chicken	Turkey	Pigeons	Waterfowl	Partridge	Birds of prey	Songbirds
M. gallisepticum	●	●	—	—	—	—	—
M. synoviae	●	●	—	—	—	—	—
M. iowae	●	●	—	—	—	—	—
M. gallopavonis	—	●	—	—	—	—	—
M. cloacale	—	●	—	—	—	—	—
M. gallinarum	●	—	—	—	—	—	—
M. gallinaceum	—	●	—	—	—	—	—
M. pullorum	●	—	—	—	—	—	—
M. iners	●	—	—	—	—	—	—
M. lipofaciens	●	—	—	—	—	—	—
M. glycophilum	●	—	—	—	—	—	—
M. columbinasale	—	—	●	—	—	—	—
M. columbinum	—	—	●	—	—	—	—
M. columborale	—	—	●	—	—	—	—
M. anatis	—	—	—	●	—	—	—
M. anseris	—	—	—	●	—	—	—
M. imitavis	—	—	—	●	●	—	—
M. sturni	—	—	—	—	—	—	●
M. buteonis	—	—	—	—	—	●	—
M. falconis	—	—	—	—	—	●	—
M. gypis	—	—	—	—	—	●	—

rently assess the significance of these organisms as a disease factor, although the house finch situation clearly illustrates the potential for clinical disease to occur. Of added significance is the suppression of reproduction through lowered egg production that commonly affects poultry. Reproduction has also been suppressed during natural MG infections of captive chukar partridge, pheasants, peafowl, and other species and during experimental studies with MM in wild turkey. Preliminary studies at the National Wildlife Health Center (NWHC) with *M. anatis* isolated from a wild duck resulted in reduced hatchability of mallard eggs inoculated with that isolate and decreased growth of the infected hatchlings.

Mycoplasmas have been recovered from domestic or semidomestic ducks since 1952, but the bacteria have not been reported from wild North American waterfowl before a 1988–1990 waterfowl survey by scientists from the NWHC. *M. anatis* has more recently been isolated from wild shoveler ducks and coot and from a captive saker falcon during surveys conducted in southern Spain. The finding of *M. anatis* in three different major groups of wild birds (Falconiformes, Gruiformes, Anseriformes) demonstrates how the ability of a single strain to infect different avian groups could facilitate interspecies transmission.

Distribution

Avian mycoplasmas cause disease in poultry and other captive-reared birds worldwide. The current reported distribution of mycoplasma-caused conjunctivitis in wild songbirds roughly corresponds with the distribution of the eastern house finch population (Fig. 11.1).

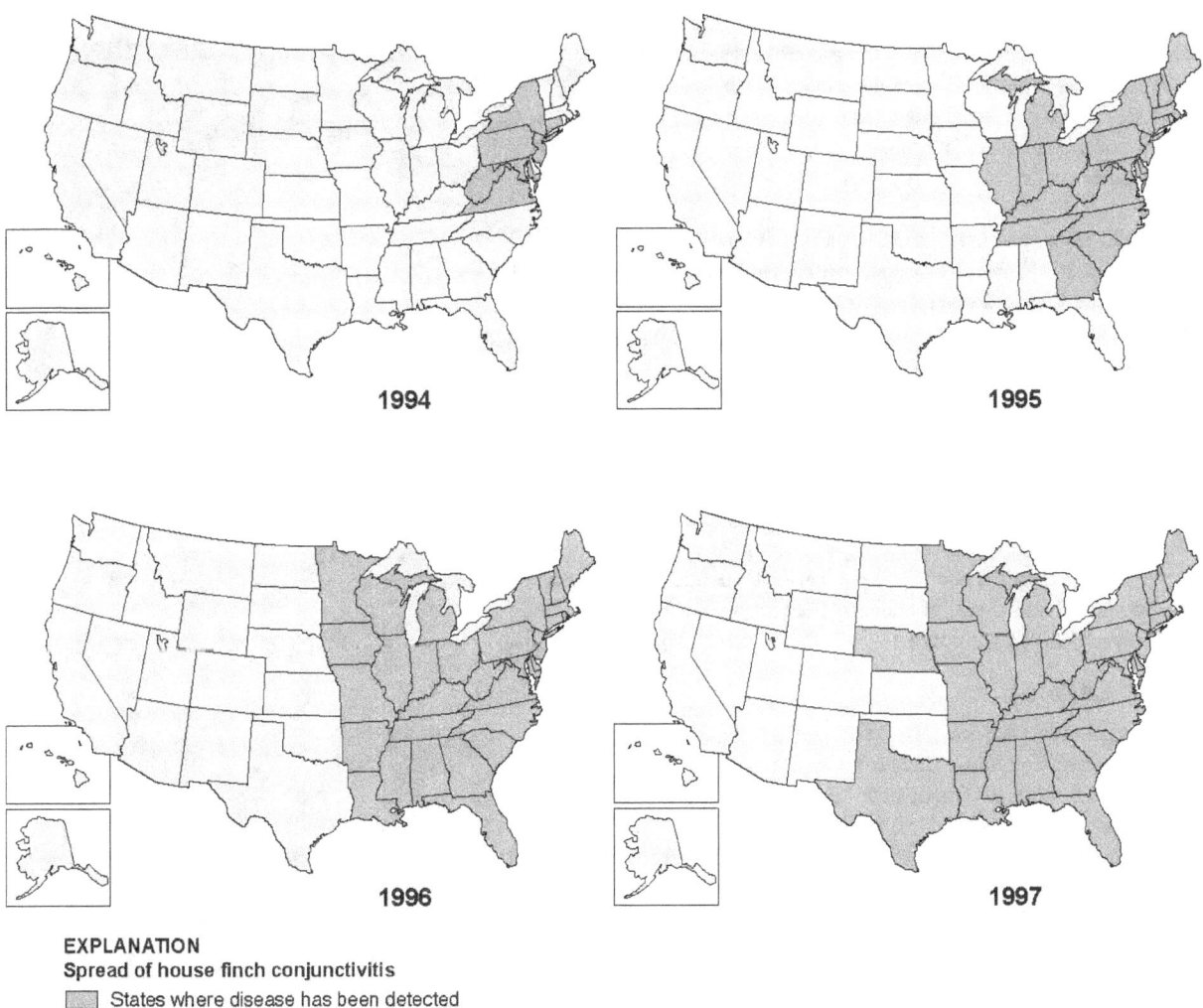

EXPLANATION
Spread of house finch conjunctivitis
States where disease has been detected

Figure 11.1 Reported geographic spread of house finch inner eyelid inflammation (conjunctivitis) since the initial 1994 observation. (Data adapted from reports in the scientific literature and personal communications between the National Wildlife Health Center and other scientists.)

Seasonality

Because mycoplasmas in poultry are commonly transmitted through the egg and are present in carrier birds, there is no distinct seasonality associated with disease in those species. Observations of house finch conjunctivitis are most frequent when birds are using birdfeeders during the colder months of the year.

Field Signs

Mycoplasma infections in poultry are generally more severe than those reported for house finches, the only wild bird for which any substantial field observations of clinical disease have been made. The prominent field signs are puffy or swollen eyes and crusty appearing eyelids (Fig. 11.2). A clear to somewhat cloudy fluid drainage from the eyes has been reported for some birds. Birds rubbing their eyes on branches and birdfeeder surfaces have also been reported. Other observations of infected birds include dried nasal discharge, severely affected birds sitting on the ground and remaining at feeders after other birds have departed, and birds colliding with stationary objects due to impaired vision. The European starling recently diagnosed to have been infected by *M. sturni* had similar clinical signs and was apparently blind.

Initial field signs observed during a natural outbreak of MG in a backyard gamebird operation included foamy eyes, excessive tearing, and severely swollen sinuses in chukar partridge and ring-necked pheasant, along with reduced egg production. As the disease progressed, severe depression, lethargy, and weight loss preceded respiratory distress and death. Eye inflammation was the only sign observed in Indian blue peafowl that became infected.

A captive saker falcon from Spain infected with *M. anatis* displayed signs of respiratory illness in addition to involvement of the eyes. Irregular breathing, wheezing, and a mucous discharge from the nose and beak were seen in this bird along with anorexia or loss of appetite. These signs are typical of mycoplasmosis in poultry.

Gross Lesions

Mycoplasmosis lesions in wild birds reflect the observed field signs. Infected house finches typically have a mild to severe inflammation of one or both eyes and the surrounding area including swollen, inflamed eyelids; a clear to a cloudy, thickened discharge from the eye; and drainage from the nares of the bill (Fig. 11.2). Chukar partridge and pheasant naturally infected with MG have had moderate to severe swelling of the eyelids, mild to moderate tearing, swelling of one or both of the sinuses near the eyes, and moderate to large amounts of cheesy discharge within the sinuses.

Diagnosis

Mycoplasma are among the most difficult organisms to grow from clinical specimens because of their fastidious

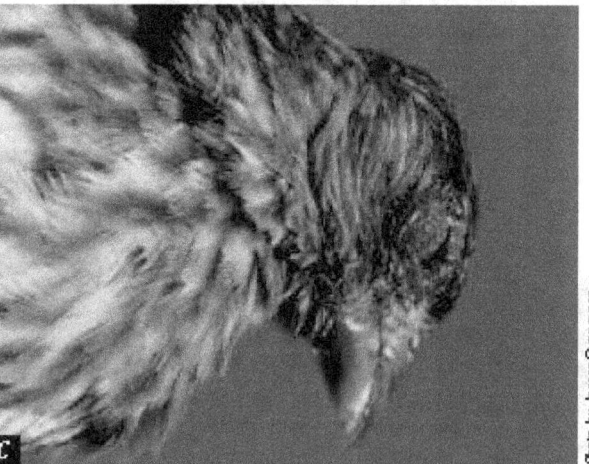

*Figure 11.2 Field signs and gross lesions of **Mycoplasma gallisepticum** infections in house finches: (A) and (B) Inflammation of the eye; (C) pasty, crusty appearance of the area surrounding the eye of a dying house finch.*

nature, intimate dependence upon the host species they colonize, and slow growth on artificial media. The greatest success in isolating MG from house finches has been when tissue swabs were obtained from live trapped, freshly killed, or fresh dead birds. There has been limited success from frozen carcasses. When mycoplasma is suspected, contact with a disease diagnostic laboratory is recommended to obtain guidance on how to handle specimens. If field conditions permit, selective media provided by a diagnostic laboratory should be inoculated with swabs from the inner eyelids, sinus, the funnel-shaped area at the back of the sinuses where they split right and left (choanal cleft), and trachea of suspect birds and shipped to the laboratory with the freshly killed or dead birds from which those swabs were made. If birds can be submitted, they should be chilled, rather than frozen, and immediately transported to a qualified disease diagnostic laboratory.

Control

Routine cleaning and disinfection of birdfeeders with household bleach is recommended to prevent mycoplasmosis and other diseases that can be transmitted at birdfeeders. A 10 percent solution of household bleach applied weekly for feeders with high bird use will reduce the potential for contaminated surfaces to transmit disease. Close observation of birds using feeders and the prompt reporting of suspect cases of mycoplasmosis to authorities will provide the opportunity for early intervention based on timely diagnosis and for initiating an appropriate disease-control strategy specific to the location and population involved. Special consideration needs to be given to the fact that house finch conjunctivitis is a new and emerging disease problem that has been documented in two additional species of songbirds. One of these included a case where a blue jay being rehabilitated in a cage previously occupied by an infected house finch became infected. That case demonstrates the need for adequate cleaning and disinfection of cages used in wildlife rehabilitation. Birds that survive infection can become disease carriers that serve as a source for initiating new outbreaks. Also, aerosol and egg transmission of mycoplasmosis is common for poultry. Similar transmission is likely for wild birds and must be taken into consideration during the rehabilitation of wild birds infected with mycoplasmosis.

The potential for interspecies transmission of MG from poultry to upland gamebirds being reared in captivity for sporting purposes must also be considered. This same consideration exists for raptors that may be fed poultry carcasses and waste.

Human Health Considerations

None. Mycoplasmas that infect birds are not known to be hazards for humans.

Milton Friend

Supplementary Reading

Cookson, K.C., and Shivaprasad, H.L., 1994, Mycoplasma gallisepticum infection in chukar partridges, pheasants, and peafowl: Avian Diseases, v. 38, p. 914–921.

Dhondt, A.A., Tessaglia, D.L., and Slothower, R.L., 1998, Epidemic mycoplasmal conjunctivitis in house finches from eastern North America: Journal of Wildlife Diseases, v. 34, p. 265–280.

Fischer, J.R., Stallknecht, D.C., Luttrell, P., Dhondt, A.A., and Converse, K.A., 1997, Mycoplasmal conjunctivitis in wild songbirds: The spread of a new contagious disease in a mobile host population: Emerging Infectious Diseases, v. 3, p. 69–72.

Poveda, J.B., Carranza, J., Miranda, A., Garrido, A., Hermoso, M., Fernandez, A., and Comenech, J., 1990, An epizootiological study of avian mycoplasmas in southern Spain: Avian Pathology, v. 19, p. 627–633.

Chapter 12
Miscellaneous Bacterial Diseases

Disease in free-ranging birds is caused by many other pathogenic bacteria in addition to those illustrated within this section. These other diseases are currently considered less important because of their infrequent occurrence, the small numbers of birds generally lost annually, or because they primarily result from infection by opportunistic pathogens and they require concurrent disease processes for them to become apparent. The following brief highlights about the more important of these diseases are included to acquaint readers with their existence and provide some basic information about their ecology.

Erysipelas

Erysipelas is caused by infection with the bacterium *Erysipelothrix rhusiopathiae*. This disease is primarily associated with swine and domestic turkeys, but it has been diagnosed in many groups of birds (Fig. 12.1) and in mammals. The causative agent has also been isolated from the slime layer of marine and freshwater fish and from crocodiles. Erysipelas is found worldwide. Little is known of the ecology of this disease in birds. Most reports of erysipelas in free-ranging birds involve individuals or small numbers of birds, but major die-offs can occur. The largest recorded die-off killed an estimated 5,000 birds, primarily eared grebes, during 1975 on the Great Salt Lake, Utah. Small numbers of waterfowl (green-winged teal, northern shoveler, and common mergansers) and a few herring gulls also died. Erysipelas has also been diagnosed as the cause of a die-off of brown pelicans in southern California during the late 1980s. Other free-ranging birds diagnosed with erysipelas include hawks, crows, raven, wood pigeon, starling, doves, finches, and European blackbird. The causative bacterium is able to survive in the environment for prolonged periods of time, and it was isolated from grebe carcasses approximately 18 weeks after their death during the Great Salt Lake mortality event. The bacteria probably are transmitted through ingestion, such as when gulls feed on carcasses, or entry of the organism through cuts and abrasions. Humans are susceptible to infection. Most human cases involve localized infections resulting from entry through a cut in the skin. Human cases have been fatal when the disease progressed to an infection of the blood and spreads throughout the body (a septicemic infection).

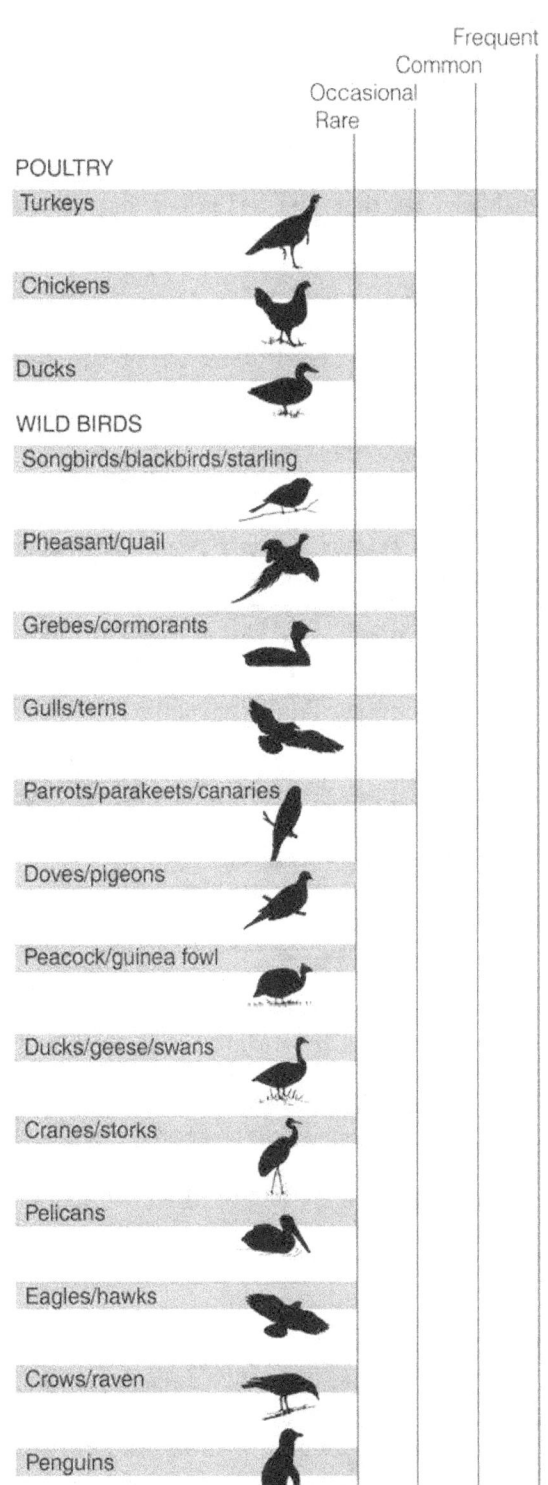

Figure 12.1 Reported occurrences of erysipelas in birds.

New Duck Disease

Pasteurella anatipestifer causes an important disease of domestic ducks that has infrequently caused the deaths of wild birds. This disease has also killed domestic turkeys and chickens and captive-reared pheasants, quail, and waterfowl. Major mortality events from infection with *P. anatipestifer* have occurred in free-ranging black swans in Tasmania and in tundra swans in Canada. New duck disease has also been diagnosed as the cause of mortality in small numbers of other free-ranging birds, including lesser snow geese. In the domestic duck industry, mortality primarily involves birds 2–3 months old. The swans that died in Tasmania and Canada were primarily young-of-the-year, which is consistent with mortalities of captive wild waterfowl. Birds can die within 24–48 hours after the onset of clinical signs of listlessness, a droopy appearance, fluid discharges from the eyes and bill, greenish diarrhea, and variety of nervous system disorders. The most prominent lesion seen during postmortem examination is a fibrinous covering on the surface of various organs such as the liver and heart (Fig. 12.2).

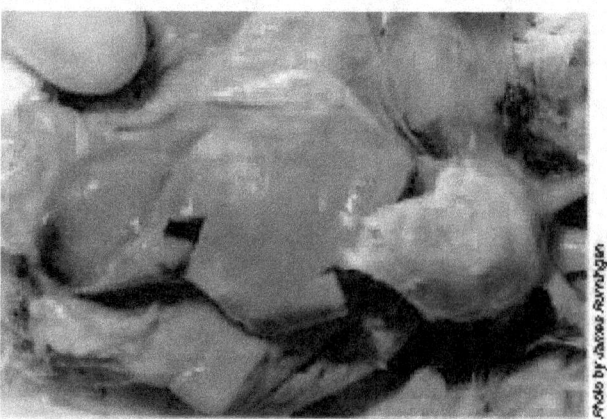

Figure 12.2 Fibrinous covering on the heart and liver of a bird with **Pasteurella anatipestifer**.

Necrotic Enteritis

Necrotizing enteritis is caused by an enterotoxemia or toxins in the blood produced in the intestine resulting from infections with *Clostridium perfringens*. This disease is found throughout much of the world where poultry are produced, and it is often an important cause of mortality for adult domestic breeder ducks. Sporadic cases have been diagnosed in waterfowl collections and in wild mallards, black ducks, and Canada geese. A die-off in Florida involved mallards and other wild ducks along with several species of shorebirds and wading birds. Wild ducks are also reported to have died from this disease in Germany.

During recent years, increasing numbers of small die-offs have been detected in snow geese, Canada geese, and white-fronted geese in Canada and the United States. An abrupt change in diet associated with seasons and bird migrations are thought to disrupt the intestinal microflora and allow *C. perfringens* to proliferate in the intestine. The toxins produced by these bacteria are the cause of death. The onset of death is generally rapid and without obvious clinical signs. Severe depression is sometimes observed in chickens along with reluctance to move, diarrhea, and ruffled feathers. Lesions generally appear as a mixture of dead cellular materials and plasma debris, tan-yellow in color, that covers much of the lower region of the intestine of affected waterfowl (Fig. 12.3).

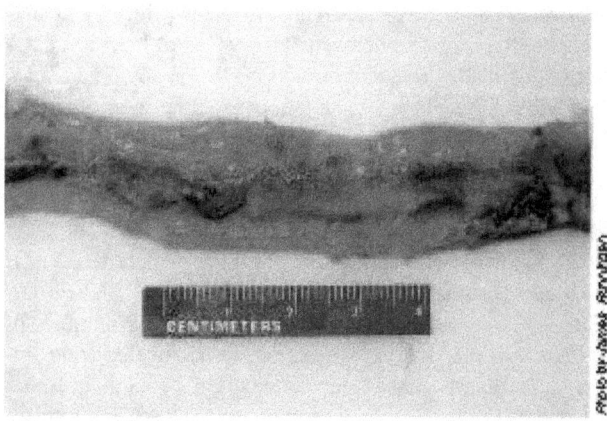

Figure 12.3 Lesions of necrotic enteritis in the intestine of a goose.

Ulcerative Enteritis

Quail are highly susceptible to infection by *Clostridium colinum*, the cause of ulcerative enteritis or "quail disease." Outbreaks of this disease in free-ranging wild birds are rare, but outbreaks have been reported for California quail in Washington State. This acute bacterial infection is charac-

Figure 12.4 Advanced lesions of ulcerative enteritis in the intestine of a chukar partridge.

terized by sudden onset followed by rapid spread through the flock. Outbreaks have been reported worldwide wherever game birds are raised in captivity under crowded conditions. In addition to upland game species such as grouse, quail, pheasant, and partridges, outbreaks have been reported in chickens, pigeons and robins. Mortality in young quail can reach 100 percent of the flock. Gross lesions vary and depend upon how long the bird lives following infection. Ulcers within the intestine originate as small yellow spots or infected areas with hemorrhagic borders and progress to circular forms that may join together as large areas of dead tissue that resemble thickened mucous membranes with raised edges (Fig. 12.4). Liver lesions include yellow areas of tissue death or necrosis along the edges of the liver and scattered grey spots or small yellow circumscribed spots within the liver itself that sometimes are surrounded by a light yellow halo effect.

Staphylococcosis

All avian species are susceptible to staphylococcal infections, and *Staphylococcus aureus* is the most common cause of disease. An often observed form of infection is a lesion that appears as an inflammation of the skin of the foot or pododermatitis, that is commonly referred to as "bumblefoot" (Fig. 12.5). Staphylococcal bacteria are ubiquitous, normal inhabitants of the skin and mucous membranes, and the bacteria require a break in those protective layers for infection to occur. Captive birds are more commonly found infected than free-ranging birds. Abrasions from rough surfaces where birds perch or stand may contribute to the occurrence of this disease. Studies in Spain with free-ranging imperial eagles demonstrated that staphylococcal infection can be transferred from humans to chicks being handled for banding. Infection was common in nestlings handled without latex gloves, whereas infection was rare in those birds handled with gloves. Mallard and redhead duck, bald and golden eagle, and ferruginous hawk have been among the species submitted to the National Wildlife Health Center (NWHC) that have been diagnosed with this condition.

Septicemic staphylococosis or staphylococcal blood poisoning can also occur, generally in birds that are immunocompromised or whose immune systems are not fully functioning. These types of infection can result in sudden death. Lesions associated with this form of infection generally consist of congestion of internal organs, including the liver, spleen, kidneys, and lungs, accompanied by areas of tissue death (Fig. 12.6). Bald eagles, American kestrels, red-tailed hawks, a duck, a mute swan, and herring and ring-billed gulls are among the species submitted to the NWHC for which septicemic staphylococcal infections have been diagnosed.

S. aureus can also cause serious disease in humans both as a wound infection and as a source of food poisoning. Good sanitation procedures should always be followed when han-

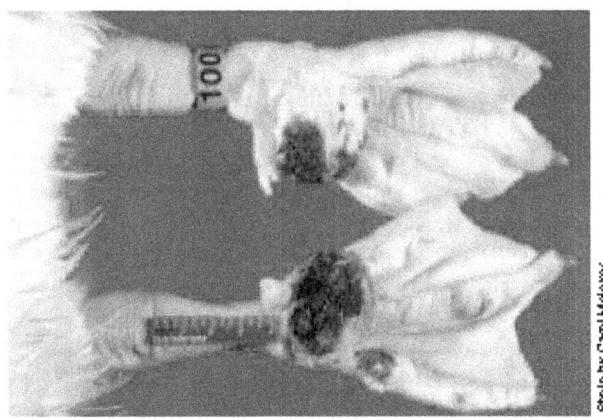

Figure 12.5 Bumblefoot in a domestic duck.

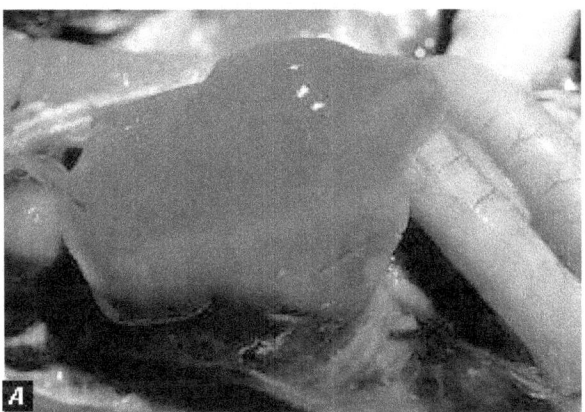

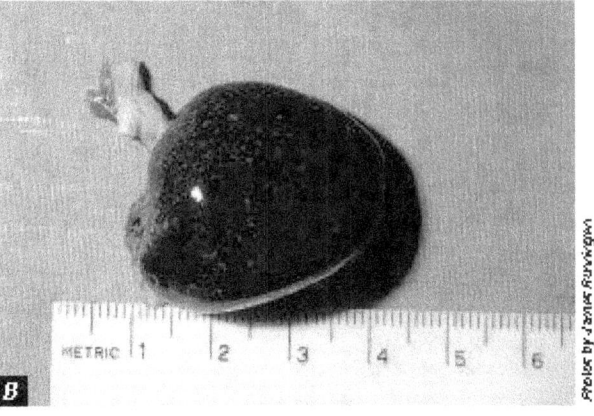

Figure 12.6 Congested liver (A) and spleen (B) from birds with staphylococosis.

dling animals, and protective gloves should be worn when handling wildlife found dead.

Tularemia

Tularemia is primarily a disease of mammals, but natural infections by *Francisella tularensis* have caused die-offs of ruffed grouse and other grouse species. A variety of avian species have been found to be susceptible to infection as a result of serological surveys that have detected antibody against tularemia, experimental studies to determine susceptibility, and by cause-of-death assessments for birds submitted for necropsy (Table 12.1). The strains of *F. tularensis* that caused natural infection of ruffed grouse are of low virulence for humans despite ruffed grouse becoming infected by the same tick (*Haemaphysalis leporispalustris*) that causes highly virulent tularemia in snowshoe hare.

Ticks are the primary source for disease transmission in natural cases of tularemia in upland game birds such as grouse and pheasants; ingestion of diseased birds and rodents is the primary source of disease transmission to raptors, gulls, and other scavenger species. Tularemia is infrequently reported as a cause of disease in wild birds. Ruffed grouse in northern climates have been the primary focus for reports in the scientific literature. The primary lesion seen is multiple, discrete spots scattered throughout the liver tissue (Fig. 12.7).

Table 12.1 *Avian species reported to be susceptible to infection by Francisella tularensis.*

Upland game species	Other birds
Ruffed grouse	Gulls and terns
Sharp-tailed grouse	Raptors (such as hawks and eagles)
Sage grouse	
Ptarmigan	Scavengers (such as shrikes)
Blue grouse	
Bobwhite quail	Ducks and geese
Pheasant	

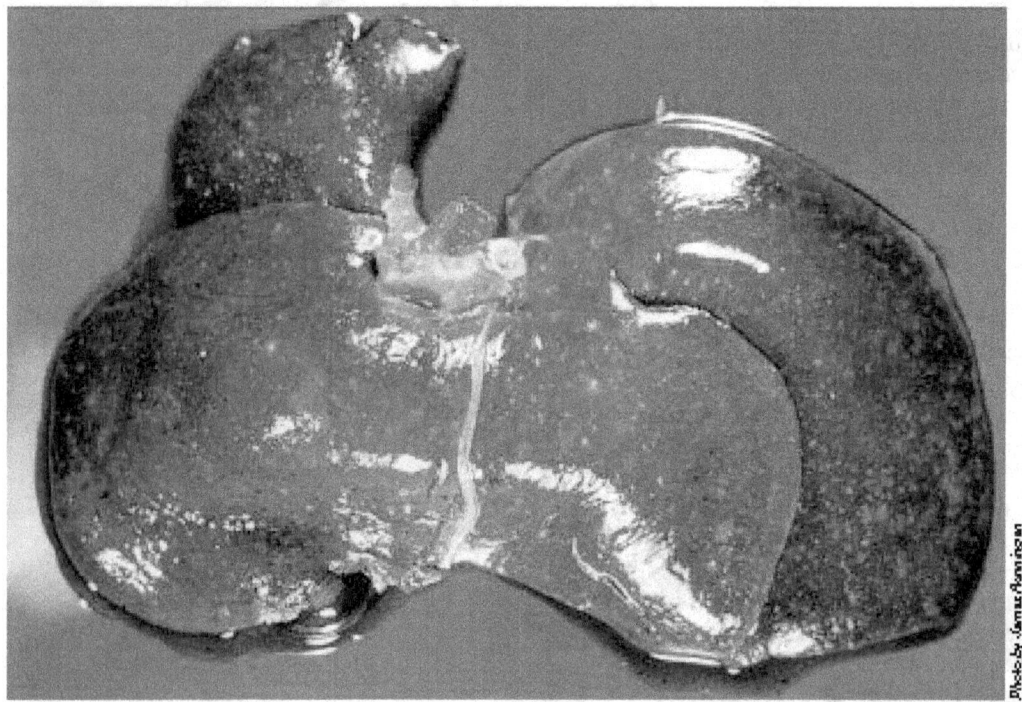

Figure 12.7 *The numerous, small, yellow and white spots on the liver of this beaver that died of tularemia are similar to the appearance of liver lesions in ruffed grouse.*

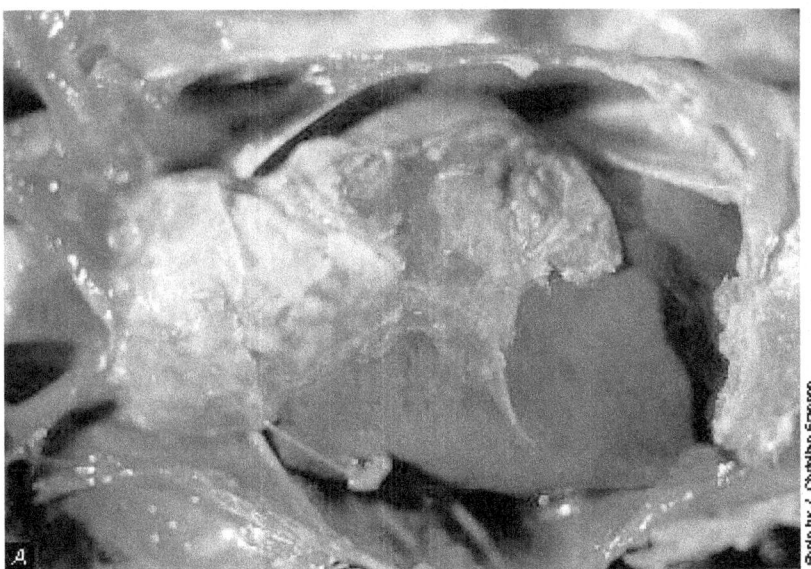

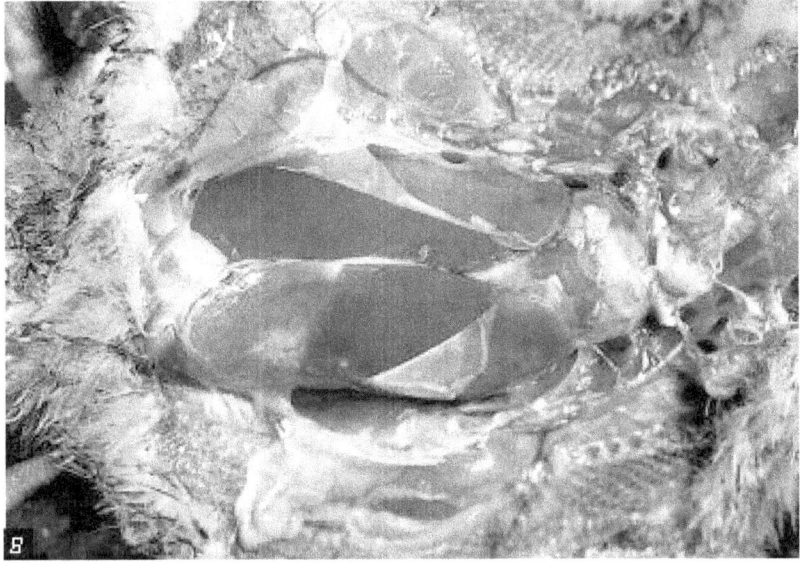

Figure 12.8 (A) Pericarditis and perihepatitis in a bird with colibacillosis. (B) Infection results in the liver being encased in a translucent covering.

Other

Colibacillosis, which is caused by infection with *Escherichia coli*, is one of several additional bacterial diseases occasionally encountered in wild birds. Avian strains of *E. coli* are generally not considered important causes of infection for humans or species other than birds. *E. coli* is a common inhabitant of the intestinal tract, but it often infects the respiratory tracts of birds, usually in conjunction with infection by other pathogens. These infections result in disease of the air sacs, and the infections are referred to as chronic respiratory disease. Lesions commonly associated with this disease include pericarditis or inflammation of the transparent membrane that encloses the heart and perihepatitis or inflammation of the peritoneal covering of the liver. These conditions make the coverings of the heart and liver look like a white or yellow mass that somewhat resembles the icing of a cake (Fig. 12.8). The livers of infected birds often appear swollen, dark in color, and may be bile stained (Fig. 12.9). Unhygienic hatcheries and other areas where young waterfowl and gamebirds are being held are often heavily contaminated with *E. coli*, and this results in infections causing acute mortality.

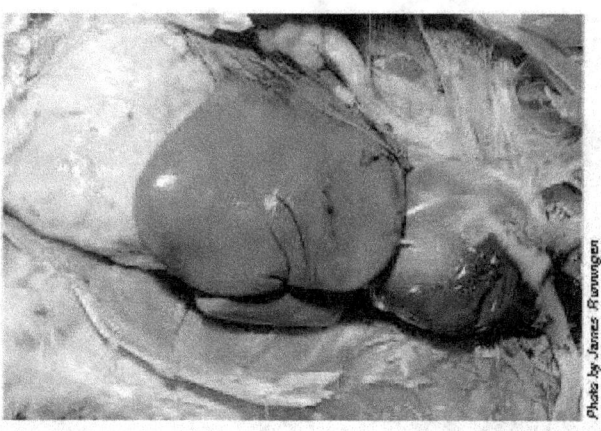

Figure 12.9 Swollen bile-stained liver in a bird with colibacillosis.

Similar to the other sections of this Manual, the bacterial diseases discussed are not comprehensive of diseases of wild birds. The similarities in clinical signs and gross lesions displayed in illustrations in this section emphasize the need for cause-of-death evaluations by qualified animal disease laboratories. Also, the environmental persistence and human health impacts noted for some of these pathogens emphasize the need to consider personal and environmental protection when handling dead birds. Assumptions that the cause of death is due to a pathogen of minor importance could have serious consequences if highly virulent infections are involved.

Milton Friend

Supplementary Reading

Calnek, B.W., and others, eds., 1997, Diseases of poultry (10th ed.): Ames, Iowa, Iowa State University Press, 929 p.

Davis, J.W., Anderson, R.C., Karstad, L., and Trainer, D.O. eds., 1971, Infectious and parasitic diseases of wild birds: Ames, Iowa, Iowa State University Press, 344 p.

Jellison, W.L., 1974, Tularemia in North America, 1930–1974: University of Montana, Missoula, p. 1–276.

Jensen, W.I., and Cotter, S.E., 1976, An outbreak of erysipelas in eared grebes (*Podiceps nigricollis*)., Journal of Wildlife Diseases, v. 12: 583–586

Wobeser, G., and Rainnie, D.J., 1987, Epizootic necrotic enteritis in wild geese: Journal of Wildlife Diseases, v. 23: 376–386

Section 3
Fungal Diseases

Aspergillosis

Candidiasis

Miscellaneous Fungal Diseases

Moldy grain may be the source of aspergillosis in wild waterfowl

Introduction to Fungal Diseases

"Fungi are of an ancient lineage and have a fossil record that extends back to the Devonian and Pre-Cambrian eras...the earliest written record of fungi are not of the fungi themselves, but of their depredations... To the physician and poet Nicander [ca. 185 B.C.], fungi were 'the evil ferment of the earth; poisonous kinds originating from the breath of vipers,'..."
(Ainsworth)

Fungi are important causes of disease in wild birds and other species. Three basic types of disease are caused by these agents: mycosis, or the direct invasion of tissues by fungal cells, such as aspergillosis; allergic disease involving the development of a hypersensitivity of the host to fungal antigens; and mycotoxicosis, which results from ingestion of toxic fungal metabolites. Mycosis and allergic disease may occur together, especially when the lung is infected. This section will address only mycosis. Mycotoxicosis is addressed in Section 6, Biotoxins. Allergic disease is not well studied in wild birds and it is beyond the scope of this Manual.

Most disease-causing fungi are commonly found within the normal environment of hosts that may become diseased. Host resistance is the main determinant of whether or not disease will occur. Opportunistic infections often result when birds and other species are immunosuppressed, when their mechanisms for inflammatory response are inhibited, or when they experience physical, nutritional, or other stress for prolonged periods of time. Newborn do not have fully functioning immune systems and are, therefore, especially vulnerable to mycosis as are very old animals that are likely to have impaired immune systems. Inhalation is the primary route for exposure to most fungi-causing mycosis.

Aspergillosis is the primary mycosis affecting wild birds. Candidiasis is a less common mycosis of wild birds and other species, but it differs greatly from aspergillosis by being transmitted by ingestion. These two diseases are the primary mycoses of wild birds and are the main subjects of this section.

Quote from:

Ainsworth, G.C., and Sussman, A.S., 1965, The fungi: An advanced treatise, v. 1 of The Fungal Cell: Academic Press, New York, p. 4, 8.

Chapter 13
Aspergillosis

Synonyms

Brooder pneumonia, pseudotuberculosis, "asper" mycosis, mycotic pneumonia

Cause

Aspergillosis is a respiratory tract infection caused by fungi of the genus Aspergillus, of which *A. fumigatus* is the primary species responsible for infections in wild birds (Fig. 13.1). Aspergillosis is not contagious (it will not spread from bird to bird), and it may be an acute, rapidly fatal disease or a more chronic disease. Both forms of the disease are commonly seen in free-ranging birds, but the acute form is generally responsible for large-scale mortality events in adult birds and for brooder pneumonia in hatching birds. *Aspergillus* sp. also produce aflatoxins (see Chapter 37, Mycotoxins), but the significance of those toxins in the ability of the fungus to cause disease in birds is unknown.

Aspergilli are saprophytic (live upon dead or decaying organic matter) molds that are closely associated with agriculture and other human activities that make nutrients available to fungi. *A. fumigatus* commonly grows in damp soils, decaying vegetation, organic debris, and feed grains. High numbers of spores (called conidia) are released into the atmosphere and are inhaled by humans, birds, and other animals. These spores travel through the upper respiratory tract to the lungs. If the spores colonize the lungs, then the fungi may be disseminated to other parts of the body and disease, often leading to death, occurs.

Acute aspergillosis has caused devastating loss of birds in hatcheries. The source of infection in some instances has been contaminated litter. Also, infection of broken eggs prior to hatching provides an ideal growth medium for the fungus and the subsequent production of massive numbers of spores for infection of newly hatched birds. Inhaled spores initiate a cellular response in the lungs that results in the air passages soon becoming obliterated by cellular material and branching fungal filaments. Asphyxiation quickly follows and causes death. Acute aspergillosis has also been found in free-ranging waterfowl. The circumstances of these events are uniformly associated with birds feeding in waste grain and in silage pits during inclement weather. The mallard duck has been the primary species involved, and the events have only lasted a few days, terminating when the weather improved and allowed the ducks to resume normal feeding. Field investigations of several events disclosed heavily contaminated feed that resulted in overwhelming exposure to *A. fumigatus* (Fig. 13.2).

Chronic forms of aspergillosis have been described in wild birds since at least 1813. Typically, the lungs and air sacs are chronically infected, resulting in a gradual reduction in respiratory function. Eventual dissemination of the fungus to the liver, gut wall, and viscera is facilitated by infection of the extensive system of air sacs that are part of the avian respiratory system.

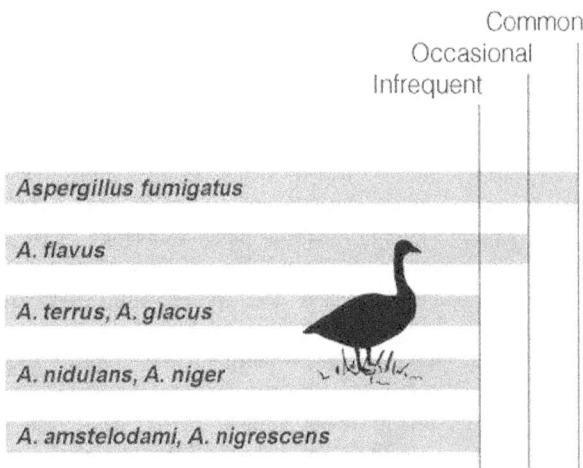

Figure 13.1 *Primary causes of aspergillosis in birds.*

Figure 13.2 *Moldy grain pile that was the source of acute aspergillosis in wild waterfowl.*

Species Affected

A wide variety of birds have died of aspergillosis and probably all birds are susceptible to it. Aspergillosis was one of the first diseases described for wild birds; it was noted in a scaup in 1813 and in a European jay in 1815. Loons and marine birds that are brought into rehabilitation, captive raptors, and penguins being maintained in zoological parks and other facilities commonly die from aspergillosis. This disease also develops at birdfeeding stations and it causes waterfowl die-offs. Young birds appear to be much more susceptible than adults. Most reported mortalities of free-ranging wild birds involve isolated mortalities found during postmortem evaluations rather than mortalities found during major die-offs (Fig. 13.3).

Distribution

Aspergillosis in birds is reported nearly worldwide.

Seasonality

Most aspergillosis outbreaks in waterfowl happen in fall to early winter; individual cases can occur at any time, particularly among birds stressed by crippling, oiling, malnutrition, recent capture, and concurrent disease conditions. This disease can cause serious losses among seabirds in rehabilitation programs after oil spills. Aspergillosis is a frequent complication in hunter-crippled waterfowl, among birds on nutritionally deficient diets, and in Canada geese whose immune systems have been compromised by exposure to environmental contaminants such as lead.

Environmental factors also contribute to the time of year when aspergillosis is seen. Scattered outbreaks of this disease occurred among American coot, diving ducks, tundra swan, and passerine birds throughout California one winter at the end of a 3-year drought. Severe dust conditions associated with this weather pattern are thought to have interfered with respiratory clearance mechanisms by reducing the amount of mucous and other body secretions that coat the cellular lining of the throat and air passages to the lungs, thereby increasing bird susceptibility to aspergillosis. A fall outbreak in Steller's jays in British Columbia was associated with a particularly dry and warm summer.

Brooder pneumonia, a specialized springtime form of aspergillosis, infects chicks or ducklings that are placed in *Aspergillus*-contaminated brooders. Catastrophic losses have occurred on game farms under these circumstances. Chicks have also been lost during captive-rearing of endangered species. Aspergillosis is also an important cause of mortality in winter roosts of blackbirds in Maryland and Pennsylvania.

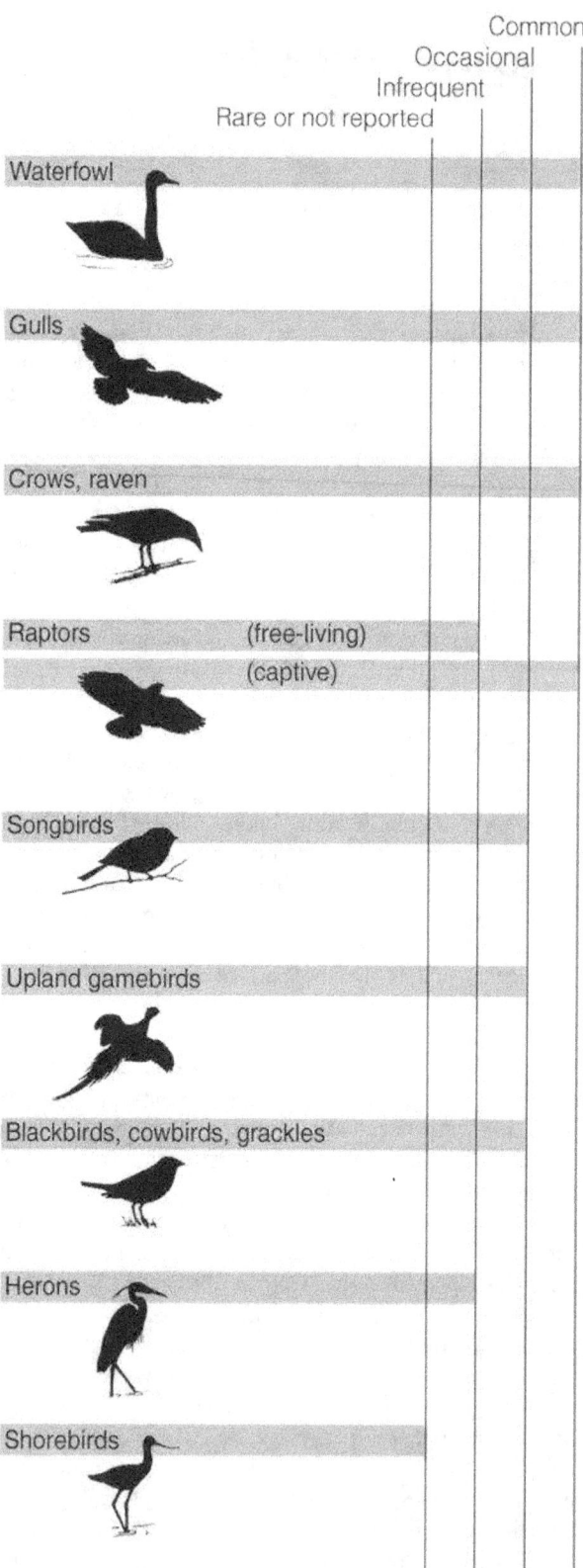

Figure 13.3 Relative occurrence of aspergillosis in free-ranging wild birds.

Field Signs

The typical aspergillosis-affected bird is emaciated, and it frequently exhibits severe and progressive difficulty in breathing by gaping or rapid opening and closing of the bill (Fig. 13.4A). Birds often appear to be unthrifty, and their wings may droop (Fig. 13.4B). Infected birds are usually weak and may fail to try to escape. With the exception of visible evidence of breathing difficulties, these signs are similar to those for lead poisoning. Infection that reaches the brain can result in obvious loss of muscular coordination and twisting of the head and neck so that the head is held in unnatural positions. Inflammation of the covering of the brain or meningoencephalitis with associated areas of brain tissue death has been reported for eider ducklings dying from aspergillosis.

Epizootic aspergillosis and brooder pneumonia outbreaks are often characterized by sudden deaths of previously healthy birds. Sick birds show acute respiratory distress and failure.

Figure 13.4 (A) Respiratory distress and gaping (note the open bill) in a herring gull suffering from aspergillosis. (B) Wing droop also occurs. Note that the wing on the near side of this bird is drooping well below the body.

Gross Lesions

Birds infected with the more typical chronic form of aspergillosis usually have variously sized lesions in their lungs and air sacs. Typically, these lesions appear as flattened, yellow plaques with a cheesy appearance and consistency (Fig. 13.5). Continuous masses of these lesions may completely line the air sac. There may also be an extensive fungus growth on tissue and air sac surfaces that appears similar to bread mold. This velvety, blue-green or grey fungal mat is striking in appearance (Fig. 13.6).

In cases of acute aspergillosis, the birds are usually in good flesh and have good-to-moderate deposits of fat. Air sacs are usually thickened, but the most striking lesion is a dark red, firm lung that is often studded or peppered with small, 1–2 millimeter, yellow nodules (Fig. 13.7).

Other, less common lesions that have been described include necrotic skin granulomas or semifirm growths of granular consistency in chickens and pigeons. Cheesy plaques that form in the eye beneath the nictitating membrane, which is the transparent membrane that forms a rapidly moving third eyelid that keeps the eye clean and moist, or on the surface of the eye have also been observed.

Diagnosis

Whole carcasses should be submitted for necropsy by qualified diagnosticians. Diagnosis is based on finding the typical lesions and on isolating the fungus from the tissues. *Aspergillus* sp. can be identified by microscopically examining material from fungal mats and from tissue sections that have been specially stained. However, the specific species of *Aspergillus* cannot be identified by these means.

Control

The spores of the mold *A. fumigatus* are widely distributed and are often present in moldy feeds, unclean brooders and incubators, moldy straw, and rotting agricultural waste. *Aspergillus* grows best on decomposing organic matter left in a warm, dark, moist environment. Failure to maintain a clean environment often leads to severe outbreaks.

Aspergillosis has broken out in mallards feeding in fields that were previously covered by discarded moldy corn and silage. Dusty straw hay placed as litter in the bottoms of wood duck nest boxes has resulted in losses of wood duck ducklings. Avoid using moldy or dusty straw, silage, or feed, and dumping moldy waste grain in areas where waterfowl and other birds feed. Birds should be denied the use of fields where moldy agricultural waste products such as waste corn, peanuts, straw, or hay have accumulated. Monitoring for such situations in waterfowl concentration areas and establishing contingency plans that can be implemented at the onset of inclement weather can minimize the potential for waterfowl deaths if the concentrated sources of *Aspergillus* spores can not be dealt with in other ways. People who feed birds should be educated to periodically clean their feeding stations.

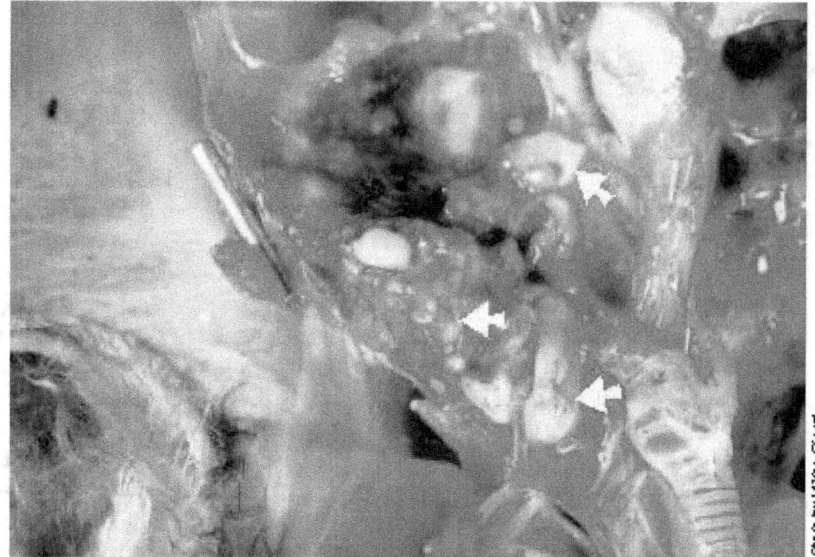

Figure 13.5 "Cheesy" plaques in the lungs and air sacs of a bird with aspergillosis.

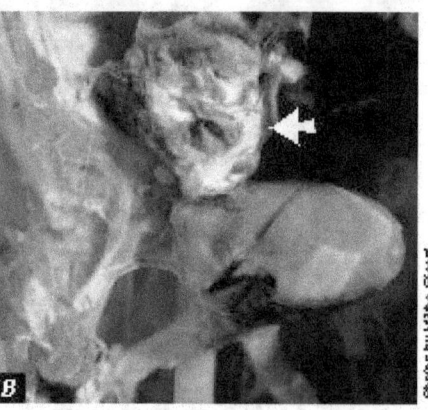

Figure 13.6 Lung of a bird with chronic aspergillosis showing (A) "cheesy" fungal plaques, and (B) "bread mold" fungal mat totally involving the air sac adjacent to the heart of this bird.

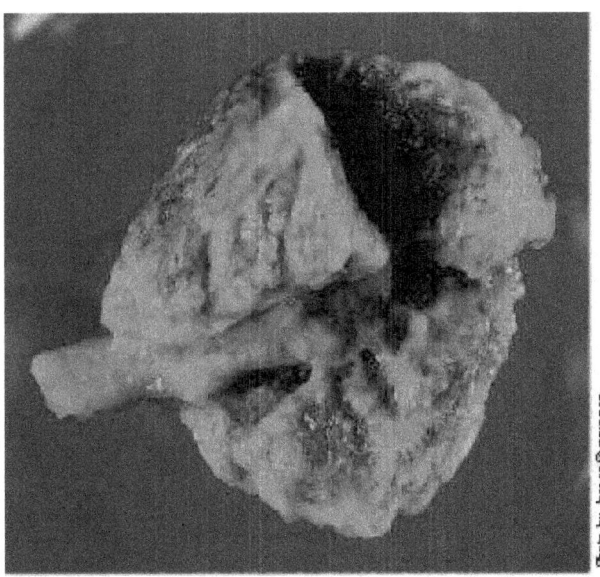

Figure 13.7 *Acute aspergillosis or "brooder pneumonia" in a lung of a wood duck duckling. Note dark red, "studded" (granular) appearance of lung.*

Human Health Considerations

Aspergillosis is not contagious. However, when human resistance to infection is impaired, aspergilli can cause rapidly developing acute infection following environmental exposure. Invasive aspergillosis in humans involving dissemination of fungi to organs other than the lungs is often associated with the person being immunocompromised and, if the disease is not properly diagnosed, it may be life threatening. A few individuals who have worked with *A. fumigatus* have become allergic to it. Allergic response can result in an acute, life-threatening reaction to this fungus. It is unlikely that infected bird carcasses would provide sufficient exposure to result in either of these outcomes.

Milton Friend

Supplementary Reading

Adrian, W.J., Spraker, T.R., and Davies, R.B., 1978, Epornitics of aspergillosis in mallards (*Anas platyrhynchos*) in north central Colorado: Journal of Wildlife Diseases, v. 14, p. 212–217.

Barden, E.S., Chute, H.L., O'Meara, D.C., and Wheelwright, H.T., 1971, A bibliography of avian mycosis (partially annotated): Orono, Me., College of Life Sciences and Agriculture. Univ. of Maine, 193 p.

O'Meara, D.C., and Witter, J.F., 1971, Aspergillosis, *in* Davis, J.W., and others, eds., Infectious and parasitic diseases of wild birds: Ames, Iowa, Iowa State University Press, p. 153–162.

Powell, K.A., Renwick, A. and Peberdy, J.F., 1994, The genus *Aspergillus* from taxonomy and genetics to industrial application: Plenum Press, New York, N.Y., 380 p.

Wobeser, G.A., 1997, Aspergillosis, *in* Diseases of wild waterfowl (2nd ed): New York, N.Y., Plenum Press, p. 95–101.

Chapter 14
Candidiasis

Synonyms
Moniliasis, candidiasis, thrush, sour crop

Cause
Candida albicans, a yeast-like fungi, is the primary cause of candidiasis or candidiosis. *C. albicans* is a normal inhabitant of the human alimentary canal, as well as that of many species of lower animals. Ingestion in food or in water is the usual means for its transmission. Contaminated environments, such as litter from poultry and gamebird rearing facilities, refuse disposal areas, discharge sites for poultry operations, and areas contaminated with human waste have all been suggested as sources for *Candidia* exposure for birds.

Species Affected
There have been few reports of candidiasis causing disease in free-ranging wild birds and few investigations of its prevalence. Therefore, little can currently be said about its occurrence in wild species. Candidiasis is an occasional disease of importance within some poultry flocks, and it has been reported as a disease or an intestinal infection in numerous species of wild birds being raised in captivity. It has also been an occasional cause of disease in wild species being transported within the pet bird industry (Fig. 14.1).

Distribution
Candidiasis is found worldwide.

Seasonality
There is no known seasonal occurrence. Life-cycle patterns for bird populations are likely to influence any temporal occurrence for this disease because young birds are generally more susceptible to infection.

Field Signs
There are no unique signs of disease. Affected poultry have retarded growth, stunted appearance, are listless, and have ruffled feathers.

Gross Lesions
Lesions are generally confined to the upper areas of the digestive tract. The mouth, esophagus, and, primarily, the crop, may have grayish-white, loosely attached, plaque-like areas on their internal surfaces. Circular, raised, ulcerative

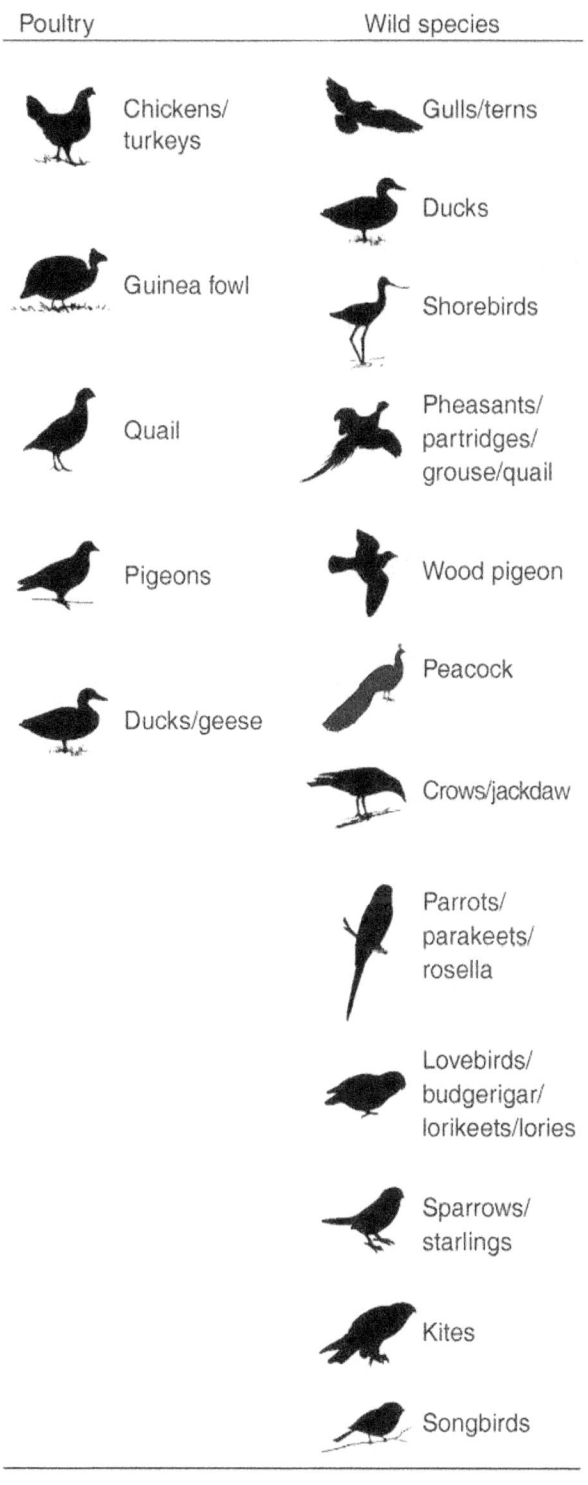

Figure 14.1 *Avian groups reported to have been infected with candidiasis.*

nodules that appear as rose-like clusters may be within the crop, and the crop surface is often so unevenly thickened that it appears to have the texture of a Turkish bath towel or curds. Other areas of the upper digestive tract develop false membranes that resemble those which develop during diptheria, areas of dead tissue, and contain considerable tissue debris.

Control

The infrequent reports of this disease in free-ranging wild birds do not warrant the need for disease control. This disease is more likely to be encountered in captive-rearing situations. Disease prevention should be practiced to prevent infections. Cages, equipment, and other materials in contact with infected birds should be disinfected because of the broad host range of species that can become infected.

Human Health Considerations

Humans can be infected, and infections can result in acute or chronic disease that can involve the mucous membranes (oral thrush), skin, nails, and internal organs.

Milton Friend

Supplementary Reading

Chute, H.L., 1997, Thrush (mycosis of the digestive tract), *in* Calnek, B.W., and others, eds., Diseases of Poultry (10th ed.): Ames, Iowa, Iowa State University, p. 361–365.

Odds, F.C., 1988, Candida and candidosis: London, Baillière Tindall, 468 p.

O'Meara, D.C., and Witter, J.F., 1971, Candidiasis, *in* Davis, J.W., and others, eds., Infectious and parasitic diseases of wild birds: Ames, Iowa, Iowa State University, p. 163–169.

Chapter 15
Miscellaneous Fungal Diseases

As for other types of disease, fungal infections probably are more common causes of disease in wild birds than is currently recognized. Also, the similarity in gross lesions produced by some fungi mask the detection of less common fungi as disease agents. Numerous types of disease-causing fungi in addition to *Aspergillus fumigatus* and *Candida albicans* have been isolated from birds; most isolations have been from poultry and wild birds being maintained in captivity. Enhanced disease surveillance that is often associated with privately owned birds and greater opportunity to detect disease in confined birds are reasons for these findings rather than any known differences in the occurrence of fungal diseases in free-ranging and captive birds. Many of the reported infections appear to have been opportunistic invasions by the fungi involved. The important points are that many fungi are capable of causing disease in birds but their collective impacts do not rival *A. fumigatus* as a single cause of disease in wild birds. Nevertheless, it is important to be aware of the diversity of pathogenic or disease causing fungi.

Infectious diseases caused by fungi have been grouped into categories that represent their involvement within the host.

Types of mycosis
(direct invasion of tissue by fungal cells)

Category	Area of the body affected
Superficial	Found on the outermost layers of the body covering; are generally of cosmetic impact rather than causes of illness or death; have not been reported in birds.
Cutaneous (dermatophytosis)	Found on the skin and appendages.
Subcutaneous	Usually found in the fat-containing tissues underneath the skin and in the skin.
Systemic	Result in infection of internal organs as well as other tissues.

Aspergillosis and candidiasis are diseases characteristic of systemic mycosis. Candidiasis can also be a cutaneous mycosis.

Trichophyton gallinae is the primary cause of ringworm, or fowl favus, in birds, and has been reported in poultry and several species of wild birds in addition to companion animals, humans, and other mammalian species. *T. gallinae* is widely distributed geographically, and infection by this fungus is a striking example of a cutaneous mycosis (Fig. 15.1). Ringworm in birds is highly contagious, and it is transmitted by direct bird-to-bird contact or by contact with a contaminated environment. The fungus can remain viable at room temperature in infected scales or skin lesions that slough from the body for up to 1 year. *Microsporum gallinae* is another widely distributed fungus that is a significant cause of ringworm in birds and mammals.

Dactylaria gallopova causes a subcutaneous mycosis reported for poultry. This fungus is found in warm habitats such as hot springs and thermal soils. The fungi generally enter the body at a traumatized or injured site and may then invade other sites following fungal establishment and growth. *D. gallopova* is not contagious, but it can invade the brain following its spread from the site of infection. Death is the outcome when the brain is invaded.

Aspergillus niger is another fungus within the genus *Aspergillus* that has caused bird deaths.

As noted in the Introduction of this Section, disease due to infection of tissues is only one aspect of the potential impacts of fungi. The added issues of mycotoxins (see Section 6, Biotoxins), allergic responses, and other aspects of fungal diseases make fungi an important area for consideration in the management and stewardship of free-ranging bird populations.

Figure 15.1 *Extensive loss of feathers of the head of a loon believed to have been caused by ringworm resulting from infection by Trichophyton sp.*

Supplementary Reading

Hubbard, G.B., Schmidt, R.E., Eisenbrandt, D.L., Witt, W.M., and Fletcher, K.C., 1985, Fungal infections of ventriculi in captive birds: Journal of Wildlife Diseases, v. 21, p. 25–28

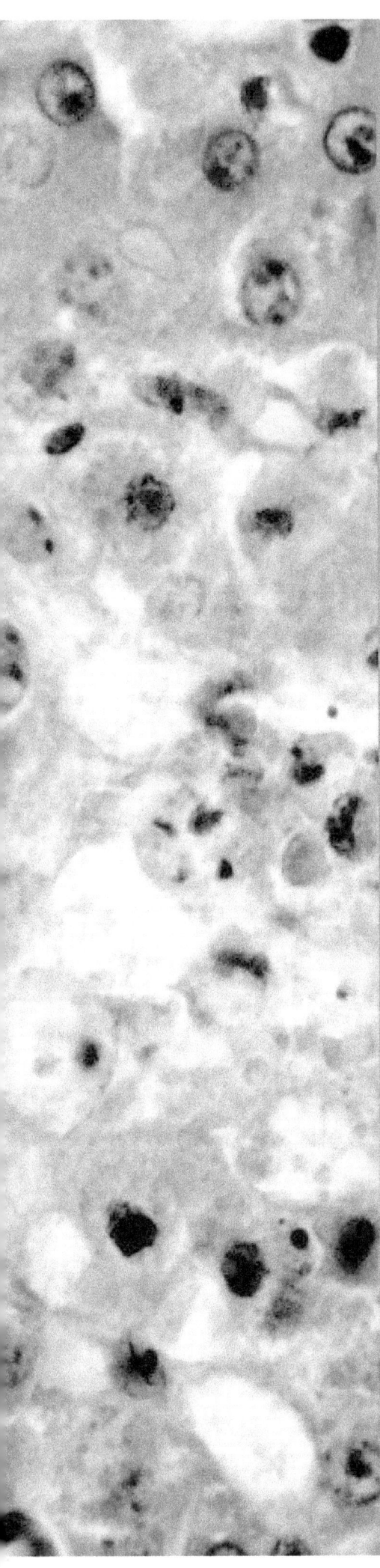

Section 4
Viral Diseases

Duck Plague

Inclusion Body Disease of Cranes

Miscellaneous Herpesviruses of Birds

Avian Pox

Eastern Equine Encephalomyelitis

Newcastle Disease

Avian Influenza

Woodcock Reovirus

Inclusion bodies in the liver of a bird that died of herpesvirus infection
Photo by Jim Runningen

Introduction to Viral Diseases

"The viruses almost surely antedate our species." (Johnson)

"...viral emergence is essentially a two-step process: (1) introduction of the virus (whatever its origin) into a new host, followed by (2) dissemination within that new host population...That second step might not occur at all...However, changing conditions might increase the chances of this second step occurring." (Morse)

Historically, viral diseases have not been recognized as major causes of illness and death in North American wild birds. Until relatively recently, this may have been due to inadequate technology to culture and identify these organisms. Unlike bacteria, viruses are too small to be seen under the light microscope and they cannot be grown on artificial media. Nevertheless, studies of infectious diseases caused by viruses have often predated discovery of the causative agents by many years as evidenced by smallpox immunizations being used centuries before that virus was identified. The isolations of a tobacco mosaic virus in 1892 and foot and mouth disease viruses in 1898 mark the development of virology as a distinct biological science. The era of modern virology began in the post-World War II years of 1945–50 with the application of cell culture techniques to the study of animal viruses.

For centuries, gross and microscopic pathology associated with tissue alterations caused by viral infections have been recorded for species of domestic birds, captive wild birds, and, occasionally, for free-living wild birds. However, significant concern about viral diseases in wild birds has primarily occurred since the 1970s. This timeframe is consistent with an apparent increase of emerging infectious diseases and emerging viruses in other species, including humans. It is noteworthy that this pattern exists for the diseases included in this section. Duck plague first appeared in the United States in 1967 and the first major loss of wild waterfowl from duck plague occurred in 1973. Eastern equine encephalitis erupted in a captive breeding flock of whooping cranes in 1984; a highly virulent form of Newcastle disease virus has appeared several times among double-crested cormorants in Canada since 1990 and in the United States since 1992; and a previously undescribed reovirus was the cause of death for woodcock in 1989 and again in 1993. In 1978, inclusion body disease of cranes appeared in a captive crane breeding colony in the Midwestern United States; that outbreak was the first identification of this herpesvirus infection. In 1978 also, avian pox viruses were first isolated from free-living waterfowl and from bald eagles the following year.

Avian influenza has been included in this section to give wildlife resource managers basic information about this group of generally avirulent viruses that exchange genetic material to create new forms of the virus, some of which are capable of causing disease. Interest in influenza is primarily focused on the role of migratory birds as a source of viruses that infect domestic poultry and humans.

It seems likely that viral diseases will assume even greater future importance as causes of disease in wild birds. Greater attention needs to be given to the study of this source of disease, especially in captive-propagation programs intended for supplementing and enhancing wild stocks of birds.

Quotes from:

Johnson, K.M., 1993, Emerging viruses in context: an overview of viral hemmorhegic fevers, *in* Morse, S.S. [editor], Emerging Viruses: Oxford University Press, New York, p. 46.

Morse, S.S., 1993 Examining the origins of emerging viruses, *in* Morse, S.S. [editor], Emerging Viruses: Oxford University Press, New York, p. 16–17.

Chapter 16
Duck Plague

Synonyms

Duck virus enteritis, DVE

Cause

Duck plague is caused by a herpesvirus. Infection often results in an acute, contagious, and fatal disease. As with many other herpesviruses, duck plague virus can establish inapparent infections in birds that survive exposure to it, a state referred to as latency. During latency, the virus cannot be detected by standard methods for virus isolation. Studies of domestic species of waterfowl have detected multiple strains of the virus that vary in their ability to cause disease and death. Little is known about the response of wild waterfowl to strain differences.

Duck plague outbreaks are thought to be caused when birds that carry the virus shed it through fecal or oral discharge, thus releasing the virus into food and water with which susceptible birds may have contact. Experimental studies have demonstrated spontaneous virus shedding by duck plague carriers during spring. Changes in the duration of daylight and onset of breeding are thought to be physiological stresses that stimulate virus shedding at this time of year. The carriers are immune to the disease, but the virus shed by them causes infection and disease among susceptible waterfowl. Bird-to-bird contact and contact with virus that has contaminated the environment perpetuate an outbreak. Scavenging and decomposition of carcasses of infected birds also contaminate the environment by releasing viruses from tissues and body fluids. Virus transmission through the egg has been reported, but the role of the egg in the disease cycle remains to be resolved.

Species Affected

Only ducks, geese, and swans are susceptible to duck plague. Other aquatic birds do not become infected, and the absence of mortality of American coot, shorebirds, and other waterbirds that may be present during a waterfowl die-off can be an important indication that duck plague may be involved. Susceptibility varies greatly among waterfowl species (Fig. 16.1). In one study with a highly virulent virus, it took 300,000 times more virus material to infect northern pintail than to infect blue-winged teal.

Distribution

The first reported duck plague outbreak in North America struck the white Pekin duck industry of Long Island, New

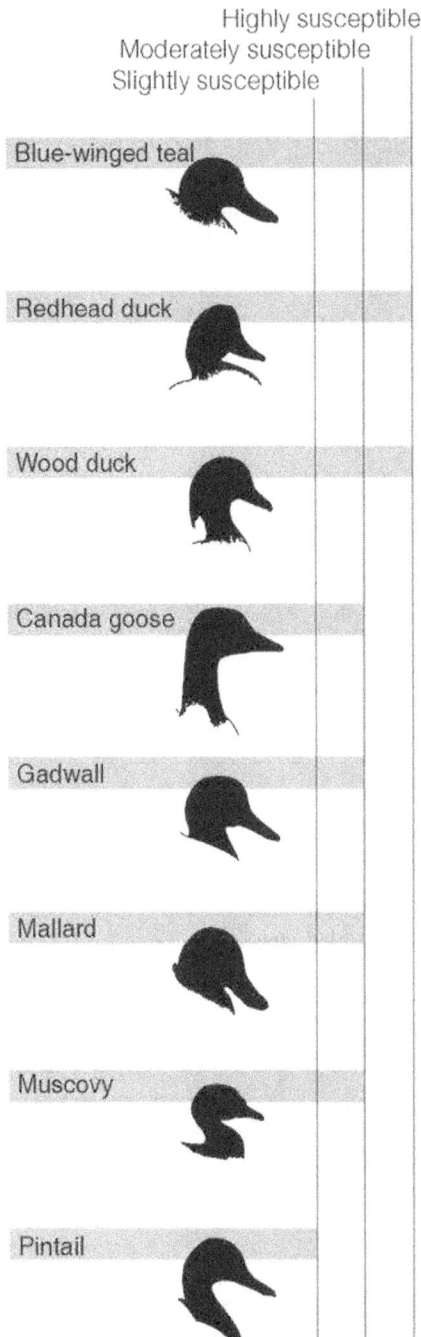

Figure 16.1 Comparative susceptibility of eight waterfowl species to duck plague virus.

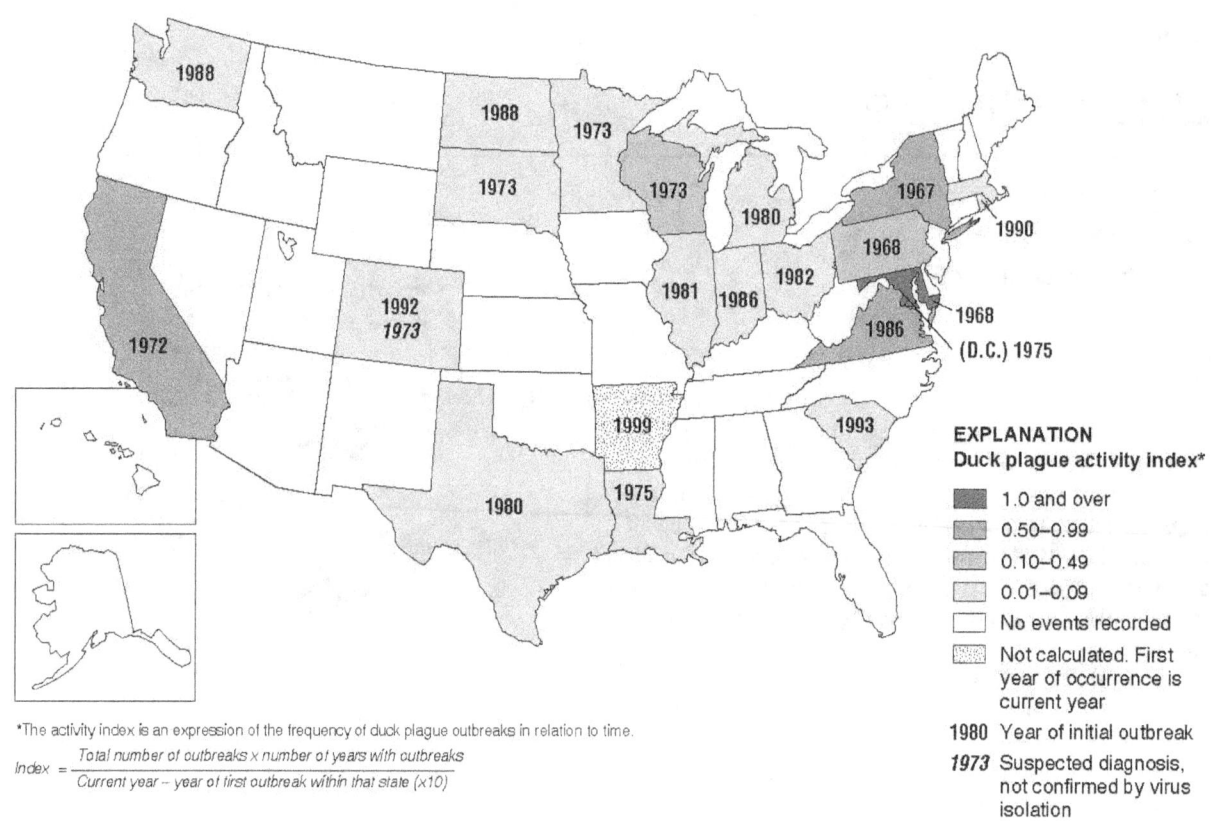

Figure 16.2 Frequency of duck plague since year of first outbreak (1967–1996).

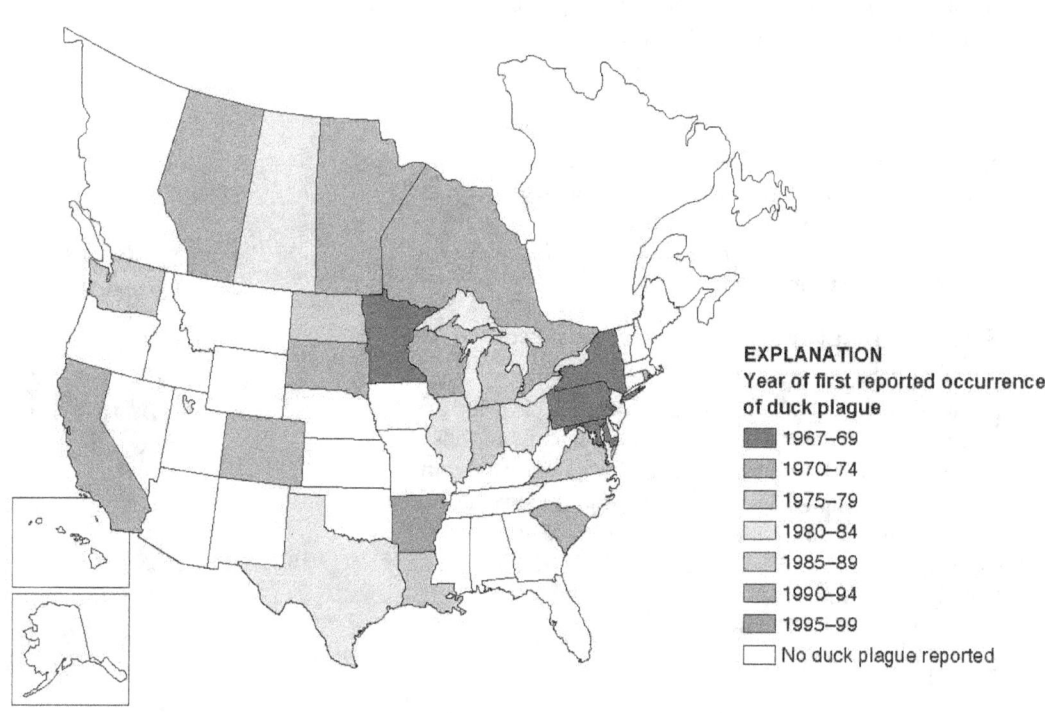

Figure 16.3 Reported North American distribution of duck plague by period of first occurrence.

York in 1967. Since then, duck plague has broken out from coast to coast and from Canada to Texas. The frequency of duck plague outbreaks has varied considerably geographically. The greatest frequency of duck plague activity has been reported in Maryland, followed by California, Virginia, and New York (Fig. 16.2). The disease has also been reported in several Canadian Provinces since it first was observed in the United States (Fig. 16.3). First reported in the Netherlands in 1923, duck plague has also been reported in several other countries in Europe and in Asia since 1958. The frequency of duck plague varies within different types of waterfowl, and failure to respond to these differences complicates disease prevention and control efforts. The different types of waterfowl aggregations involved and the relative frequency of duck plague activity within these different populations are highlighted in Tables 16.1 and 16.2.

Despite the cumulative widespread geographic distribution and frequent occurrence of duck plague in captive and feral waterfowl in North America, wild waterfowl have been affected only infrequently. The only major outbreaks in migratory waterfowl have happened in South Dakota and New York. In January 1973, more than 40,000 of 100,000 mallards and a smaller number of Canada geese and other species died at Lake Andes National Wildlife Refuge in South Dakota while they were wintering there (Fig. 16.4). The only other duck plague event that caused substantial loss of wild waterfowl occurred during February 1994 in the Finger Lakes region of western New York State. Approximately 1,200 carcasses were recovered, primarily American black duck and mallard, with nearly three times as many black duck as mallard carcasses. The carcasses that were recovered were approximately 24 percent of the black duck and 3 percent of the mallard populations present at the outbreak location. During the initial 1967 outbreak in white Pekin ducks on Long Island, several hundred wild waterfowl carcasses (primarily mallard and American black duck) were recovered

Table 16.1 Types of waterfowl involved in outbreaks of duck plague in the United States.

Waterfowl classification	Population composition
Commercial	Birds raised for consumptive markets; for example, white Pekin ducks.
Captive collections	Zoological and other collections of birds for display and research.
Game farm	Birds raised for release for sporting programs; for example, mallard ducks.
Feral	Nonmigratory, nonconfined waterfowl of various species.
Nonmigratory	Resident populations of native wild species; for example, mallard ducks and Canada geese.
Migratory	North American waterfowl that breed in one geographic area and winter in another before returning to their Northern breeding grounds.

Table 16.2 Relative frequency of duck plague in different types of waterfowl within the United States.

Waterfowl classification	Occurrence of disease	
	Mortality events	Trends, 1967–1996
Commercial	Rare	Was the primary virus source, but is currently rare
Captive collections	Occasional	None; sporadic outbreaks
Game farm	Occasional	None; sporadic outbreaks
Feral	Common	Increasing outbreaks, and currently prime virus source
Nonmigratory	Occasional	None; sporadic outbreaks
Migratory	Rare	None; rare

from adjacent Flanders Bay, apparently as a result of disease transmission from white Pekin ducks. Those carcasses represented approximately 5 percent of the wild mallard and black duck populations on Flanders Bay during the duck plague outbreak. Mortality in the white Pekin duck flocks was much greater, averaging 45 percent in mature ducks (2-year olds) and 17 percent in immature ducks (younger than 5 months of age). Equally important was the 25–40 percent decrease in egg production by mature breeder ducks that were present during the outbreak. With the exception of the Lake Andes, Finger Lakes, and Flanders Bay outbreaks, duck plague in migratory waterfowl has been limited to a small number of birds. All confirmed outbreaks have also involved commercial, avicultural, captive-raised, or feral waterfowl.

The pattern of duck plague within North America is that of an emerging disease. The number of outbreaks being diagnosed is increasing each decade (Fig. 16.5). The great majority of outbreaks occur within the Atlantic Flyway (Fig. 16.6) and nearly all of those events are within Maryland and Virginia (Fig. 16.7). The factors responsible for the continued emergence and geographic spread of duck plague within North America are unknown, as is the distribution of duck plague among free-living North American waterfowl populations.

Some individuals believe that a large number of surviving wild waterfowl exposed to this disease at Lake Andes became disease carriers, that these disease carriers have perpetuated infections in other wild waterfowl, and that duck plague is now widespread among migratory waterfowl. However, surveys of wild waterfowl conducted by the U.S. Department of Agriculture in 1967 and by the National Wildlife Health Center (NWHC) from 1978 to 1986 and in 1982–1983 did not detect any evidence of duck plague carriers. In the latter NWHC survey, more than 4,500 waterfowl across the United States were sampled (Fig. 16.8). Sampling sites included major waterfowl concentration areas and areas where duck plague has been a recurrent disease problem in captive and feral waterfowl. Although none of the birds sampled during either NWHC survey were shedding detectable duck plague virus, the previously described problem of inapparent carriers complicates interpretation of these results. New technology that was not yet developed at the time of that survey provides increased ability to detect duck plague carriers and resolve the question of sources for infection.

The absence of duck plague as a cause of mortality in the thousands of wild waterfowl necropsied by the NWHC provides additional evidence that duck plague is not an established disease in wild North American waterfowl. These examinations, performed since 1975, were of waterfowl found dead on National Wildlife Refuges and other major waterfowl concentration areas.

Seasonality

Duck plague outbreaks have been reported during every month except August and September. Approximately 86 percent of these outbreaks occurred from March through June (Fig. 16.9). This pattern of spring outbreaks has also been reported for captive waterfowl collections in England, and it may be associated with the physiological changes referred to above.

Figure 16.4 During the 1973 outbreak of duck plague at Lake Andes National Wildlife Refuge in South Dakota, more than 40,000 mallards died.

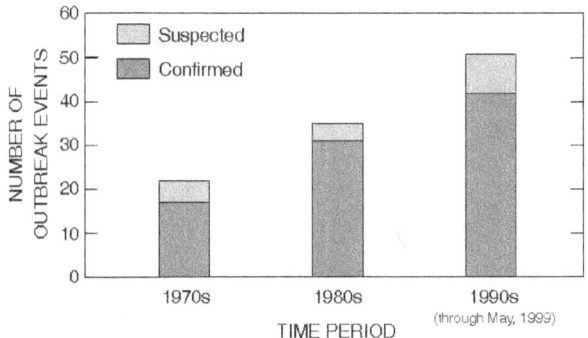

Figure 16.5 Duck plague outbreaks in the United States, 1970s to 1999.

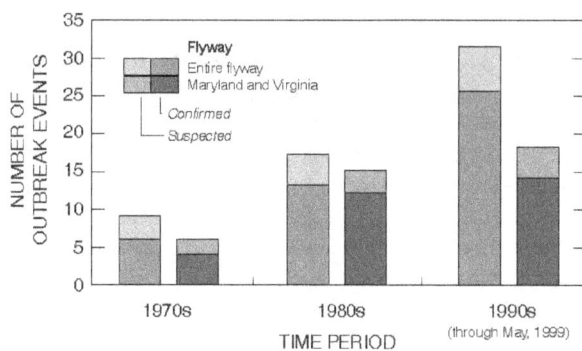

Figure 16.7 Duck plague outbreaks in the Atlantic Flyway, 1970s to 1999.

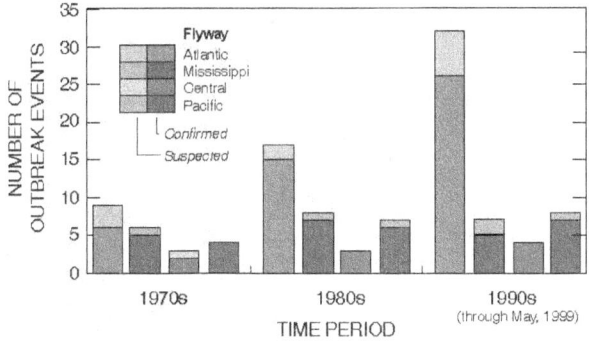

Figure 16.6 Duck plague outbreaks in the United States by flyway, 1970s to 1999.

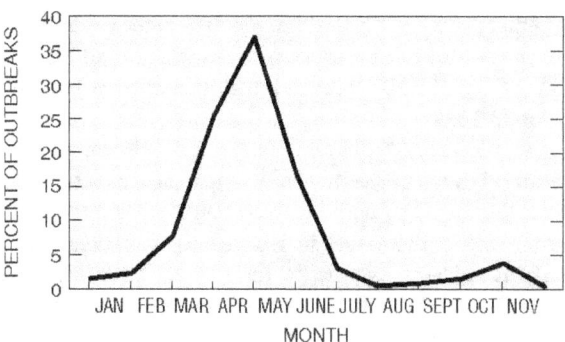

Figure 16.9 Month of onset of duck plague outbreaks, 1967–1996.

Figure 16.8 Sampling locations for 1982–1983 duck plague survey.

Duck Plague 145

Field Signs

There is no prolonged illness associated with duck plague; therefore, sick birds are seldom seen in the field, and birds that are healthy one day may be found dead the next. The incubation period between virus exposure and death is generally 3–7 days in domestic ducks, and experimental studies have found that it is as long as 14 days in wild waterfowl. Wing-clipped mallards released to monitor the Lake Andes duck plague outbreak died 4–11 days after their release.

Sick birds may be hypersensitive to light, causing them to seek dense cover or other darkened areas. They may exhibit extreme thirst, droopiness, and bloody discharge from the vent (Fig. 16.10A) or bill (Fig. 16.10B). The ground may be blood-stained where sick birds have rested (Fig. 16.10C). Therefore, duck plague should be suspected when blood-soiled areas are seen following the flushing of birds, where blood splotches that do not appear to be related to predation or other plausible explanations are seen in the environment, or where bloody discharges are seen where dead birds are lying (Fig. 16.10D). In males, the penis may be prolapsed (Fig. 16.10E).

An ulcerative "cold sore" lesion under the tongue from which virus can be shed has been seen in some infected waterfowl (Fig. 16.11). Routine examination of apparently healthy waterfowl for this lesion during banding operations may be helpful in identifying inapparent carriers. Birds with these lesions should be euthanized (see Chapter 5, Euthanasia) and submitted to a qualified disease diagnostic laboratory for examination.

Death may be preceded by loss of wariness, inability to fly, and finally by a series of convulsions that could be misinterpreted as pesticide poisoning or other diseases such as avian cholera (Fig. 16.12).

Gross Lesions

Duck plague virus attacks the vascular system, and can result in hemorrhaging and free blood throughout the gastrointestinal tract (Fig. 16.13A). At the Lake Andes outbreak, the most prominent lesions were hemorrhagic or necrotic bands circumscribing the intestine in mallards (Figs. 16.13B, C, and D) and disk-shaped ulcers in Canada geese (Figs. 16.13E and F). Sometimes there were "cheesy," raised plaques along the longitudinal folds of the esophagus and proventriculus (Fig. 16.14A) and on the mucosal surface of the lower intestine (Fig. 16.14B). Areas of tissue death (spots) were also evident in the liver (Fig. 16.14C), as was hemorrhaging on the heart surface of some birds (Fig. 16.14D).

It is important to recognize that the appearance of lesions may differ somewhat from species to species and that not all lesions are present in all birds at all times. Outbreaks of duck plague in captive and nonmigratory waterfowl have often resulted in infected birds with less distinct lesions. Of all the lesions illustrated, those of greatest value in diagnosing duck

Figure 16.10 Field signs associated with duck plague include: (A) blood staining of the vent area; (B) blood dripping from the bill or a blood-stained bill; (C) blood-stained environment from which a resting mallard has just taken flight; (D) blood-stained ice from the nasal discharge of a mallard dying from duck plague; and (E) prolapse of the penis.

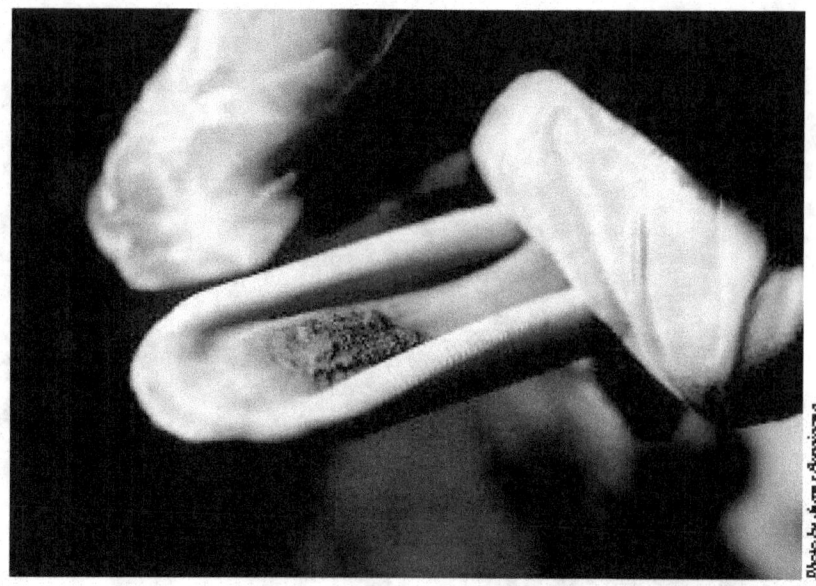

Figure 15.11 A "cold sore" under the tongue.

Figure 15.12 Death sequence observed during terminal stages of duck plague infection at Lake Andes National Wildlife Refuge began with (A) the head of the bird dropping forward, wings becoming partially extended from the sides, and tail becoming fanned and rigid. This was followed by (B) the bird swimming in a tight circle while rapidly beating the water with its wings and with the head pulled back and twisted to the side. (C) At times, birds would fall over on their side, be unable to regain a normal body position, and drown. (D) Other birds would simply stop swimming, relax, and quietly die. This entire sequence generally lasted only a few minutes.

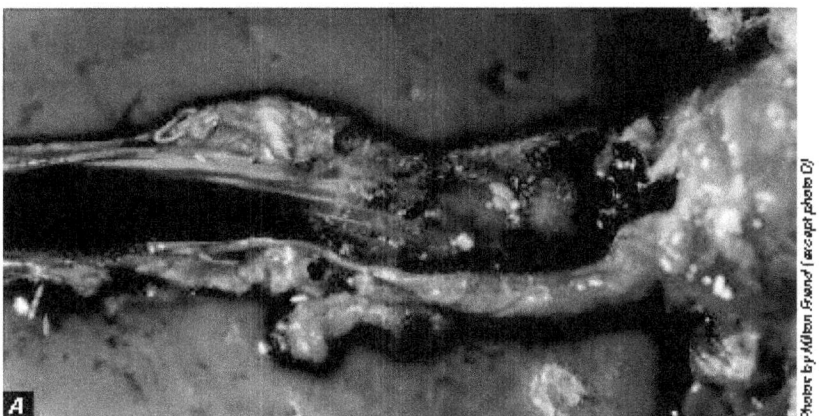

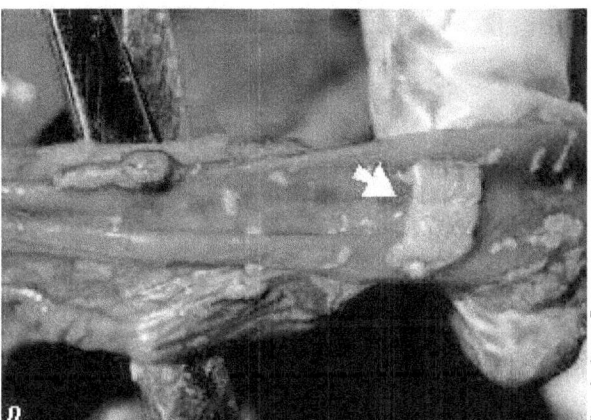
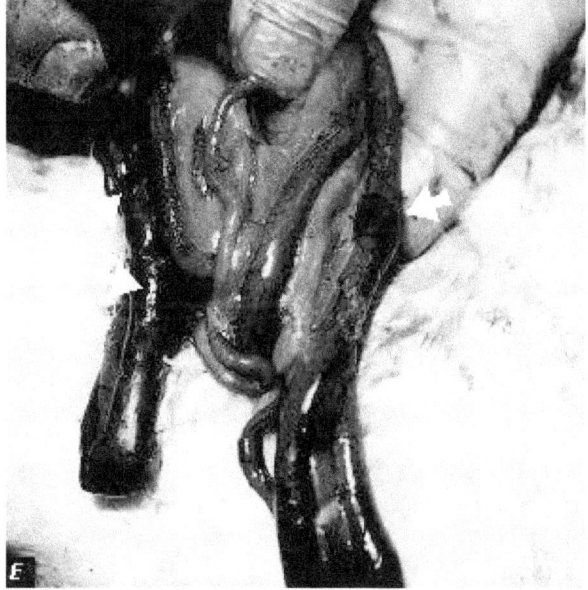
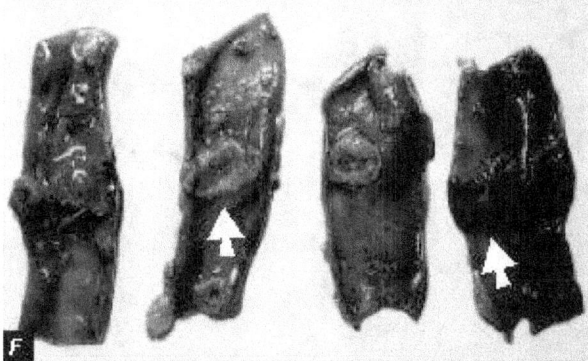

Figure 16.13 Appearance of major lesions of duck plague; **(A)** hemorrhage and free blood in the lumen of the gastrointestinal tract; **(B and C)** external appearance of hemorrhagic bands in mallard intestine; and **(D)** appearance of bands when intestine is opened; **(E)** external appearance of similar lesions in intestine of a Canada goose; and **(F)** buttonlike rather than bandlike appearance of lesions when intestine is opened.

Duck Plague 149

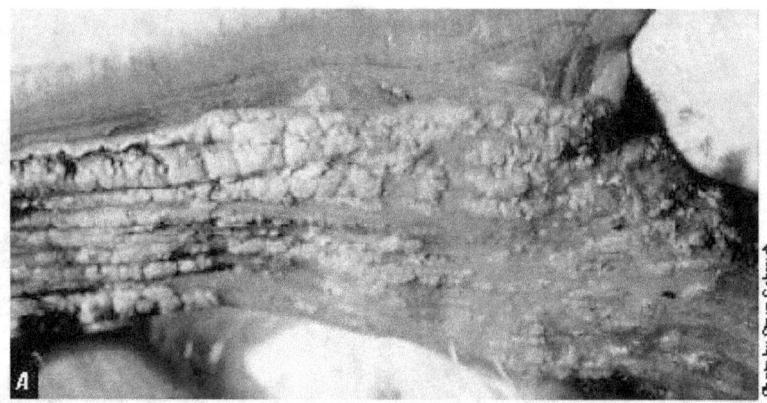

Figure 16.14 Other internal lesions of duck plague include: **(A)** cheesy, raised plaques along the longitudinal folds of the esophagus, proventriculus, and **(B)** inside (mucosal) surface of the lower intestine. **(C)** Necrotic spots may occur in the liver, and **(D)** varying degrees of hemorrhage on the heart surface.

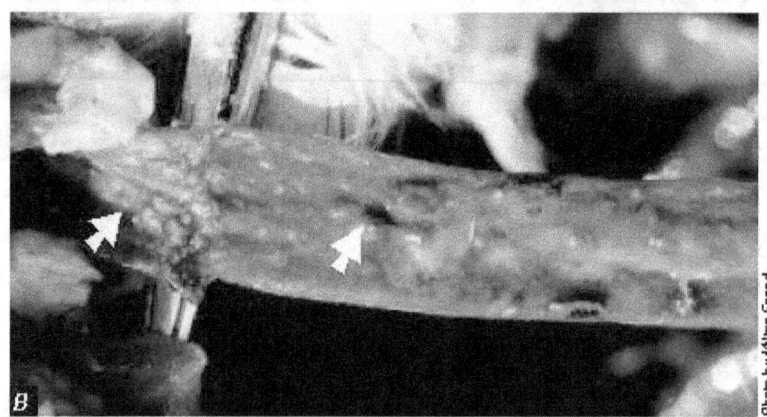

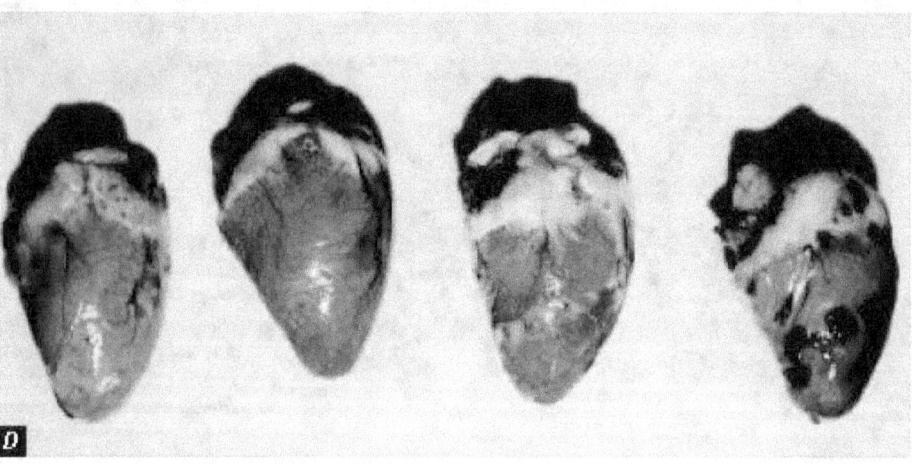

plague are hemorrhagic or necrotic bands or disks within the intestine, large amounts of free blood in the digestive tract, and cheesy plaques in the esophagus and cloaca. Liver and heart lesions of duck plague are grossly similar to those of avian cholera, and they cannot be used to distinguish between these two diseases.

Diagnosis

Although a presumptive diagnosis of duck plague may be made on the basis of characteristic internal lesions, final diagnosis can only be made by virus isolation and identification. Ducks, geese, and swans that have characteristic signs or lesions should be euthanized and shipped to a qualified diagnostic laboratory as quickly as possible. Submit whole birds rather than tissues. When this is not possible, the liver should be removed, wrapped in clean aluminum foil, and then placed in a plastic bag and frozen for shipment. The remainder of the carcass should be incinerated if possible and the area and instruments used to process the carcass disinfected. Take particular care in preserving and packaging specimens to avoid their decomposition during transit and contamination of the shipping containers (see Chapter 2, Specimen Collection and Preservation, and Chapter 3, Specimen Shipment).

Control

The primary objectives for duck plague control activities are to minimize exposure of the population-at-risk at the outbreak site and to minimize the amount of virus present in the environment as a source for potential exposure of waterfowl that may use the site in the near future. Control of duck plague outbreaks requires rapid response and aggressive actions to prevent disease spread and establishment.

Birds with inapparent duck plague infections are probably the major reservoir of this disease and they pose the greatest problem for disease prevention and control. Clinically ill birds actively shed the virus and are recognized as sick birds. However, asymptomatic healthy duck plague carriers can shed the virus periodically, but they are not overtly identifiable. Therefore, destruction of infected flocks, including eggs, is recommended whenever possible because infected birds that survive are likely to become carriers and can initiate subsequent outbreaks. New technology provides promise for determining whether or not there are carriers in a flock. The success of new technology for detecting carriers will allow selective euthanization of those birds and not the remainder of the flock.

Duck plague virus is hardy, and it can remain viable for weeks under certain environmental conditions; for example, the virus could be recovered from Lake Andes water held at 4 °C for 60 days under laboratory conditions. Duck plague virus is instantly inactivated at pH 3 and below and at pH 11 and above. Therefore, rigorous decontamination of infected waters (for example, by chlorination) and grounds (that is, by raising pH) and burning or decontamination of physical structures, litter, and other materials at outbreak sites should be carried out to the extent practical. Carcass collection should be thorough and incineration used for disposal. Personnel and equipment used at outbreak sites should be decontaminated before leaving the site to prevent mechanical spread of the virus to other waterfowl areas; chlorine bleach and phenol base disinfectants are suitable for this (see Chapter 4, Disease Control Operations).

A low virulence live-virus vaccine has been developed for combating duck plague in the domestic white Pekin, but this vaccine has not been proven entirely reliable in protecting other species of ducks and geese. It should not be considered as a means of controlling or preventing outbreaks in migratory birds.

The close association between duck plague outbreaks and captive waterfowl, especially muscovy and mallard, needs to be considered. Waterfowl release programs should not use birds or eggs from flocks with a history of this disease unless the flock has subsequently been shown by adequate testing and other technical assessments to be free of duck plague. Birds scheduled for release should be confined for at least 2 weeks before release. Birds that die during this period should be submitted to a qualified disease diagnostic laboratory. If duck plague is found to be the cause of death in any of these birds, none of the remaining birds should be released. Also, managers of areas for wild waterfowl should not permit the maintenance of domestic waterfowl, especially muscovy ducks, on the area or waterfowl display flocks that have not been certified free of duck plague.

Human Health Considerations

None.

Milton Friend
(Modified from an earlier chapter by Christopher J. Brand)

Supplementary Reading

Brand, C.J., and Docherty, D.E., 1984, A survey of North American migratory waterfowl for duck plague (duck virus enteritis) virus: Journal of Wildlife Disease, v. 20, p. 261–266.

Hansen, W.R., Brown, S.E., Nashold, S.W., and Knudson, D.L., 1999, Identification of duck plague virus by polymerase chain reaction: Avain Diseases v. 43, p. 106–115.

Leibovitz, L., 1971, Duck plague, *in* Davis, J. W., and others, eds., Infectious and parasitic diseases of wild birds: Ames, Iowa, Iowa State University Press, p. 22–33.

Wobeser, G.A., 1997, Duck plague, *in* Diseases of wild waterfowl (2nd ed): New York, N.Y., Plenum Press, p. 15–27.

Chapter 17
Inclusion Body Disease of Cranes

Synonym

Crane herpes

Cause

In March 1978, a previously unidentified herpesvirus was isolated at the National Wildlife Health Center (NWHC) from a die-off of captive cranes housed at the International Crane Foundation (ICF) in Baraboo, Wisconsin. Serological testing of this virus against other previously isolated avian herpesviruses does not result in cross-reactions, thereby supporting this agent's status as a distinctly new virus. The NWHC assigned the descriptive name, "inclusion body disease of cranes" (IBDC) to this disease when reporting the outbreak in the scientific literature, because the disease is characterized by microscopic inclusions in cell nuclei throughout the liver and spleen.

Very little is known about how this disease is transmitted. As with duck plague and avian cholera, outbreaks are thought to be initiated by disease carriers within a population of birds. The disease likely spreads by direct contact between infected birds and other susceptible birds and by contact with a virus-contaminated environment. Findings of antibody in sera of cranes bled nearly 3 years before the deaths at ICF indicates that the IBDC virus can be maintained in a captive crane population for at least 2 years and 8 months without causing mortality. The IBDC virus has been isolated from the cloaca of antibody-positive cranes, which indicates the potential for fecal shedding of the virus.

Species Affected

Spontaneous infections have developed in several species of captive cranes whose ages ranged from immature to adult (Fig. 17.1). Laboratory-induced infections and death occurred in adult cranes and in white Pekin ducklings between 3–17-days old, but not in 64-day-old Muscovy ducks. Adult coot were also susceptible, but white leghorn chicks were not (Fig. 17.2). These findings demonstrate that at least several species of cranes may become infected by this virus (virus replication develops in the bird following exposure), but the occurrence of illness and death is highly variable among different crane species. Too little is known about IBDC to assess other species' susceptibility to it based solely on the experimental infection of ducklings and adult coot. However, those findings need to be considered as a potential for this disease to involve more species than cranes. Further studies are needed to determine the true significance of IBDC as a threat to waterbirds.

Distribution

Herpesviruses have been associated with captive crane die-offs in several countries. Die-offs have occurred in Austria (1973), the United States (1978), France (1982), China (1982), the Commonwealth of Independent States [formerly the Soviet Union (1985)], and Japan (1992).

The relation between the herpesviruses from these die-offs has not been determined; however, the lesions and general pathological findings are similar. Serologic data indicates that captive cranes in the Commonwealth of Inde-

Cranes	Response	
	Mortality	Antibody
Stanley	●	○
Sandhill	●	●
Manchurian	●	●
Hooded	●	●
Sarus	○	●
Common	○	●
White-naped	○	○
Demoiselle	○	●
Brolga	○	○
East African crowned	○	●

Positive response ●
Negative response ○

Figure 17.1 Results of natural exposure to IBDC at the International Crane Foundation, Baraboo, Wisconsin.

pendent States and Japan have been exposed to the IBDC virus or to a very closely related herpesvirus.

Since the ICF die-off, many zoological collections have submitted crane sera for testing by the NWHC. Nine collections in the United States contained cranes that were found to have been exposed to the virus because they tested positive for antibodies to it. Testing of endangered species of cranes that were imported into the United States detected four additional exposed cranes. All of the antibody positive cranes came from Asia. Serological testing by the NWHC has found the antibody to the IBDC virus in 11.3 percent of 452 samples from 14 species of captive cranes in the United States. Results from other laboratories are not available; however, it is known that some antibody-positive cranes have been detected in United States zoological collections in addition to samples tested by the NWHC.

There is no evidence that wild North American crane populations have been exposed to IBDC. None of 95 sandhill crane sera collected in Wisconsin and Indiana during 1976 and 1977 had antibody to this virus. Additional testing would provide more information about the status of IBDC in wild cranes.

Seasonality

There have not been enough known outbreaks of IBDC to indicate whether or not the disease has seasonal trends. The outbreak of IBDC in Wisconsin happened in March. The other herpesvirus-associated die-offs in Austria, the Commonwealth of Independent States, and Japan happened in December. There is not enough information currently available to determine the season of the die-off in China.

Field Signs

During the ICF die-off, signs such as lethargy and loss of appetite persisted for 48 hours, with occasionally bloody diarrhea just before death. Critically ill cranes often died when they were handled.

Gross Lesions

Cranes that died from IBDC at the ICF had swollen livers and spleens. These organs contained many pinpoint-to-pinhead-size lesions that appeared as yellow-white spots throughout the tissue (Fig. 17.3). Other notable gross lesions included hemorrhages in the thymus gland and intestines. The acute nature of the disease was evident by abundant subcutaneous fat in the carcasses that were examined.

Diagnosis

A presumptive diagnosis can be made on the basis of gross lesions in the liver and spleen (Fig. 17.3). However, laboratory confirmation of this diagnosis is essential and it requires virus isolation from affected tissues. Submit whole carcasses to a disease diagnostic laboratory (see Chapter 3, Specimen Shipment). When this is not possible, remove the liver and spleen (see Chapter 2, Specimen Collection and Preservation), place them in separate plastic bags, and ship them frozen. Because this disease causes characteristic intranuclear inclusion bodies in the liver and spleen, it is also useful to place a piece of the liver and spleen in 10 percent buffered formalin when whole carcasses cannot be submitted. Care must be taken not to contaminate tissue samples being taken for virus isolation when taking a portion of these tissues for formalin fixation.

Control

Any outbreak of IBDC in North America should be considered a serious event requiring the immediate involvement of disease control specialists; destroying the infected flock and decontaminating the site of the outbreak currently are the only means of controlling the disease. This extreme response is complicated because endangered species of cranes may be involved and it may be difficult to sacrifice them for the benefit of other species. Nevertheless, failure to take aggressive action could result in IBDC being established as a significant cause of mortality in free-living North American cranes, jeopardize captive breeding programs for endangered

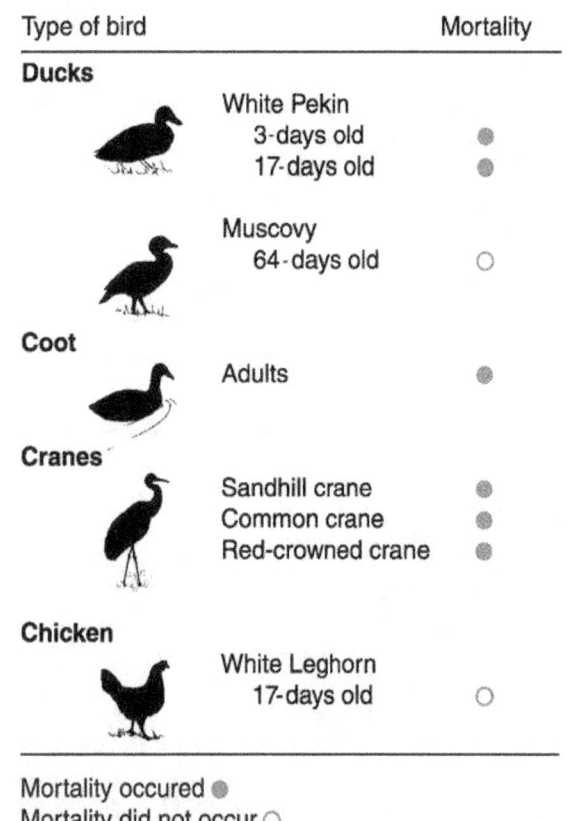

Type of bird		Mortality
Ducks		
	White Pekin 3-days old	●
	17-days old	●
	Muscovy 64-days old	○
Coot		
	Adults	●
Cranes		
	Sandhill crane	●
	Common crane	●
	Red-crowned crane	●
Chicken		
	White Leghorn 17-days old	○

Mortality occured ●
Mortality did not occur ○

Figure 17.2 Known susceptibility of avian species to experimental infection of IBDC.

species of cranes, and result in this disease becoming a serious mortality factor among zoological collections.

When captive infected flocks cannot be destroyed, it is important to make every effort to permanently isolate the survivors from other birds. Birds that survive infection can become carriers of the virus and infect other birds by intermittently discharging virus into the environment. Care must also be taken to prevent spread of the virus to susceptible birds by contact with potentially contaminated materials such as litter, water, feed, and feces from the confinement area. Clothes and body surfaces of personnel who were in contact with diseased birds are other potential sources of contamination.

There is no evidence that the IBDC virus can be transmitted through the egg. However, until more is known about this disease, eggs from birds surviving infection should be disinfected and hatched elsewhere. Young from these eggs should be reared at a facility free of IBDC, tested, and found free of exposure to IBDC before they are allowed to have contact with other birds.

Infection with the IBDC virus elicits an antibody response that persists for several years. This is a useful indicator of exposure to this virus. All captive cranes that are being transferred to other facilities or released into the wild should be tested for exposure to the IBDC virus. Birds found to have antibodies to IBDC should be considered potential carriers of this virus and either be destroyed or confined under the conditions specified above.

Good husbandry practices are important for reducing the potential for transmitting IBDC and for minimizing conditions favorable to virus shedding. Crowding, inclement weather, interspecies interactions, and poor sanitation were all possible contributing factors to the die-off at the ICF. IBDC has not reappeared at the ICF since corrective actions

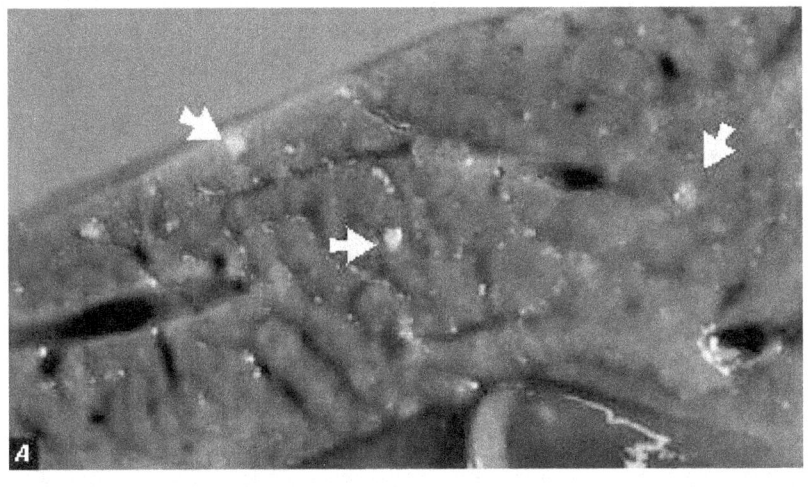
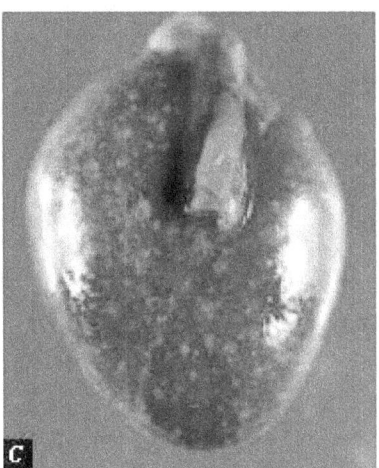
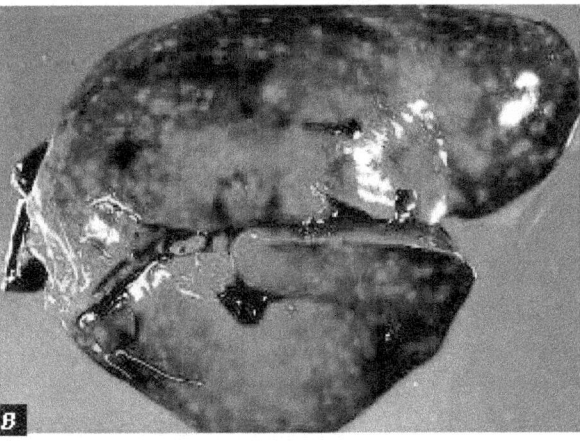
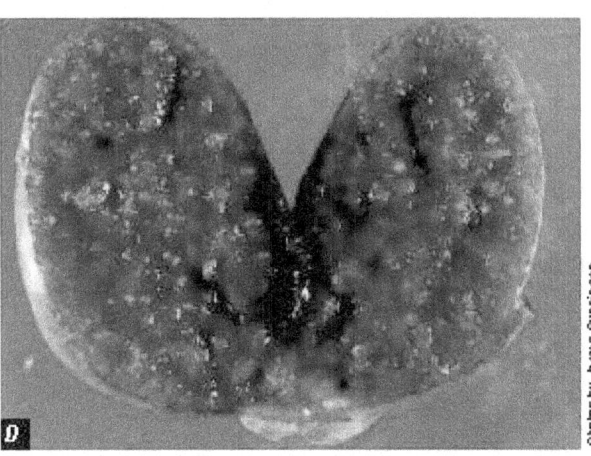

Figure 17.3 Gross lesions of IBDC: *(A)* small, yellow-white spots throughout the cut surface of the liver; *(B)* abundance of spots create mottled appearance of the liver surface; *(C)* external surface of the spleen; *(D)* cut surface of the spleen.

were taken, which include isolating the survivors of the die-off and initiating and maintaining an aggressive flock health-surveillance program.

Human Health Considerations
None known.

Douglas E. Docherty

Supplementary Reading

Docherty, D.E., and Henning, D.J., 1980, The isolation of a herpesvirus from captive cranes with an inclusion body disease: Avian Diseases, v. 24, p. 278–283.

Docherty, D.E., and Romaine, R.I., 1983, Inclusion body disease of cranes: a serological follow-up to the 1978 die-off: Avian Diseases, v. 27, p. 830–835.

Schuh, J.C.L., and Yuill, T.M., 1985, Persistence of inclusion body disease of cranes virus: Journal of Wildlife Disease, v. 21, p. 111–119.

Schuh, J.C.L., Sileo, L., Siegfried, L.M., and Yuill, T.M., 1986, Inclusion body disease of cranes: Comparison of pathologic findings in cranes with acquired versus experimentally induced disease: Journal of American Veterinary Medical Association, v. 189, p. 993–996.

Chapter 18
Miscellaneous Herpesviruses of Birds

Synonyms
Inclusion body disease of falcons, owl herpesvirus, pigeon herpes encephalomyelitis virus, psittacine herpesvirus

Cause
Herpesviruses other than duck plague and inclusion body disease of cranes (see Chapters 16 and 17 in this Section) have been isolated from many groups of wild birds. The diseases that these viruses cause have been described, but their comparative taxonomy and host ranges require additional study. All of these DNA viruses are classified in the family Herpesviridae, but they belong to various taxonomic subfamilies. The mechanisms for transmitting avian herpesviruses appear to be direct bird-to-bird contact and exposure to a virus-contaminated environment. The virus is transmitted to raptors and owls when they feed on infected prey that serve as a source of virus exposure. The development of disease carriers among birds that survive infection is typical of herpesvirus. Stress induced by many different factors is often associated with the onset of virus shedding by carrier birds resulting in the occurrence and spread of clinical disease.

Species Affected
Herpesviruses infect a wide variety of avian species (Fig. 18.1). Many virus strains appear to be group-specific in the bird species they infect and sometimes only infect a limited range of species within a group. A few of these viruses infect a wide species range. For example, although duck plague only affects ducks, geese, and swans, it affects most species within this taxonomic grouping (see Chapter 16). However, inclusion body disease of cranes has been shown under experimental conditions to infect birds of several families (see Chapter 17). Viruses included in the falcon-owl-pigeon complex resulted from experiments to cross-infect birds in these different groups. Herpesviruses as a group have been isolated from almost every animal species in which they have been sought and the viruses also cause disease in humans. In nature, the ability of these viruses to transmit to new hosts is governed by species behavior and host susceptibility to specific types of herpesviruses.

Distribution
To date, avian herpesviruses have been reported from North America, Europe, the Middle East (Iraq), Asia, Russia, Africa, and Australia and they are probably distributed worldwide (Table 18.1). Knowledge of their distribution in wild bird populations is limited to occasional isolated disease events in the wild, isolation of the viruses in association with other disease events, and from surveys of healthy birds. Unfortunately, there are few followup laboratory or field studies to expand information on those viruses that have been isolated. Most of the information on avian herpesvirus comes from disease events that affect or are found in captive flocks. The presence of this group of viruses in wild bird populations is probably more extensive than current data would indicate.

Seasonality
Little is known about the seasonality of disease caused by avian herpesviruses. Late spring appears to be the peak season for duck plague outbreaks (see Chapter 16), but less information about other herpesvirus infections of wild birds is available. The ability of this virus group to establish latent or persistently infected birds reduces the requirement for continual virus transmission to survive in an animal population (see Chapter 16, Duck Plague, and Chapter 17, Inclusion Body Disease of Cranes). Breeding season probably provides the best time of the year for bird-to-bird virus transmission in solitary species. Transmission of herpesviruses via the egg has been shown for some species, but more research is required to determine the importance of egg transmission for virus perpetuation. Seasonality probably plays a more important role for virus transmission in and among bird species that assemble for migration between summer breeding and wintering grounds.

Field Signs
The general signs of disease include depression of normal activity and sudden mortality in a group of birds. Respiratory distress may also be seen. Captive pigeons may show pronounced neurological signs such as extremity paralysis, head-shaking, and twisting of the neck.

Gross Lesions
Birds dying from infection with this group of viruses can have tumors (chicken and pigeon), hemorrhagic lesions (chicken, pheasants, ducks, cranes, peafowl, and guinea fowl), or, more commonly, hepatitis, and disseminated focal necrosis or visual areas of localized tissue death that appear as spots within the normal tissue in the liver, spleen, (Fig. 18.2) and bone marrow along with occasional intestinal necrosis. This broad array of lesions complements and extends those seen for duck plague and inclusion body disease of cranes (Chapters 16 and 17).

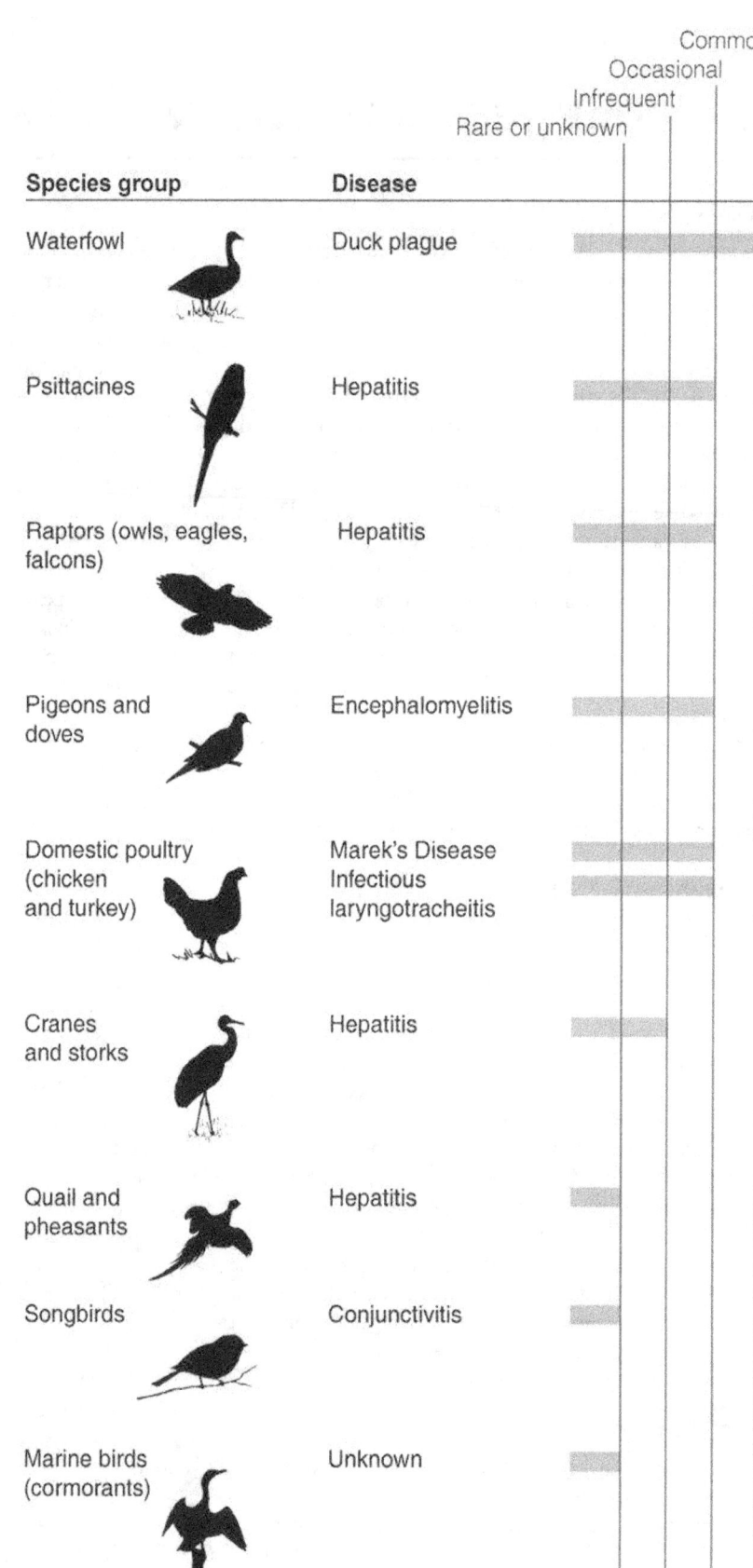

Figure 18.1 Relative frequency of disease from herpesvirus infections in birds of North America.

Table 18.1 Geographical distribution of avian herpesvirus infections.

	Continents				Other
	North America	Europe	Africa	Australia	
Raptors					
Booted eagle		●			
Bald and golden eagles	●				
Common buzzard (Old World)		●			
Falcons					
Prairie	●				
Red-headed	●				
Peregrine	●				
Gyrfalcon	●				
Kestrels	●				
Owls					
Eagle owl		●			
Long-eared owl		●			
Great horned owl	●				
Snowy owl	●	●			
Pigeon	●	●	●	●	Egypt
Ringed turtle dove	●				
Storks		●			
Cranes	●	●			China Japan Russia
Wild turkey	●				
Psittacines (several species)	●	●			Japan
Bobwhite quail		●			
Waterfowl (non-duck plague)	●		●	●	
Black-footed penguin	●				
Passeriforms					
Exotic finches		●			
Weavers		●			
Finches, including canary		●			
Cormorants				●	

Gallinaceous birds such as chicken, pheasants, peafowl, and guinea fowl raised in captivity have also been infected.

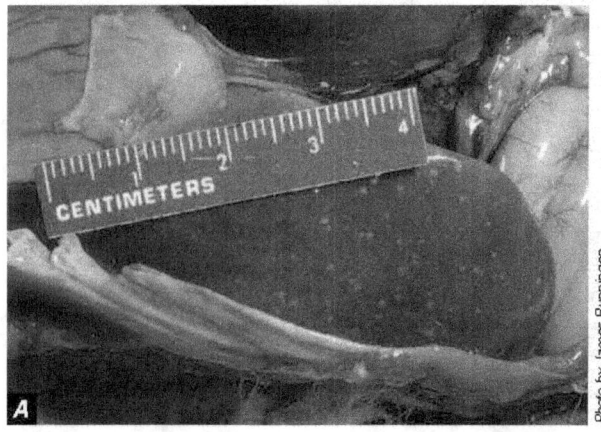 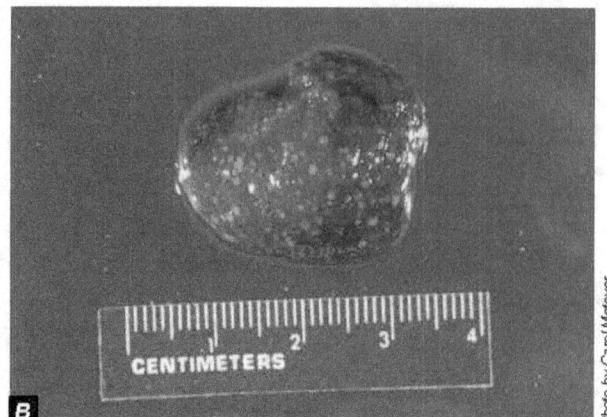

Figure 18.2 Herpesviruses can produce areas of tissue necrosis, appearing as white spots, such as in this peregrine falcon liver **(A)** and this great horned owl spleen **(B)**.

Diagnosis

The primary methods for diagnosing herpesvirus as a cause of disease are virus isolation from infected tissues and finding, during microscopic examination of infected cells, the characteristic accumulations of cellular debris referred to as intranuclear (Cowdry type A) inclusion bodies (Fig. 18.3). These lesions are most often seen in the liver, spleen, and bone marrow. The virus can usually be isolated in chicken or duck embryo fibroblast tissue culture or in embryonated chicken eggs.

Control

Control actions warranted for outbreaks of herpesvirus infections are dependent upon the type of herpesvirus infection and the prevalence of disease in the species or populations involved (see Chapter 4, Disease Control Operations). Euthanasia of infected flocks should be considered for exotic viruses and viruses that are likely to cause high mortality within the population at risk. When depopulation is not appropriate because of the ubiquitous nature of the disease, or for other reasons, disease-control steps should still be taken. Sick birds, as well as those in the preclinical stages of illness, will be shedding virus into their environment; therefore, birds that are suspected of being infected should be segregated from other birds and quarantined for 30 days. Any birds noticeably ill should be isolated from the rest of the contact group. A high level of sanitation should be imposed and maintained for the full quarantine period where birds are housed. Decontamination procedures are needed to minimize disease transmission via virus that is shed in feces and by other means.

Dead birds should be removed immediately and submitted for disease evaluations. Standard bagging and decontamination procedures should be used to avoid off-site transfer of the virus. Personnel should follow good hygiene methods and should not have any contact with other birds for 7 days to prevent mechanically carrying contamination from the quarantine site.

Surviving birds should be tested for virus and virus specific antibody. All birds with antibody are probably virus carriers and they pose a risk as a source for future virus infection. Future use of these birds should take this into consideration. This is especially important when endangered species are involved and for wildlife rehabilitation activities because survivors of herpesvirus infections are potential sources for the initiation of new outbreaks and further spread of the disease.

Human Health Considerations

Avian herpesviruses have not been associated with any disease of humans.

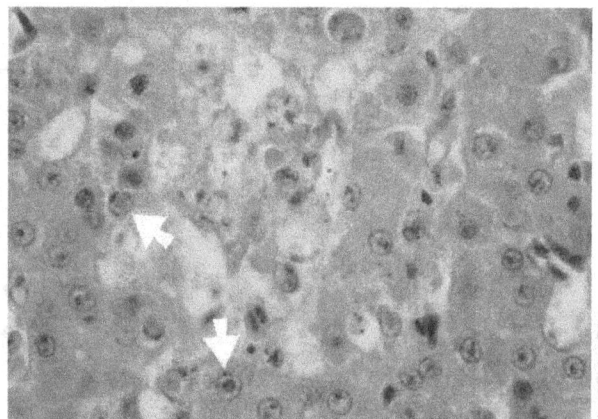

Figure 18.3 Inclusion bodies (arrows) in liver cell nuclei of a great horned owl that died of herpesvirus infection.

Wallace Hansen

Supplementary Reading

Burtscher, H., and Sibalin, M., 1975, *Herpesvirus strigis*: Host spectrum and distribution in infected owls: Journal of Wildlife Diseases, v. 11, p. 164–169.

Graham, D.L., Mare, C.J., Ward, F.P., and Peckham, M.C., 1975, Inclusion body disease (herpesvirus infection) of falcons (IBDF): Journal of Wildlife Diseases, v. 11, p. 83–91.

Kaleta, E.F., 1990, Chapter 22; Herpesviruses of free-living and pet birds, *in* A laboratory manual for the isolation and identification of avian pathogens, American Association of Avian Pathologists. p. 97–102.

Chapter 19
Avian Pox

Synonyms

Fowl pox, avian diphtheria, contagious epithelioma, and poxvirus infection

Cause

Avian pox is the common name for a mild-to-severe, slow-developing disease of birds that is caused by a large virus belonging to the avipoxvirus group, a subgroup of poxviruses. This group contains several similar virus strains; some strains have the ability to infect several groups or species of birds but others appear to be species-specific. Mosquitoes are common mechanical vectors or transmitters of this disease. Avian pox is transmitted when a mosquito feeds on an infected bird that has viremia or pox virus circulating in its blood, or when a mosquito feeds on virus-laden secretions seeping from a pox lesion and then feeds on another bird that is susceptible to that strain of virus. Contact with surfaces or exposure to air-borne particles contaminated with poxvirus can also result in infections when virus enters the body through abraded skin or the conjunctiva or the mucous membrane lining that covers the front part of the eyeball and inner surfaces of the eyelids of the eye.

Species Affected

The highly visible, wart-like lesions associated with the featherless areas of birds have facilitated recognition of avian pox since ancient times. Approximately 60 free-living bird species representing about 20 families have been reported with avian pox. However, the frequency of reports of this disease varies greatly among different species (Fig. 19.1). Avian pox has rarely been reported in wild waterfowl, and all North American cases have been relatively recent (Table. 19.1). The first case was in a free-living green-winged teal in Alaska. Single occurrences have also been documented in a Canada goose in Ontario, Canada, a mallard duck in Wisconsin, a feral mute swan cygnet in New York, and a tundra swan in Maryland. Three cases in American goldeneye have been reported in Saskatchewan, Canada, and New York. Avian pox also appeared in Wisconsin among captive-reared trumpeter swans that were part of a reintroduction program. Zoological garden cases include common scoter in the Philadelphia Zoo and a Hawaiian goose in the Honolulu Zoo.

Avian pox in a bald eagle was first diagnosed in 1979 in Alaska and it was a lethal infection. Since then, additional bald eagles in Alaska and at other locations have been diagnosed with this disease (Fig. 19.2). The severity of infection resulted in several of these cases being lethal. Poxvirus infections have been reported in other raptors, most recently

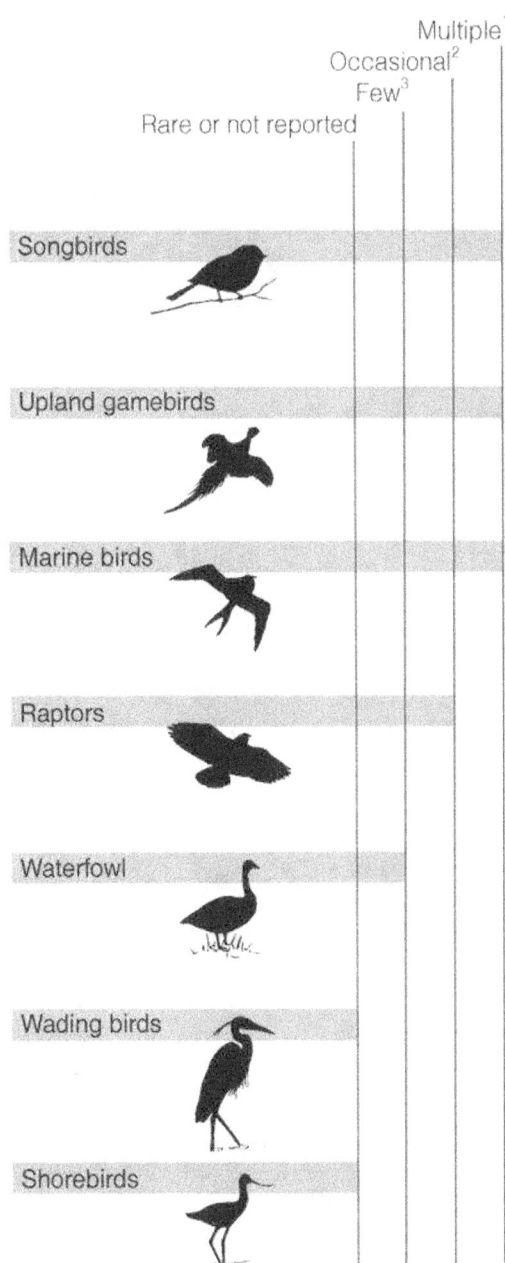

[1] Reports often involve a number of birds in a single event.
[2] Reports tend to involve individuals rather than groups of birds.
[3] Small number of reports, generally involving individual birds.

Figure 19.1 *Reported avian pox occurrence in wild birds in North America.*

Table 19.1 Waterfowl in North America reported to have avian pox.

Species	Locations	Year of first report in species
Harlequin duck	Alaska	1994
Blue-winged teal	Wisconsin	1991
Wood duck	Wisconsin	1991
Redhead duck	Wisconsin	1991
Trumpeter swan	Wisconsin	1989
Common goldeneye	New York	1994
	Saskatchewan	1981
Tundra swan	Maryland	1978
Green-winged teal	Alaska	1978
Mallard	Wisconsin	1978
Canada goose	Ontario	1975
Common scoter	Pennsylvania	1967
Mute swan	New York	1964

Figure 19.2 Number of bald eagles with cutaneous pox by State, 1979–97. (From National Wildlife Health Center records.)

in the eastern screech owl and barred owl in Florida (Table 19.2).

On Midway Atoll, large numbers of colonial nesting birds, such as the Laysan albatross, have become infected with avian pox. Red-tailed tropicbirds on Midway Atoll previously had been affected by avian pox. The shift in predominant species infected relates to the dramatic shift in population densities for the two species over time (1963–1978). Mourning dove, finches, and other perching birds using backyard feeders (Fig. 19.3) frequently have been reported to have been struck by avian pox epizootics. Avian pox is suspected as a factor in the decline of forest bird populations in Hawaii and northern bobwhite quail in the southeastern United States, where it is also an important disease of wild turkey.

Distribution

Avian pox occurs worldwide, but little is known about its prevalence in wild bird populations. The increased frequency of reported cases of this highly visible disease and the involvement of new bird species during recent years suggests that avian pox is an emerging viral disease. Birds can become disease carriers and spread avian pox among local populations, such as between birdfeeding stations, and along migratory routes used by various bird species. Mosquitoes that feed on birds play the most important role for both disease transmission and long term disease maintenance. However, contamination of perches and other surfaces used by captive birds can perpetuate disease in captivity. Pox outbreaks are commonly reported at aviaries, rehabilitation centers, and other places where confinement provides close contact among birds. The disease can spread rapidly when avian pox is introduced into such facilities. Species that would not ordinarily have contact with avian pox virus in the wild often become infected in captivity if the strain of virus present is capable of infecting a broad spectrum of species. Common murres rescued from an oil spill in California developed poxvirus lesions while they were in a rehabilitation center. Endangered avian species also have been infected during captive rearing.

Seasonality

Although wild birds can be infected by pox virus year-round (Fig. 19.4), disease outbreaks have been associated with the environmental conditions, the emergence of vector populations, and the habits of the species affected. Environmental factors such as temperature, humidity, moisture, and protective cover all play a role in the occurrence of this disease by affecting virus survival outside of the bird host. Avian pox virus can withstand considerable dryness, thereby remaining infectious on surfaces or dust particles. Mosquitoes that feed on birds are the most consistent and efficient transmitters of this disease. Mosquito populations are controlled by breeding habitat and annual moisture.

The time of appearance and magnitude of vector populations varies from year to year, depending on annual weather conditions. This influences the appearance and severity of the disease in any given year. Only limited studies have been carried out to assess the relations between avian pox and insect vector populations. Studies on the Island of Hawaii disclose a close relation between the prevalence of poxvirus infections in forest birds and seasonal mosquito cycles. The lowest prevalence of pox virus infection in California quail in Oregon was reported in the dry summer months and the highest was reported during the wetter fall and winter months. In Florida, reports of avian pox in wild turkey correspond to the late summer and early fall mosquito season. On Sand Island of the Midway Atoll, avian pox was first reported in September 1963 in the nestlings of the red-tailed tropicbird. In March and April of the late 1970s, this disease was found in nestling Laysan albatross on Sand Island. This is an example of disease seasonality influenced by dramatic shifts in predominant species populations.

Birdfeeding stations have been the source of numerous poxvirus outbreaks in the continental United States (Fig. 19.3). Contact transmission of the virus through infected surfaces and close association of birds using those feeders is the likely means of transmission during cooler periods of the year when mosquitoes are not a factor, and birdfeeders provide additional sources of infection when mosquitoes are present.

Field Signs

Birds with wart-like nodules on one or more of the featherless areas of the body, including the feet, legs, base of the beak, and eye margin should be considered suspect cases of avian pox (Fig. 19.5). The birds may appear weak and emaciated if the lesions are extensive enough to interfere with their feeding. Some birds may show signs of labored breathing if their air passages are partially blocked. Although the course of this disease can be prolonged, birds with extensive lesions are known to completely recover if they are able to feed.

Gross Lesions

Avian pox has two disease forms. The most common form is cutaneous and it consists of warty nodules that develop on the featherless parts of the bird. This form of the disease is usually self-limiting; the lesions regress and leave minor scars. However, these nodules can become enlarged and clustered, thus causing sight and breathing impairment and feeding difficulty (Figs. 19.6A and B). Secondary bacterial and other infections are common with this form of the disease, and these infections can contribute to bird mortality. In some birds, feeding habits result in the large warty nodules becoming abraded and then infected by bacterial and fungal infections (Figs. 19.6C and D).

The internal form of disease is referred to as wet pox and it is primarily a problem of young chickens and turkeys. This diphtheritic form appears as moist, necrotic lesions on the mucus membranes of the mouth and upper digestive and respiratory tracts (Fig. 19.7), and it has occasionally been reported in wild birds (Fig. 19.8). This form of avian pox probably occurs more frequently in wild birds than it is reported because it is less observable than the cutaneous form. Also, the more severe consequences of wet pox undoubtedly causes greater morbidity and mortality, thereby leading to removal of infected birds by predators and scavengers.

Table 19.2 Birds of prey from North America reported to have contracted avian pox.

Species	Locations	Year of first report in species
Barred owl	Florida	1995
Bald eagle	Maine	1995
	Ohio	1995
	Rhode Island	1993
	Michigan	1992
	Minnesota	1989
	California	1987
	Nebraska	1987
	Maryland	1986
	Massachusetts	1986
	South Dakota	1986
	Wisconsin	1986
	Pennsylvania	1985
	Arkansas	1984
	New York	1983
	Florida	1982
	Virginia	1981
	Washington	1981
	Alaska	1978
Eastern screech owl	Florida	1994
Peregrine falcon	New York	1994
Ferruginous hawk	Texas	1993
Golden eagle	Missouri	1989
	Kansas	1986
	California	1976
	British Columbia	1970
Red-tailed hawk	Nebraska	1988
	Wisconsin	1985
	Washington	1981
	Missouri	1970
Rough-legged hawk	North Dakota	1971

Diagnosis

A presumptive diagnosis of avian pox can be made from the gross appearance of the wart-like growths that appear on body surfaces. However, these observations must be confirmed by examining lesions microscopically for character-

Figure 19.3 Number of avian pox outbreaks involving passerines at birdfeeding stations by State, 1975–79. (National Wildlife Health Center Database.)

Species group	Season			
	Spring	Summer	Fall	Winter
Marine birds	●	○	●	
Raptors	○	○	○	○
Upland gamebirds		○	●	●
Songbirds	○	○		●
Hawaiian forest birds		○	●	●

High prevalence of infection ●
Low prevalence of infection ○

Figure 19.4 Seasonal avian pox outbreaks in wild birds.

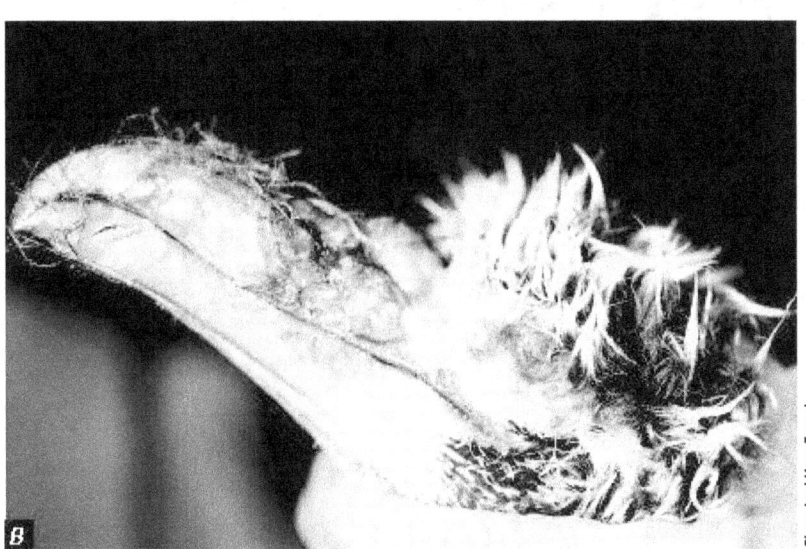
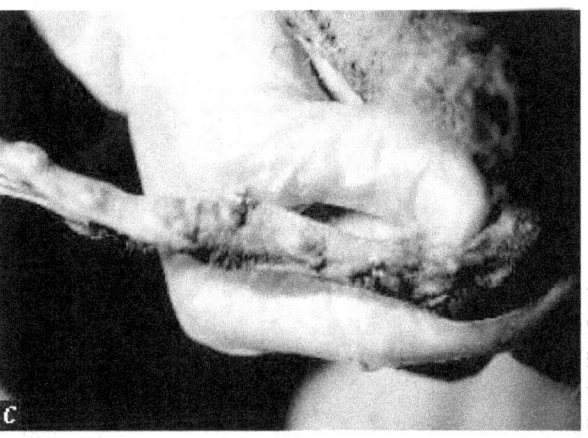

Figure 19.5 (A) Avian pox lesions typically are found on featherless parts of the body. This Laysan albatross chick has small pox nodules on the face and eyelid. (B) As the disease progresses, these lesions become more extensive. (C) Lesions also are commonly seen on the legs and (D) feet.

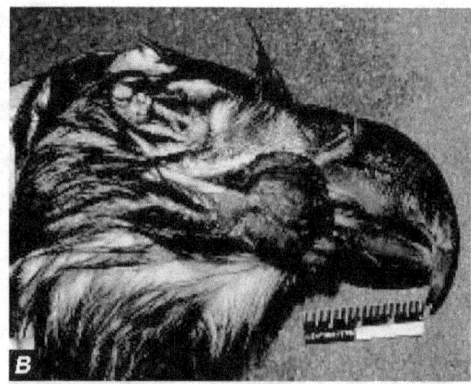

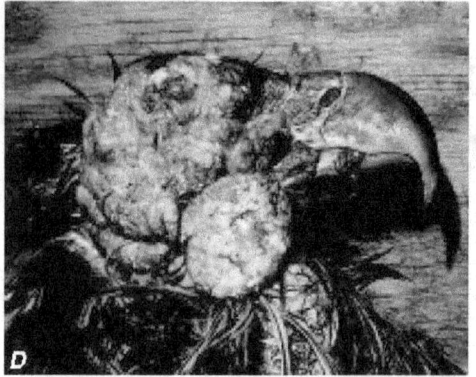

Figure 19.6 Pox lesions can be so extensive that they impair breathing, sight, and feeding as seen in these bald eagles: **(A)** extensive infection of both sides of the face, **(B)** obstruction to feeding due to the size and location of these lesions at the base of the bill, and **(C)** obstruction of sight due to complete occlusion of the eye. **(D)** Massive facial lesions often become abraded and subject to secondary infections.

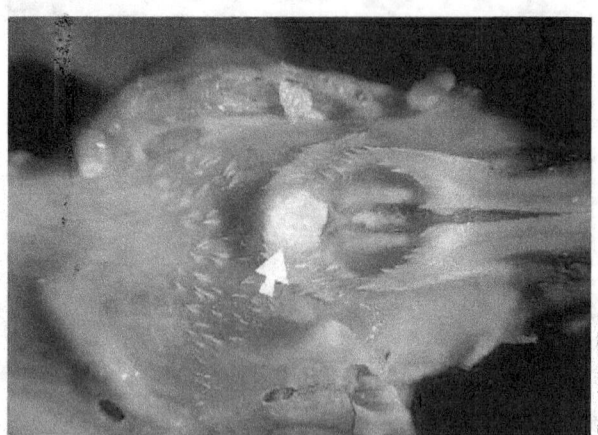

Figure 19.7 Diphtheritic form of avian pox (arrow) in a Laysan albatross at Midway Atoll.

Species group	Species
Marine birds	White-tailed tropicbird Laysan albatross Common murre
Upland gamebirds	Morning dove Ruffed grouse Northern bobwhite quail
Songbirds	Bullfinch House finch

Figure 19.8 Wild bird species in which wet pox has been reported.

istic cellular inclusion bodies. Avian pox is confirmed by virus isolation and serological identification. Submit the whole bird or the affected body part (for example, the feet or head) to a disease diagnostic laboratory that has virus isolation capabilities. Immediately freeze samples that must be held for more than 24 hours before shipment. Virus isolation can be attempted from a live bird by collecting samples from the affected area. However, consult with the diagnostic laboratory staff before collecting samples.

Control

The fundamental principle for controlling avian pox is to interrupt virus transmission. The difficulty in applying control procedures is related to the type of transmission taking place, the mobility of the infected birds, and the size of the affected area. The more confined a population at risk, the more effective the control procedures will be. Therefore, prevention is the first method for controlling this disease. Vector control (primarily mosquitoes) in and around the disease area should be considered first. Identifying and eliminating vector breeding and resting sites together with controlling adult mosquito populations are most desirable. Removing heavily infected animals is also helpful because it diminishes the source of virus for vector populations. This also reduces the opportunity for contact transmission between infected and noninfected birds.

Special vigilance of captive birds is needed, especially when threatened and endangered species are involved. Because poxvirus is resistant to drying, disease transmission by contaminated dust, food, perches, cages, and clothing can pose a continuing source of problems. Therefore, these items need to be decontaminated with disinfectant, such as a 5 percent bleach solution, before they are disposed of or reused.

The poultry industry uses modified live vaccines to prevent avian pox, but their safety and effectiveness in wild birds have not been determined. In addition, strain differences in the virus, host response to those different strains, and logistical problems of a vaccination program further complicate using vaccines for wild birds. The greatest potential use of vaccination is for protecting captive-breeding populations of threatened and endangered species and for providing immunity in birds that are to be released into areas where pox is a problem.

Human Health Considerations

Avian poxvirus is part of a larger family of poxviruses that includes the human disease known as variola or smallpox. However, there is no evidence that avipoxviruses can infect humans.

Wallace Hansen

Supplementary Reading

Karstad, L., 1971, Pox, *in* Davis, J.W. and others, eds., Infectious and parasitic diseases of wild birds: Ames, Iowa, Iowa State University Press, p. 34–41.

Kirmse, P., 1967, Pox in wild birds: an annotated bibliography: Journal of Wildlife Diseases, v. 3, p. 14–20.

Chapter 20
Eastern Equine Encephalomyelitis

Synonyms

EEE, eastern encephalitis, EE, eastern sleeping sickness of horses

Cause

Eastern equine encephalitis (EEE) is caused by infection with an RNA virus classified in the family Togaviridae. The virus is also referred to as an "arbovirus" because virus replication takes place within mosquitoes that then transmit the disease agent to vertebrate hosts such as birds and mammals, including humans. The term arbovirus is shortened nomenclature for arthropod (insect) borne (transmitted) viruses. *Culiseta melanura* is the most important mosquito vector; it silently (no disease) transmits and maintains the virus among birds. However, several other mosquito species can transmit this virus, including the introduced Asian tiger mosquito. New hosts become infected when they enter this endemic natural cycle and are fed upon by an infected mosquito. Therefore, the presence of mosquito habitat, the feeding habits of different mosquito species, and the activity patterns of vertebrate hosts are among the important factors for disease transmission.

Distribution

This disease is primarily found in eastern North America especially along the Atlantic and Gulf Coasts, and the disease range extends into Central and South America. The causative virus has been isolated from eastern Canada to Argentina and Peru, and it is maintained in a mosquito-wild bird cycle as an endemic (enzootic) focus of infection in nature that is usually associated with freshwater marshes. Wild bird die-offs from EEE have been limited to captive-rearing situations. Die-offs have occurred in pheasants in coastal States from New Hampshire to Texas, where they have been raised, in chukar partridge and whooping cranes in Maryland, and in emus and ostriches in Louisiana, Georgia, Florida, and Texas.

Species Affected

EEE virus produces inapparent or subclinical infections in a wide range of wild birds (Fig. 20.1). However, EEE virus has caused mortality in glossy ibis and in several bird species that are exotic to the United States, including pigeon, house sparrow, pheasants, chukar partridge, white Peking ducklings, and emu. The infection rate in penned emus in the United States has reached 65 percent with a case mortality rate of 80 percent. In the past, extensive losses have

Figure 20.1 Relative frequency of EEE virus isolation or presence of antibodies in birds.

In September and November of 1984, EEE virus was associated with the deaths of 7 of 39 captive whooping cranes at the Patuxent Wildlife Research Center in Laurel, Maryland. Sandhill cranes coexisting with the whooping cranes did not become clinically ill or die.

Passerines (perching songbirds), some small rodents, and bats are highly susceptible to infection and they often die from experimental infections. Horses are highly susceptible and they often die from natural infections.

Seasonality

EEE is associated with the early summer appearance of *C. melanura* mosquito populations (Fig. 20.2). Nestling birds, such as passerines and other perching birds, are the amplification hosts for the virus, producing high concentrations of virus in their blood or viremia following mosquito infection. New populations of emerging mosquitoes become infected when they feed on the viremic birds. *C. melanura* and other species of infected mosquitoes can transmit the virus to other species of birds susceptible to disease (Fig. 20.3).

The summer-fall transmission cycle is followed by little virus transmission during the winter and spring months. The overwintering mechanism for virus survival is not known. Infected mosquitoes, other insects, cold-blooded vertebrate species, or low levels of virus transmission by mosquitoes are among current theories for virus cycle maintenance in milder climates. It is also believed that bird migration spreads the virus to higher latitudes in the spring.

Field Signs

Clinical signs do not develop in most native species of wild birds infected with EEE virus. Clinical signs for nonindigenous birds (including pheasants) include depression, tremors, paralysis of the legs, unnatural drowsiness, profuse diarrhea, voice changes, ataxia or loss of muscle coordination, and involuntary circular movements (Fig. 20.4). Some of the EEE-infected whooping cranes became lethargic and incoordinated or ataxic, with partial paralysis or paresis of the legs and neck 3–8 hours prior to death; other cranes did not develop clinical signs before they died.

Gross Lesions

Gross lesions in whooping cranes included fluid accumulation in the abdominal cavity or ascites, intestinal mucosal discoloration, fat depletion, enlarged liver or hepatomegaly, enlarged spleen or splenomegaly, and visceral gout (Fig. 20.5).

Diagnosis

Because of human health hazards, field personnel should not dissect birds suspected of having died from EEE. Whole carcasses should be submitted to diagnostic laboratories capable of safely handling such specimens. EEE can be diag-

Figure 20.2 *Seasonality of virus isolation from birds and mosquitoes.*

occurred in ring-necked pheasant being reared in captivity for sporting purposes, including one outbreak in a South Dakota pheasant farm. Large-scale mortalities in captive pheasants are perpetuated by bird-to-bird disease spread through pecking and cannibalism after EEE has been introduced by mosquitoes, usually of the genus *Culiseta*. Outbreaks have not been reported during recent years.

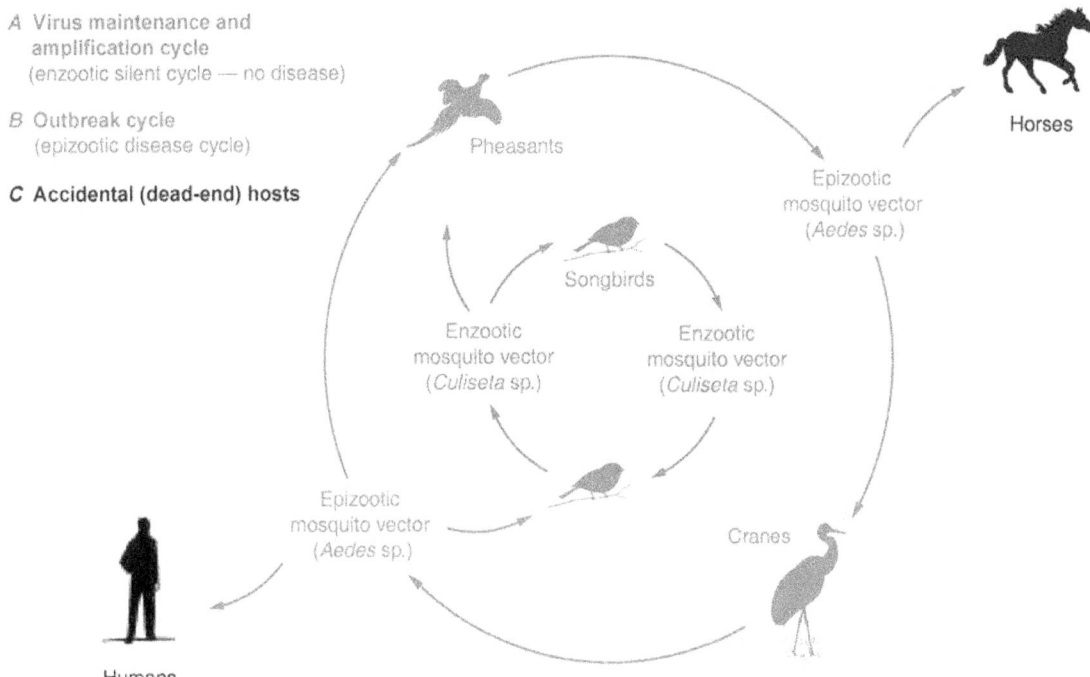

Figure 20.3 Transmission of eastern equine encephalomyelitis. **(A)** Virus circulates in songbird populations by being transmitted by mosquitoes. Those birds are susceptible to infections, but they do not become clinically ill or die. **(B)** The outbreak cycle is started either when an infected mosquito from the enzootic cycle feeds on highly susceptible birds such as pheasants or cranes, or when another species of mosquito, that primarily feeds on these same birds, becomes infected after feeding on songbirds in the enzootic cycle and transmits the virus. The epizootic cycle is maintained by the second mosquito species. **(C)** The broader host feeding range of the second mosquito results in exposure of horses and humans. No disease cycle is maintained between these species by mosquitoes.

Figure 20.4 A hen pheasant with EEE exhibiting neurologic signs.

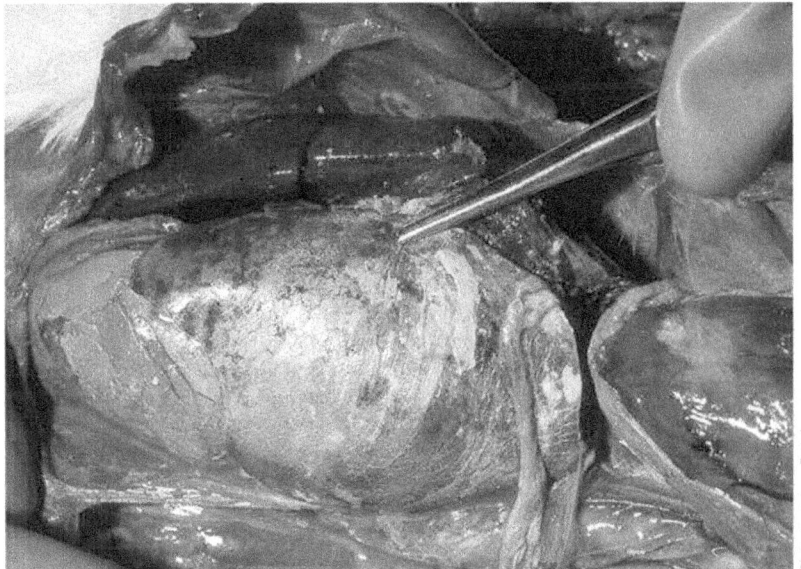

Figure 20.5 The white, grainy material on the liver of this whooping crane is evidence of visceral gout.

Eastern Equine Encephalomyelitis 173

nosed by virus isolation from infected whole blood or brain and other tissues from dead birds. Diagnosis of virus activity can be made from surviving birds because they will have virus neutralizing antibodies in their blood serum. The rise and fall in serum antibodies that occurs after virus exposure can be used to assess infection rates and the relative timing of exposure before antibody levels reach nondetectible levels. Most native birds do not suffer clinical infections.

Control

There are two approaches to protecting susceptible animals from infection from vector-borne diseases. The first approach includes separating mosquitoes from animals at risk. This requires eliminating mosquito breeding and resting sites in an endemic area or protecting animals from mosquito contact by maintaining them in an insect-proof enclosure. In the second approach, vaccination is used to render the animal immune. A killed-virus vaccine was used in captive whooping cranes to protect the rest of the breeding flock following the 1984 outbreak. Vaccination has also been used to protect whooping cranes released into an area where EEE is prevalent in mosquito populations. Field data suggest that immunity in those cranes is being boosted by natural infections after their release.

Human Health Considerations

Humans are susceptible to EEE and human cases typically arise after the disease has appeared in horses. EEE is a significant disease in humans with a case fatality rate of between 30–70 percent and it often causes severe permanent neurological disorders among survivors. Aerosol infection is possible but rare. Laboratory personnel have been infected with this virus. Human pre-exposure vaccination is recommended for people who may handle infected tissues.

Wallace Hansen and Douglas E. Docherty

Supplementary Reading

Dein, J.F., Carpenter, J.W., Clark, G.G., Montali, R.J., Crabbs, C.L., Tsai, T. F., and Docherty, D.E., 1986, Mortality of captive whooping cranes caused by eastern equine encephalitis virus: The Journal of American Veterinary Medical Association, v. 189, p. 1,006–1,010.

Gibbs, E.P.J., and Tsai, T.F. 1994, Eastern encephalitis, *in* Beran, G.W., and Steele, J.H., eds., Handbook of zoonoses (2d ed.), Section B: Viral: Boca Raton, Fla., CRC Press, p. 11–24.

Kissling, R.E., Chamberlain, R.W., Sikes, R.K., and Eidson, M.E., 1954, Studies on the North American arthropod-borne encephalitides III. eastern equine encephalitis in wild birds: American Journal of Hygiene, v. 60, p. 251–265.

Chapter 21

Newcastle Disease

Synonyms

ND, paramyxovirus-1, NDV, VVND, NVND

Newcastle Disease (ND) in domestic poultry is a focus for concern throughout much of the world's agricultural community because of severe economic losses that have occurred from illness, death, and reduced egg production following infection with pathogenic or disease causing strains. Prior to 1990, this disease had rarely been reported as a cause of mortality in the free-living native birds of the United States or Canada. Repeated large-scale losses of double-crested cormorants from ND in both countries has resulted in a need for enhanced awareness of ND as a disease of wild birds and, therefore, its inclusion within this Manual. Background information about ND in poultry is needed to provide a perspective for understanding the complexity of the disease agent, Newcastle disease virus (NDV). Some general information about ND in other avian species is also provided, but the primary focus for this chapter is the effect of NDV on double-crested cormorants.

Cause

Newcastle disease is caused by infection with an RNA virus within the avian paramyxovirus-1 group. NDV is highly contagious and there is great variation in the severity of disease caused by different strains of this virus. A classification system for the severity of disease has been established to guide disease control efforts in poultry because of the economic damage of ND.

The most virulent ND form causes an acute, lethal infection of chickens of all ages with mortality in affected flocks often reaching 100 percent. These strains produce hemorrhagic lesions of the digestive tract, thus resulting in the disease being referred to as viscerotropic or having an affinity for abdominal tissue, and velogenic or highly virulent Newcastle disease or VVND. This form of ND is rare in the United States, and it is primarily introduced when exotic species of birds are trafficked in the pet bird industry. Another acute, generally lethal infection of chickens of all ages affects respiratory and neurologic tissues and is referred to as neurotropic velogenic Newcastle disease or NVND. Morbidity or illness from NVND may affect 100 percent of a flock, but mortality is generally far less with extremes of 50 percent in adult birds and 90 percent in young chickens. The NVND form of ND was essentially eradicated from the United States in about 1970, but it has occasionally been reintroduced via pet birds and by other means. A less pathogenic form of ND causes neurologic signs, but usually only young birds die and, except for very young susceptible chicks, mortality is low. These strains are classified as mesogenic or moderately virulent. NDV strains that cause mild or inapparent respiratory infections in chickens are classified as lentogenic or low virulence. Lentogenic strains do not usually cause disease in adult chickens, but these forms can cause serious respiratory disease in young birds. Some strains of lentogenic NDV cause asymptomatic-enteric infections without visible disease (Table 21.1).

The virus classification standard applies to ND in poultry and the standard is not directly transferrable to wild birds. Experimental studies have demonstrated differences in bird response to the same strain of NDV. Thus, a highly pathogenic strain isolated from wild birds may be less hazardous for poultry and vice-versa. ND may be transmitted among birds by either inhalation of contaminated particulate matter or ingestion of contaminated material.

Species Affected

NDV is capable of infecting a wide variety of avian species. In addition to poultry, more than 230 species from more than one-half of the 50 orders of birds have been found to be susceptible to natural or experimental infections with avian paramyxoviruses. Experimental infections in mallard ducks exposed to large amounts of a highly virulent form of NDV for chickens disclosed that ducklings were more susceptible than adults, and that mortality of 6-day-old ducklings was higher than in 1-day-old and 3-day-old ducklings. Captive-reared gamebirds, such as pheasants and Hungarian partridge, have died of ND. However, large-scale illness and death from NDV in free-ranging wild birds has only occurred in double-crested cormorants in Canada and the United States. White pelicans, ring-billed gulls, and California gulls were also reported to have died from NDV in association with cormorant mortalities in Canada.

The 1990 epizootic of ND in Canada killed more than 10,000 birds, mostly double-crested cormorants. Mostly sub-adult cormorants died in these cormorant colonies. Losses in the United States have been primarily in nestlings and other young of the year. The total mortality attributed to ND during 1992 exceeded 20,000 birds. Mortality in Great Lakes cormorant colonies ranged from 2 to 30 percent, while that in Midwestern colonies was estimated to be 80 to 90 percent. In 1997, nesting failure of a cormorant colony at the Salton Sea in California was attributed to NDV. The total mortality in 1997 was about 2,000 cormorants. During the 1992 epizootic, a domestic turkey flock in the Midwestern United States was infected at the same time NDV occurred in cormorants near that poultry flock.

Distribution

Different strains of NDV exist as infections of domestic poultry and within other species of birds throughout much of the world. Highly pathogenic strains of NDV have spread throughout the world via three panzootics or global epizootics since ND first appeared in 1926. The first of these highly pathogenic strains appears to have arisen in Southeast Asia; it took more than 30 years to spread to chickens worldwide, and it was primarily spread through infected poultry, domestic birds, and products from these species. The virus responsible for the second panzootic involving poultry appears to have arisen in the Middle East in the late 1960s; it reached most countries by 1973, and it was associated with the importation and movement of caged psittacine species. The most recent panzootic also appears to have its origin in the Middle East, and it began in the late 1970s. This panzootic differs in that pigeons and doves kept by bird fanciers and raised for food are the primary species involved. This NDV spread worldwide primarily through contact between birds at pigeon races, bird shows, and through international trade in these species. It has spread to chickens in some countries. A current question is whether or not the ND outbreaks that have occurred in double-crested cormorants are the beginning of a fourth panzootic.

In North America, NDV has caused disease in double-crested cormorants from Quebec to the West Coast (Fig. 21.1). Most cormorant mortality has occurred in the Upper Midwest and the Canadian prairie provinces, although smaller outbreaks have occurred at Great Salt Lake, in southern Cali-

Table 21.1 Disease impacts on chickens resulting from exposure to different strains of Newcastle disease virus. [Pathotype refers to the severity of disease in susceptible, immunologically naive chickens. Velogenic is the most severe; lentogenic is the least severe.]

Pathotype	Disease impacts
Velogenic	
Viscerotrophic velogenic ND (VVND)	Acutely lethal, kills chickens of all ages, often with lesions in the digestive tract. Flock mortality approaches 100 percent.
Neurotrophic velogenic ND (NVND)	Acutely lethal, kills chickens of all ages, often with signs of neurological disease. Flock mortality approaches 50 percent in adults and 90 percent in young birds. Sharp decrease in egg production.
Mesogenic	Moderate infection rates as indicated by clinical signs. Mortality generally only in young birds, but for very young chicks the death rate is low. Sharp and persistent decrease in egg production by adults.
Lentogenic	Mild or inapparent respiratory infections occur. Disease seldom seen in adults, but serious illness (generally nonlethal) can occur in young chickens.
Asymptomatic lentogenic	Infects the intestine but causes no forms of visible disease in chickens of any age.

fornia, and on the Columbia River between Washington and Oregon. Cormorants, the closely related shag, and gannets, which are another species of marine bird that has close associations with cormorants, were believed to be an important source of NDV for the poultry outbreaks along the coast of Britain during the 1949–51 epizootic in that country.

Seasonality

All of the North American cormorant die-offs from ND have occurred in breeding colonies. Mortality has occurred during the months of March through September.

Field Signs

Clinical signs, observed only in sick juvenile double-crested cormorants, include torticollis or twisting of the head and neck, ataxia or lack of muscular coordination, tremors, paresis or incomplete paralysis including unilateral or bilateral weakness of the legs and wings, and clenched toes (Fig. 21.2). Paralysis of one wing is commonly observed in birds surviving NVD infection at the Salton Sea in southern California (Fig. 21.3).

Experimental inoculations in adult mallard ducks with a highly virulent form of NDV from chickens resulted in onset of clinical signs 2 days after inoculation. Initially, mallards would lie on their sternum with their legs slightly extended to the side. As the disease progressed, they were unable to rise when approached and they laid on their sides and exhibited a swimming motion with both legs in vain attempts to escape. Breathing in these birds was both rapid and deep. Other mallards were unable to hold their heads erect. By day 4, torticollis and wing droop began to appear, followed by paralysis of one or both legs (Fig. 21.4). Muscular tremors also became increasingly noticeable at this time.

EXPLANATION

· Sites of mortality in double-crested cormorants caused by Newcastle disease in North America

Figure 21.1 Locations in North America where Newcastle disease has caused mortality in double-crested cormorants.

Figure 21.2 Clinical signs of Newcastle disease in cormorants include **(A)** torticollis or twisting of head and neck in these two nestlings, and **(B)** wing droop and abnormal posture in this subadult.

Gross Lesions

Dead cormorants examined at necropsy have had only nonspecific lesions. Mildly enlarged livers and spleens and mottled spleens have been noted, but these may be the result of other concurrent diseases, such as salmonellosis.

Diagnosis

Virus isolation and identification, supported by characteristic microscopic lesions in tissues, is necessary to diagnose ND as the cause of illness or death. Whole carcasses should be submitted, and the samples should be representative of all species and age-classes affected. Clinically ill birds should be collected, euthanized by acceptable methods (see Chapter 5, Euthanasia), and, if possible, a blood sample should be collected from euthanized birds and the sera submitted with the specimens. Contact with the diagnostic laboratory is recommended to obtain specific instructions on specimen collection, handling, and shipment. A good field history describing field observations is of great value (see Chapter 1) and should be included with the submission.

Control

An outbreak of ND is a serious event requiring immediate involvement of disease control specialists. NDV infections can be devastating for the domestic poultry industry and an immediate objective in the diagnosis is to determine if the strain of virus involved poses a high risk for poultry. As soon as ND is suspected, strict biosecurity procedures should be followed to contain the outbreak as much as possible and to prevent disease from spreading to other sites.

Figure 21.3 A double-crested cormorant fleeing from observers during the Newcastle disease outbreak at the Salton Sea, California. Note that only the right wing is functional. This bird is typical of juvenile birds surviving infection. The same condition was also observed in adults prior to the breeding season; these birds were presumably survivors from a previous Newcastle disease outbreak.

Figure 21.4 Clinical signs of Newcastle disease in adult mallards that were experimentally infected with a velogenic form of NDV: **(A)** leg paralysis and inability of two of the birds to hold their heads erect, **(B)** torticollis, and **(C)** wing droop.

Large amounts of virus are often shed in the excrement of infected birds and these can contaminate the surrounding environment. Also, NDV is relatively heat-stable and, under the right conditions, it can remain infectious in a carcass for weeks.

The spread of ND in poultry epizootics has occurred via several means including human movement of live birds such as pet or exotic species or both, gamebirds, poultry and other types of birds; other animals; movement of people and equipment; movement of poultry products; airborne spread; contaminated poultry feed; water; and vaccines. Humans and their equipment have had the greatest role because contaminated surfaces provide mechanical transportation for the virus to new locations and to susceptible bird populations.

The critical points are to recognize the outbreak site as a contaminated area, regardless of whether or not poultry or wild birds are involved, to be sensitive to the wide variety of ways that NDV can be moved from that site, and to take all reasonable steps to combat the disease and minimize its spread to other sites and to additional birds at that site.

Control efforts can become complicated by wildlife rehabilitation interests, the presence of strains of NDV that are highly virulent for domestic poultry, and the proximity of the wildlife involved to domestic poultry operations. Collaboration involving all concerned parties is essential in these situations.

Human Health Considerations

NDV is capable of causing a self-limiting conjunctivitis or inflammation of the membrane covering the eyeball and a mild flu-like disease in humans. Most reported cases in humans have occurred among poultry slaughterhouse workers, laboratory personnel, and vaccinators applying live virus vaccines. Aerosols, rather than direct contact, are most often involved as the route for transmission to humans.

Douglas E. Docherty and Milton Friend

Supplementary Reading

Alexander, D.J., 1997, Newcastle disease and other paramyxovirus infections *in* Calnek, B.W., and others, eds., Diseases of Poultry, (10th ed.): Ames, Iowa, Iowa State University Press, p. 541–569.

Friend, M., and Trainer, D.O., 1972, Experimental Newcastle disease studies in the mallard: Avian Diseases, v. 16, no. 4, p. 700–713.

Meteyer, C.U., Docherty, D.E., Glaser, L.C., Franson, J.C.F., Senne, D.A., and Duncan, R., 1997, Diagnostic findings in the 1992 epornitic of neurotropic velogenic Newcastle disease in double-crested cormorants from the upper midwestern United States: Avian Diseases, v. 41, p. 171–180.

Wobeser, G.F., Leighton, F.A., Norman, R., Meyers, D.J., Onderka, D., Pybus, M.J., Neufeld, J.L., Fox, G.A., and Alexander, D.J., 1993, Newcastle disease in wild water birds in western Canada, 1990: Canada Veterinary Journal, v. 34, p. 353–359.

Chapter 22
Avian Influenza

Synonyms

Fowl pest, fowl plague, avian influenza A.

Wild birds, especially waterfowl and shorebirds, have long been a focus for concern by the poultry industry as a source for influenza infections in poultry. Human health concerns have also been raised. For these reasons, this chapter has been included to provide natural resource managers with basic information about avian influenza viruses.

Cause

Avian influenza is usually an inapparent or nonclinical viral infection of wild birds that is caused by a group of viruses known as type A influenzas. These viruses are maintained in wild birds by fecal-oral routes of transmission. This virus changes rapidly in nature by mixing of its genetic components to form slightly different virus subtypes. Avian influenza is caused by this collection of slightly different viruses rather than by a single virus type. The virus subtypes are identified and classified on the basis of two broad types of antigens, hemagglutinan (H) and neuraminidase (N); 15 H and 9 N antigens have been identified among all of the known type A influenzas.

Different combinations of the two antigens appear more frequently in some groups of birds than others. In waterfowl, for example, all 9 of the neuraminidase subtypes and 14 of the 15 hemagglutinin subtypes have been found, and H6 and H3 are the predominant subtypes. In shorebirds and gulls, 10 different hemagglutinin subtypes and 8 neuraminidase subtypes have been found. Many of the antigenic combinations of subtypes are unique to shorebirds. H9 and H13 are the predominate subtypes. More influenza viruses from shorebirds infect waterfowl than chickens. Hemagglutinin subtypes H5 and H7 are associated with virulence or the ability to cause severe illness and mortality in chickens and turkeys. However, two viruses with the same subtype antigens can vary in virulence for domestic birds.

Species Affected

Avian influenza viruses have been found in many bird species, but are most often found in migratory waterfowl, especially the mallard duck (Fig. 22.1). However, the only mortality event known in wild birds killed common terns in South Africa in 1961. This was the first influenza virus from marine birds and it was classified as subtype H5N3. Other wild birds yielding influenza viruses include various species of shorebirds, gulls, quail, pheasants, and ratites (ostrich and rhea). Experimental infections of domestic birds with viruses

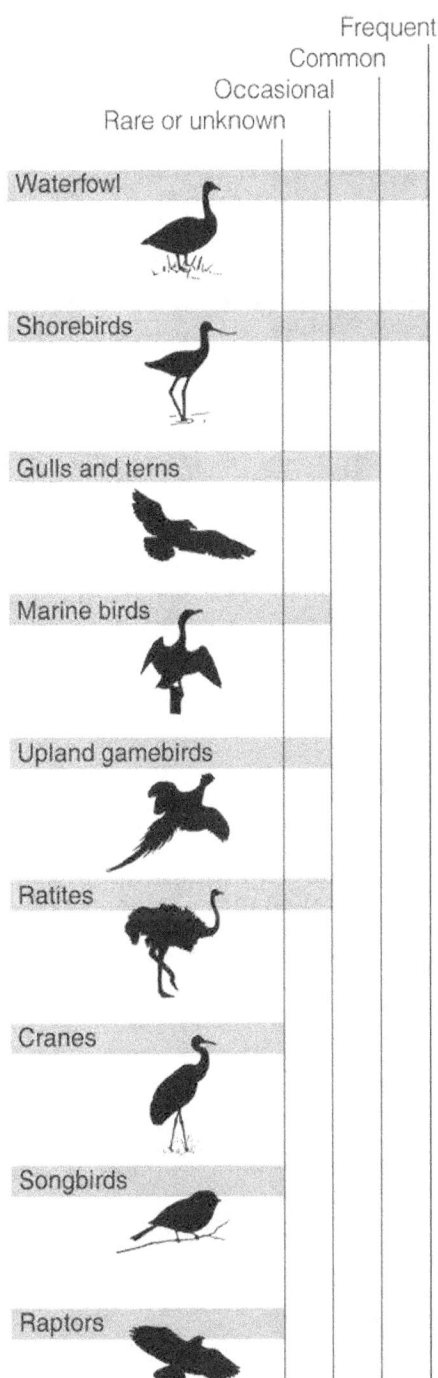

Figure 22.1 Relative occurrence of avian influenza virus in various bird groups.

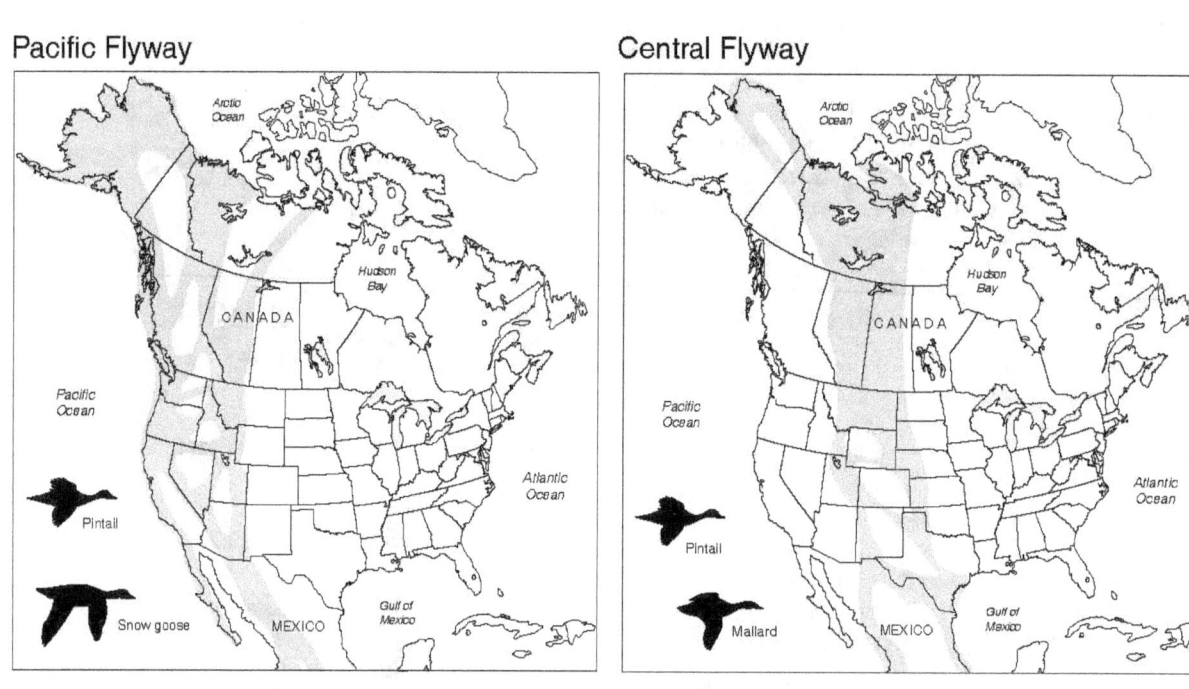
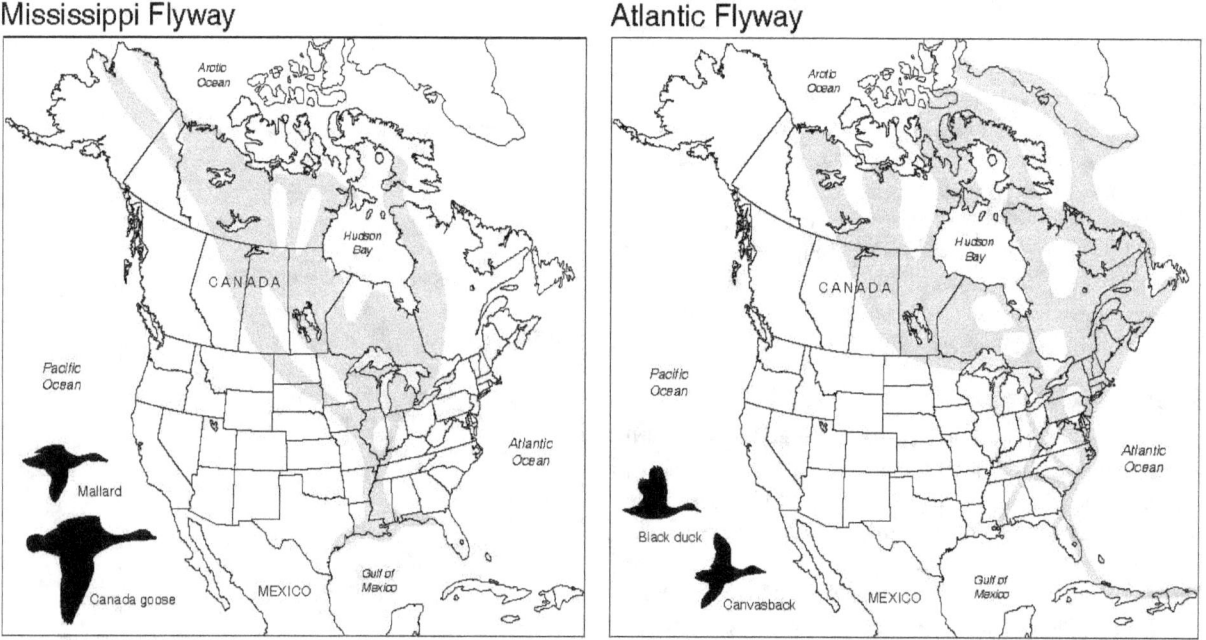

Figure 22.2 *General migratory pathways followed by North American waterfowl. Species shown are typical of these flyways (Modified from Hawkins and others, 1984).*

from wildlife do not cause mortality. Likewise, virulent viruses or viruses that cause disease in domestic fowl do not cause mortality in wild waterfowl.

Distribution

Although influenza tends to be most commonly detected in birds that use the major waterfowl flyways, these viruses are found throughout North America and around the world. The majority of North American waterfowl migration takes place within four broad geographic areas (Fig 22.2). Many species other than waterfowl follow these same migratory pathways from their breeding grounds to the wintering grounds and return to the breeding grounds. The virus subtype that are found in birds in adjacent flyways will differ, especially if the birds from each flyway do not mix during migration. In any given year the percentage of waterfowl and shorebirds carrying influenza viruses will vary by flyway. Likewise, the percentage of birds carrying virus in an flyway will vary in consecutive years. The virus subtype found in birds that use a flyway are rarely the same in consecutive years.

Seasonality

Influenza virus has been found in wild birds throughout the year, but waterfowl are the only group in which these viruses are found year round (Fig. 22.3). The highest occurrence of infection is in the late summer months in juvenile waterfowl when they assemble for their first southward migration. The number of infected waterfowl decreases in the fall as birds migrate toward their southern wintering grounds and is lowest in the spring, when only one bird in 400 is infected during the return migration to the north. In contrast, the number of birds infected is highest in shorebirds (primarily ruddy turnstone) and gulls (herring) during spring (May and June). Infection in shorebirds is also high in September and October. Influenza viruses have not been found in shorebird and gull populations during other months of the year. Influenza viruses have been found in marine birds such as murres, kittiwakes, and puffins while they have been nesting, but the pelagic habits of these species preclude sampling during other periods of the year.

Field Signs

In domestic birds, the signs of disease are not diagnostic because they are highly variable and they depend on the strain of virus, bird species involved, and a variety of other factors including age and sex. Signs of disease may appear as respiratory, enteric, or reproductive abnormalities. Included are such nonspecific manifestations as decreased activity, food consumption, and egg production; ruffled feathers; coughing and sneezing; diarrhea; and even nervous disorders, such as tremors. Observable signs of illness have not been described for wild birds. In domestic chickens and turkeys, certain virus subtypes like H5N2 and H7N7, respectively, are usually highly virulent and may cause up to 100 percent mortality in infected flocks. Another major impact of influenza viruses in domestic birds is decreased egg production. Too little is known about the impact of influenza viruses on the reproductive performance of wild birds to assess whether or not they are affected in the same manner as poultry.

Gross Lesions

Avian influenza virus infection in wild birds is not indicated by gross lesions. Common terns that died in South Africa did not have gross lesions, but a few birds had microscopic evidence of meningoencephalitis or inflammation of the membrane that covers the brain. These lesions were not reproducible during experiments. Mallards experimentally infected with a virulent influenza virus developed discrete purple areas of lung firmness and cloudy lung coverings. However, virulent viruses are rarely found in wild birds, and these lesions may not appear in natural infections.

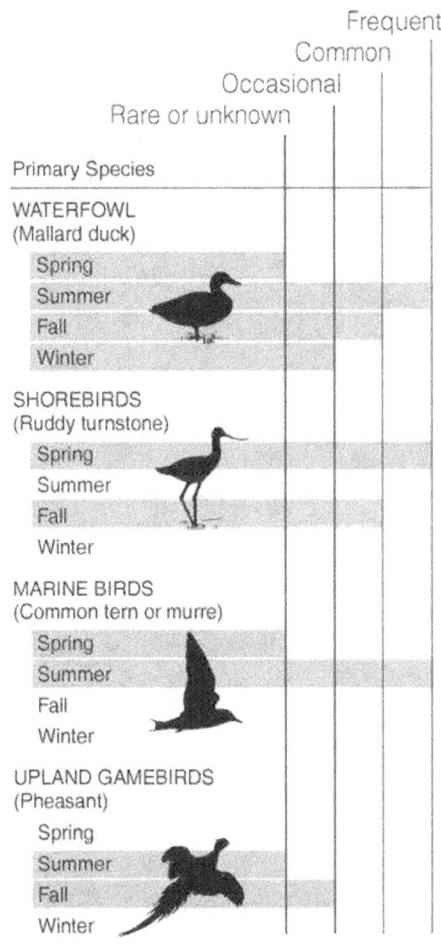

Figure 22.3 Relative seasonal occurrence of influenza A in birds.

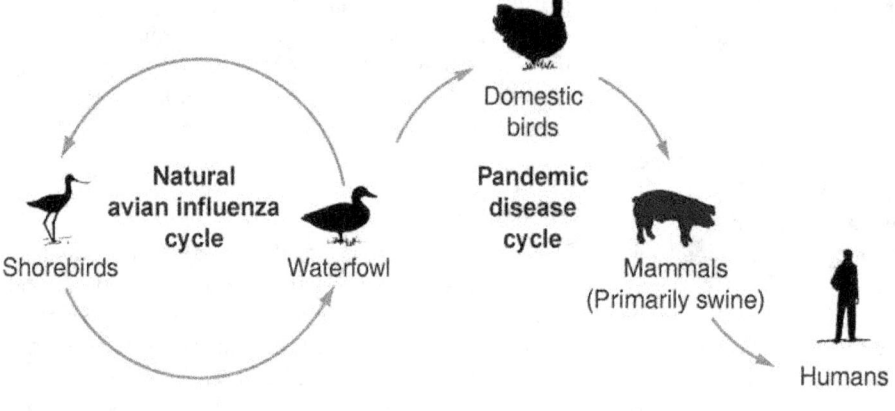

Figure 22.4 Global cycle of avian influenza viruses in animals.

Diagnosis

Infected birds are detected by virus isolation from cloacal swabs in embryonated chicken eggs, and by serological testing of blood for antibody. The last test indicates that a bird was exposed to these viruses rather than if it is infected or carries the disease. Reference antisera to all of the subtype antigen combinations are used to determine the identity of the virus; however, the virulence of a virus cannot be determined by the antigenic subtype. Virulent and avirulent strains of the same subtype can circulate in nature. Laboratory and animal inoculation tests are required to establish the virulence of strains based on an index established for domestic birds.

Control

Avian influenza viruses in wild birds cannot be effectively controlled because of the large number of virus subtypes and the high frequency of virus genetic mixing resulting in new virus subtypes. Also, virus has been recovered from water and fecal material in areas of high waterfowl use. During experiments, influenza virus was recovered from infected waterfowl fecal material for 8 days and from fecal contaminated river water for 4 days when both were held at 22 °C. Poultry manure is a primary residual source of virus for domestic flocks. The virus has been recovered from poultry houses more than 100 days after flock depopulation for markets.

In the domestic bird industry, preventing the entry of the virus into poultry flocks is the first line of defense. Killed vaccines are selectively used to combat less virulent forms of this disease. Antibody present in the blood of recovered and vaccinated birds prevents virus transmission. Therefore, these birds pose little risk to other birds. Flocks are generally killed when they are infected with highly virulent viruses.

In the past, the poultry industry and the wildlife conservation community have been in conflict regarding wildlife refuge development and other waterbird habitat projects. The fear that waterbirds are a source of influenza viruses for infection of poultry has resulted in strong industry opposition that has negatively impacted some projects. This issue should be considered when land use near wetlands is planned and when wildlife managers plan for development for wildlife areas. Open communication during project development and sound plans that are developed in a collaborative manner may help industry and conservation groups avoid confrontation and support each others' interests.

Human Health Considerations

Although this group of viruses includes human influenza viruses, the strains that infect wild birds do not infect humans. It is believed that waterfowl and shorebirds maintain separate reservoirs of viral gene pools from which new virus subtypes emerge. These gene pools spill over into other animals (mammals) and may eventually cause a new pandemic (Fig. 22.4).

Wallace Hansen

Supplementary Reading

Easterday, B.C., and Hinshaw, V.S., 1991, Influenza, *in* Calnek, B.W., and others, eds., Diseases of Poultry (9th ed.): Ames, Iowa, Iowa State University Press, p. 532–552.

Hinshaw, V.S., Wood, J.M., Webster, R.G., Deible, R., and Turner, B., 1985, Circulation of influenza viruses and paramyxoviruses in waterfowl originating from two different areas of North America: Bulletin of the World Health Organization 63, 711–719.

Kawoka, Y., Chambers, T.M., Sladen, W.L., and Webster, R.G., 1988, Is the gene pool of influenza viruses in shorebirds and gulls different from that in wild ducks?: Virology, v. 163, p. 247–250.

Stallknecht, D. E., Shane, S. M., Zwank, P. J., Senne, D. A., and Kearney, M. T., 1990, Avian influenza viruses from migratory and resident ducks of coastal Louisiana: Avian Diseases, v. 34, p. 398–405.

Chapter 23
Woodcock Reovirus

Synonyms

None

This chapter provides information on a recently identified disease of the American woodcock. Little is known about the disease or the virus that causes it. It has been included in this Manual to enhance awareness that such a disease exists and to stimulate additional interest in further investigations to define the importance of woodcock reovirus. More information about this disease is needed because it is not known whether or not this virus is a factor in the decline of woodcock populations within the United States.

Cause

The first virus isolated from the American woodcock is a reovirus that was found during woodcock die-offs during the winters of 1989–90 and 1993–94. Avian reovirus infections have been associated with numerous disease conditions including viral arthritis/tenosynovitis or inflammation of the tendon sheath; growth retardation; pericarditis or inflammation of the sac surrounding the heart; myocarditis or inflammation of the heart muscle; hydropericardium or abnormal accumulation of fluid in the pericardium; enteritis or inflammation of the intestine; hepatitis or inflammation of the liver; bursal and thymic atrophy or wasting away; osteoporosis or rarefaction of the bone; and respiratory syndromes. The infections are generally systemic, transmitted by the fecal-oral route, and are often associated with nutritional factors or concurrent infections with other agents.

Species Affected

American woodcock are the only species known to be infected with this particular reovirus. Investigations have not been conducted to determine whether or not other species are susceptible to infection, which species are not susceptible, and which species become diseased.

Distribution

The virus was isolated from woodcock that were found dead at the Eastern Shore of Virginia National Wildlife Refuge (Fig. 23.1). To determine the prevalence of woodcock reovirus in the eastern and central regions of the United States, virus isolation was attempted from woodcock samples collected from the breeding and wintering populations in 1990–92 (Fig. 23.1). No viruses were isolated from 481 tissue samples or 305 cloacal swabs that were obtained from live-trapped and hunter-killed woodcock.

Figure 23.1 *Site of woodcock mortality from reovirus infection and field sampling sites where other woodcock were tested and found to be negative for exposure to this disease.*

Seasonality

Both die-offs occurred during the winter months. Nothing more is known about the seasonality of this disease.

Field Signs

Sick woodcock have not been observed. Therefore, field signs are unknown.

Gross Lesions

Most of the birds found dead were emaciated. Little or no food was found in their digestive tracts, and no obvious gross lesions were noted upon necropsy of the carcasses. Healthy "control" woodcock collected during the same time did not yield virus.

Diagnosis

Diagnosis requires laboratory isolation and identification of the causative virus. Winter concentrations of woodcock should be monitored and carcasses picked up and submitted for diagnosis. Whole carcasses of woodcock found dead should be shipped to a diagnostic laboratory where pathological assessments and virus isolation can be made (see Chapter 3, on Specimen Shipment). Although the virus has been isolated from a variety of tissues including intestine, brain, cloacal swab, heart, and lung, the majority of isolates were obtained from intestines and cloacal swabs. These findings suggest that a fecal-oral route of transmission is likely.

Control

Field carcasses not needed for diagnostic study should be collected for disposal to minimize environmental contamination. Too little is known about this disease to recommend response actions.

Human Health Considerations

There are no known human health considerations.

Douglas E. Docherty

Supplementary Reading

Docherty, D.E., Converse, K.A., Hansen, W.R., and Norman, G.W., 1994, American woodcock *(Scolopax minor)* mortality associated with a reovirus: Avian Diseases, v. 38, p. 899–904.

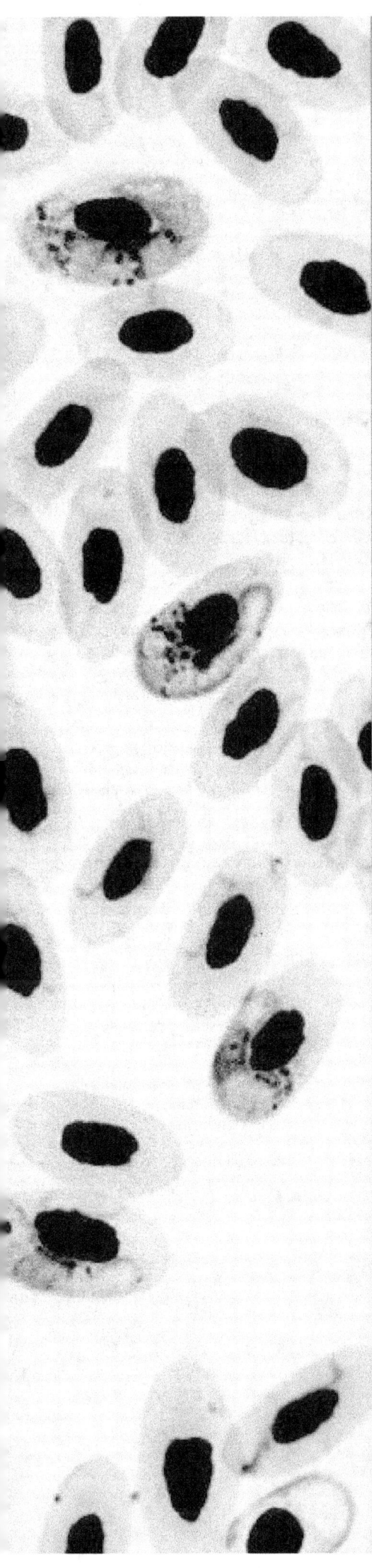

Section 5
Parasites and Parasitic Diseases

Hemosporidiosis

Trichomoniasis

Intestinal Coccidiosis

Renal Coccidiosis

Sarcocystis

Eustrongylidosis

Tracheal Worms

Heartworm of Swans and Geese

Gizzard Worms

Acanthocephaliasis

Nasal Leeches

Miscellaneous Parasitic Diseases

Stained blood smear from a turkey infected with the parasite Haemoproteus meleagridis
Photo by Carter Atkinson

Introduction to Parasitic Diseases

"Parasites form a large proportion of the diversity of life on earth."
(Price)

Parasitism is an intimate relationship between two different species in which one (parasite) uses the other (host) as its environment from which it derives nourishment. Parasites are a highly diverse group of organisms that have evolved different strategies for infecting their hosts. Some, such as lice and ticks, are found on the external parts of the body (ectoparasites), but most are found internally (endoparasites). Some are microscopic, such as the blood protozoans that cause avian malaria; however, many are macroscopic. Life cycles differ greatly between major types of parasites and are generally classified as direct or indirect (Table 1). Direct life cycles do not require an intermediate host (Fig. 1A). For direct life cycles, only a definitive host is required: the species in which the parasite reaches sexual maturity and produces progeny. Indirect life cycles may involve one or more intermediate hosts (Fig. 1B and C). Intermediate hosts are required by the parasite for completion of its life cycle because of the morphological and physiological changes that usually take place in the parasite within those hosts. Wild birds can serve as the definitive hosts for most of the parasites that are discussed in the following chapters. In addition, paratenic or transport hosts are present in some parasite life cycles. The parasites generally do not undergo development in paratenic hosts. Instead, paratenic hosts provide both an ecological and temporal (time) bridge for the parasite to move through the environment and infect the definitive host. Typically, in these situations one or more intermediate hosts are required for development of the parasite but they are not fed upon by the bird. Instead, the bird feeds on the paratenic hosts, which in turn have fed on the intermediate host(s), thereby, "transporting" the parasite to the bird (Fig. 2).

The presence of parasites in birds and other animals is the rule, rather than the exception. Hundreds of parasite species have been identified from free-ranging wild birds; however, the presence of parasites does not necessarily equate with disease. Most of the parasites identified from wild birds cause no clinical disease. Others cause varying levels of disease, including death in the most severe cases. The pathogenicity or the ability to cause disease, of different species of parasites varies with 1) the species of host invaded (infected or infested), 2) the number or burden of parasites in or on the host, and 3) internal factors impacting host response. For example, when birds are in poor nutritional condition, have concurrent infections from other disease agents (including other species of parasites), or are subject to other types of stress, some parasites that do not normally cause disease do cause disease. Lethal infections may result from parasites that generally only cause mild disease.

This section highlights some of the parasitic diseases such as trichomoniasis that are associated with major mortality events in free-ranging wild birds and those that because of the gross lesions they cause (*Sarcocystis* sp.), their visibility (nasal leeches), or general interest (heartworm) are often the subject of questions asked of wildlife disease specialists.

Quote from:

Price, P.W., 1980, Evolutionary biology of parasites: Princeton University Press, Princeton, NJ, p. 3.

Table 1 General characteristics of major groups of internal parasites (endoparasites) of free-ranging birds.

Type of parasite	Common name	Type of life cycle	Characteristics
Nematodes	Roundworms	Indirect and direct	Most significant group relative to number of species infecting birds and to severity of infections. Unsegmented cylindrical worms. Found throughout the body. Generally four larval stages. Sexes are separate. Most are large in size (macroscopic).
Cestodes	Tapeworms	Indirect	Flattened, usually segmented worms with a distinct head, neck and body. Found primarily in the lumen of the intestines. Lack a mouth or an alimentary canal; feed by absorbing nutrients from the host's intestinal tract. Most are hermaphroditic (self-fertilization; have both male and female reproductive tissues). Attachment is by suckers, hooks. Large size (macroscopic).
Trematodes	Flukes	Indirect	Flatworms, generally leaf-shaped (some almost cylindrical). Generally found in the lower alimentary tract, respiratory tract, liver, and kidneys. Complex life cycles; usually require two intermediate hosts, one of which is usually a snail. Hermaphroditic except for blood flukes, which have separate sexes. Attachment is usually by suckers.
Acanthocephalans	Thorny-headed worms	Indirect	Cylindrical, unsegmented worms. Found in the digestive tract. No intestinal tract; nutrients absorbed through the tegument (similar to tapeworms). Sexes are separate. Attachment by means of a retractable proboscis that has sharp recurved hooks or spines.
Protozoans	Coccidians, malarias, trichomonads, others	Direct and indirect	Microscopic. Different types are found in different parts of the body. Asexual and sexual multiplication.

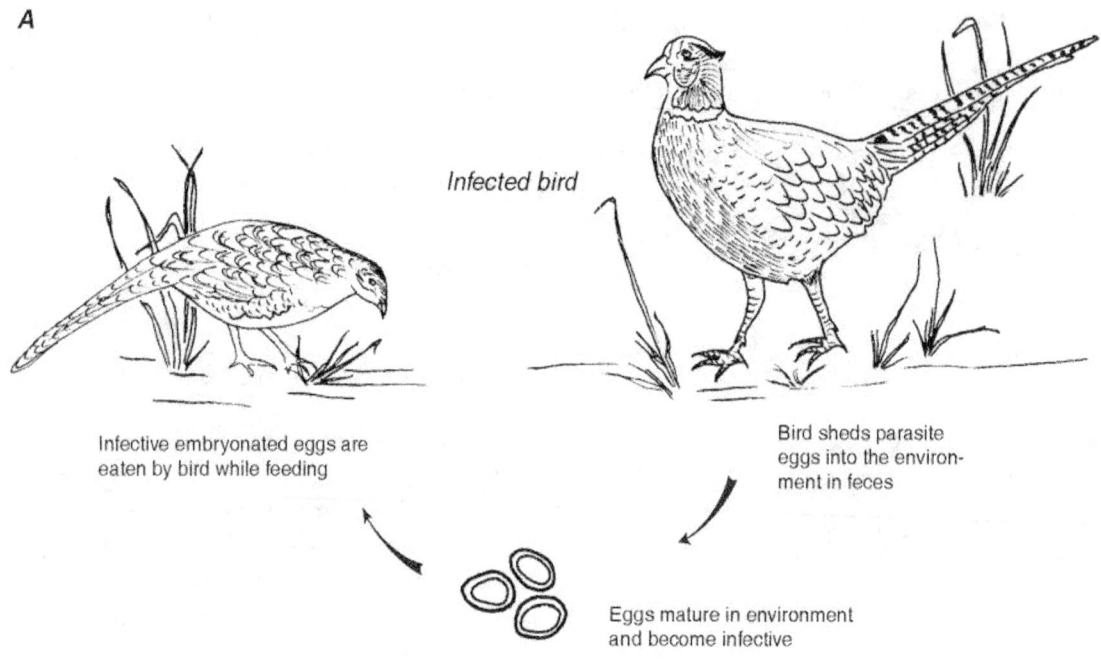

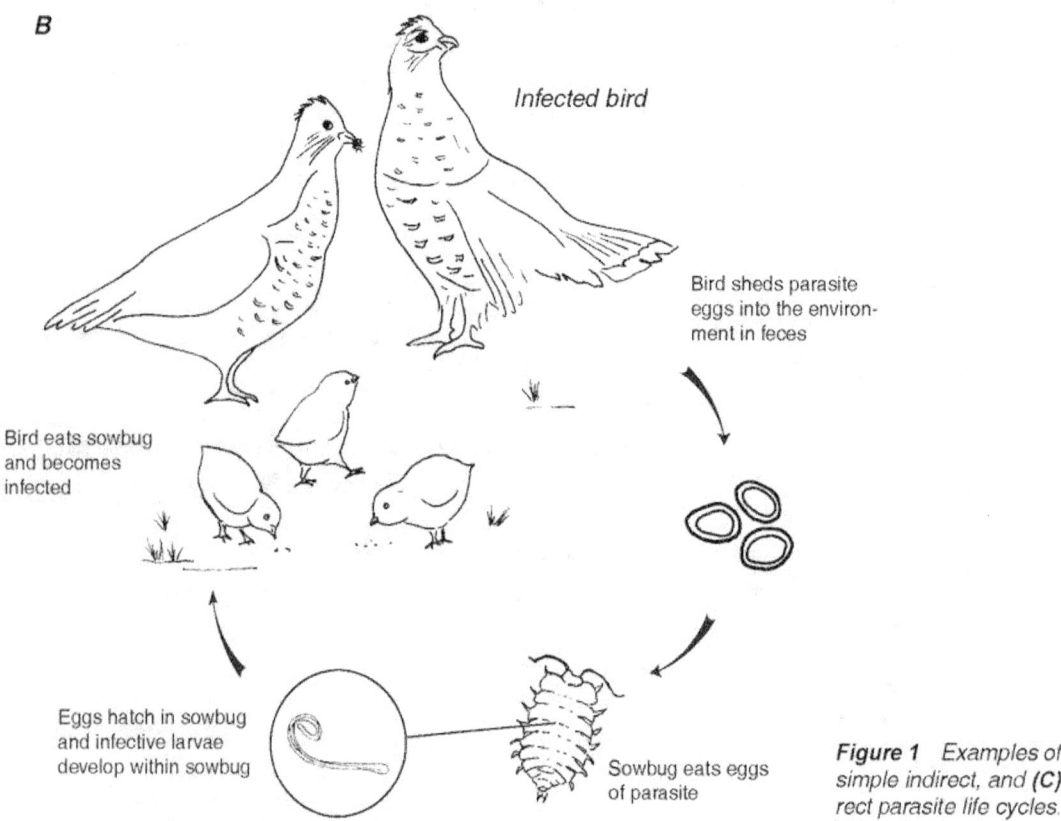

Figure 1 Examples of **(A)** direct, **(B)** simple indirect, and **(C)** complex indirect parasite life cycles.

c

Infected bird

Parasite eggs are passed in feces and hatch in water

First intermediate host

Larvae (miracidium) swims to a snail and penetrates it, undergoing further larval development within the host

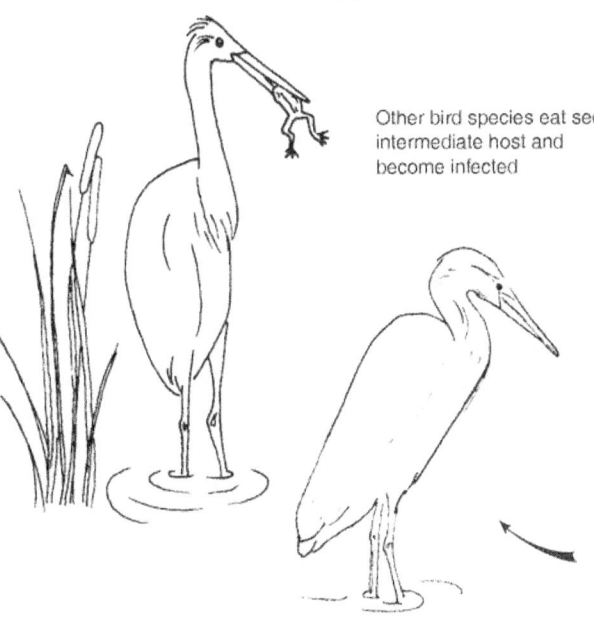

Other bird species eat second intermediate host and become infected

Second intermediate host

New larval stage emerges from snail (cercaria) and swims to new host where it penetrates and encysts

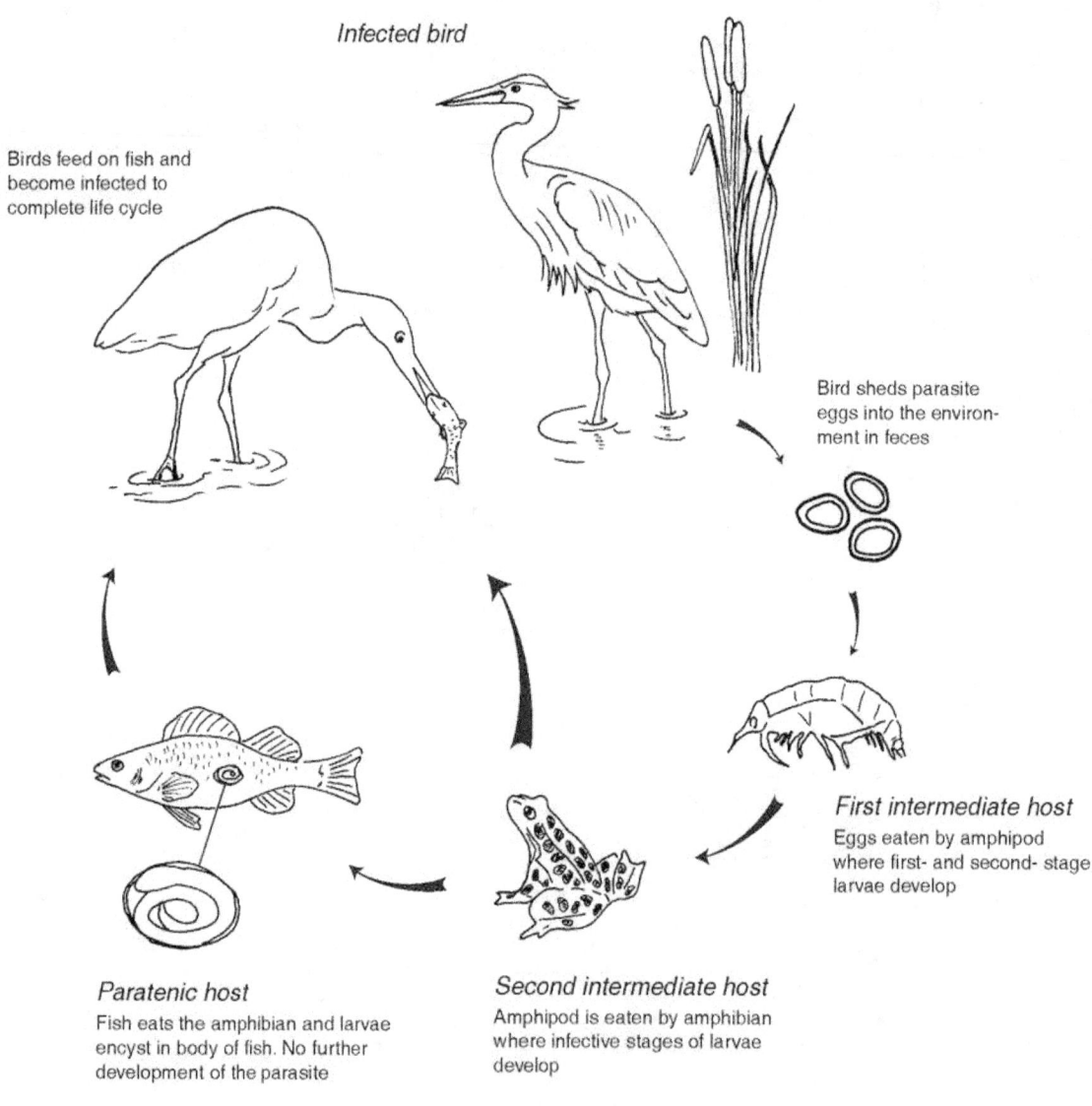

Figure 2 Hypothetical parasite life cycle illustrating the role of paratenic (transport) hosts.

Chapter 24
Hemosporidiosis

Synonyms
Avian malaria

Cause
Hemosporidia are microscopic, intracellular parasitic protozoans found within the blood cells and tissues of their avian hosts. Three closely related genera, *Plasmodium*, *Haemoproteus*, and *Leucocytozoon*, are commonly found in wild birds. Infections in highly susceptible species and age classes may result in death.

Life Cycle
Hemosporidia are transmitted from infected to uninfected birds by a variety of biting flies that serve as vectors, including mosquitoes, black flies, ceratopogonid flies (biting midges or sandflies) and louse flies (Fig. 24.1) (Table 24.1). When present, infective stages of the parasites (sporozoites) are found in the salivary glands of these biting flies. They gain entry to the tissues and blood of a new host at the site of the insect bite when these vectors either probe or lacerate the skin to take a blood meal. Insect vectors frequently feed

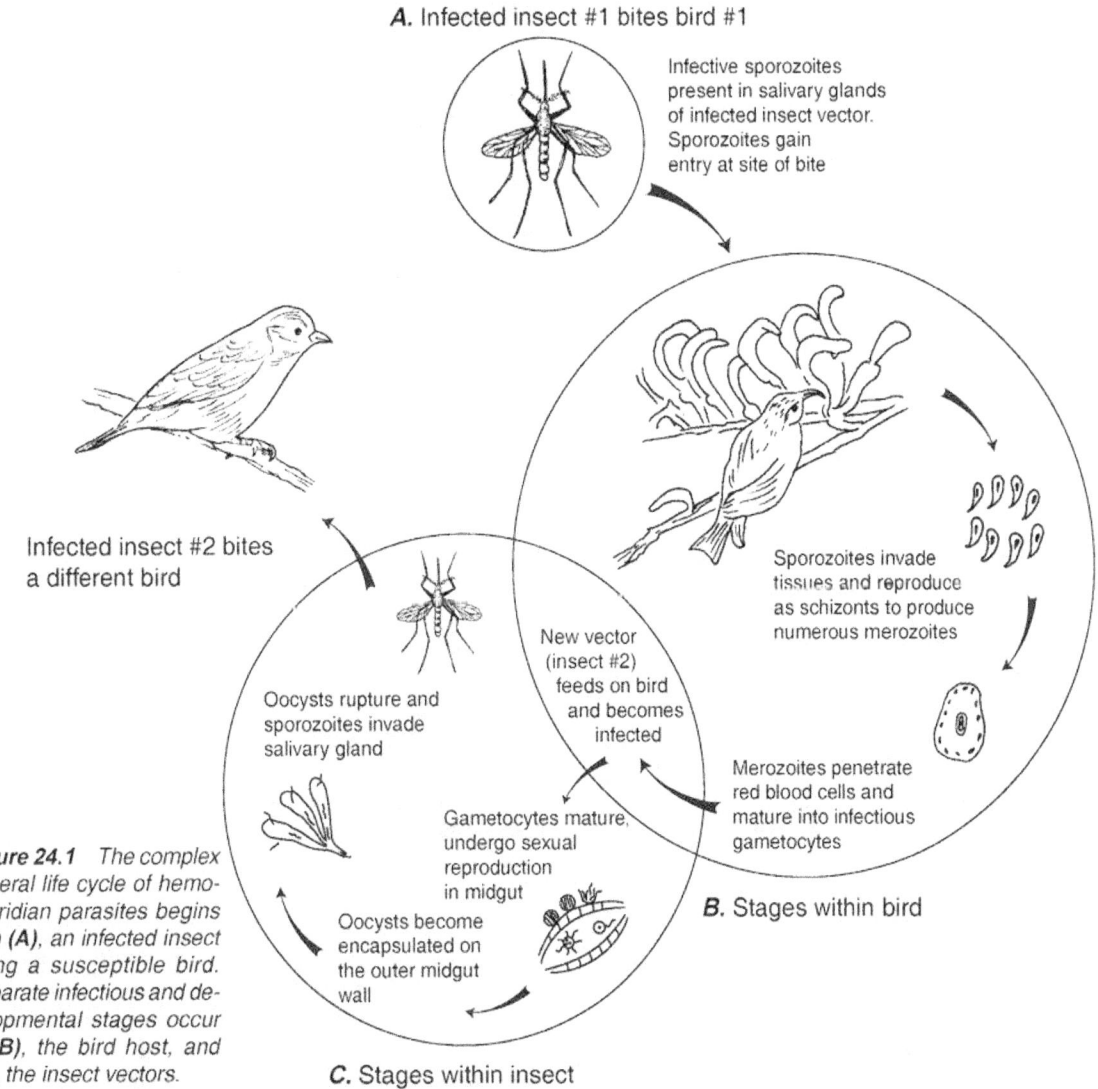

Figure 24.1 The complex general life cycle of hemosporidian parasites begins with (A), an infected insect biting a susceptible bird. Separate infectious and developmental stages occur in (B), the bird host, and (C), the insect vectors.

Table 24.1 Avian hemosporidia parasites and their documented vectors.

Parasite	Vector type	Common name
Haemoproteus	Ceratopogonidae (*Culicoides* sp.)	Punkies, no-see-ums, sand flies
	Hippoboscidae (*Ornithomyia* sp.)	Hippoboscid or louse flies
Plasmodium	Culicidae (*Culex*, *Aedes* sp.)	Mosquitoes
Leucocytozoon	Simulidae (*Simulium* sp.)	Black flies

Figure 24.2 A Culex mosquito feeding on the unfeathered area surrounding the eye of an apapane, a native Hawaiian honeycreeper.

on exposed flesh around the eyes (Fig. 24.2), the beak, and on the legs and feet, although black flies, ceratopogonid flies, and louse flies can crawl beneath the bird's feathers to reach the skin surface. Immediately after they infect a bird, sporozoites invade the tissues and reproduce for one or more generations before they become merozoites. Merozoites penetrate the red blood cells and become mature, infectious gametocytes. The cycle is completed when the gametocytes in the circulating blood cells of the host bird are ingested by another blood-sucking insect, where they undergo both sexual and asexual reproduction to produce large numbers of sporozoites. These invade the salivary glands of the vector and are transmitted to a new host bird during the vector's next blood meal.

Species Affected

The avian hemosporidia are cosmopolitan parasites of birds, and they have been found in 68 percent of the more than 3,800 species of birds that have been examined. Members of some avian families appear to be more susceptible than others. For example, ducks, geese and swans are commonly infected with species of *Haemoproteus*, *Leucocytozoon*, and *Plasmodium*, and more than 75 percent of waterfowl species that were examined were hosts for one or more of these parasites. Wild turkeys in the eastern United States are also commonly infected by these parasites. Pigeons and doves have similar high rates of infection, but members of other families, such as migratory shorebirds, are less frequently parasitized.

Differences in the prevalence, geographic distribution, and host range of hemosporidia are associated with habitat preferences of the bird hosts, the abundance and feeding habits within those habitats of suitable insect vectors, and innate physiological differences that make some avian hosts more susceptible than others. For example, some species of black flies (*Simulium* sp.) prefer to feed on waterfowl within a limited distance of the shoreline. Ducks and geese that spend more of their time in this zone will be more likely to be exposed to bites that carry infective stages of *Leucocytozoon simondi*. Biting midges or no-see-ums (*Culicoides* sp.) that transmit species of *Haemoproteus* are more active at dusk in the forest canopy. Birds that roost here, for example, increase their chances for being infected with this parasite. Finally, some avian hosts are more susceptible to hemosporidian parasites than others, but the physiological basis for this is still poorly understood.

Species of *Plasmodium* and *Leucocytozoon* are capable of causing severe anemia, weight loss, and death in susceptible birds. Young birds are more susceptible than adults, and the most serious mortality generally occurs within the first few weeks of hatching. This is also the time of year when increasing temperatures favor the growth of the populations of insect vectors that transmit hemosporidia. Major outbreaks of *L. simondi* that caused high mortality in ducks and geese in Michigan and subarctic Canada have been documented. Species of *Haemoproteus* are generally believed to be less pathogenic, with only scattered reports of natural mortality in wild birds.

Penguins and native Hawaiian forest birds are highly susceptible to *Plasmodium relictum*, a common parasite of songbirds that is transmitted by *Culex* mosquitoes. This parasite causes high mortality in both captive and wild populations of these hosts, and it is a major factor in the decline of native forest birds in the Hawaiian Islands.

Distribution

Species of *Plasmodium*, *Haemoproteus*, and *Leucocytozoon* have been reported from most parts of the world with the exception of Antarctica, where cold temperatures prevent the occurrence of suitable insect vectors. Studies of the distribution of hemosporidia in North America have shown that areas of active transmission of the parasites coincide with the geographic distribution of their vectors. *Leucocytozoon* is most common in mountainous areas of Alaska and the Pacific Northwest where abundant fast-moving streams create suitable habitat for aquatic black fly larvae. Species of *Haemoproteus* and *Plasmodium* are more evenly distributed across the continent because their ceratopogonid and mosquito vectors are less dependent on the presence of flowing water for larval development. Migratory birds may winter in habitats that lack suitable vectors; therefore, the simple presence of infected birds may not be evidence that the parasites are being transmitted to birds at the wintering grounds.

Seasonality

Infections with *Plasmodium*, *Haemoproteus*, and *Leucocytozoon* are seasonal because transmission depends upon the availability of vector populations. In temperate North America, most birds become infected with hemosporidia during the spring when conditions for transmission become optimal. Some of these conditions include the onset of warmer weather; increases in vector populations; the reappearance or relapse of chronic, low-level infections in adult birds; and the hatching and fledging of susceptible, nonimmune juvenile birds. In warmer parts of the United States, these parasites may be transmitted at other times of the year. In Hawaii, *P. relictum* in forest bird populations may be transmitted throughout the year in warm low-elevation forests, but transmission is more seasonal at elevations above 3,000 ft. where cooler winter temperatures limit mosquito populations.

Field Signs

Birds with acute infections of *Plasmodium*, *Haemoproteus*, and *Leucocytozoon*, may exhibit similar signs in the field. These include emaciation, loss of appetite, listlessness, difficulty in breathing, and weakness and lameness in one or both legs. Survivors develop persistent, low-level infections in the blood and tissues that stimulate immunity to reinfection. These survivors do not exhibit any signs of disease, but they serve as reservoirs of infection, allowing the parasites to survive droughts and cold winter weather when vector populations have died off.

Gross Lesions

Gross lesions associated with acute infections include enlargement of the liver and spleen (Fig. 24.3) and the appearance of thin and watery blood as a result of infected blood cells being destroyed and removed from circulation (Fig. 24.4). In *Plasmodium* and *Haemoproteus* infections, parasites within the red blood cells produce an insoluble black pigment called hemozoin when they digest the host's oxygen-bearing, iron-laden red blood cell protein or hemoglobin. The hemozoin is deposited extensively in the host's spleen and liver tissue as the host's immune system responds to the infection. In very heavy infections, the kidneys may also be affected. These organs typically appear chocolate brown or black at necropsy and they may be two or more times their normal size (Fig. 24.3). Hemozoin pigment is not produced in *Leucocytozoon* infections; therefore, organs will not be as discolored and dark at necropsy, but they will still appear enlarged. Some species of *Haemoproteus* form large, cyst-like bodies in muscle tissue that superficially resemble tissue cysts produced by species of *Sarcocystis* (Fig. 24.5).

Diagnosis

Definitive diagnosis of hemosporidian infections is dependent on microscopic examination of a stained blood smear or on an organ impression smear to detect the presence and form of the parasites within the red blood cells (Figs. 24.6, 7, 8). Species of *Leucocytozoon* frequently produce dramatic changes in the host's cell structure (Fig. 24.6). Parasitized red blood cells are often enlarged and elongated so that they

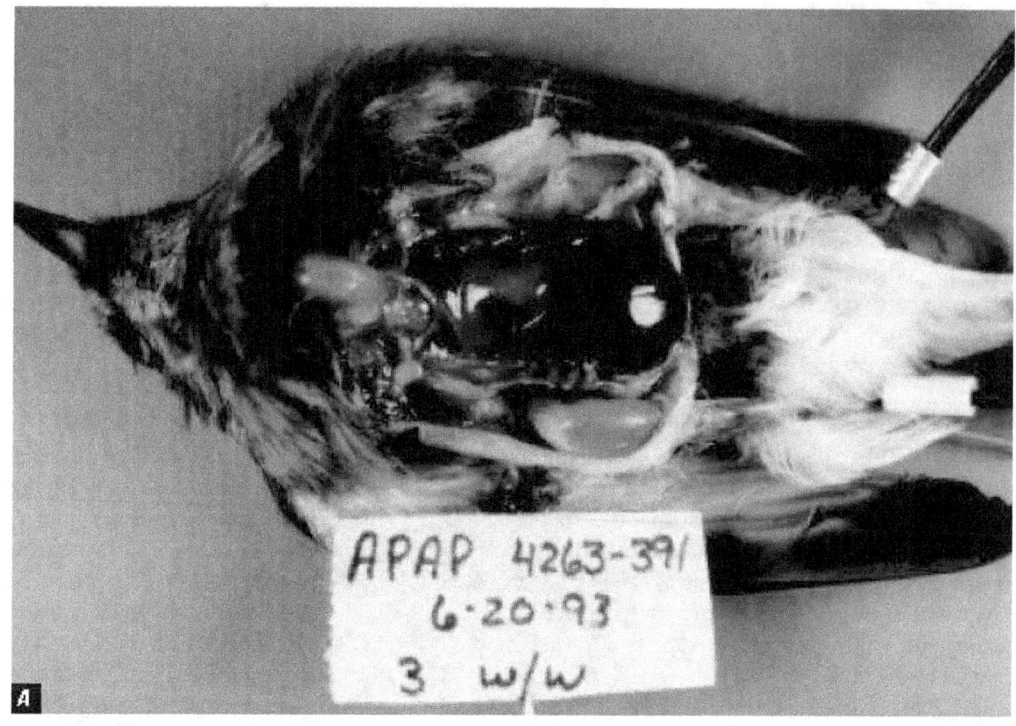

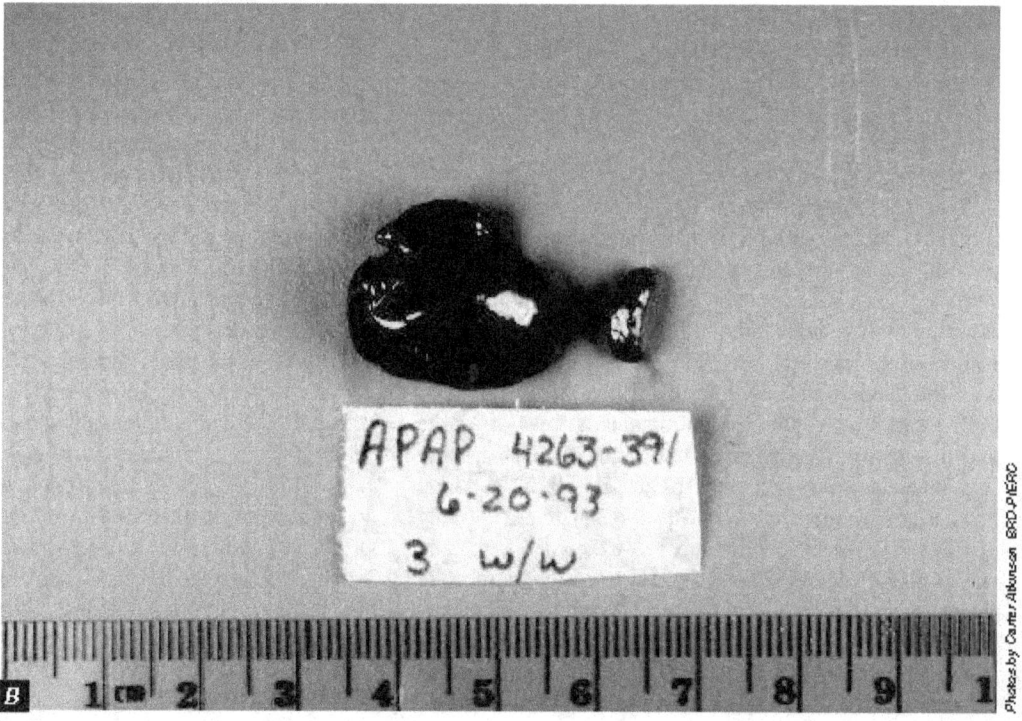

Figure 24.3 Gross lesions caused by **Plasmodium relictum** in an apapane. Enlargement and discoloration of the (A), liver and (B), spleen are typical in acute infections when large numbers of parasites are found in the circulating red blood cells.

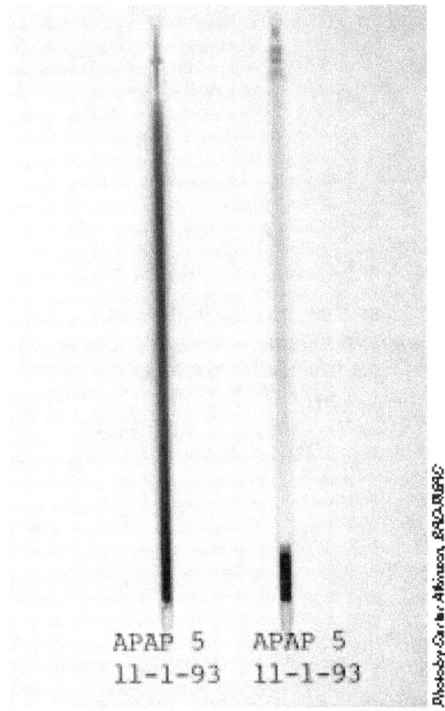

Figure 24.4 Thin and watery blood from an apapane infected with *Plasmodium relictum* before (left) and after (right) centrifugation. In uninfected songbirds, approximately half of the blood volume is occupied by red blood cells. Note that most of the blood cells have been destroyed by the parasite (right).

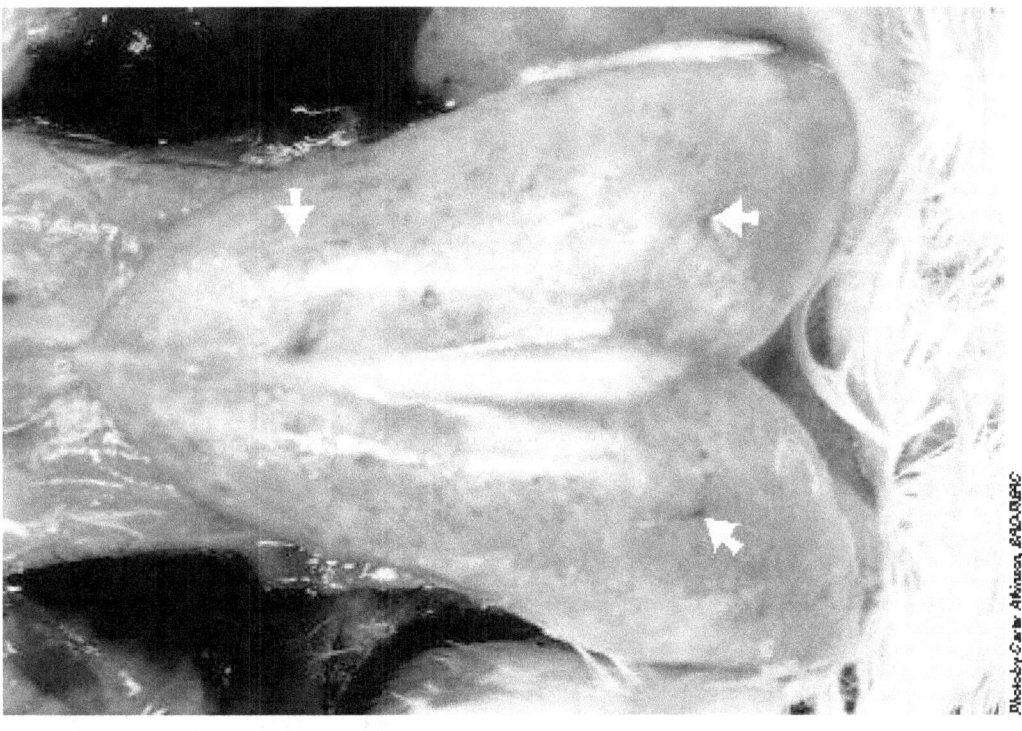

Figure 24.5 Pectoral muscles of a turkey infected with *Haemoproteus meleagridis*. Note the white streaks and bloody spots in the muscle tissue of this bird (arrows). The tissue stages of this hemosporidian form large, cyst-like bodies that may superficially resemble those caused by species of *Sarcocystis*.

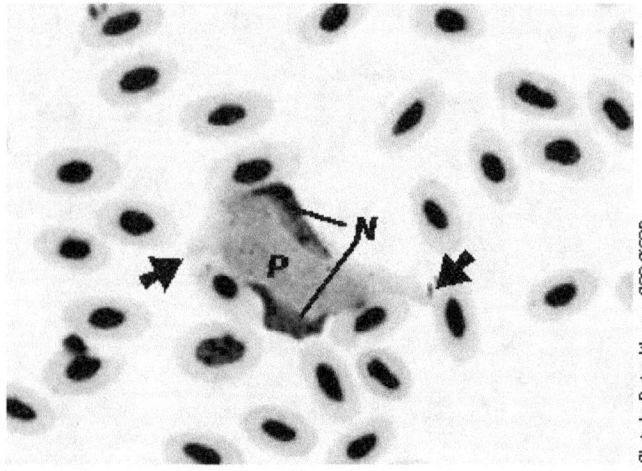

Figure 24.6 Stained blood smear from a turkey infected with **Leucocytozoon smithi**. This parasite causes enlargement and distortion of the infected blood cell. The red blood cell nucleus (**N**) is divided in two halves that lay on either side of the parasite (**P**). The membrane of the infected cell is stretched into two hornlike points (arrows).

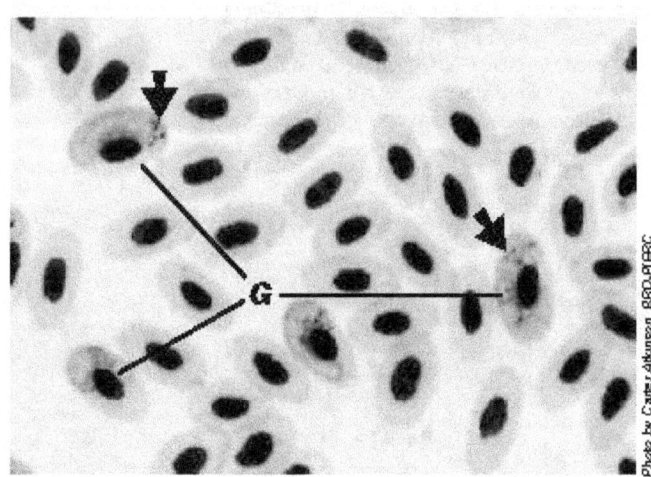

Figure 24.7 Stained blood smear from a turkey infected with **Haemoproteus meleagridis**. Gametocytes (**G**) contain a single pink-staining nucleus and contain black or golden brown pigment granules (arrows).

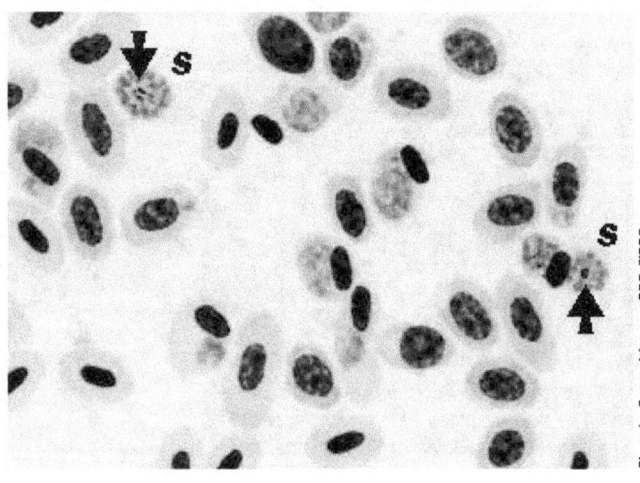

Figure 24.8 Stained blood smear from an apapane infected with **Plasmodium relictum**. Some red blood cells contain multinucleated, asexually-reproducing stages of the parasite called schizonts (**S**). These are diagnostic for Plasmodium infections and contain one or more centrally-located pigment granules (arrows).

form a pair of horn-like extensions from either end of the cell. Species of *Plasmodium* and *Haemoproteus* produce fewer changes in their host's red blood cells, but these parasites may cause slight enlargement of infected host cells and displacement of the red blood cell nucleus to one side (Figs. 24.7, 8). Unlike *Leucocytozoon*, *Plasmodium* and *Haemoproteus* produce golden brown or black deposits of hemozoin pigment in the parasite cell (Figs. 24.7, 8). Further differentiation of *Plasmodium* from *Haemoproteus* may be difficult. Diagnosis of a *Plasmodium* infection is dependent on detecting the presence of asexually reproducing stages of its life cycle (schizonts) in the red blood cells of the infected host (Fig. 24.8).

Control

Control of the avian hemosporidia is dependent on reducing transmission from infected birds to healthy birds through reduction or elimination of vector populations. Many of the same techniques that were developed for control of vector-transmitted human diseases can be used effectively, but few agencies have the resources or manpower to apply them over large areas. Most techniques rely on habitat management to reduce vector breeding sites or depend on the application of pesticides that affect larval or adult vectors to reduce vector populations. Large-scale treatment of infected survivor birds could prevent disease outbreaks by reducing sources of infection, but the logistics and practicality of treating sufficient numbers of birds to interrupt transmission are prohibitive. Although some experimental vaccines for these parasites have been developed, none are currently available for general use.

Human Health Considerations

The avian hemosporidia are closely related to the malarial parasites of humans, but are not capable of infecting people.

Carter T. Atkinson
Pacific Island Ecosystems Research Center
Kilauea Field Station

Supplementary Reading

Atkinson, C.T., 1991, Vectors, epizootiology, and pathogenicity of avian species of *Haemoproteus*: Bulletin of the Society for Vector Ecology, vol.16, p. 109–126.

Atkinson, C.T., and van Riper, C., III, 1991, Pathogenicity and epizootiology of avian haematozoa: *Plasmodium, Leucocytozoon*, and *Haemoproteus*, in Loye, J.E., and Zuk, M., eds., Bird-parasite interactions, Ecology, evolution, and behavior: New York, Oxford University Press, p. 19–48.

Bennett, G.F., Whiteway, M., and Woodworth-Lynas, C.B., 1982, Host-parasite catalogue of the avian haematozoa: Memorial University of Newfoundland Occasional Papers in Biology Number 5, p. 243.

Greiner, E.C., Bennett, G.F., White, E.M., and Coombs, R.F., 1975, Distribution of the avian hematozoa of North America, v. 53, p. 1,762–1,787.

Greiner, E.C., 1991, Leucocytozoonosis in waterfowl and wild galliform birds: Bulletin of the Society for Vector Ecology, v. 16, p. 84–93.

Chapter 25
Trichomoniasis

Synonyms

Canker (doves and pigeons), frounce (raptors), avian trichomoniasis

Cause

Avian trichomoniasis is caused by a single celled protozoan, *Trichomonas gallinae*. Avirulent *T. gallinae* strains that do not cause disease and highly virulent strains are found in nature and circulate within bird populations. The factors that make a strain virulent are not known, but they are thought to be controlled genetically within the parasite. Similarly, the reasons why an avirulent or a virulent form of the parasite is found within a bird population at any period of time also remain unknown. Virulent strains of *T. gallinae* have caused major mortality events or epizootics in doves and pigeons in addition to less visible, chronic losses (Table 25.1). Infection typically involves the upper digestive tract of doves and pigeons but other species have also been infected (Fig. 25.1).

Trichomoniasis in doves and pigeons, but not in other species, is generally confined to young birds. The parasite was introduced to the U.S. with the introduction of pigeons and doves brought by European settlers. It has been reported that 80 to 90 percent of adult pigeons are infected, but they show no clinical signs of disease. It is speculated that most of these birds became immune as a result of exposure to avirulent strains of the parasite or because they survived mild infections. In pigeons and mourning doves, the parasites are transmitted from the adults to the squabs in the pigeon milk produced in the crop of the adult. Squabs usually become infected with the first feeding of pigeon milk, which is gen-

Table 25.1 Examples of wild bird mortalities reported in the scientific literature due to trichomoniasis.

Year	Magnitude	Geographic area	Comments
1949–51	Tens of thousands of mourning doves	Southeastern United States	Trichomoniasis broke out in virtually all States in the region; the magnitude of losses focused attention on the devastation that could be caused by this disease and stimulated research on the ecology of this disease.
1950–51	25,000 to 50,000 mourning doves each year	Alabama	Breeding birds were the focus of infection; mortality was thought to have been grossly underestimated.
1972	Several hundred	Nebraska	Railroad yards and a grain elevator were focal points of infection; birds fed on spilled grain.
1985	Approximately 800 mourning doves	New Mexico	Losses at birdfeeders near Las Cruces.
1988	At least 16,000 band-tailed pigeons	California	First major epizootic of trichomoniasis in this species.
1991	Approximately 500 mourning doves	North Carolina	—

erally within minutes after hatching. The resulting infection may range from asymptomatic or mild disease to a rapidly fatal course resulting in death within 4–18 days after infection. Other modes for infection are through feed, perhaps contaminated drinking water, and feeding on infected birds (Fig. 25.2).

There is no cyst or resistant stage in the parasite's life cycle; therefore, infection must be passed directly from one bird to another, in contaminated feed or water. Feed and water are contaminated when trichomonads move from the mouth of infected birds, not from their feces. Lesions in the mouth or the esophagus or both of an infected bird (see below) often prevent the passage of ingested grain seeds and cause the bird to regurgitate contaminated food items. Water becomes contaminated by contact with the contaminated bill and mouth. Pigeons that feed among domestic poultry are often blamed for contaminating feed and water and passing the disease to the poultry. Similar transmission has been associated with dove mortality at grain elevators and at birdfeeders. Doves and pigeons cross-feed and bill during courtship, and this behavior facilitates direct transmission as does the consumption of infected birds by raptors. It has been reported that some moist grains can maintain viable *T. gallinae* for at least 5 days and that parasite survival in water can range from 20 minutes to several hours. These conditions are adequate for disease transmission at birdfeeders and waterers because of the gregarious habits of doves and pigeons.

Species Affected

Trichomoniasis is considered by many avian disease specialists to be the most important disease of mourning doves in North America. Band-tailed pigeons have also suffered large-scale losses from trichomoniasis. This disease has been reported as a cause of mortality in birds of prey for hundreds of years prior to the causative organism being identified. Songbirds are less commonly reported to be infected, but *T. gallinae* is reported to be the most important trichomonad of caged birds; it is often responsible for epizootics among captive collections. Domestic turkeys and chickens also become infected.

Distribution

It is likely that *T. gallinae* is found wherever domestic pigeons and mourning doves are found. Disease in free-ranging wild birds is grossly underreported. Outbreaks at birdfeeding stations and similar locations reported to the National Wildlife Health Center have occurred from coast-to-coast within the United States (Fig. 25.3).

Seasonality

Epizootics due to *T. gallinae* can happen yearround, but most outbreaks have been reported during late spring, summer, and fall.

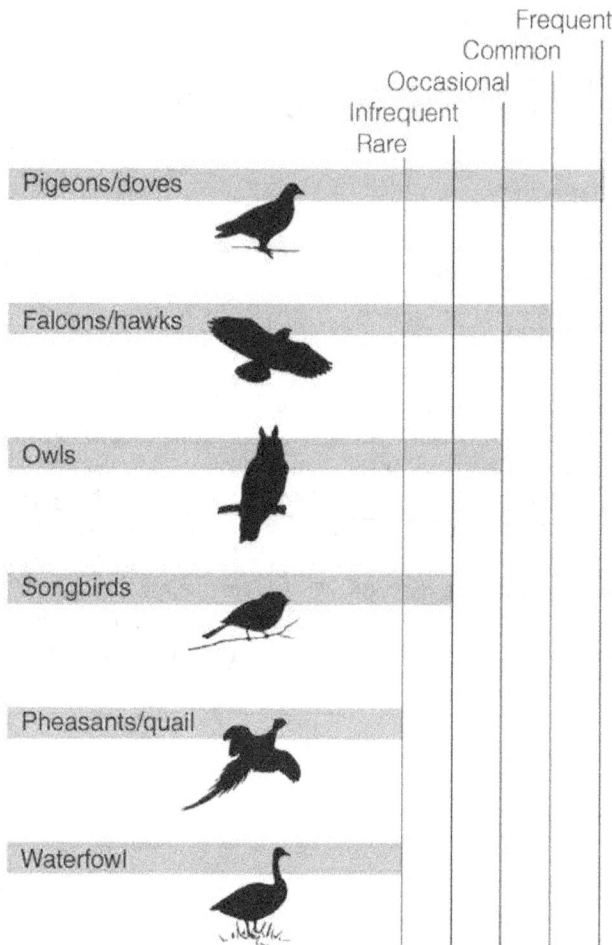

Figure 25.1 Relative frequency of trichomoniasis in free-ranging birds.

Field Signs

Because oral lesions often affect the ability of the bird to feed, infected birds lose weight, appear listless, and stand grouped together. These birds often appear ruffled. Caseous or cheesy, yellowish lesions may be seen around the beak or eyes of mourning doves and the face may appear "puffy" and distended (Fig. 25.4). Severely infected pigeons may fall over when they are forced to move.

Gross Lesions

The severity and appearance of lesions varies with the virulence of the strain of the parasite, the stage of infection, and the age of the bird. The most visible lesions from mildly pathogenic strains may simply appear as excess salivation and inflammation of the mucosa or lining of the mouth and throat. Early oral lesions appear as small, well defined, cream to yellowish spots on the mucosal surface (Fig. 25.5A). As the disease progresses the lesions become larger, thicker, and

Figure 25.2 Transmission of trichomoniasis.

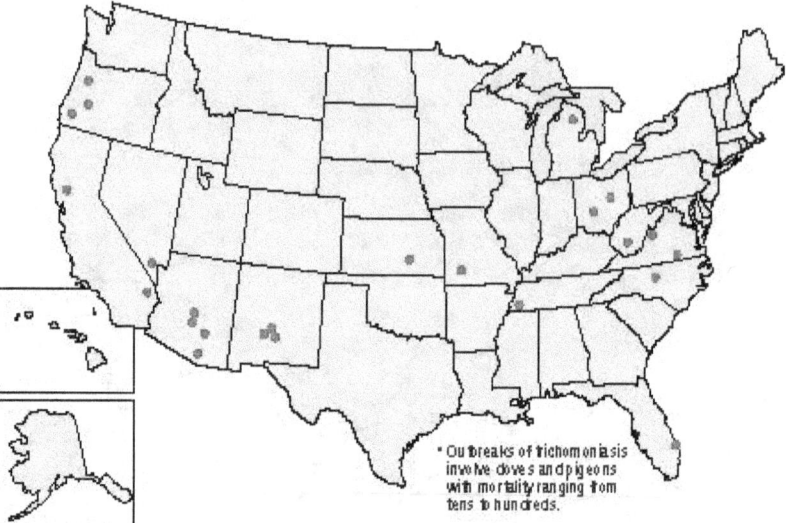

EXPLANATION
Trichomoniasis outbreak sites, 1983–97*

- Outbreak site

*Outbreaks of trichomoniasis involve doves and pigeons with mortality ranging from tens to hundreds.

Figure 25.3 Locations of outbreaks of trichomoniasis in free-ranging birds, January 1983 through March 1997.

Figure 25.4 Mourning doves at a backyard waterbath. Note the puffy appearance (arrow) of the face of a *T. gallinae* infected dove.

are caseous (consistency of cheese) in appearance (Fig. 25.5B). In more advanced lesions, a wet, sticky type of discharge and nodules within the mouth are characteristic of acute disease. Hard, cheesy lesions are most often seen in more chronic infections. Although lesions are generally confined to the inside of the mouth and esophagus, they can extend externally to the beak and eyes and be confused with avian pox (see Chapter 19).

Early lesions of the pharynx to the crop are also cream to yellow in color and caseous. As the disease progresses, these lesions may spread to the esophagus (Fig. 25.5C), and can eventually block its opening (Fig. 25.5D). A bird can suffocate if the blockage is severe enough. A bird will starve when these masses prevent it from swallowing food and water. These large, caseous masses may invade the roof of the mouth and sinuses (Fig. 25.5E) and even penetrate through the base of the skull into the brain. Also, a large amount of fluid may accumulate in the crop of severely infected birds. Lesions may extend down the alimentary tract and the parasite may invade the liver, particularly in domestic pigeons. Other organs such as the lungs, occasionally become involved. The digestive tract below the proventriculus is rarely involved.

Diagnosis

A tentative diagnosis can be made for doves and pigeons on the basis of finding caseous, obstructive lesions within the upper areas of the digestive tract. However, other disease agents such as pox virus, *Aspergillus* sp. fungi, *Candida* sp. yeasts, nematodes of the genus *Capillaria*, and vitamin A deficiency can produce similar lesions. Diagnosis is estab-

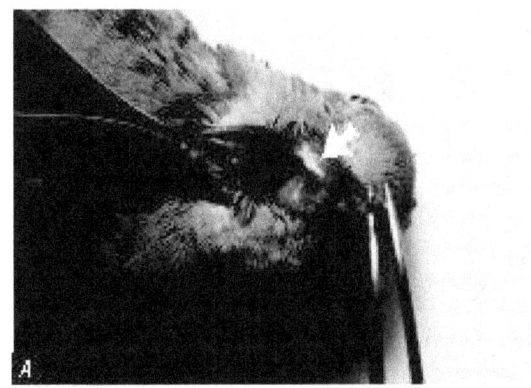

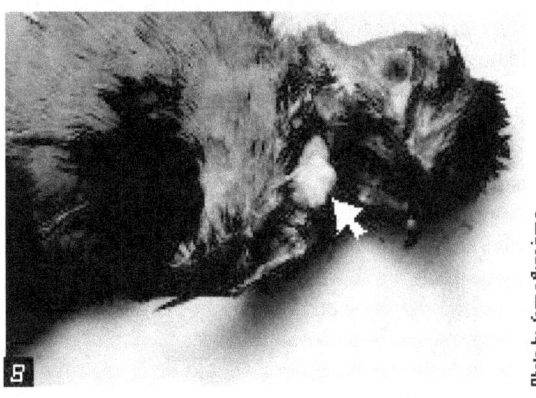

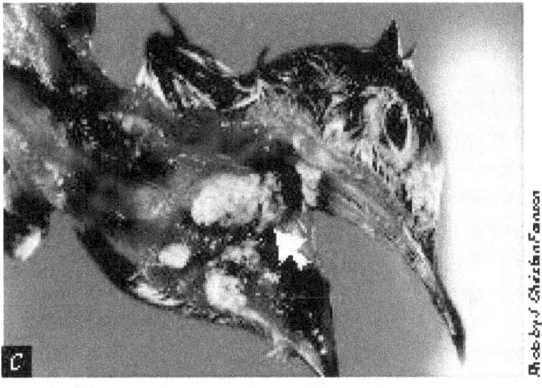

Figure 25.5 Gross lesions of trichomoniasis in mourning doves. (A) Small, cream-colored lesion on oral mucosa (arrow). (B) Large, caseous lesions in back of mouth (arrow). (C) Large lesion in upper esophagus (arrow). (D) Occlusion of esophagus by a large, caseous lesion. (E) Lesions on the roof of the mouth, in the region of the sinuses (arrow).

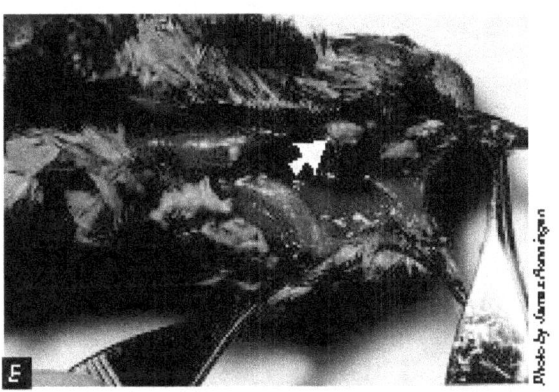

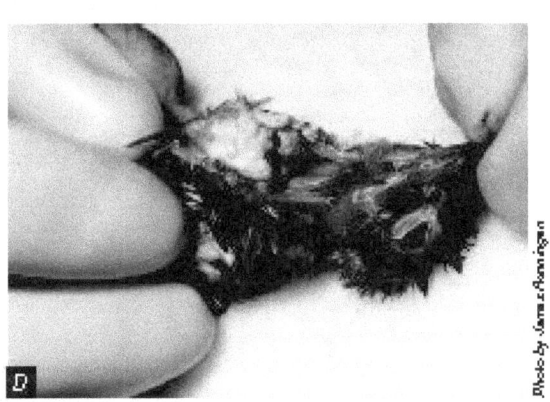

lished by finding the trichomonads in the saliva or smears of the caseous lesions of infected birds. Specimens are best taken from sick birds, or from recently dead birds that are kept chilled and reach the diagnostic laboratory within 48 hours after death. Samples of tissues with lesions preserved in 10 percent buffered formalin or frozen whole carcasses can be used if fresh carcasses cannot be provided.

Control

The removal of infected birds is recommended for combating trichomoniasis in poultry and captive pigeons and in captive collections of wild birds. The focus in both instances is on birds that harbor virulent strains of the parasite. Elimination of infection from adult birds by drug treatment has also been recommended, but this is not a practical approach for wild birds. Prevention of the build-up of large concentrations of doves at birdfeeders and artificial watering areas is recommended to minimize disease transmission in the wild. Stock tanks, livestock feedlots, grain storage facilities and clusters of urban birdfeeders should be targeted for disease prevention activities. Although the environmental persistence for *T. gallinae* is rather limited, contaminated feed is suspected as a significant source of disease transmission. Therefore, fresh feed should be placed in feeders daily, if it is practical. Platforms and other surfaces where feed may collect, including the area under feeders, should be frequently decontaminated with 10 percent solution of household bleach in water, preferably just prior to placing clean feed in the feeder. Pigeons and doves are high risk food sources for birds of prey; therefore, before they are fed to raptors, pigeons and doves should be inspected first and found to be free of trichomoniasis or other infectious diseases.

Human Health Considerations

None. *T. gallinae* has not been reported to infect humans.

Rebecca A. Cole

Supplementary Reading

Conti, J.A., 1993, Diseases, parasites, and contaminants, *in* Baskett, T.S., and others, eds., Ecology and management of the mourning dove: Harrisburg, Pa., Stackpole Books, p. 205–224.

Levine, N.D., 1985, Flagellates: the trichomonads in veterinary protozoology: Ames, Iowa, Iowa State University Press, p. 72–74.

Pokras, M.A., Wheeldon, E.B., and Sedgwick, C.J., 1993, Raptor biomedicine, *in* Redig, P.T. and others, eds., Trichomoniasis in owls: report on a number of clinical cases and a survey of the literature: Minneapolis, Minn., University Minnesota Press, p. 88–91.

Rupier, D.J., and W.M. Harmon, 1988, Prevalence of *Trichomonas gallinae* in central California mourning doves: California Fish and Game, v. 74, no. 4, p. 471–473.

Stabler, R.M., 1951, A survey of Colorado band-tailed Pigeons, mourning doves, and wild common pigeons for *Trichomonas gallinae*: Journal of Parasitology, v. 37, p. 471–473.

Chapter 26
Intestinal Coccidiosis

Synonyms
Coccidiosis, coccidiasis

Cause

Coccidia are a complex and diverse group of protozoan (single-celled organisms) parasites; the coccidia group contains many species, most of which do not cause clinical disease. In birds, most disease-causing or pathogenic forms of coccidia parasites belong to the genus *Eimeria*. Coccidia usually invade the intestinal tract, but some invade other organs, such as the liver and kidney (see Chapter 27).

Clinical illness caused by infection with these parasites is referred to as coccidiosis, but their presence without disease is called coccidiasis. In most cases, a bird that is infected by coccidia will develop immunity from disease and it will recover unless it is reinfected. The occurrence of disease depends, in part, upon the number of host cells that are destroyed by the juvenile form of the parasite, and this is moderated by many factors. Severely infected birds may die very quickly. Often, tissue damage to the bird's intestine results in interrupted feeding; disruption of digestive processes or nutrient absorption; dehydration; anemia; and increased susceptibility to other disease agents. In cranes, coccidia that normally inhabit the intestine sometimes become widely distributed throughout the body. The resulting disease, disseminated visceral coccidiosis (DVC) of cranes, is characterized by nodules, or granulomas, on the surface of organs and tissues that contain developmental stages of the parasite.

Collectively, coccidia are important parasites of domestic animals, but, because each coccidia species has a preference for parasitizing a particular bird species and because of the self-limiting nature of most infections, coccidiosis in free-ranging birds has not been of great concern. However, habitat losses that concentrate bird populations and the increasing numbers of captive-reared birds that are released into the wild enhance the potential for problems with coccidiosis.

Life Cycle

Most intestinal coccidia have a complex but direct life cycle in which the infective forms of the parasite invade a single host animal for development to sexual maturity; the life cycle is completed in 1–2 weeks (Fig. 26.1). A mature female parasite in the intestine of an infected host bird produces noninfective, embryonated eggs or oocysts, which are passed into the environment in the feces of the host bird. The oocysts quickly develop into an infective form while they are in the environment. An uninfected bird ingests the infective oocysts while it is eating or drinking, and the infective oocysts invade the bird's intestine. Within the intestine, the oocysts may or may not undergo several stages of development, depending on the parasite species, before they become sexually mature male and female parasites. The complex life cycle for *Eimeria* (Fig. 26.2) illustrates the exponential rate of infection and destruction of the intestinal epithelial cells, which are the cells that provide the covering of the intestinal lining. The mature female parasites release noninfective oocysts to the environment, and, thus, the cycle begins anew.

Species Affected

Many animal species, including a wide variety of birds (Table 26.1) may harbor coccidia. Although disease is not common in free-ranging wild birds, several epizootics due to *E. aythyae* have been reported among lesser scaup in the United States. During those events, predominantly females have died, which suggests that female lesser scaup may be more susceptible to the disease than male lesser scaup. Lesions of DVC were first seen in captive sandhill cranes in the late 1970s. Since then, mortality of captive sandhill and whooping cranes has been attributed to DVC, and the disease has been found in wild sandhill cranes, including the endangered Mississippi sandhill crane.

Characteristics of Intestinal Coccidiosis

All domestic birds carry more than one species of coccidia, and pure infections with a single species are rare.

Different coccidia species are usually found in a specific location within the intestinal tract of the host bird.

After initial exposure to the parasite, the host bird may quickly develop immunity to it but immunity is not absolute. A bird can be reinfected by the same or a different species of the parasite.

Infections do not generally cause a problem of free-ranging birds; instead, coccidiosis is considered a disease of monoculture and of the raising of birds in confinement.

Distribution

Coccidia are found worldwide. The few reported outbreaks of coccidiosis in free-ranging waterfowl have all occurred in the Midwestern United States (Fig. 26.3). Recurrent epizootics have broken out at a single reservoir in eastern Nebraska, and coccidiosis is also believed to be the cause of waterfowl die-offs in Wisconsin, North Dakota, Illinois, and Iowa. DVC has been found in migratory sandhill cranes at several locations, and it is a recurring problem in the only free-ranging population of the nonmigratory Mississippi sandhill crane. These birds reside at the Mississippi Sandhill Crane National Wildlife Refuge in Mississippi.

Seasonality

Birds may be infected with coccidia at any time. Although little is known about the conditions that may lead to the development of clinical disease in wild birds, birds may become diseased more frequently during periods of stress. Most epizootics of intestinal coccidiosis in waterfowl in the Upper Midwest have broken out in early spring, during a stressful staging period of spring migration. Mississippi sandhill cranes also die from DVC most frequently during the spring.

Field Signs

Field signs for free-ranging wild birds have not been reported. Nonspecific clinical signs reported for captive birds include inactivity, anaemia, weight loss, general unthrifty appearance, and a watery diarrhea that may be greenish or bloody. Tremors, convulsions, and lameness are also occasionally seen. Rapid weight loss may lead to emaciation and dehydration followed by death. Young birds that survive severe infections may suffer retardation of growth.

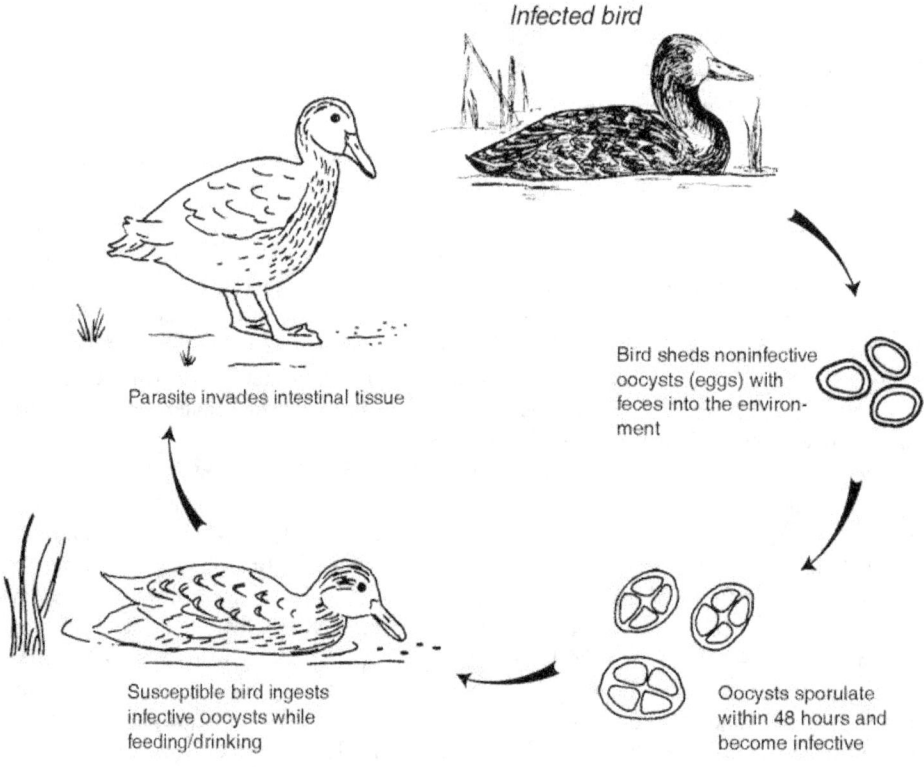

*Figure 26.1 Direct life cycle of **Eimeria** infection in birds.*

A. Noninfective parasite oocysts (eggs) containing a single cell referred to as the sporont are passed via feces into the environment.

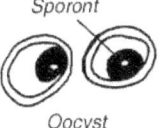
Sporont
Oocyst

B. Oocysts become infective after 2 days in the environment at ordinary temperatures through sporolation (sporogony), which is a developmental process that results in the sporont dividing and forming four sporocysts each containing two infective sporozoites.

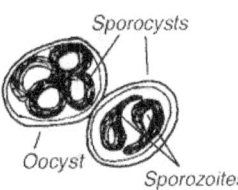
Sporocysts
Oocyst
Sporozoites

C. Infective oocysts are ingested by birds in contaminated feed, water, soil, or other ingesta.

D. The oocyst wall breaks within the gizzard of the bird and releases the sporocysts.

E. The sporozoites escape from the sporocysts in the small intestine and enter the epithelial cells, which are cells that line the internal and external surfaces of the body of the intestine.

F. The sporozoites develop within the epithelial cells, and asexual multiple fission results in the formation of first-generation meronts, each of which produces about 900 first-generation merozoites.

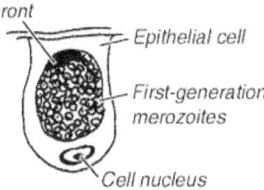
Meront
Epithelial cell
First-generation merozoites
Cell nucleus

G. Merozoites break out of the epithelial cells into the intestinal canal about 2.5–3 days after infection. The merozoites enter new host cells and undergo developmental processes resulting in the formation of second-generation meronts. By dividing many times, each of these meronts produce about

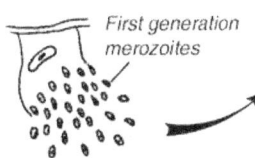
First generation merozoites

200–350 second-generation merozoites that are 4–8 times larger in size than the first-generation merozoites and that are produced about 5 days after initial oocyst ingestion.

H. The cycle may continue with a third generation of a small number (4–30) of merozoites of intermediate size (between those of the first and second generation). However, many of the second-generation merozoites enter new host cells and begin the sexual phase of the life cycle referred to as gamogony.

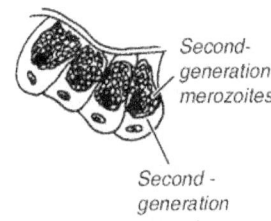
Second-generation merozoites
Second-generation meront

I. Most of the second-generation merozoites develop into female gametes or macrogamonts and some become males or microgamonts. The females grow until they reach full size while a large number of tiny microgametes are formed within each of the microgamonts. The macrogamonts are fertilized by the microgametes and new oocysts result.

J. Seven days after ingestion of infected coccidia, the oocysts break out of their host cells and enter the intestinal canal to be passed from the body via feces to continue the cycle.

Figure 26.2 A typical life cycle of ***Eimeria*** sp. in birds. (Adapted from ***Eimeria tenella*** in chickens.)

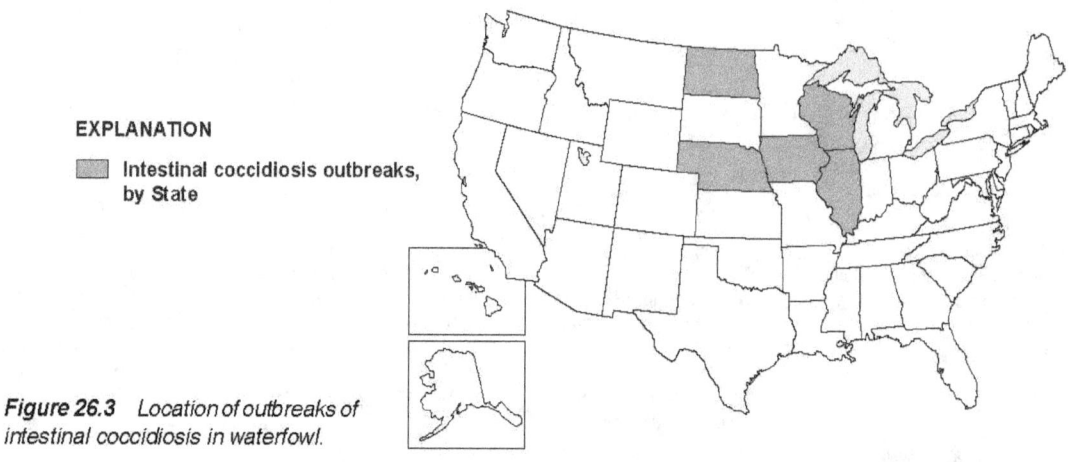

Figure 26.3 Location of outbreaks of intestinal coccidiosis in waterfowl.

Table 26.1 Relative occurrence of coccidia in different groups of birds. [Frequency of occurrence: ● occasional, ● common, — not reported]

Bird types (and examples)	Coccidia species				
	Eimeria sp.	*Isospora* sp.	*Tyzzeria* sp.	*Cryptosporidium* sp.	*Wenyonella* sp.
Poultry (Chicken, turkey)	●	—	—	●	—
Anseriformes (Ducks, geese)	●	●	●	●	●
Charadriiformes (Gulls, shorebirds)	●	—	—	—	—
Columbiformes (Pigeons, doves)	●	—	—	—	●
Coraciiformes (Kingfishers)	—	●	—	—	—
Falconiformes (Hawks, falcons)	—	●	—	—	—
Galliformes (Pheasant, quail)	●	●	—	●	—
Gruiformes (Cranes, rails)	●	—	—	—	—
Passeriformes (Songbirds)	—	●	—	—	—
Pelicaniiformes (Pelicans)	●	—	—	—	—
Piciformes (Woodpeckers)	—	●	—	—	—
Psittaciformes (Parrots)	●	●	—	●	—
Strigiformes (Owls)	—	●	—	—	—
Struthioniformes (Ostriches)	—	●	—	—	—

Gross Lesions

The location of lesions varies with the species of coccidia and the severity and intensity of infection. In acutely-affected lesser scaup, bloody inflammation or enteritis is commonly seen in the upper small intestine (Fig. 26.4A). In scaup that survive for longer periods, dry crusts form on the mucosal (internal) surface of the intestinal tract. The severity of this lesion decreases from the small intestine to the large intestine (Fig. 26.4B). Chronic lesions of intestinal coccidiosis take other forms in different species, sometimes appearing as rather distinct light-colored areas within the intestinal wall (Fig. 26.5).

Lesions of DVC in cranes typically consist of small (usually less than 5 millimeters in diameter), raised, light-colored granulomas. These nodules may be found on any surface within the body cavity, but they are commonly seen on the lining of the esophagus near the thoracic inlet area and on the inner surface of the sternum (Fig. 26.6A–C). Light-colored patches may also appear on and within organs such as the heart and liver (Fig. 26.7A, B).

Diagnosis

When large numbers of oocysts are found in the feces of live birds concurrent with diarrhea, emaciation, and pallor or pale skin color, coccidiosis should be suspected as the cause of illness. However, a diagnosis of coccidiosis as cause of death requires a necropsy evaluation combined with identification of the causative coccidia. Fecal evaluations are not adequate for a diagnosis of coccidiosis because disease may develop before large numbers of oocysts are present in feces and because oocysts seen in the feces may not be those of pathogenic species. As with other diagnostic evaluations, submit chilled, whole carcasses for necropsy by qualified specialists. When carcasses cannot be provided, remove intestinal tracts and submit them chilled. If submissions will be delayed for several days or longer and carcasses cannot be preserved by freezing, remove the entire intestinal tract and preserve it in an adequate volume of neutral formalin (see Chapter 3).

Control

Oocysts can rapidly build up in the environment when birds are overcrowded and use an area for a prolonged period of time. The disease risk increases significantly when these conditions result in oocyst contamination of food and drinking water. In captive situations, good husbandry and sanitation, including continual removal of contaminated feed and litter, can minimize the potential for coccidiosis. Captive birds can be treated with therapeutic agents that control, but that do not eliminate, the level of infection. Therefore, oocyst shedding by those birds after they are removed from therapy should be considered if they are to be released or mixed with other birds. Light infections result in a substantial level of immunity to that species of coccidia and are use-

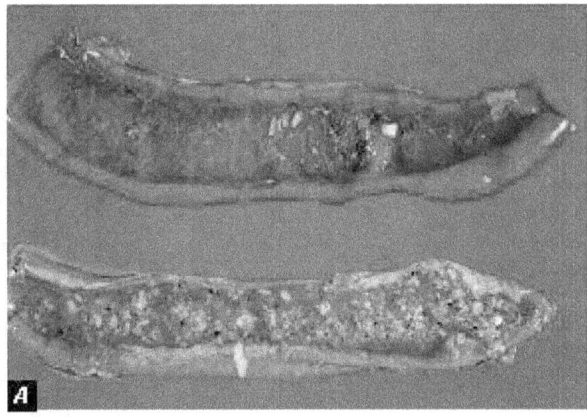
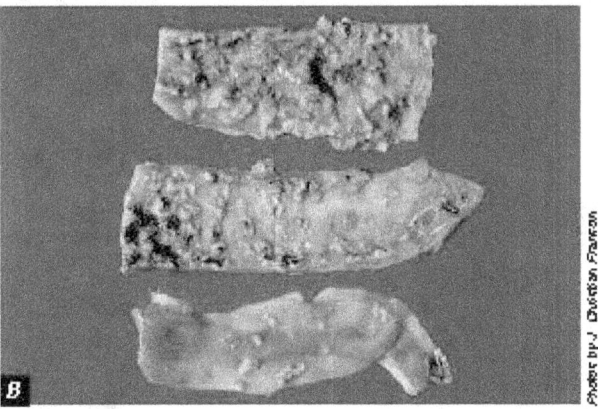

Figure 26.4 (A) Hemorrhage in the small intestine of a lesser scaup with acute intestinal coccidiosis (upper part of photo), compared with normal small intestine (lower part of photo). (B) Dry, crust-like lesions in the intestinal tract of a lesser scaup with chronic intestinal coccidiosis. The lesions are most severe in the upper small intestine (top section in photo). The severity decreases in lower parts of the intestine (middle and bottom sections in photo).

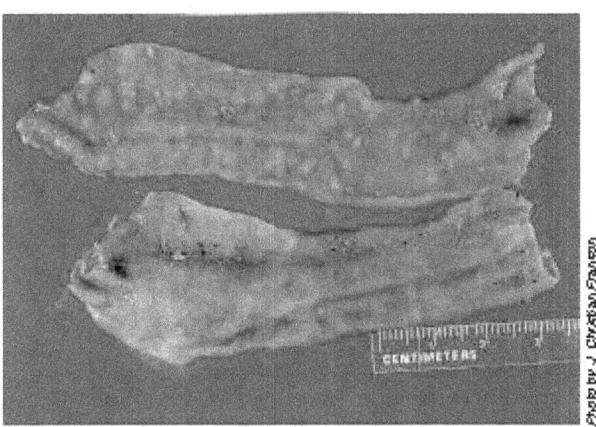

Figure 26.5 Intestinal coccidiosis in a common eider from Alaska, showing distinct light-colored areas within the wall of the intestine.

Intestinal Coccidiosis 211

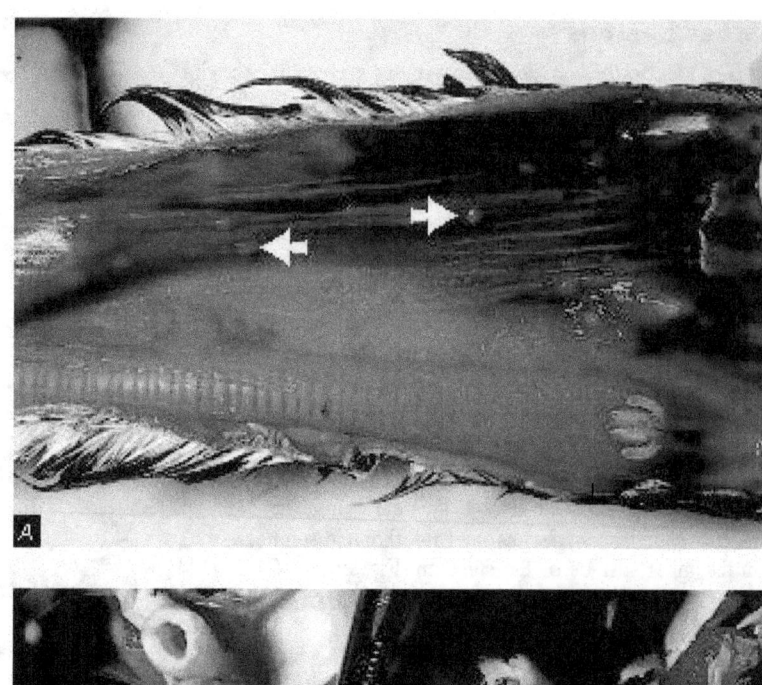
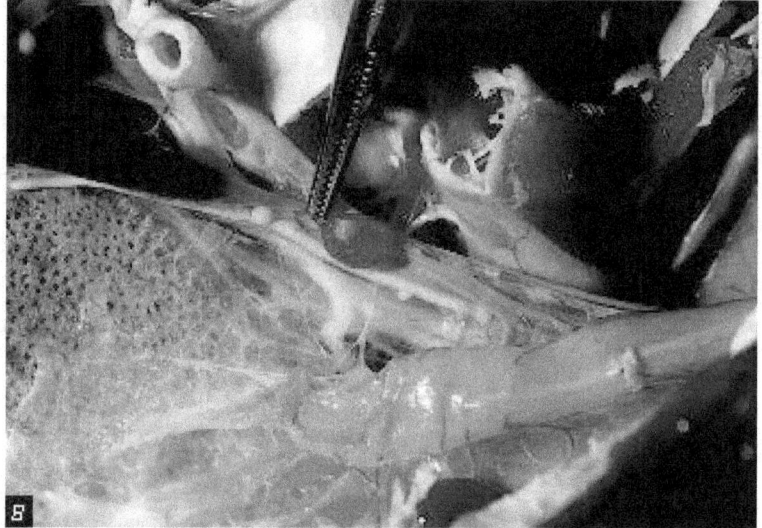

Figure 26.6 Gross lesions of disseminated visceral coccidiosis of cranes. (A) Granulomas on the lining of the esophagus (arrows); and (B) in the area of the thoracic inlet [the tip of the forceps is between granulomas on the surface of a vessel and nerve (left) and on the thyroid gland (right)]; and (C) on the inside surface of the sternum (arrow).

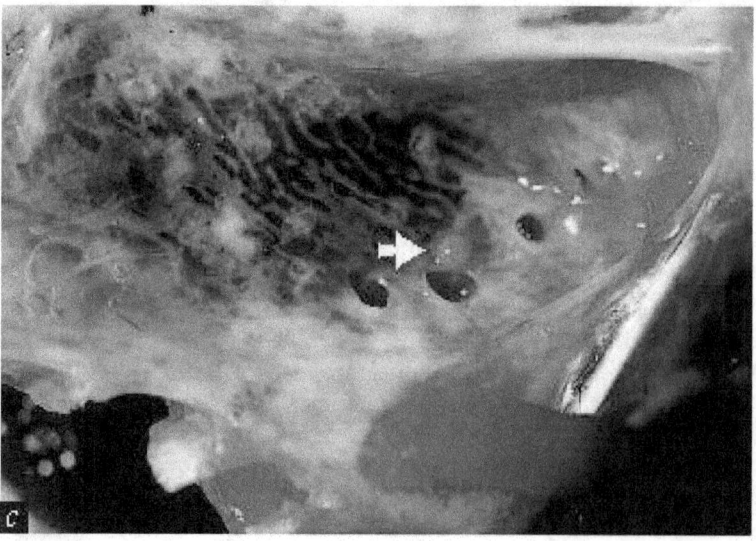

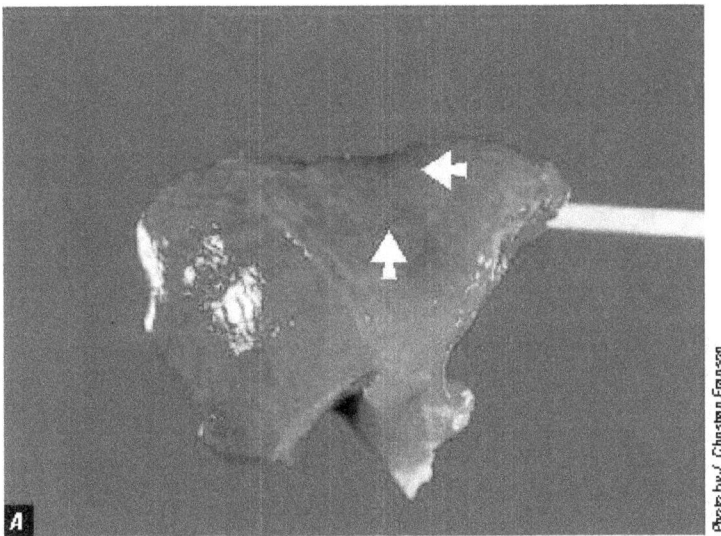
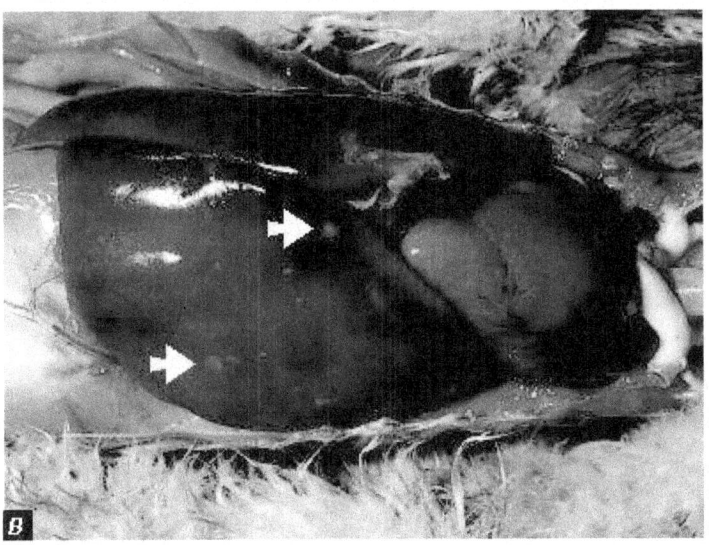

Figure 26.7 Lesions of disseminated visceral coccidiosis also may include light patches as seen here on the **(A)**, surfaces of the heart muscle and **(B)** on the liver (arrows).

ful in preventing epizootics from this disease. Therefore, the objective is not to completely eliminate infection with coccidia; instead, the focus should be on preventing heavy infections and the establishment and persistence of high levels of environmental contamination with coccidia. For free-ranging birds, flock dispersal may be warranted when overcrowding continues for prolonged periods of time.

Human Health Considerations

None. Coccidia of birds are not infectious for humans.

Milton Friend and J. Christian Franson

Supplementary Reading

Carpenter, J.W., Novilla, M.N., Fayer, R., and Iverson, G.C., 1984, Disseminated visceral coccidiosis in sandhill cranes: Journal of the American Veterinary Medical Association, v. 185, no. 11, p. 1,342–1,346.

Courtney, C.H., Forrester, D.J., Ernst, J.V., and Nesbitt, S.A., 1975, Coccidia of sandhill cranes, *Grus canadensis*: The Journal of Parasitology, v. 61, no. 4, p. 695–699.

Novilla, M.N., Carpenter, J.W., Spraker, T.R., and Jeffers, T.K., 1981, Parenteral development of Eimerian coccidia in sandhill and whooping cranes: Journal of Protozoology, v., 28, no. 2, p. 248–255.

Parker, B.B., and Duszynski, D.W., 1986, Coccidiosis of sandhill cranes (*Grus canadensis*) wintering in New Mexico: Journal of Wildlife Diseases, v. 22, no. 1, p. 25–35.

Windingstad, R.M., McDonald, M.C., Locke, L.N., Kerr, S.M., and Sinn, J.A., 1980, Epizootic of coccidiosis in free-flying lesser scaup: Avian Diseases 24, p. 1,044–1,049.

Chapter 27
Renal Coccidiosis

Cause

Renal coccidiosis is caused by protozoal parasites that infect the kidneys and associated tissues. Most of the coccidia that infect the tissues in birds are *Eimeria* sp. As with most other parasitic infections, this infection is not synonymous with clinical or apparent disease. Asymptomatic infections are far more common than those that are severe and cause mortality.

Life Cycle

Typical *Eimeria*-type life cycles have an internal or endogenous phase of development within the host. A bird becomes a host when it feeds or drinks from a source that is contaminated with oocysts (cystic, infectious stage) that have become infectious following multiple fission of the sporont (zygote) to form four sporocysts, each containing two infec-

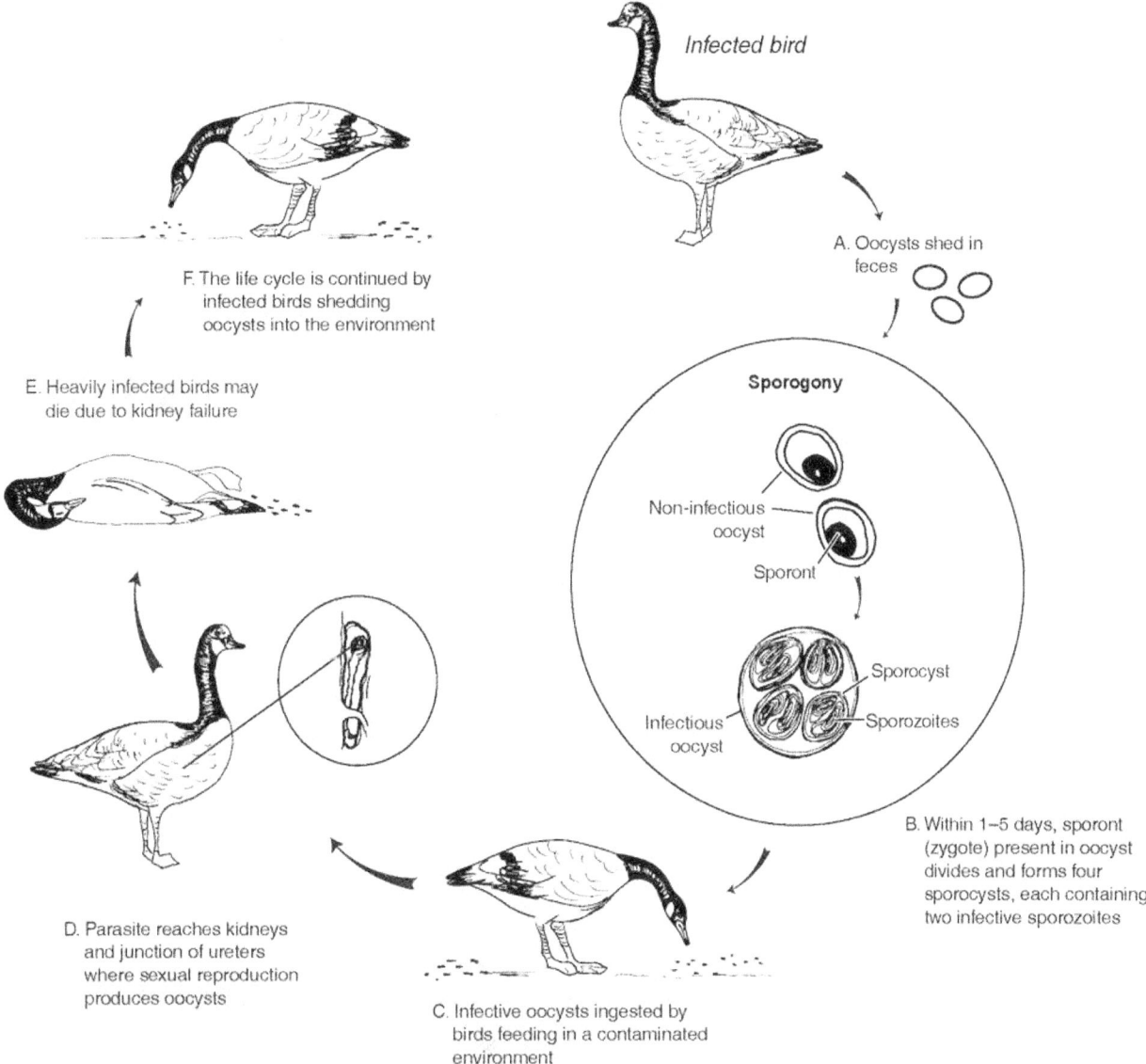

Figure 27.1 Life cycle of **Eimeria truncata**, which is one of the parasites that causes renal coccidiosis.

tious sporozoites (sporogony) within each oocyst. The infective sporozoites within the sporocysts of the oocysts invade the bird's intestinal lining, where they may undergo several developmental stages depending on the *Eimeria* species. *E. truncata*, the most well known of the renal coccidia, matures and reproduces only in the kidneys and in the cloaca near its junction with the ureter (Fig. 27.1). It is not known how the *E. truncata* sporozoites get from the intestine to the kidneys; the sporozoites probably undergo asexual reproduction or multiple fission before they reach the kidneys. The sexual phase of the *E. truncata* life cycle, or gamogony, takes place in the kidneys, producing noninfectious oocysts which are voided with the host bird's feces into the environment. Sporulated oocysts are resistant to environmental extremes, and their sporozoites can remain infectious for months.

The life cycles of the coccidia that cause renal coccidiosis are similar to those that cause intestinal coccidiosis (see Chapter 26). However, less is known about the species of *Eimeria* that cause renal coccidiosis than about those that cause intestinal coccidiosis.

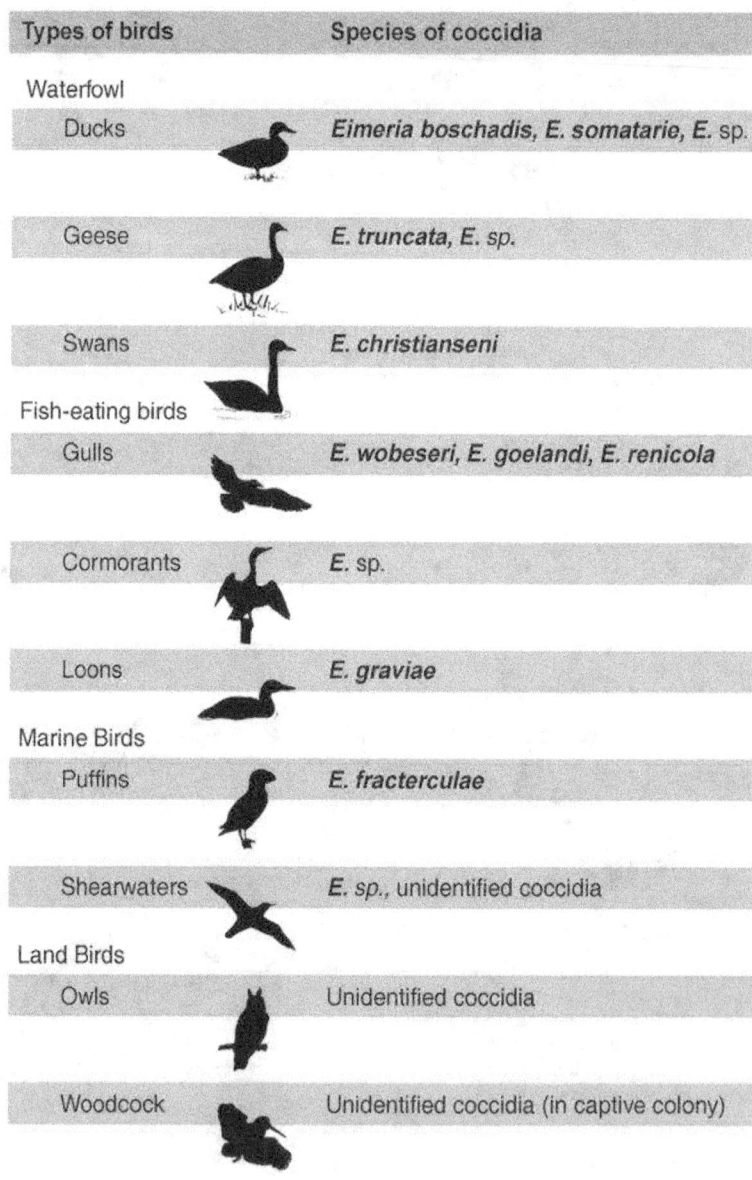

Figure 27.2   Reported occurrences of renal coccidia in wild birds.

Species Affected

Avian coccidiosis was first reported in France. Canadian investigators have reported that virtually all species of wild ducks they examined are susceptible to renal coccidiosis. Different species of renal coccidia are found in different species of birds (Fig. 27.2). Most reports of renal coccidiosis are of asymptomatic birds or birds that show minor physiological or pathological changes due to the parasite. Young birds and those that have been stressed by various conditions are most likely to have clinical cases of renal coccidiosis. Mortality has occurred in free-ranging wild geese, eider ducklings, and double-crested cormorants. Disease in domestic geese is usually acute, lasts only 2–3 days, and can kill large segments of the flock.

Distribution

Renal coccidiosis is found in birds worldwide.

Seasonality

Mortality from renal coccidiosis is most common during periods of the year when birds are densely aggregated on their breeding grounds or wintering areas.

Field Signs

There are no specific field signs that indicate that a bird is infected with renal coccidia. Young birds will often be emaciated and weak, but many other diseases cause similar clinical signs.

Gross Lesions

Infected birds may be emaciated and have a prominent keel. In severe infections, kidneys may become enlarged and pale, containing multiple spots or foci of infection that coalesce into a mottled pattern (Fig. 27.3). Cutting through these white foci may reveal material that has the consistency of chalk due to the buildup of uric acid salts (Fig. 27.4).

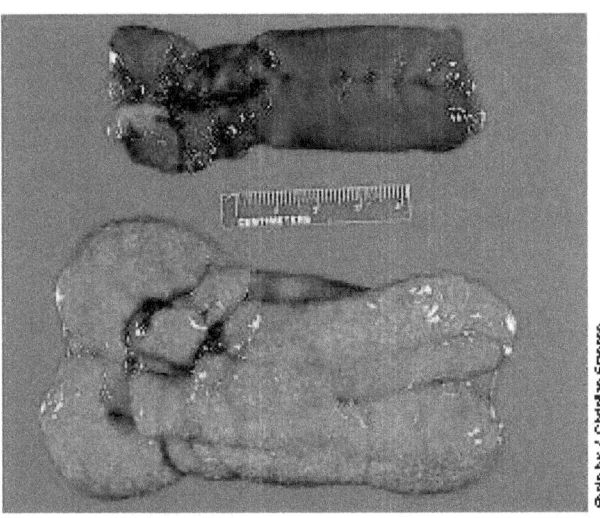

Figure 27.3 Kidneys from double-crested cormorants. Top: normal size and color. Bottom: enlarged kidneys with diffuse pale areas from a bird infected with renal coccidia.

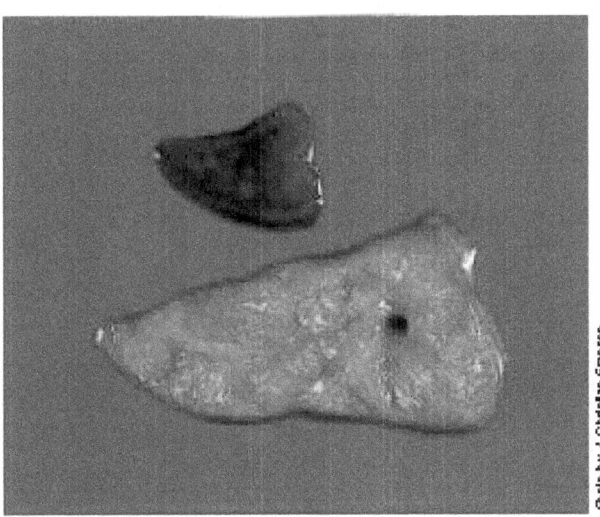

Figure 27.4 Cut surfaces from the same two kidneys as in Fig. 27.3. Bottom kidney shows chalky material from buildup of uric acid salts.

Diagnosis

Confirmation of renal coccidiosis requires microscopic examination of tissue by the trained staff of a diagnostic laboratory. Whole carcasses are generally needed to determine the cause of death unless kidney damage is so severe that it unquestionably would have caused death. When whole refrigerated carcasses cannot be provided for evaluation because of field circumstances, the kidneys should be removed, preserved in a 10:1 volume of 10 percent buffered neutral formalin and submitted for diagnosis (see Chapter 2).

Control

Control of renal coccidiosis in free-ranging birds is not feasible. Crowded conditions facilitate transmission of the parasite through fecal contamination of the environment. Prevention of degradation of habitat quantity and quality on breeding grounds and wintering areas is needed to minimize disease risks.

Human Health Considerations

There are no reports of human health concerns with this disease.

Rebecca A. Cole

Supplementary Reading

Gajadhar, A.A., and Leighton, F.A., 1988, *Eimeria wobeseri* sp. n. and *Eimeria goelandi* sp. n. (Protozoa: Apicomplexa) in the kidneys of herring gulls (*Larus argentatus*): Journal of Wildlife Diseases, v. 24, p. 538–546.

Oksanen, A., 1994, Mortality associated with renal coccidiosis in juvenile wild greylag geese (*Anser anser anser*): Journal of Wildlife Disease, v. 30, p. 554–556.

Wobeser, G., and Stockdale, P.H.G., 1983, Coccidia of domestic and wild waterfowl (Anseriformes): Canadian Journal of Zoology, v. 61, p. 1–24.

Chapter 28
Sarcocystis

Synonyms
Rice breast disease, sarcosporidiosis, sarcocystosis

Cause
Sarcocystis is a nonfatal, usually asymptomatic infection that is caused by a parasitic protozoan. Various species of this parasite affect mammals, reptiles, and birds. The most commonly reported species of the parasite in North America is *Sarcocystis rileyi*, the species most commonly found in waterfowl.

Life Cycle
The *Sarcocystis* sp. parasites have an indirect life cycle (Fig. 28.1) that requires a paratentic or transport host animal (a bird), in which they live for a time before they are transported to a definitive host animal (a carnivore), in which they reach maturity. Birds ingest the eggs or oocysts of the mature parasite in food or water that is contaminated by carnivore feces, which contain the oocysts. The oocysts develop in the intestine of the bird into an intermediate form, the sporozoites, that enter the bird's bloodstream and infect specific cells of the blood vessels. Multiplication of these cells gives rise to a second intermediate form, merozoites, that are carried by the blood to the voluntary muscles, where elongated cysts or macrocysts are eventually produced (Fig. 28.2). The life cycle is completed when a carnivore ingests the infected muscle tissue of a bird and the parasite reaches maturity and releases oocysts in the intestines of the carnivore. The carnivore is infected only in its intestine. Macrocysts do

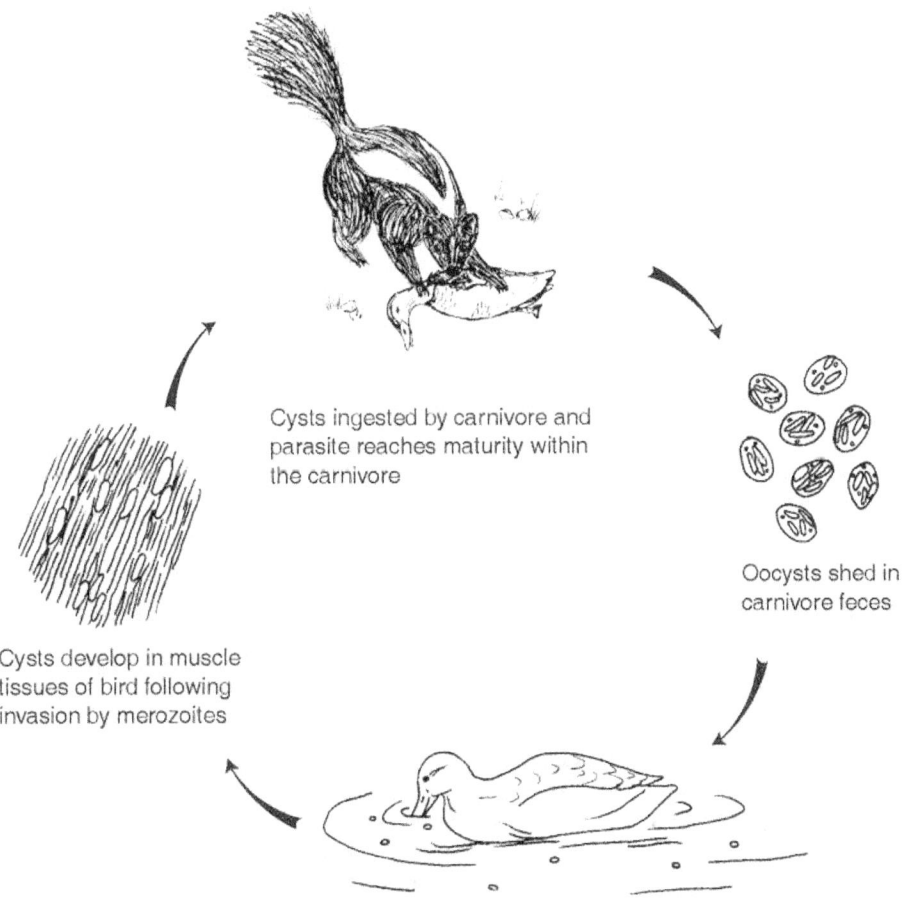

Figure 28.1 General life cycle of **Sarcocystis** *sp.*

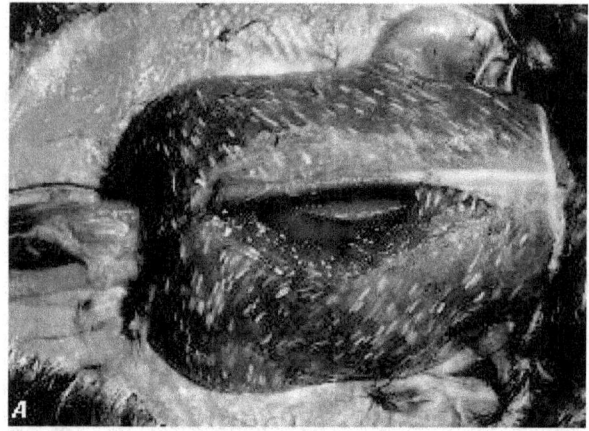 

Figure 28.2 Rice-grain sized cysts of **Sarcocystis** sp. evident in parallel streaks in **A**, breast muscle fibers of a mallard and **B**, thigh and leg muscle of an American black duck.

not develop in the carnivore, and the *Sarcocystis* sp. parasite rarely causes the carnivore illness or other forms of disease.

Species Affected

Dabbling ducks (mallard, northern pintail, northern shoveler, teal, American black duck, gadwall, and American wigeon) commonly have visible or macroscopic forms of *Sarcocystis* sp.; these forms are far less frequently found in other species of ducks and are infrequently found in geese and swans. Recent studies of wading birds in Florida have disclosed a high prevalence of *Sarcocystis* sp.; similar findings have previously been reported from South Africa. Land birds, such as grackles and other passerine birds, as well as mammals and reptiles can have visible forms of sarcocystis, but it is unlikely that *S. rileyi* is the species of parasite involved. With the exception of waterfowl, this parasite has received little study in migratory birds. This must be taken into account when considering the current knowledge of species affected (Fig. 28.3).

Distribution

Sarcocystis is a common parasitic infection of some waterfowl species, and it is found throughout the geographic range of those species in North America. Less is known about *Sarcocystis* sp. in other species of wild birds, but this parasite has been reported from waterbirds in South Africa, Australia, Canada, and Mexico in addition to the United States.

Seasonality

Infected birds can be found year round, but waterfowl that are infected with *Sarcocystis* sp. are usually observed during the hunting season. Infection is not seen in prefledgling waterfowl, nor is it often seen in juveniles. Two possible reasons for these differences between the age classes may be that the development of visible forms of the parasite requires time or that birds may not be infected until after they have left their breeding grounds. Because visible forms of sarcocystis are more frequently developed in older birds, hunter detection tends to be greatest during years of poor waterfowl production when the bag contains a greater proportion of adult birds. A moderate percentage of juvenile mottled ducks that were collected in Louisiana primarily after the hunting season were recently found to have light sarcocystis infections. Because this species does not migrate, this suggests that the birds were infected within the general geographic area where they were collected and that the later collection date allowed the macrocyst lesions to be visible.

Too little is known about sarcocystis in other groups of wild birds to evaluate its seasonality.

Field Signs

Usually, there is no externally visible sign of this disease nor is it recognized as a direct cause of migratory bird mortality. Severe infections can cause loss of muscle tissue and result in lameness, weakness, and even paralysis in rare cases. The debilitating effects of severe infections could increase bird susceptibility to predation and to other causes of mortality.

Gross Lesions

Visible forms of infection are readily apparent when the skin is removed from the bird. In waterfowl and in many other species, infection appears as cream-colored, cylindrical cysts (the macrocysts) that resemble grains of rice running in parallel streaks through the muscle tissue. The cysts are commonly found in the breast muscle (Fig. 28.2A), but they are also found in other skeletal and cardiac muscle (Fig. 28.2B). Calcification of the muscle tissue around these cysts

makes them obviously discrete bodies. The degree of calcification is often sufficient to give a gritty feeling to the tissue when it is cut with a knife.

Lesions that were observed in wading birds differed in appearance; the cysts were white and opaque, and they generally extended throughout the entire length of the infected muscle fiber. Cysts were present in the heart muscle and they were confined to striated muscles.

Diagnosis

The visible presence of sarcosporidian cysts in muscle tissue is sufficient to diagnose this disease. Visible cysts may vary in size and shape in different bird species. Good quality color photographs (prints or 35 millimeter slides) of the external surface of infected muscle are generally sufficient for a disease specialist to recognize this disease if tissues or a whole carcass cannot be provided. Whole birds should be submitted if possible. If only tissues can be submitted, then a portion of the infected muscle should be fixed in a 10 percent formalin solution. Frozen muscle tissue is also suitable for diagnosis, and the distinctive appearance of these cysts allows a diagnosis from even partially decomposed carcasses.

Control

There are no known control methods for this disease, nor do any seem to be needed or are any being developed. Control of sarcocystis would require interruption of the life cycle of the parasite. Although the life cycles of the *Sarcocystis* sp. that affect wild birds are not precisely known, they are probably similar to the two-host, indirect life cycle known for some other *Sarcocystis* sp. (Fig. 28.1). The predator-prey relationship between the intermediate bird hosts and the definitive carnivore hosts may be the primary reason that juvenile birds or some bird species are seldom found to be infected. The appropriate carnivores may not be present on the breeding grounds.

Different species of carnivores seem to be involved in the infection of different bird species, which suggests that birds are infected by more than one species of the genus *Sarcocystis* sp. If the carnivore-bird cycle is species-specific, that is, if a specific species of bird can only be infected by oocysts that are produced by a parasite in a specific carnivore species, then selective control of sarcocystis might be feasible. However, current knowledge of the disease does not indicate a need to initiate control because there is little evidence that bird health is often compromised by infection. Nevertheless, the role of carnivores in the life cycle of *Sarcocystis* sp. infections should be considered when feeding

Figure 28.3 Relative frequency of grossly visible forms of sarcocystis in selected groups of North American migratory birds.

uncooked, infected waterfowl to house pets and to farm animals such as hogs.

Human Health Considerations

Sarcocystis sp. presents no known health hazard to humans. The primary importance to humans of sarcocystis in waterfowl is the loss of infected birds for food; the unaesthetic appearance of parasitized muscle may prompt hunters to discard the carcass. Limited evaluations of hunter responses to infected carcasses indicate no reduction in carcass consumption in areas where the infection is commonly seen. Also, the recognized high prevalence of infection in northern shovelers in some areas results in this species often being left unretrieved by some hunters and focuses additional hunting pressure on other species.

Benjamin N. Tuggle and Milton Friend
(Modified from and earlier chapter by Benjamin N. Tuggle)

Supplementary Reading

Cawthorn, R.J., Rainnie, D., and Wobeser, G.A., 1981, Experimental transmission of *Sarcocystis* sp. (Protozoa: Sarcocystidae) between the shoveler (*Anas clypeata*) duck and the striped skunk (*Mephitis mephitis*): Journal of Wildlife Diseases, v. 17, p. 389–394.

Cornwell, G, 1963, New waterfowl host records for *Sarcocystis rileyi* and a review of sarcosporidiosis in birds: Avian Disease, v. 7, p. 212–216.

Moorman, T.E., Baldassarre, G.A., and Richard, D.M., 1991, The frequency of *Sarcocystis* spp. and its effect on winter carcass composition of mottled ducks: Journal of Wildlife Disease, v. 27, no. 3, p. 491–493.

Spalding, M.G., Atkinson, C.T., and Carleton, R.E., 1994, *Sarcocystis* sp. in wading birds (Ciconiiformes) from Florida: Journal of Wildlife Disease, v. 30, no. 1, p. 29–35.

Chapter 29
Eustrongylidosis

Synonyms
Verminous peritonitis

Cause

Eustrongylidosis is caused by the nematodes or roundworms *Eustrongylides tubifex, E. ignotus,* and *E. excisus. Eustrongylides* sp. can cause large die-offs of nestlings in coastal rookeries, especially of egrets and other wading birds.

Life Cycle

The three species of *Eustrongylides* that cause disease in birds have similar indirect life cycles that require two intermediate hosts (Fig. 29.1). Four developmental stages of the parasite are required from egg to sexually mature worm. The first larval stage develops within the eggs that are shed in the feces of the bird host and are eaten by freshwater oligochaetes or aquatic worms. The oligochaetes serve as the first intermediate host. The eggs hatch within the oligochaetes, where they develop into second- and third-stage larvae. Minnows and other small fish, such as species of *Fundulus* and *Gambusia*, feed upon the infected oligochaetes and serve as the second intermediate host. The third-stage larvae become encapsulated on the internal surface areas of the fish, develop into infective fourth-stage larvae, and await ingestion by birds. Predatory fish, which consume infected fish, can serve as paratenic or transport hosts when they are fed upon by birds. Amphibians and reptiles have also been reported as second-stage intermediate hosts and serve as paratenic hosts. Larvae that are infective for birds can penetrate the ventriculus (stomach) within 3–5 hours after a bird ingests an intermediate or paratenic host, and the larvae quickly become sexually mature worms that begin shedding eggs 10–17 days postinfection.

Species Affected

E. tubifex has been reported from four different bird families, *E. ignotus* from three, and *E. excisus* from three (Fig. 29.2). Young wading birds are the most common species to have large mortalities from eustrongylidosis (Table 29.1). *Eustrongylides* sp. have also been reported in birds of prey.

Distribution

Eustrongylides sp. have been reported from birds throughout much of the world. *E. tubifex* and *E. ignotus* are the species reported within the United States (Table 29.2). Eustrongylid infections within the United States have been reported from many areas (Fig. 29.3). Typical rookeries where birds are infected with *Eustrongylides* sp. are found in coastal areas and consist of dense populations of birds nesting on low islands, often surrounded by canals or ditches. Nesting habitat often includes stands of low trees, such as willows, with an understory that may be submergent, semisubmergent, or upland mixed-prairie species. Inland rookeries are usually adjacent to lakes or rivers, and nesting trees, particularly those used by great blue herons, may be much higher than those in coastal rookeries. Several wading bird species may nest in these areas, but typically one or two species account for most of the birds in the rookery (Fig. 29.4).

Seasonality

Birds can harbor infections yearround. Mortality usually is reported in spring and summer and birds less than 4 weeks old are more likely to die than adults. Disease in older birds tends to be of a more chronic nature and infection may be seen at any time of the year.

Field Signs

Disease results in a variety of clinical or apparent signs that are not specific to eustrongylidosis. However, consideration of the species affected, the age class of birds involved, and the full spectrum of signs may suggest that eustrongylidosis is the cause of mortality. Very early in the infection as the worm is penetrating the ventriculus, some birds will shake their heads, have difficulty swallowing, have dyspnea or difficult or labored breathing and, occasionally, regurgitate their food. Anorexia or loss of appetite has been noted in experimentally infected nestlings. It has been speculated that anorexia in combination with sibling competition for food may contribute to the emaciation seen in naturally infected birds. Infected nestlings also may wander from the nest predisposed to predation or trauma or both. Affected nestlings observed during one mortality event became progressively weakened and showed abdominal swelling. Palpation of worms on the ventriculus has been useful for detecting infection in live nestlings.

Gross Lesions

Birds that have been recently infected often have large, tortuous, raised tunnels that are visible on the serosal surface of the proventriculus, ventriculus, or intestines (Fig. 29.5A). The nematodes reside within these tunnels, which are often encased with yellow, fibrous material, and maintain openings to the lumen of the organ so that parasite eggs may be passed out with feces into the environment. A fibrinoperitonitis or fibrin-coated inflammation of the surfaces of the peritoneal cavity (the area containing the organs below

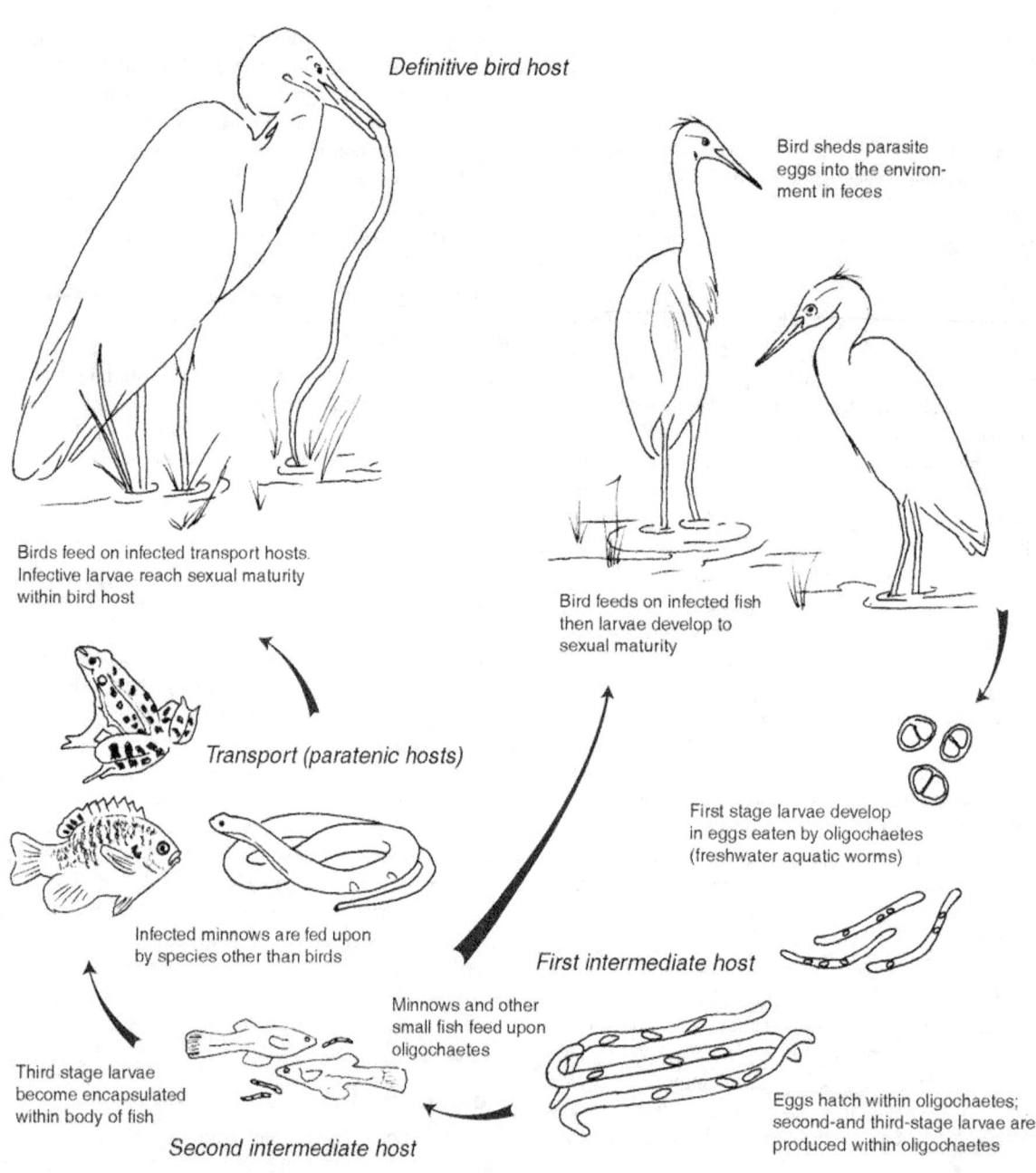

Figure 29.1 Life cycle of *Eustrongylides* sp.

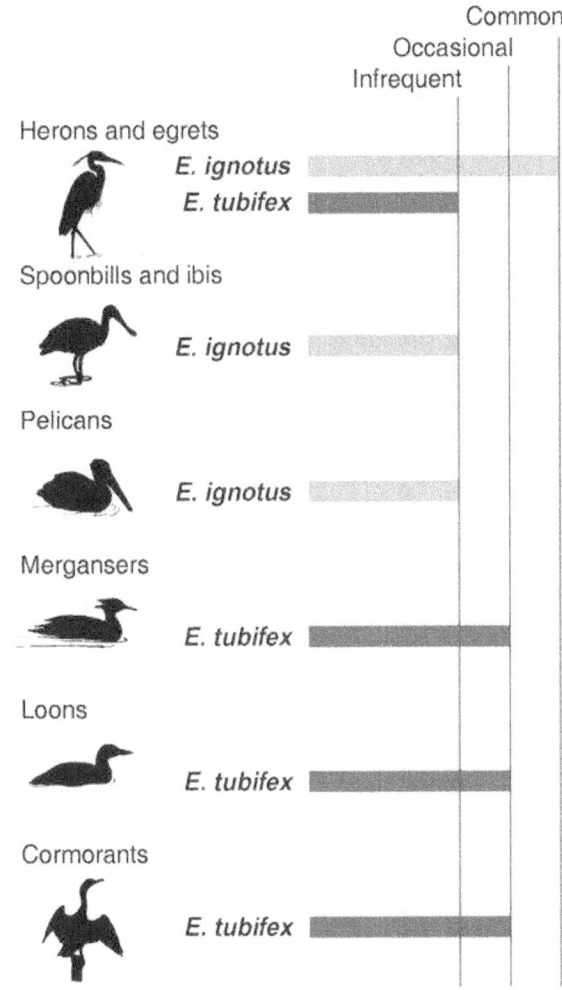

Figure 29.2 Groups of water birds reported to be infected with **Eustrongylides** sp.

the heart and lungs) and the intestinal surfaces may be present when larvae or adult worms have perforated the surface of the intestines (Fig. 29.5B). Movement of bacteria from the lumen of the digestive tract to the body cavity results in bacterial peritonitis and secondary infections that can cause the death of an infected bird. Thick-walled granulomas, which are firm nodules consisting of fibrous tissue that forms in response to inflammation with necrotic (dead) centers, caseous (cheesy) airsaculitis or inflammation of the air sacs and intestinal blockages have also been reported. The presence of the parasite is also striking when carcasses are examined. Adult worms can be quite large (up to 151 millimeters in length and 4.3 millimeters in width) and are reddish.

Lesions in chronic or resolving infections are less remarkable and appear as raised, yellow or tan-colored tunnels filled with decomposed worms or worms encased with yellow fibrous material. Some lesions will not have recognizable worm structures intact. Lesions seen in bald eagles that were examined at the National Wildlife Health Center were in the esophagus and were much less severe than those in other fish-eating birds.

Diagnosis

Large tortuous tunnels on the surface of the proventriculus, ventriculus, or intestine of fish-eating birds are most likely due to *Eustrongylides* sp. However, the presence of eustrongylid worms is not diagnostic of the cause of death, especially in older nestlings and adult birds. Therefore, entire carcasses should be provided for disease diagnosis. If interest is limited to confirming the presence of *Eustrongylides* sp., then infected organs and the gastrointestinal tract should be removed and shipped chilled on cool packs to an appropriate laboratory. If shipment is not possible within 24–48 hours, the organs can be frozen or preserved in 10 percent neutral formalin and shipped. Speciation of worms requires a diagnostician who has appropriate training.

Control

Control of eustrongylidosis depends on the difficult task of disrupting the parasite life cycle, which is further complicated by the length of time that the eggs can remain viable and that intermediate hosts can remain infective. Under experimental conditions, *Eustrongylides* sp. eggs have remained viable up to 2.5 years and freshwater fish and oligochaetes have been reported to remain infected for more than 1 year. Also, the rather quick maturation of the parasite (once it is inside the bird definitive host), along with the long time period that intermediate and paratenic hosts can remain infected, are a perfect parasite strategy for infecting transient or migratory birds. Thus, the birds in a rookery can quickly infect intermediate/paratenic hosts, which can maintain the parasite until next season's nesting.

It is known that eutrophication and warm water temperatures (20–30 °C) create optimal conditions for the parasite. It has been reported that infection among fish is highest where external sources of nutrients or thermal pollution alter natural environments. Therefore, water quality is an important factor that in some situations is subject to actions that may decrease transmission of the parasite. Water-quality improvement as a means of disease prevention should be taken into consideration relative to land-use practices and wastewater discharges that may negatively impact egret and heron rookeries and feeding areas for wading birds.

Food sources used for birds being reared in captivity or being rehabilitated for return to the wild should be free of infection with *Eustrongylides* sp. The types of fish and the sources of those fish should be considered before they are used to feed birds.

Table 29.1 Examples of reported wild bird mortality attributed to eustrongylidiosis.

Geographic location	Primary species affected	Time of Year	Parasite species	Comments
Virginia Beach, Va.	Red-breasted merganser	Dec.	*E.* sp.	50 dead, 95 moribund; mature birds were affected.
Madison County, Ind.	Great blue heron	May	*E. ignotus*	25 dead and moribund; most birds had fledged the previous year.
Pea Patch Island, Del.	Snowy egret	May-July	*E. ignotus*	Approximately 300 hatchlings in one outbreak; most deaths occurred within the first 4 weeks after hatching; other outbreaks have been reported for this location.
Avery Island, La.	Common egret	May	*E.* sp.	Minimum loss of 400 hatchlings at just prefledging age.
Goat Island, Texas	Snowy egret Great egret	Not reported	*E.* sp.	Nestlings and young of undetermined numbers; high infection prevalence in colony.
Several colonies in central and southern Florida.	Snowy egret Great egret	Not reported	*E. ignotus*	More than 250 nestlings during one event; this geographic area has recurring losses from this parasite.

Table 29.2 Reported geographic occurrence in wild birds of *Eustrongylides* sp.

	Eustrongylides sp.		
Geographic area	*E. tubifex*	*E. ignotus*	*E. excisus*
United States	●	●	—
Canada	●	—	—
Brazil	●	●	—
Europe	●	—	●
Russia	●	—	●
Middle East	—	—	●
Taiwan	—	—	●
India	—	—	●
Australia	—	—	●
New Zealand	—	●	—

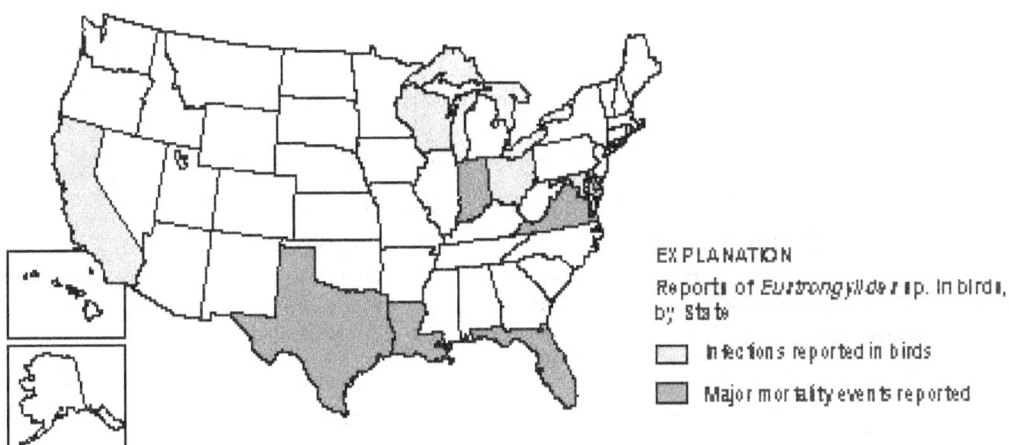

Figure 29.3 States where *Eustrongylides* sp. infections in wild birds have been reported.

Figure 29.4 Although many species may nest in wading bird rookeries, one or two species are often predominant.

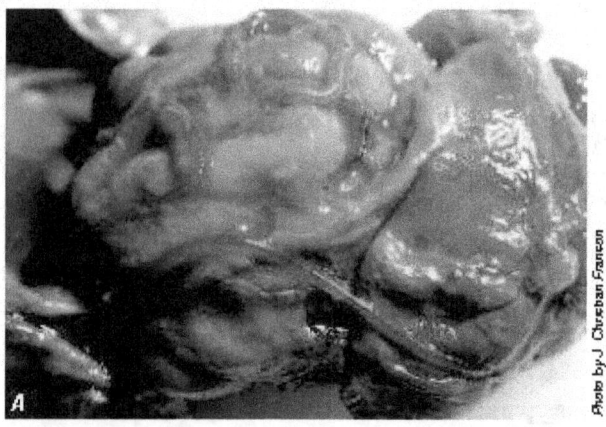

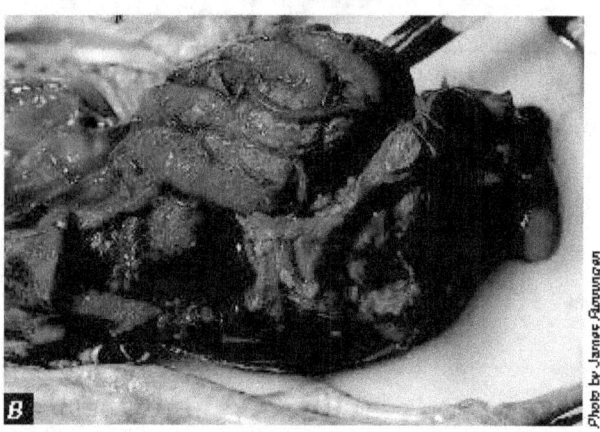

Figure 29.5 *(A) Raised tunnels caused by **Eustrongylides** sp. on intestines of a snowy egret. (B) The debris on the intestinal surfaces of this snowy egret is characteristic of the peritonitis often caused by **Eustrongylides** sp. infection.*

Human Health Considerations

Humans who have consumed raw or undercooked fish that carry the larval stages of the parasite have experienced gastritis or inflammation of the stomach and intestinal perforation requiring surgical removal of worms.

Rebecca A. Cole

Supplementary Reading

Franson, J.C. and Custer, T.W., 1994, Prevalence of eustrongylidosis in wading birds from colonies in California, Texas, and Rhode Island, United States of America: Colonial Waterbirds, v. 17, p. 168–172.

Measures, L.N., 1988, Epizootiology, pathology and description of *Eustrongylides tubifex* (Nematoda: Dioctophymatoidea) in fish: Canadian Journal of Zoology, v. 66: p. 2212–2222.

Spalding, M.A., Bancroft, G.T., and Forrester, D.J., 1993, The epizootiology of eustrongylidosis in wading birds (Ciconiiformes) in Florida: Journal of Wildlife Diseases, v. 29, p. 237–249.

Spalding, M.A., and Forrester, D.J., 1993, Pathogenesis of *Eustrongylides ignotus* (Nematoda: Dioctophymatoidea) in Ciconiiformes: Journal of Wildlife Diseases, v. 29, p. 250–260.

Wiese, J.H., Davidson, W.R., and Nettles, V.F., 1977, Large scale mortality of nestling ardeids caused by nematode infection: Journal of Wildlife Diseases, v. 13, no. 4, p. 376–382.

Chapter 30
Tracheal Worms

Synonyms
Gape worm, syngamiasis, gapes

Cause
Infection by tracheal worms often results in respiratory distress due to their location in the trachea or bronchi and their obstruction of the air passage. Infections by these parasitic nematodes or roundworms in waterbirds, primarily ducks, geese, and swans, are usually due to *Cyathostoma bronchialis* and infection of land birds are usually due to *Syngamus trachea*. However, both genera infect a variety of species, including both land and waterbirds. Infections with *S. trachea* have been more extensively studied than infections with *Cyathostoma* sp. because of its previous importance as a disease-causing parasite of poultry in many parts of the world. Changes in husbandry practices to modern intensive methods for poultry production have essentially eliminated *S. trachea* as an agent of disease in chickens, but it is an occasional cause of disease in turkeys raised on range.

Life Cycle
Tracheal worms have an indirect life cycle (Fig. 30.1) that requires a paratenic or transport host which transmits the infectious larvae to the definitive host bird, where they reach adulthood and reproduce. Adult *S. trachea* reside within the trachea. The female releases fertilized eggs, which are swallowed by the bird and voided with the feces into the soil. Eggs may also be directly expelled onto the ground from the trachea. After embryonation (1–2 weeks), infective larvae develop within the egg. Birds can become infected by eating invertebrate paratenic hosts such as earthworms, snails, slugs, or fly larvae that have consumed the eggs. Infective larvae are released from the egg and become encysted within the bodies of these invertebrates and can remain infective for up to three and one-half years. Upon ingestion by birds, the larvae are believed to penetrate the intestinal wall. Some larvae enter the abdominal cavity but most enter the bloodstream, where they are carried to the lungs. After further development in the lungs, the young worms migrate up the bronchi to the trachea. Larvae can reach the lungs within 6 hours after ingestion and eggs are produced by worms in the trachea about 2 weeks after ingestion of those larvae. *C. bronchialis* is very similar in that earthworms transmit the infective stage to the bird. Infection of birds with *C. bronchialis* by direct consumption of fully embryonated eggs has been documented experimentally; however, worm burdens were extremely low.

Species Affected
Disease caused by tracheal worms is not commonly reported for free-ranging birds within the United States and Canada, but it is common within the United Kingdom and in some other countries. High infection rates within wild birds in England attest to the potential for this parasite to be a serious pathogen. More than 50 percent of nestling and fledgling starlings, more than 85 percent of jackdaws, and 100 percent of young rooks were found to be infected in one study. Infection rates in adult birds were considerably lower, but they still exceeded 30 percent for starlings and rooks. Within the United States, *S. trachea* infections have been reported from wild turkeys, other gamebirds, a variety of passerines, (songbirds), and occasionally from other bird species. Large-scale mortalities have occurred among pheasants and other gamebirds being propagated for sporting purposes. Findings from captive bird collections have led to the conclusion that almost any species of cage or aviary bird is susceptible to infection.

S. trachea has been reported infrequently in waterfowl, but members of the genus *Cyathostoma* sp. are "characteristic" or common parasites of waterfowl. Mortality has been reported for several species of young geese, leading some investigators to suggest that *C. bronchialis* are potentially important pathogens for geese. Juvenile free-ranging sandhill cranes have also been reported to have died from *Cyathostoma* sp. infection.

Distribution
S. trachea and *Cyathostoma* sp. are found worldwide.

Seasonality
Infected birds can be found yearround. Young birds are most commonly affected and, therefore, disease is associated with breeding cycles in the spring to summer months for free-ranging birds.

Field Signs
Most birds that are infected show no signs of disease. In general, the severity of disease is dependent upon the degree of infection and the size of the bird. Small birds are more severely affected than larger birds because their narrower tracheal openings result in greater obstruction by the worms. Respiratory distress is the primary clinical sign of disease. Birds with severe infections open their mouth widely and at the same time stretch out their necks, assuming a "gaping" posture. The adult worms that are attached to the lining of

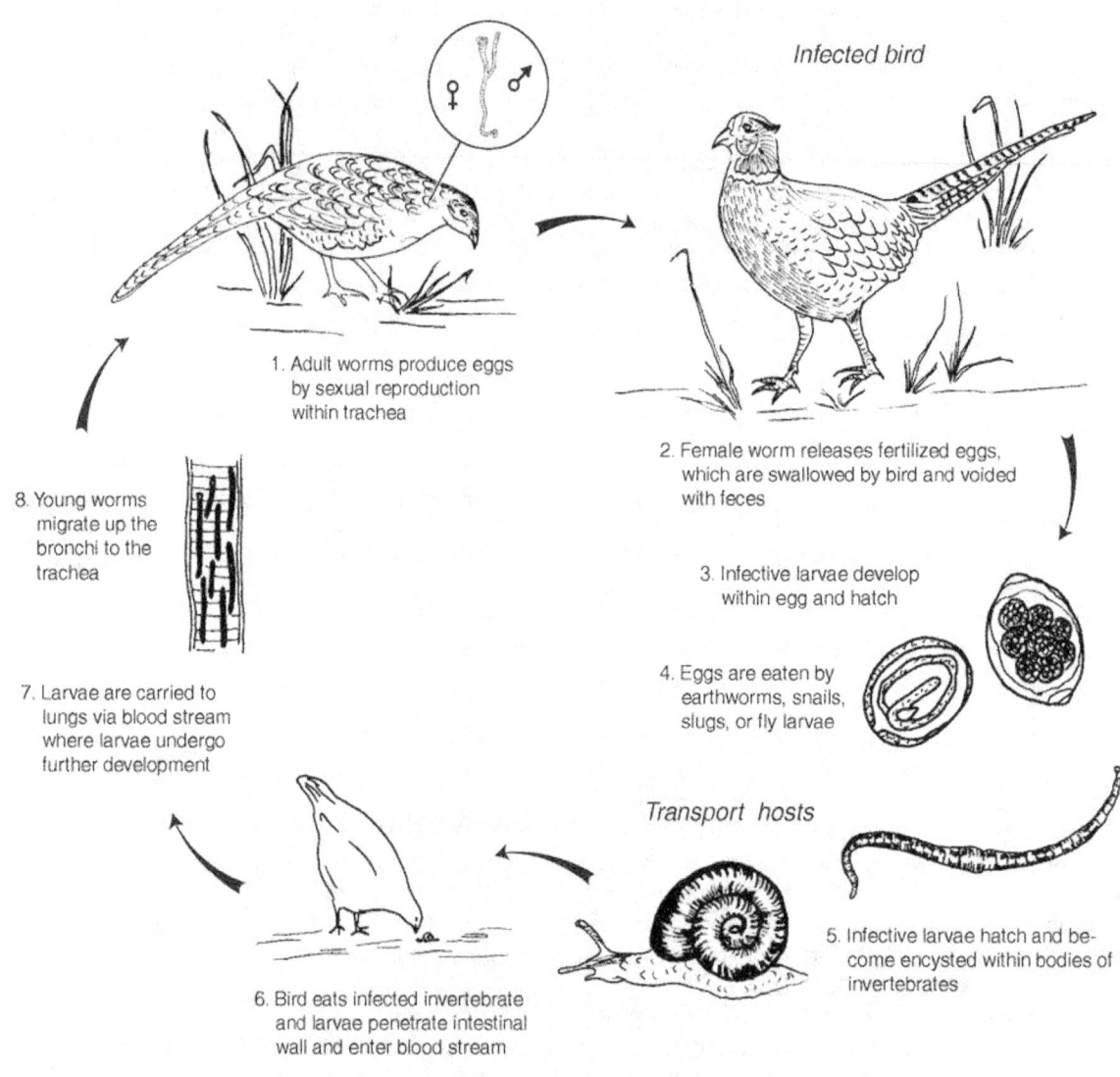

Figure 30.1 *Tracheal worm life cycle.*

the trachea cause irritation and excess mucus production. This often results in agitated bouts of coughing, head shaking, and sneezing as the birds attempt to dislodge the parasites. Severely infected birds may have most or all of the tracheal opening obstructed by worms, may stop feeding, and may rapidly lose body condition.

Gross Lesions

Severely affected birds experience severe weight loss and have poorer development of body mass than uninfected birds, and they often die from starvation (Fig. 30.2). Anemia may also be present due to the blood-feeding habits of the parasites.

Diagnosis

Identification of the worms (Fig. 30.3) and evaluation of any associated disease signs are required for a diagnosis. Clinical signs are not diagnostic because similar signs can be seen with some mite infections, aspergillosis, and wet pox.

Control

There is no feasible method for controlling tracheal worms in free-ranging birds. Disease prevention should be practiced by minimizing the potential for captive-propagation and release programs to infect invertebrates that are then fed upon by free-ranging birds. Land-use practices that provide direct contact between poultry rearing and wild birds and the disposal of bird feces and litter should also be considered because environmental contamination with infective larvae is a critical aspect of the disease cycle.

Human Health Considerations

There are no reports of these nematodes infecting humans.

Rebecca A. Cole

*Figure 30.2 Comparative breast muscle mass of noninfected pheasants (left) and those with **Syngamus trachea** infections (right).*

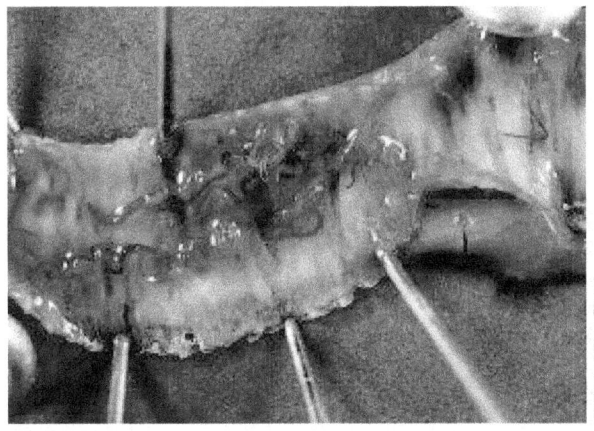

*Figure 30.3 **Syngamus trachea** in the trachea of a ring-necked pheasant.*

Supplementary Reading

Anderson, R.C., 1992, Nematode parasites of vertebrates: Their development and transmission: Wallingford, England, CAB International, 578 p.

Cram, E.B., 1927, Bird parasites of the nematode suborders Strongylata, Ascaridata, and Spirurata: U.S. National Museum Bulletin No. 140, 465 p.

Fernando, M.A., Hoover, I.J., and Ogungbade, S.G., 1973, The migration and development of *Cyathostoma bronchialis* in geese: Journal of Parasitology, v. 59, p. 759–764.

Threlfall, W., 1965, Life-cycle of *Cyathostoma lari*: Nature, v. 206, p. 1,167–1,168.

Chapter 31
Heartworm of Swans and Geese

Synonyms
Filarial heartworm, Sarconema, **Sarconema eurycerca**

Cause

Heartworm in swans and geese is caused by a filarial nematode or a roundworm of the superfamily Filarioidea which is transmitted to the bird by a biting louse. The nematode and the louse both are parasites. *Sarconema eurycerca* is the only one of several species of microfilaria or the first stage juvenile of the parasite found in the circulating blood of waterfowl that is known to be pathogenic or cause clinical disease.

Life cycle

Sarconema eurycerca has an indirect life cycle (Fig. 31.1) that requires the parasite larvae to develop in an intermediate host before they can become infective for and be transmitted to a definitive host, where they mature and reproduce. Female adult heartworms release microfilariae into the bloodstream of the definitive host bird. The microfilariae infect a biting louse, *Trinoton anserinum*, that subsequently feeds upon the bird. The larvae go through three stages of development within the louse, and the third stage is infectious to birds. A new host bird becomes infected when the louse bites it to feed on its blood and the third-stage larvae move into

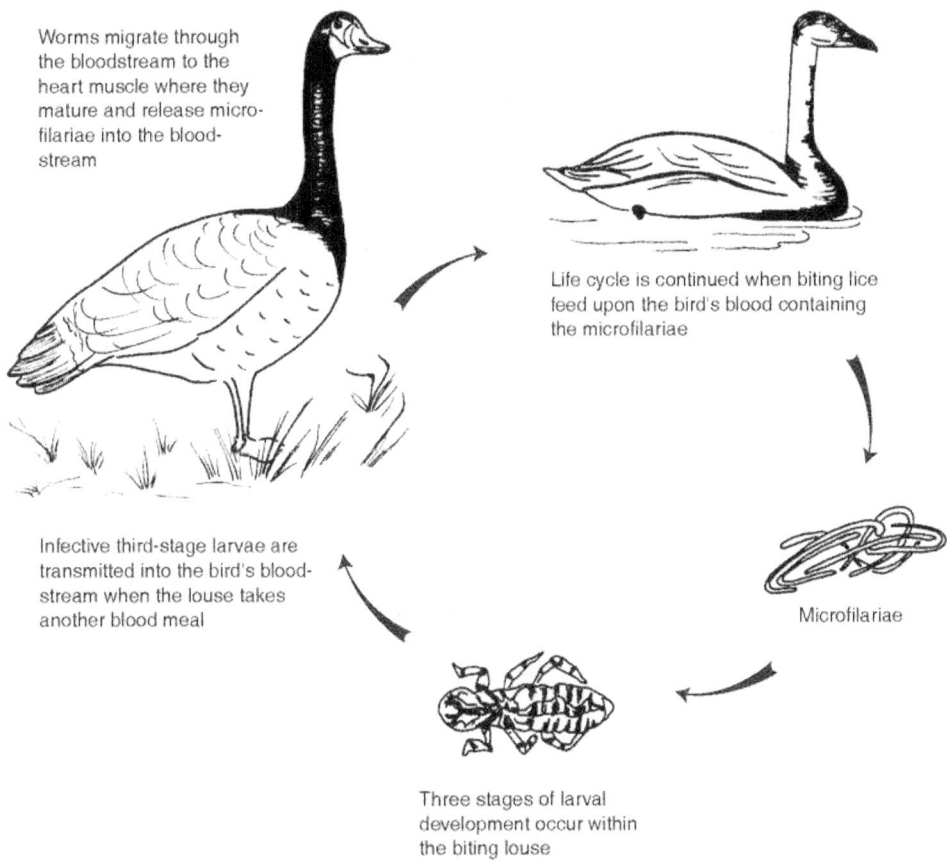

Figure 31.1 Indirect life cycle of **Sarconema eurycerca**.

the bird's bloodstream. The larvae migrate through the bloodstream to the myocardium, which is the middle and thickest layer of the heart wall composed of cardiac muscle. They are nourished by and develop to sexual maturity within the myocardium. The cycle continues as this next generation of mature heartworms release microfilariae into the bloodstream.

Infection with the parasite is not synonymous with disease; that is, the parasite may infect and develop within the bird but not debilitate it.

Species Affected

Sarconema eurycerca was first identified from a tundra swan (whistling swan) in the late 1930s. It has since been reported from trumpeter, Bewick's, and mute swans and, from Canada, snow, white-fronted, and bean geese. Varying percentages of swans (4–20 percent) have been found to be infected on the basis of blood smears that were taken from apparently healthy birds during field surveys. Canadian investigators have reported a prevalence of approximately 10 percent of snow geese that were examined at necropsy and which had died from other causes. This parasite has not received sufficient study for its full host range, its relative frequency of occurrence in different species, or its significance as a mortality factor for wild birds to be determined.

Distribution

Heartworm is found throughout the range of its swan and goose hosts.

Seasonality

It is suspected that while swans and geese are on the breeding grounds, louse infestation and colonization on birds is prevalent. Therefore, the possibility of infection by heartworm is highest while birds are on the breeding grounds.

Field Signs

Field signs are not always present in infected birds, and infection cannot be determined by the presence of clinical signs alone. Chronic types of debilitating diseases, such as lead poisoning, may exacerbate louse infestation because birds become lethargic and do not preen. No specific field sign is diagnostic for infection.

Gross Lesions

The severity of infection dictates the lesions that are seen at necropsy. Birds may be emaciated or in comparably good flesh. The heart may be enlarged and have pale foci or spots within the myocardium. The thin, long thread-like worms may be visible under the surface layer or epicardium of the heart or the worms may be embedded within the deeper muscle tissue of the myocardium (Fig. 31.2).

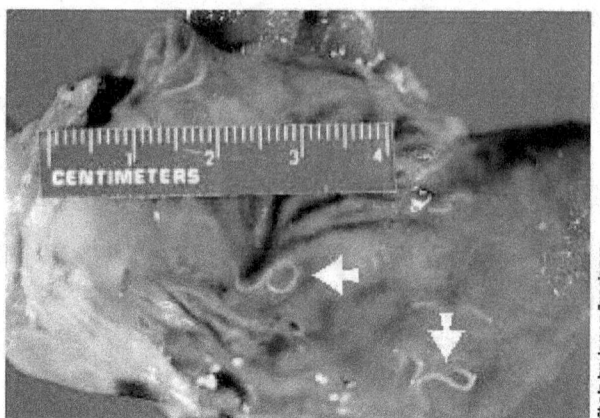

Figure 31.2 Heartworms (arrows) on the inner surface of the heart of a tundra swan.

Diagnosis

A diagnosis of heartworm as the cause of death must be supported by pathologic lesions seen during examination of the heart tissues with a microscope and consideration of other causes. Therefore, whole carcasses should be submitted for diagnostic assessments. If the transit time is short enough to avoid significant decomposition of the carcass and if the carcass can be kept chilled during transit, then chilled whole carcasses should be submitted to qualified disease diagnostic laboratories. If those conditions cannot be met, then carcasses should be submitted frozen.

Control

Control of heartworm is not practical for free-ranging birds. Decreasing the opportunity for heavy infestation of the louse intermediate host will result in reduced opportunity for heartworm infection.

Human Health Considerations

Sarconema eurycerca has not been reported to infect humans.

Rebecca A. Cole

Supplementary Reading

Cohen, M., Greenwood, T., and Fowler, J.A., 1991, The louse *Trinoton anserinum* (Amblycera: Pthiraptera), an intermediate host of *Sarconema eurycerca* (Filarioidea: Nematoda), a heartworm of swans: Medical and Veterinary Entomology, v. 5, p. 101–110.

Scheller, E.L., Sladen, W.L., and Trpis, M., 1976, A Mallophaga, *Trinoton anserium*, as a cyclodevelopmental vector for a heartworm parasite of waterfowl: Science, v. 194, p. 739–740.

Seegar, W.S., 1979, Prevalence of heartworm, *Sarconema eurycerca*, Wehr, 1939 (Nematoda), in whistling swan, *Cygnus columbianus columbianus*: Canadian Journal of Zoology, v. 57, p. 1,500–1,502.

Chapter 32
Gizzard Worms

Synonyms
Stomach worm, ventricular nematodiasis, amidostomiasis

Cause

Gizzard worms are comprised of several species of parasitic nematodes or roundworms of birds. Severe infections can result in birds becoming unthrifty and debilitated to the extent that they are more susceptible to predation and to infection by other disease agents. The two gizzard worms that are emphasized here are trichostrongylid nematodes that belong to the genera *Amidostomum* sp. and *Epomidiostomum* sp. These long (10–35 millimeter), sometimes coiled, threadlike roundworms are found just beneath the surface lining and the grinding pads of the gizzard, and they are most frequently found in waterfowl. Other species of gizzard worms are found in upland gamebirds such as grouse, in psitticine birds such as parakeets, and in passerine or perching birds such as robins in various parts of the world.

Life Cycle

Amidostomum sp. and *Epomidiostomum* sp. have a direct life cycle in which the infective parasite larvae invade a single host animal for development to reproductive maturity (Fig. 32.1). Embryonated eggs are passed in the feces of an infected host bird. First-stage larvae hatch from the eggs into the surrounding environment in about 24–72 hours, depending on the ambient temperature. These larvae molt twice after they hatch, and the time between molts also depends on the temperature. Larvae are quite resilient, surviving low temperatures and even freezing; they do not, however, survive drying.

After a bird ingests the larvae, most commonly when a bird feeds or drinks, they enter the gizzard and burrow into its surface lining where they molt again before they become adult worms. Adult worms become sexually mature in about 10–15 days after the final molt, and females shed eggs within 15–20 days. The development from egg to adulthood may take as few as 20 days or as many as 35 days depending on environmental conditions. Once a bird is infected, it can harbor gizzard worms for several years.

In contrast to the direct parasite life cycle, other gizzard worms such as *Cheilospirura spinosa* have indirect life cycles (Fig. 32.2) in which they undergo one or more stages of development in an arthropod (insect) intermediate host. *C. spinosa* is a common gizzard worm of North American ruffed grouse that also infects partridges, pheasants, quail, and wild turkey. Embryonated *C. spinosa* eggs that are discharged in the feces of grouse and other infected upland gamebirds are

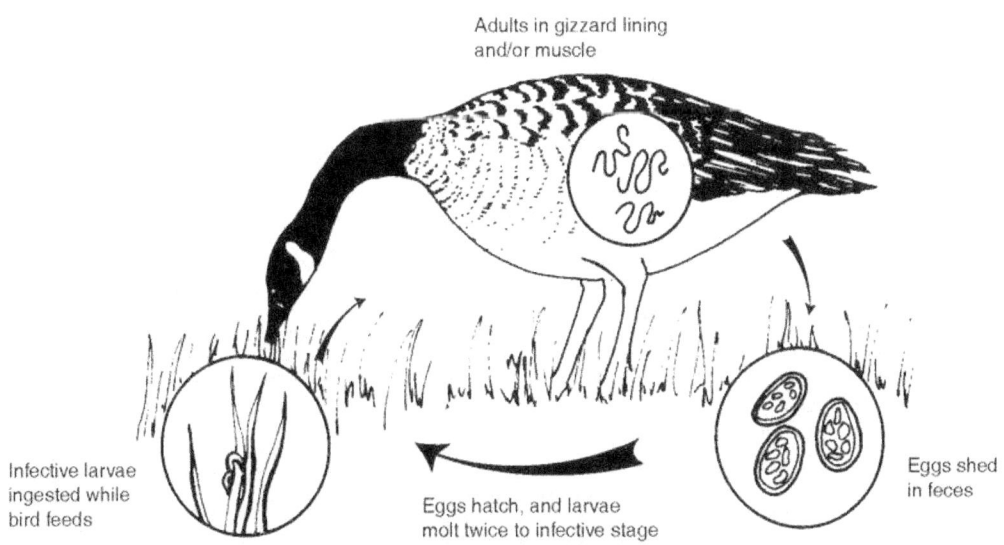

Figure 32.1 Direct life cycle of gizzard worms such as **Amidostomum** sp. and **Epomidiostomum** sp.

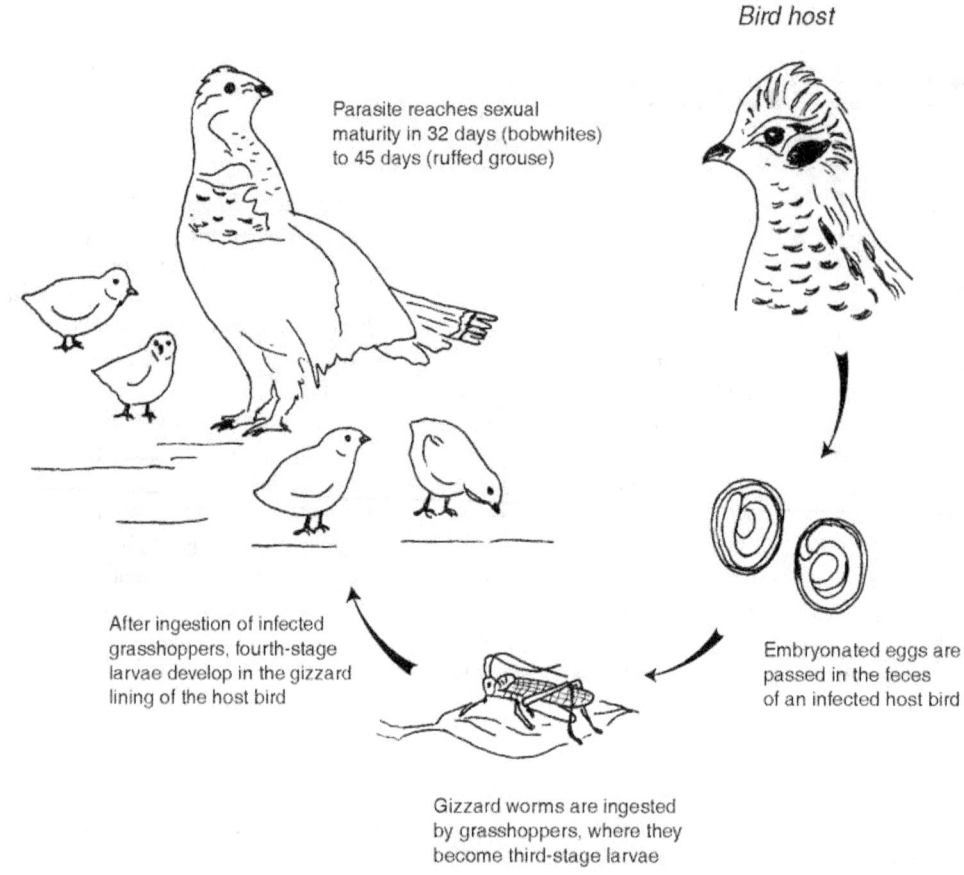

Figure 32.2 *Indirect life cycle of gizzard worms such as **Cheilospirura spinosa**.*

ingested by grasshoppers, the intermediate host, and the eggs hatch within the body of the grasshopper. Experimental studies indicate that the larvae then migrate into the body cavity of the grasshopper, where they become loosely encysted or where they invade the muscles. They then become third-stage larvae that are infective for birds; this infective stage is reached about three or three and one-half weeks after the grasshopper ingests the parasite eggs. Fourth-stage larvae (immature adult worms) have been found underneath the gizzard lining of bobwhite quail 14 days after ingestion of infected grasshoppers. Sexual maturity of the parasite is reported to be reached in bobwhites 32 days following ingestion of infected grasshoppers and in 45 days for ruffed grouse

Species Affected

Amidostomum sp. and *Epomidiostomum* sp. can be found in a variety of migratory birds, and gizzard worms have been reported in ducks, geese, swans, American coot, grebes, and pigeons (Fig. 32.3). Birds can die from gizzard worm infection, and death of very young birds is more common than death of adult birds. These worms are among the most common parasites of waterfowl, and they generally are more common in geese than in ducks or swans. However, a very high prevalence of infection of canvasback ducks with *Amidostomum* sp. (80 percent) was reported in one study. Infection is most severe in snow geese and Canada geese.

Seasonality

Migratory birds first become exposed to gizzard worms on breeding grounds, and they can continue to be exposed throughout their lives. Therefore, no seasonality is associated with this parasitism. The loss of young birds may be particularly high during the fall and winter months because of the combined effects of large worm burdens, the stresses of migration, and competition for food.

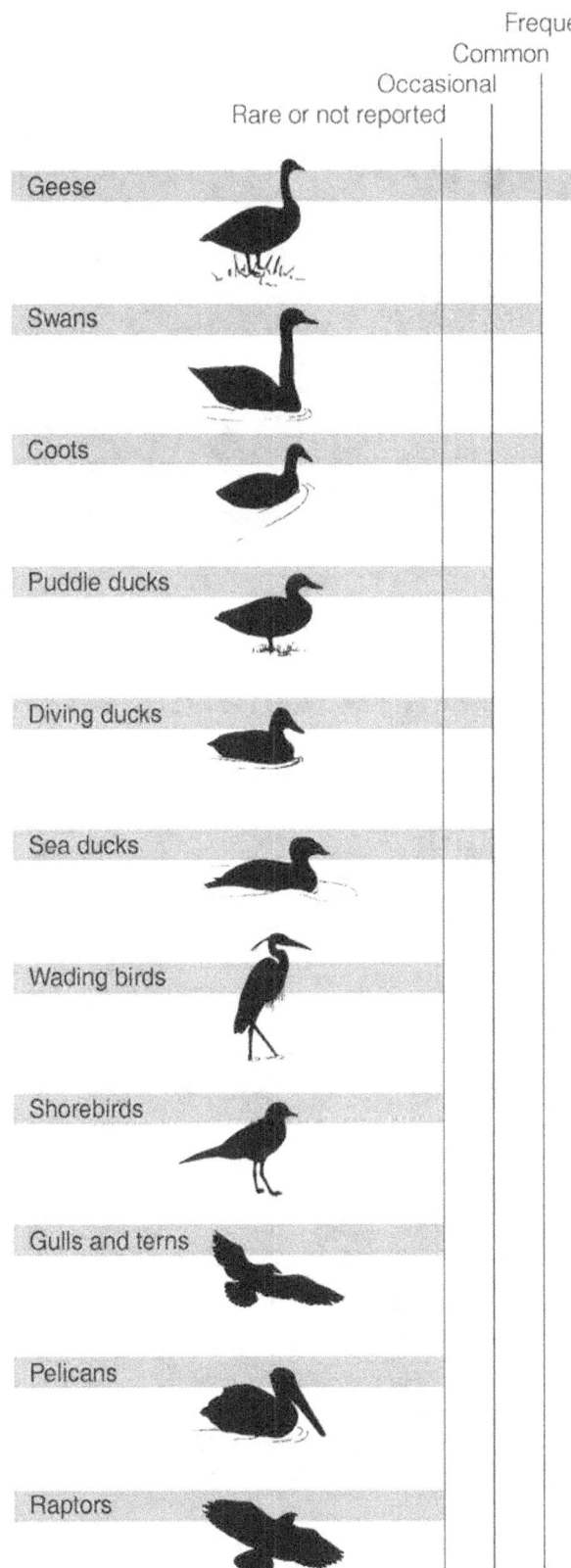

Field Signs

There are no field signs that indicate gizzard worm infection. Heavy worm burdens can result in poor growth of young birds, and birds of all ages are subject to emaciation and general weakness. Severe infections can interfere with food digestion by the bird as a result of extensive damage to the gizzard lining and muscle.

Gross Lesions

Obvious changes from the normal appearance of the gizzard result from the development, migration, and feeding of gizzard worms in that organ. The gizzard lining can slough off, become inflamed, hemorrhagic, and become ulcerated as a result of erosion of the grinding pads (Fig. 32.4). Large numbers (greater than 35) of worms can denude the surface lining of the gizzard, causing the edges of the grinding pads to degenerate and separate the pads from the underlying tissue (Fig. 32.5). In geese, portions of the gizzard muscle can die due to the presence of variable numbers of *Epomidiostomum* sp., which migrate through the tissue. Oblong tissue cavities 1–4 centimeters long can also be present (Fig. 32.6), and they can contain granular material that results from tissue reaction to worm migration through the muscle.

Diagnosis

Gizzard worm infection can be determined in live birds by finding and identifying gizzard worm eggs in the feces. The eggs of *Amidostomum* sp. and *Epomidiostomum* sp. are similar in size and appearance, and they require speciation by trained personnel.

Large numbers of worms and lesions in the gizzard lining or gizzard muscle of carcasses are highly suggestive of death caused by gizzard worms. Submit whole carcasses to disease diagnostic laboratories for more thorough evaluation. If it is not possible to submit a whole carcass and you suspect gizzard worms as the cause of mortality, then remove the gizzard (see Chap. 2) and ship it chilled or frozen. If the gizzard has been opened, remove with forceps as many whole worms as possible and place them in a 10 percent formalin solution or a 70 percent ethanol solution; do not freeze these worms. Submit the opened gizzard with the worms or preserve slices of the gizzard muscle in 10 percent formalin and forward them for microscopic examination (see Chap. 2).

Figure 32.3   Relative frequency of gizzard worms in selected groups of North American migratory birds.

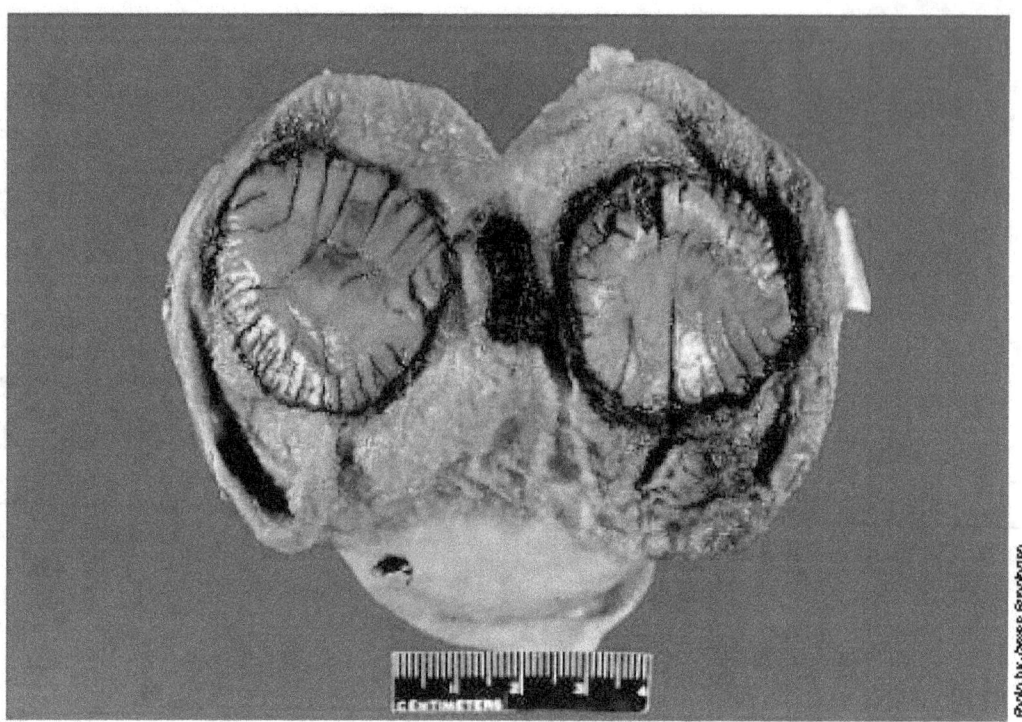

Figure 32.4 Canada goose gizzard showing ulcerations in the gizzard lining caused by gizzard worm (Amidostomum sp.) infection.

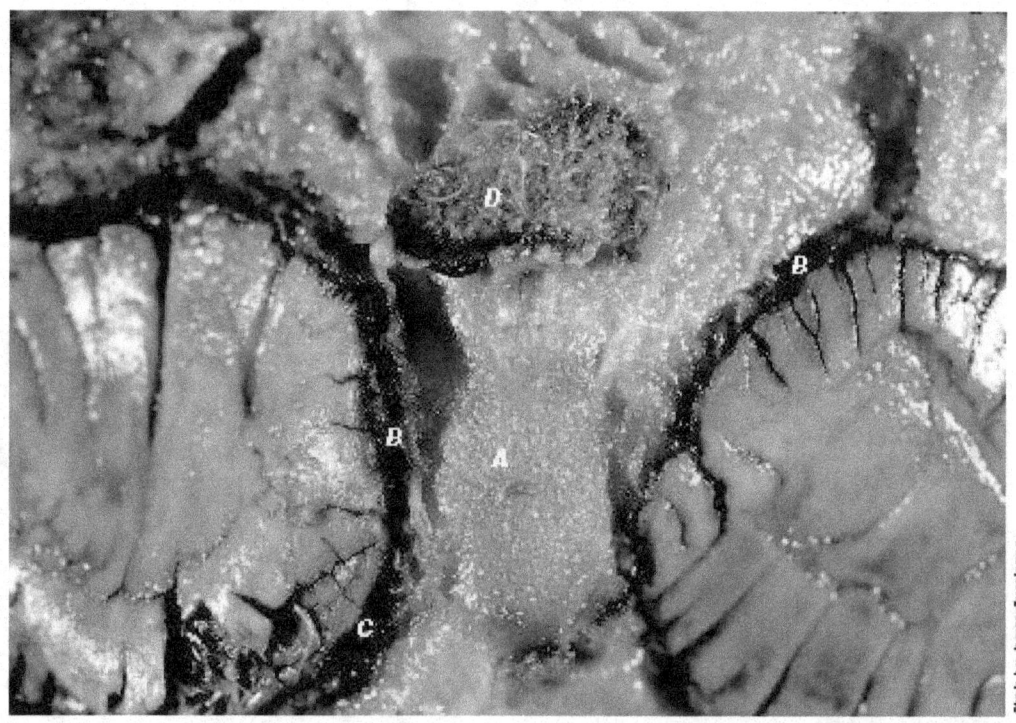

Figure 32.5 Closeup of Canada goose gizzard showing A, denuded surface lining, and B, degeneration of the edges of the grinding pads. Note also C, the separation of the pads from the gizzard lining and D, the presence of worms.

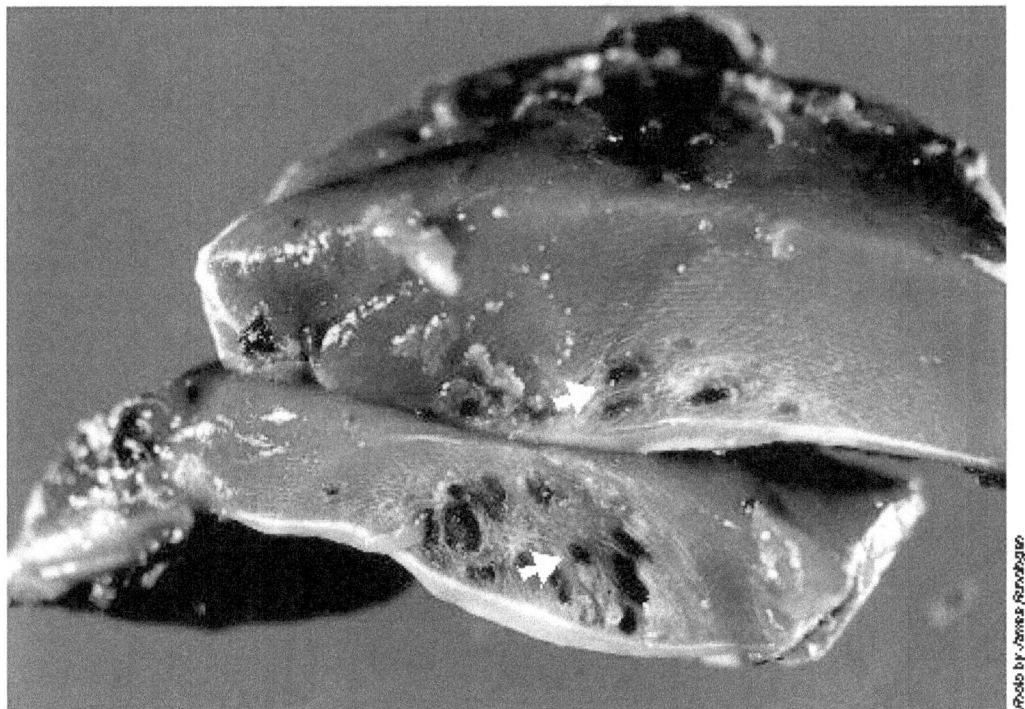

Figure 32.6 Areas of tissue destruction and reaction to migrating Epomidiostomum sp. in the gizzard muscle of a snow goose.

Control

Methods of controlling gizzard worms in free-ranging birds have not been developed. Attempts to do so would involve disruption of the parasite's life cycle. *Amidostomum* sp. and *Epomidiostomum* sp. have a direct life cycle (Fig. 32.1), and this suggests that transmission potential is greatest in crowded and continuously used habitat because of accumulative fecal contamination, provided that ambient temperatures are warm enough (68–77 °F) for larval development. Newly hatched birds are least resistant to infection, and birds of all ages are susceptible to reinfection.

Gizzard worms such as *C. spinosa* that have indirect life cycles could, theoretically, be controlled by reducing the availability of intermediate hosts to a number that is less than that which would allow transmission to be frequent enough to maintain the parasites. However, such actions, which would require habitat control or the use of insecticides, are generally not warranted because the parasite does not cause a significant number of bird deaths. Also, intermediate hosts, such as grasshoppers, have high food value for birds.

Human Health Considerations

Gizzard worms are not a threat to humans. Nevertheless, people who eat waterfowl gizzards should cook them thoroughly and should discard those that appear unhealthy because other infections may also be present.

Benjamin N. Tuggle and Milton Friend
(Modified from an earlier chapter by Benjamin N. Tuggle)

Supplementary Reading

Bump, R., Darrow, R.W., Edmister, F.C., and Crissey, W.F., 1947, The ruffed grouse, life history, propagation, management: Buffalo, N.Y., New York State Conservation Department, 915 p.

Herman, C.M., and Wehr, E.E., 1954, The occurrence of gizzard worms in Canada geese: Journal of Wildlife Management, v. 18, p. 509–513.

Herman, C.M., Steenis, J.H., and Wehr, E.E., 1955, Causes of winter losses among Canada geese: Transactions of the North American Wildlife Natural Resources Conference, v. 20, p. 161–165.

Leiby, P.D., and Olsen, O.W., 1965, Life history studies on nematodes of the genera Amidostomum and Epomidiostomum occurring in the gizzards of waterfowl: Washington, D.C., Proceedings of the Helminthological Society, v. 32, p. 32–49.

Tuggle, B.N., and Crites, J.L., 1984, The prevalence and pathogenicity of gizzard nematodes of the genera *Amidostomum* and *Epomidiostomum* (Trichostrongylidae) in the lesser snow goose (*Chen caerulescens caerulescens*): Canadian Journal of Zoology, v. 62, p. 1,849–1,852.

Chapter 33
Acanthocephaliasis

Synonyms
Thorny-headed worms, acanths

Cause

The phylum Acanthocephala contains parasitic worms referred to as thorny-headed worms because both the larval and adult parasites have a retractable proboscis or a tubular structure at the head, which has sharp, recurved hooks or spines. Much like the cestodes or tapeworms, they lack digestive tracts and absorb nutrients from the bird's intestinal canal. This may weaken the bird and may make it more susceptible to other diseases and to predation.

Adult acanthocephalans are found in a variety of bird species and in other vertebrates. More than 50 species of acanthocephalans have been reported in waterfowl, but reevaluations of acanthocephalan taxonomy are resulting in revised speciation. Nevertheless, numerous species within the phylum are found in birds.

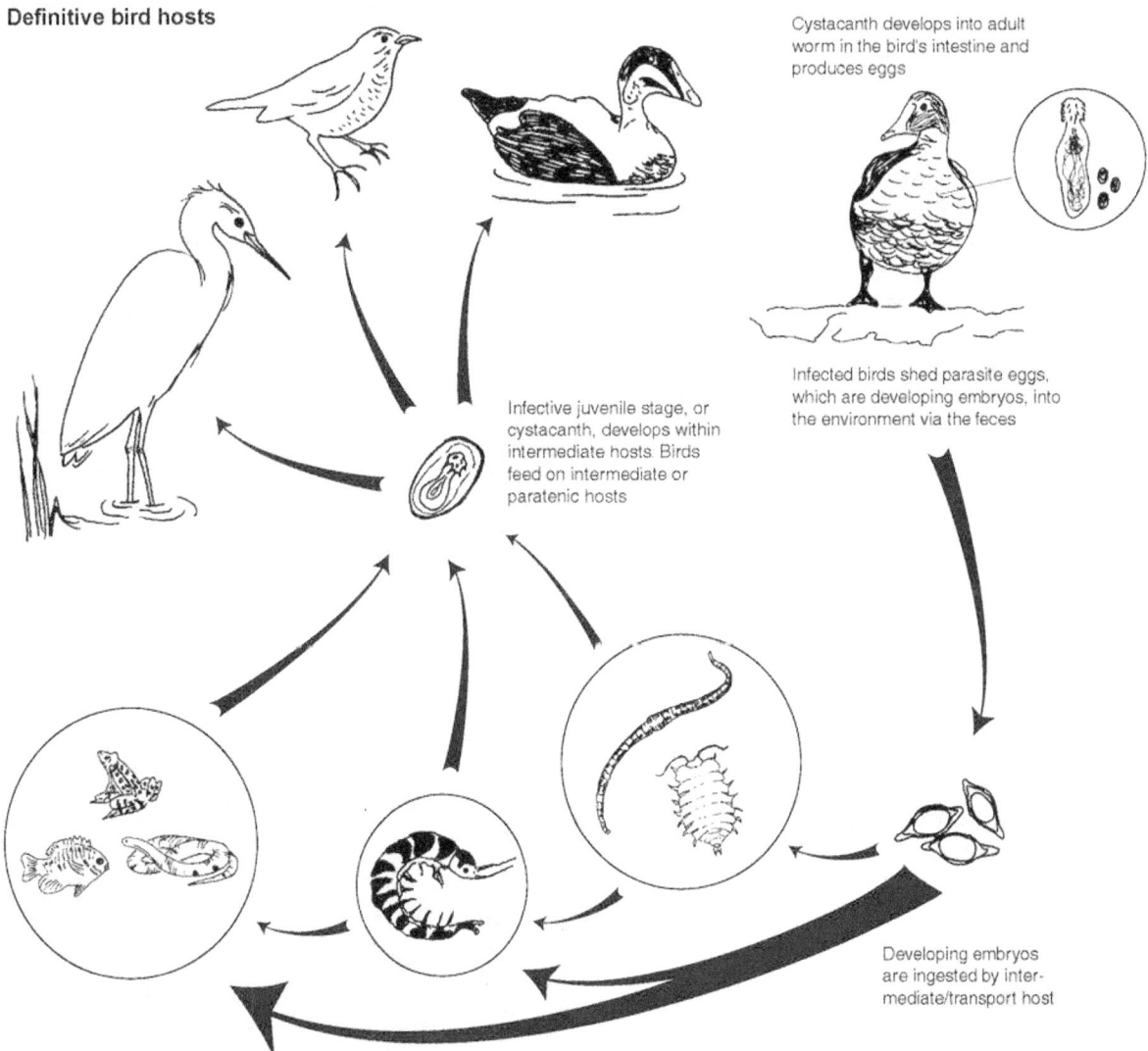

Figure 33.1 Indirect life cycle of acanthocephalan worms.

Life Cycle

All acanthocephalan species thus far examined have an indirect life cycle (Fig. 33.1) that requires at least one intermediate host. Intermediate hosts tend to be preferred food items of the definitive host; thus, the parasite also uses the intermediate host as a means of transport to the definitive host. Crustacea of the orders Amphipoda, Isopoda, and Decapoda have been identified as common intermediate hosts of acanthocephalans that infect waterfowl. Some acanthocephalans that affect passerines or perching birds are reported to use terrestrial insects as intermediate hosts. Fish, snakes, and frogs have been identified as paratenic hosts in the life cycle of some acanthocephalans that infect birds.

The adult female parasite within the definitive bird host produces eggs that are passed with the bird's feces into the environment. When the egg is ingested by the intermediate host (insect, crustacean, or centipedes and millipedes), the infective juvenile stage or cystacanth develops within the intermediate host. In many life cycles of acanthocephalans, if this intermediate host is eaten by a vertebrate host which is unsuitable as a definitive host, the cystacanth will penetrate the vertebrates gut, encyst and cease development. This vertebrate is now a paratenic host. If the paratenic or intermediate host is eaten by a suitable definitive host the cystocanth will attach to the definitive host's intestinal mucosa via the spined proboscis, mature, mate and produce eggs. A change in body coloration has been noted in some crustaceans infected with certain species of acanthocephalan. It is thought that this change in color increases predation by definitive hosts. This might be an evolutionary adaptation which increases the chances of life cycle completion by the acanthocephalan.

Species Affected

Acanthocephalans infect all classes of vertebrates and are common in birds. Ducks, geese, and swans are considered to be the most commonly infected birds along with birds of prey, and some species of passerines. All age classes can become infected. Severe disease outbreaks have been repeatedly reported from common eiders. Eider mortality from acanthocephalans has been documented throughout the arctic areas of their range and has been attributed to food habits rather than to any increased susceptibility of their species. Historical U.S. Fish and Wildlife Service disease diagnostic records reported heavy infections of acanthocephalans and mortality in trumpeter swans from Montana.

Distribution

Worldwide.

Seasonality

Birds can be infected with acanthocephalans year around. Epizootics usually correspond with food shortages, exhaustion (resulting from migration or breeding), or stressful circumstances. Mortality in immature and adult male eiders is commonly seen in late winter and early spring. Adult females experience mortality during or after brooding. Eider ducklings often suffer from both acanthocephalans and renal coccidia. Swan cygnets are also susceptible to lethal infections.

Field Signs

Lethargy and emaciation are nonspecific but common clinical signs associated with severe infections.

Gross Lesions

Gross lesions include white nodules on the serosal or external surface of the intestine (Fig. 33.2). Dissection of the nodules will reveal the proboscis of an acanthocephalan. Examination of the intestinal mucosa or internal lining will reveal white-to-orange colored parasites that are firmly attached to the mucosa (Fig. 33.3). Some parasites can penetrate the gut wall and project into the abdominal cavity. Adhesions between the loops of intestine are not uncommon in the severe cases where the intestine has been penetrated.

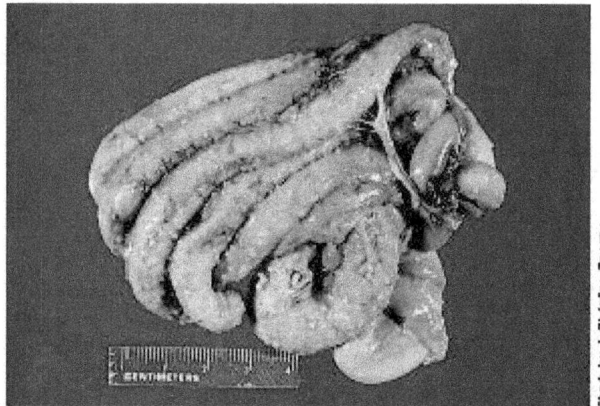

Figure 33.2 Intestinal loops of a bird infected with acanthocephalans. Parts of the worms protrude through the intestinal wall.

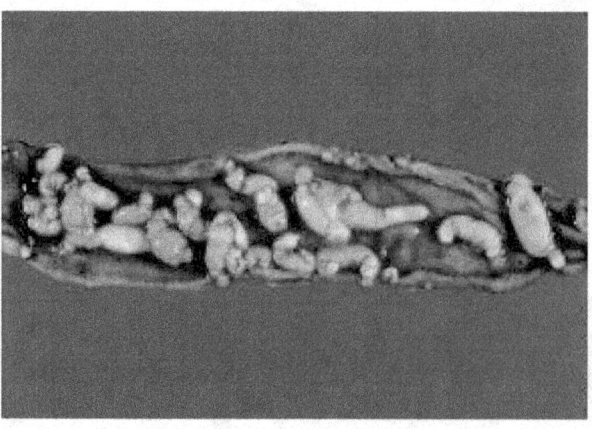

Figure 33.3 Acanthocephalans attached to the inner surface of the intestine of a bird.

Diagnosis

Postmortem examinations are required to reach a diagnosis of cause of death. When possible, submit chilled carcasses for evaluation. Fecal evaluations can be used for determining the presence of infection, but they do not provide a definitive diagnosis of disease. Evaluation by a parasitologist is required to differentiate the acanthocephalan species. If chilled carcasses cannot be submitted, the following alternatives in order of preference are: whole refrigerated intestine, frozen intestine, and formalin-fixed intestine. The formalin-fixed intestine may prohibit species identification but it will allow identification to the genus level as well as assessments of worm burden and tissue response to the parasites.

Control

Control of acanthocephaliasis in free-ranging birds is not practical. Captive flocks can be managed so that aggregations of birds and crustaceans are minimized. One approach that has been suggested for captive flocks is to limit infection in young birds by segregating them on water areas that are not used by other birds. These segregated water areas will presumably have much lower numbers of infected intermediate hosts for the birds to be exposed to. Acanthocephalans that infect mammals have been successfully treated with ivermectin or fenbendazole. Thiabendazole has been recommended for use in birds, but treatment is acknowledged to be difficult and success low.

Human Health Considerations

None

Rebecca A. Cole

Supplementary Reading

Bowman, D., 1995, Parasitology for veterinarians (6th ed.): Philadelphia, Pa., W.B. Saunders, 430 p.

Crompton, D.W.T. and Nichol, B.B., 1985, Biology of the Acanthocephala: Cambridge, England, Cambridge University Press, 519 p.

McDonald, M.E., 1988, Key to acanthocephala reported in waterfowl: U.S. Fish and Wildlife Service, Resource Publication 173, Washington, D.C., 45 p.

Wobeser, G.A., 1997, Acanthocephala, *in* Diseases of wild waterfowl (2nd ed): New York, N.Y., Plenum Press, p. 145–146.

Chapter 34
Nasal Leeches

Synonyms

Duck leeches

Cause

Bloodsucking leeches of the genus *Theromyzon* sp. are the only leeches in North America known to feed directly in the nasal passages, trachea, and beneath the nictitating membrane of the eyes of migratory birds. Three species of nasal leeches have been reported from North America, *T. rude*, *T. tessulatum*, and *T. biannulatum*. Other genera of leeches feed on the exposed surfaces of waterfowl.

Species Affected

Nasal leeches affect many aquatic bird species (Fig. 34.1). Affected waterfowl include northern pintail, teal, American wigeon, northern shoveler, ring-necked duck, canvasback, redhead, lesser scaup, bufflehead, gadwall, ruddy duck, white-winged scoter, surf scoter, trumpeter swan, and tundra swan. Geese may also be parasitized but they are parasitized less frequently than ducks and swans.

Distribution

Nasal leech infestations of waterfowl and other migratory birds have not been reported south of the 30th parallel and are most commonly observed in northern areas because these parasites are better adapted to cold-water lakes (Fig. 34.2).

Seasonality

Peak parasitism usually occurs during the spring and summer months when leeches are actively seeking potential hosts and reproducing. During the winter months, the ambient temperatures in frozen ponds and marshes considerably slow their metabolic rate and, thus, their activity. In wetlands kept free of ice during the winter, bird activity may stimulate opportunistic feeding by leeches.

Field Signs

Birds that have leeches protruding from the nares or attached externally to the mucous membranes of the eyes are easily recognized from a distance with the aid of binoculars (Fig. 34.3). Leeches may be so blood-engorged that they resemble small sacks of blood (Fig. 34.4). Infested birds may be seen vigorously shaking their heads, scratching at their bills with their feet, or sneezing in an effort to dislodge the leeches and to force air through blocked nasal passages. These efforts are usually unsuccessful. Nasal and respiratory tract

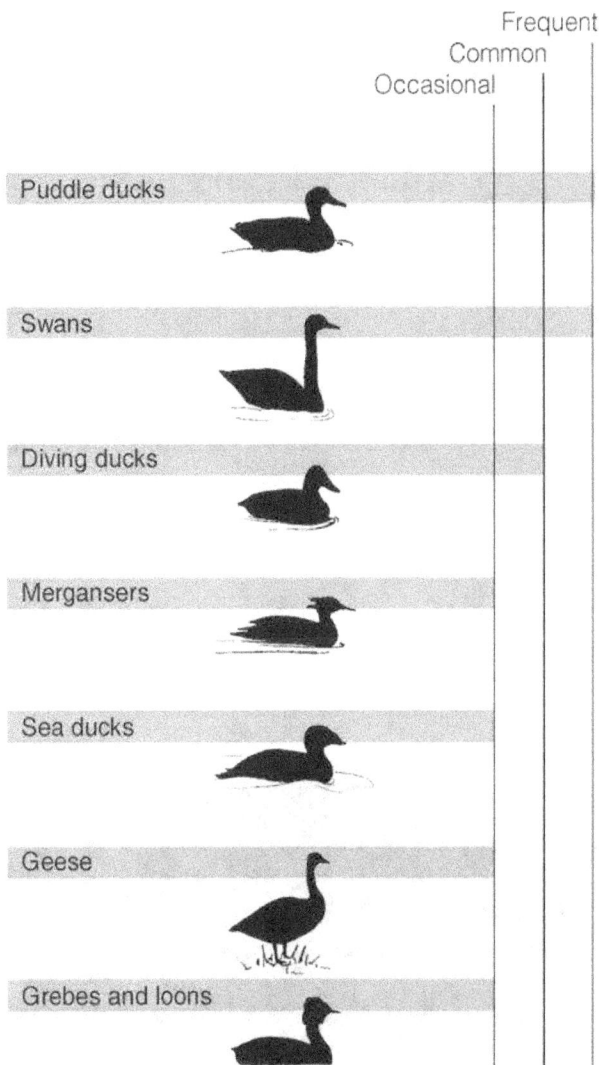

Figure 34.1 Relative frequency of nasal leech infestations in selected groups of migratory birds.

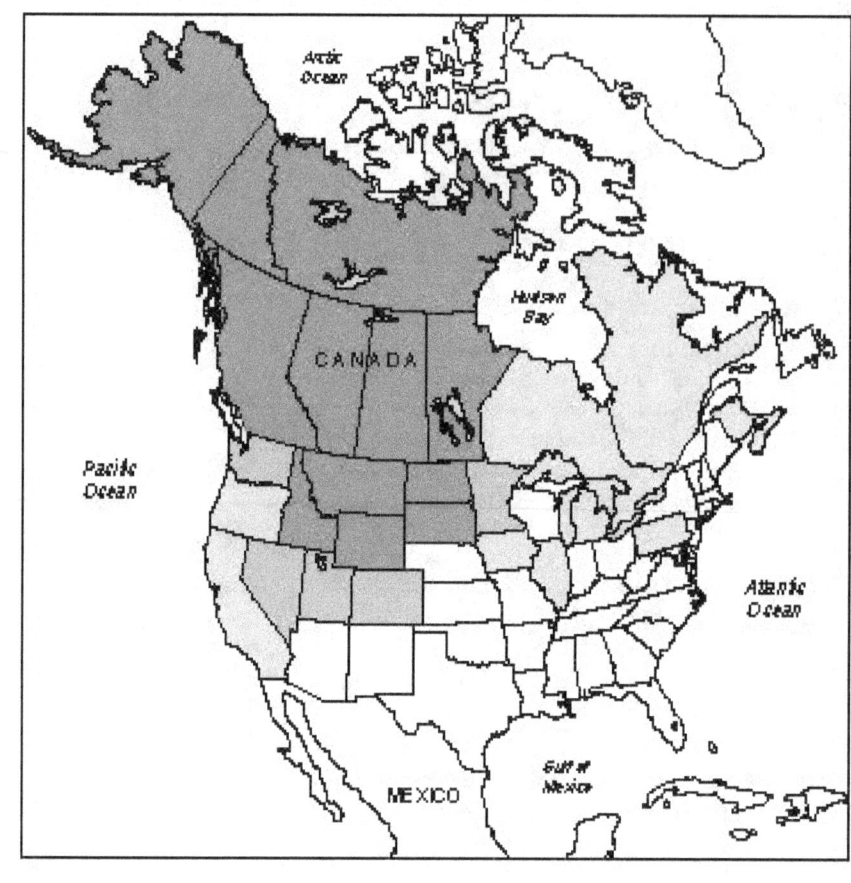

Figure 34.2 Reported distribution of nasal leeches in North America.

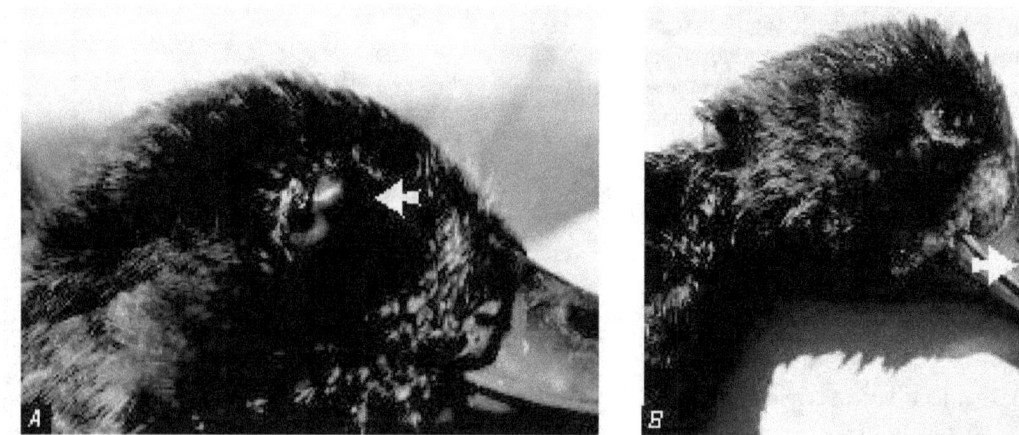

Figure 34.3 External nasal passage infestations of nasal leeches on (A), the eyes and (B), on the nares of a female redhead duck.

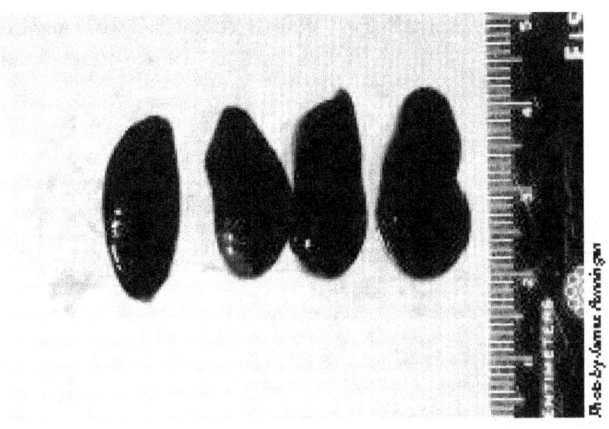

Figure 34.4 Blood-engorged nasal leeches removed from the nares of a trumpeter swan.

leech infestations can cause labored breathing and gaping similar to that seen among birds suffering from aspergillosis.

Gross Lesions

Severe leech infestations of the eye can result in temporary blindness. Eye damage may be seen as an accumulation of a stringy, cheeselike material beneath the nictitating membrane, as clouding of the cornea, and, in some instances, as collapse of the globe of the eye. Nasal passages (Fig. 34.5), throat, and trachea can become blocked by engorged leeches. The feeding action of *Theromyzon* sp. can cause inflammation and extensive damage to the lining of the nasal cavity.

Diagnosis

Nasal leeches are 10–45 millimeters long when they are blood-engorged, are amber or olive colored, and have four pairs of eyes. Those found in the free-living state are green, with variable patterns of spots on the top surface. Diagnosis of parasitism is usually made by seeing blood-engorged leeches protruding from the nares or attached to the eyes (Fig. 34.3), especially in birds that cannot be handled. Birds with internal leech infestations cannot be diagnosed by observation. In cases where the suspected cause of death is nasal leech parasitism, submit the entire carcass for examination. Leeches may depart a dead bird, making diagnosis difficult, or they may move to other areas of the body where they may be overlooked. Therefore, leeches found on carcasses should be collected and submitted with the carcass. They can be shipped alive in pond water and can be maintained in that condition for several months if they are kept refrigerated.

If leeches are to be killed before shipment, they must be preserved in a relaxed state so that species identification can be made. Straighten specimens between two glass slides (Fig. 34.6A) and flood them with a 10 percent formalin solution for 3–5 minutes while applying pressure to the top micro-

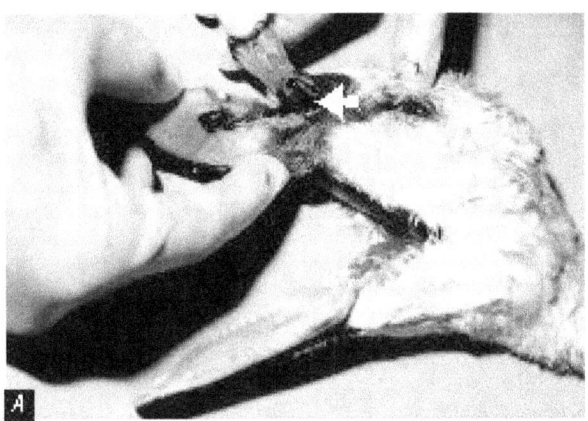

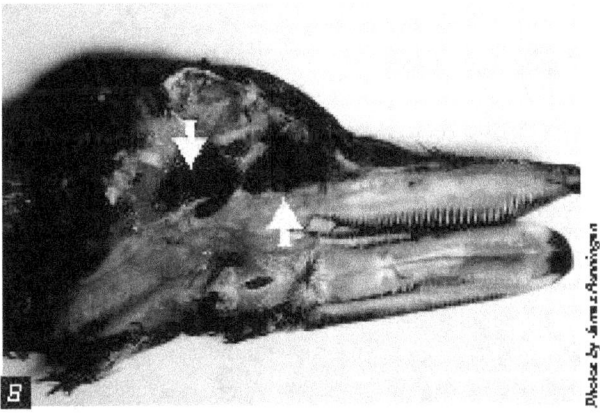

Figure 34.5 (A), Nasal passages of trumpeter swan infested with blood-engorged *Theromyzon* sp. leeches. (B), Dissected nasal passages of a mallard showing leech infestation in sinus passages.

scope slide (Fig. 34.6B); then place them in a 10 percent formalin solution for about 12 hours to complete the fixation. Afterwards, transfer leeches to a 5 percent formalin solution for preservation (Fig. 34.6C).

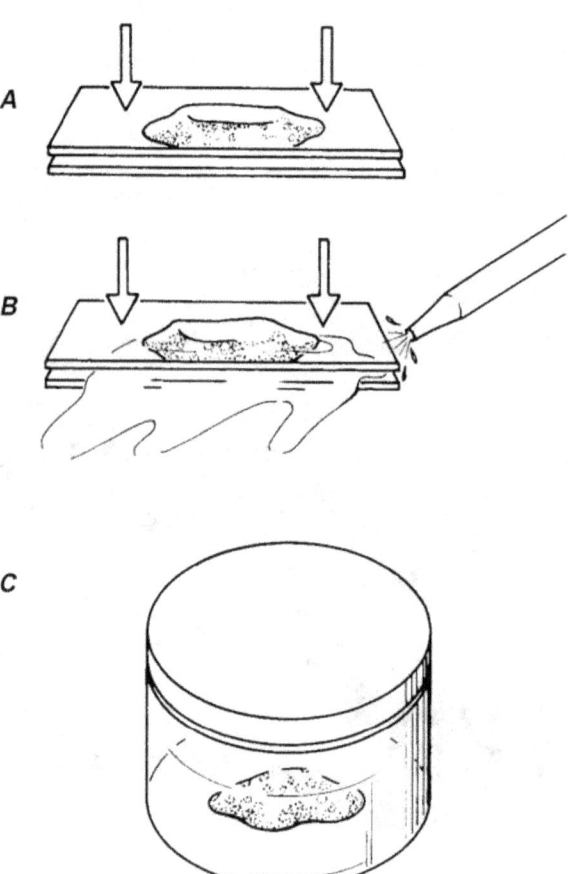

Figure 34.6 Leech fixation procedure. **(A)**, Straighten (flatten) leech between two glass slides; **(B)**, flood area between slides with 10 percent formalin for 3–5 minutes while maintaining downward pressure on top slide; and **(C)**, transfer leech to container of 10 percent formalin (see text for additional details). (Illustration by Randy Stothard Kampen)

Control

Leech infestations in waterfowl are common and can be fatal, especially in cygnets and ducklings. Tissue damage resulting from heavy infestations can facilitate secondary bacterial infections. However, no preventive measures have been developed for combating infestations in wild birds. When they are not feeding on birds, nasal leeches occur as free-living organisms in aquatic environments. Control measures to reduce leech populations might be possible if nasal leeches could be selectively killed. However, leech control must be weighed against the value of leeches as an aquatic bird food item.

Leeches protruding externally from the nares or attached to eyes can be removed with forceps. Leeches in the nasal passages can be removed by submerging the bird's bill for 5–10 seconds in a 10–20 percent salt solution, and this can be repeated several times if necessary.

Human Health Considerations

Theromyzon sp. feed exclusively on avian hosts and are not considered a threat to humans.

Benjamin N. Tuggle

Supplementary Reading

Davies, R.W., 1984, Sanguivory in leeches and its effects on growth, survivorship and reproduction of *Theromyzon rude*: Journal of Canadian Zoology, v. 62, p. 589–593.

Trauger, D.L., and J.C. Bartonek, 1977, Leech parasitism of waterfowl in North America: Wildfowl, v. 28, p. 142–152.

Tuggle, B.N., 1985, The occurrence of *Theromyzon rude* (Annelida:Hirudinea) In association with mortality of trumpeter swan cygnets (*Cygnus buccinator*): Journal of Wildlife Diseases, v. 22, p. 279–280.

Chapter 35
Miscellaneous Parasitic Diseases

Free-ranging wild birds are afflicted with numerous other parasites that occasionally cause illness and death. Some of these parasites, such as two of the trematodes or flukes highlighted below, can cause major die-offs. This section about parasitic diseases concludes with descriptions of some additional parasites that field biologists may encounter in wild birds. This listing is by no means complete and it is intended only to increase awareness of the diversity of types of parasites that might be encountered during examinations of wild birds. One should not assume that the parasites found during the examination of bird carcasses caused their death. Because parasites of birds vary greatly in size from a protozoa of a few microns in length to tapeworms of several inches in length and because they can be found in virtually all tissues, body cavities and other locations within the bird, the observation of the parasites will depend on their visibility and the thoroughness of the examination. Therefore, it is generally beneficial to submit bird carcasses to qualified disease diagnostic laboratories to obtain evaluations of the significance of endoparasites or of ectoparasites. The methods that are used to preserve the carcass, tissues, or other specimens can enhance or compromise the ability of specialists to identify the parasite to species, and even to genera, in some instances. Therefore, whenever possible, it is best to contact the diagnostic laboratory that will receive the specimens and obtain instructions for collecting, preserving, and shipping field samples (See Chapters 2 and 3).

Endoparasites

Trematodes

Most trematodes or flukes have complex life cycles that require two intermediate hosts (Fig. 35.1) in which the parasites develop before they become infective for the definitive, final bird host. In general, a mollusc is the first intermediate host, and is often a species that lives in the aquatic environment. Therefore, the aquatic environment brings potential hosts (waterbirds) and these parasites into close proximity and results in infections of waterfowl and other waterbirds that sometimes have fatal consequences.

Sphaeridiotrema globulus is a small trematode less than 1 millimeter in length that has been reported to cause die-offs in waterfowl, especially in diving ducks and swans, and, occasionally, in coot. Birds become infected by eating snails (the second intermediate host) that are infected with the metacercarial stage of the parasite (Fig. 35.1). This trematode feeds on blood, leading to severe blood loss and anemia. Birds often die acutely from shock due to an abnormally decreased volume of circulating plasma in the body or hypovolemic shock. Die-offs usually occur in the late summer or early fall and have been reported in the United States most commonly in the Northeast and North Central States, with a few reports from Western States. Die-offs will often recur yearly at the same locations. For example, mortality from *S. globulus* is suspected as a recurring event on the St. Lawrence River in southern Quebec. Infections by *S. globulus* have also been found elsewhere in Canada. The trematode is also present in the Old World and in Australia. Field signs include lethargy and bloodstained vents, although these signs are also found with duck plague (See Chapter 16). Gross lesions can appear as an inflammation or enteritis that is characterized by obvious hemorrhages in the lower small intestine or as inflammation of the intestinal wall with areas of ulcer-like erosions and the presence of a mixture of blood and fibrin (Fig. 35.2).

Experimental infections of mute swans with *S. globulus* indicated that as few as 100 metacercariae could be lethal. Juvenile mallard ducks that had no previous exposure to *S. globulus* died when infected with 550 metacercariae. Some immunity has been shown to exist in experimental infections of mallards. Adult mallards that were given 100 metacercariae and that later were challenged with 2,500 metacercariae survived, but those that received only 2,500 metacercariae died. Depending on how heavily snail populations are infected, some birds can receive a lethal dose during less than 24 hours of feeding. Susceptible waterfowl generally die 3–8 days postinfection after ingesting a lethal dose of *S. globulus*. Younger birds are generally more susceptible than older birds.

No parasite control measures have been developed. Any attempts at control would need to take into account the fact that the intermediate hosts are different within different geographic areas of the United States. There have been no reports of this parasite infecting humans.

Cyathocotyle bushiensis is another trematode that infects waterfowl and coot in the United States and Canada. This trematode can be found in the lower intestine, most commonly in the cecae or blind pouches extending out from the beginning of the large intestine (Fig. 35.3). The worms are slightly larger than *S. globulus*, measuring 1.7–1.8 millimeters in length. The life cycle for this parasite is similar to that of *S. globulus*. Birds become infected by consuming snails harboring the metacercariae. Disease caused by this trematode has been reported in black duck, blue-winged teal, green-winged teal, and coot. This trematode was first described in England and it is most likely limited by the geographic distribution of the snail, *Bithynia tentaculata*, which serves as the first intermediate host. This snail is found within the Great Lakes Basin in the United States.

There are no field signs associated with infection by *C. bushiensis* that have any diagnostic value. Common gross lesions in birds include hemorrhagic areas with plaque formation and some cheese-like or caseous core formation in the lumen of the cecae. Hemorrhage and plaque formation within the cecae are often present during early stages of infections (5–7 days after infection), whereas semisolid cellular debris that plugs the cecae are found in later infections (day 9 and later). Tissue damage within the birds is directly related to the attachment of the fluke to the mucosa and the effect of secretions by the fluke on cellular process within the intestine. Studies have indicated that there is a negative relationship between the number of flukes present and weight gain by the bird and a positive relationship between the number of flukes present and the number of white blood cells and body temperature. These findings reflect the nutritional impacts (reduced weight gain or weight loss) and body response to infection (increased white blood cells and body temperature). It is not known if weight loss in infected birds is due to cecal and lower intestine dysfunction or if the birds do not feed as much as noninfected birds. It has been suggested that morbidity and mortality are directly related to vascular leakage in the cecae from tissue damage that leads to dehydration of the bird. Experiments have shown that a fluke burden as low as 32 resulted in deaths of Pekin ducklings on day 8 postinfection, which indicates that ducks may succumb rather quickly to infection.

Control of this parasite, as for all trematode infections, would require preventing birds from feeding on infected snails or other invertebrate intermediate hosts. No reports of human infection have been reported for this trematode.

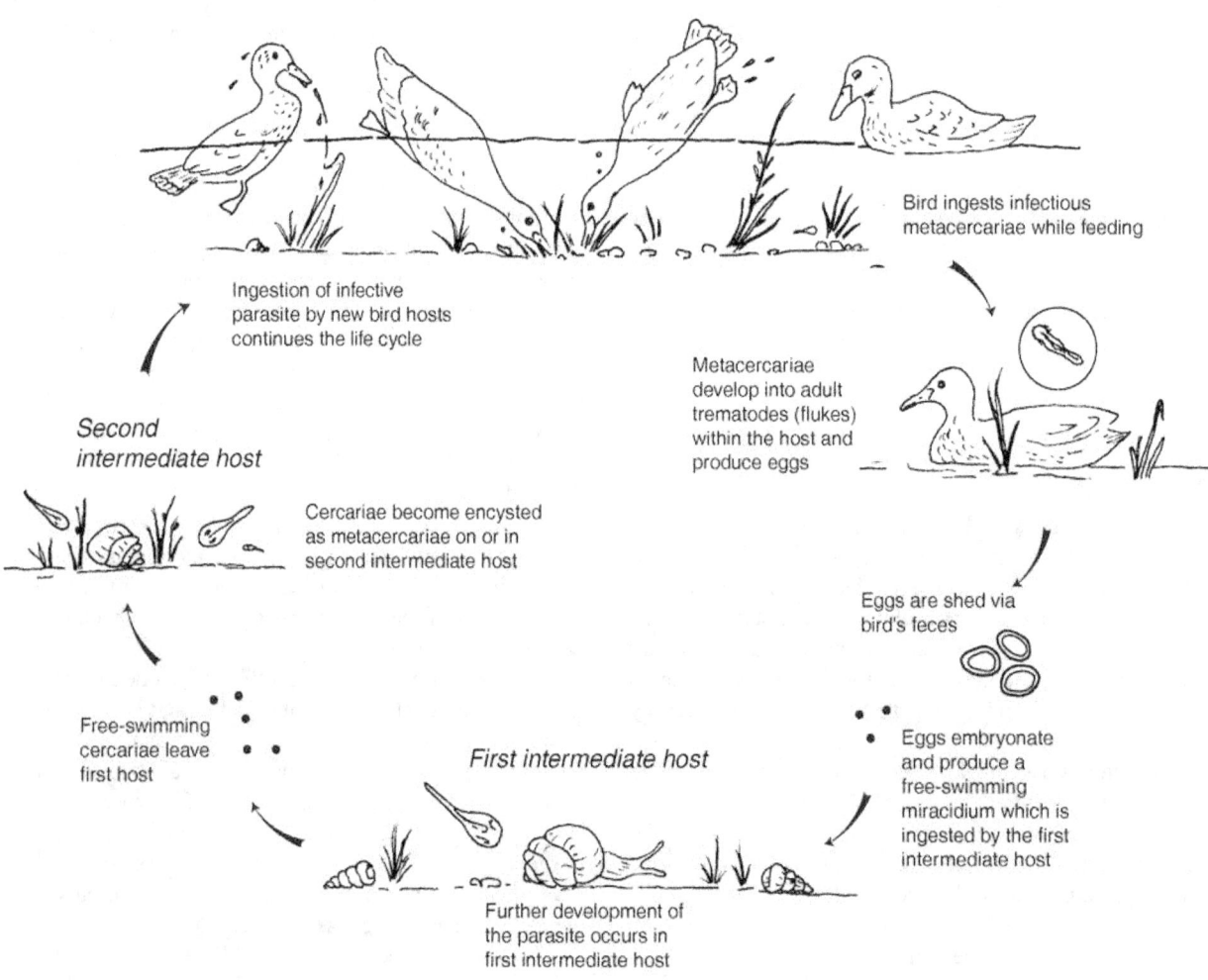

Fig. 35.1 General trematode (fluke) life cycle.

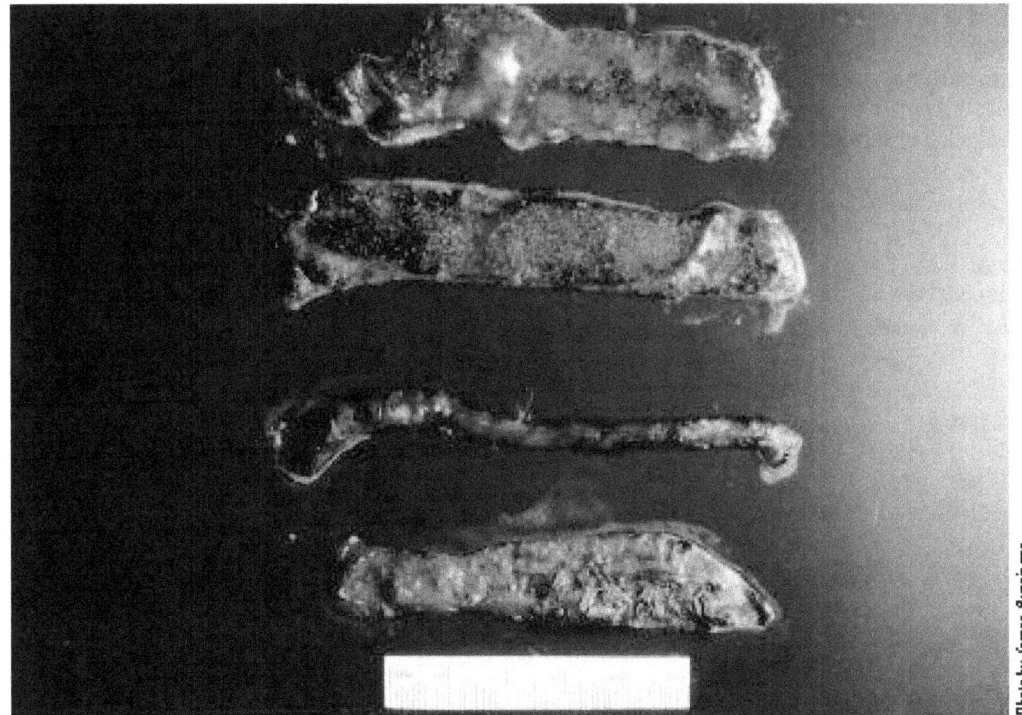

Fig. 35.2 Small intestine of lesser scaup that died from infection with *Sphaeridiotrema globulus*. Note the white flecks, which are *S. globulus*, and obvious hemorrhage.

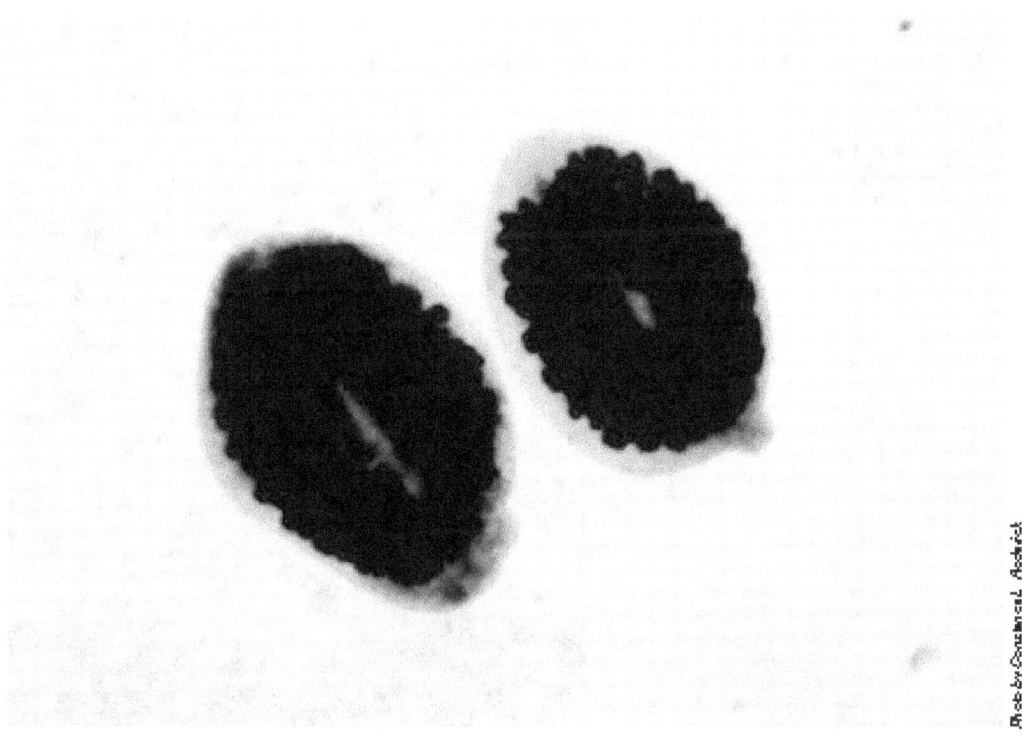

Fig. 35.3 *Cyathocotyle bushiensis* from the cecae of an American coot.

Leyogonimus polyoon (Fig. 35.4) is a newly reported trematode in North America that caused the deaths of over 1,500 and 11,000 coots in Northeastern Wisconsin during the falls of 1996 and 1997, respectively (Fig. 35.5). This trematode is of similar small size (0.7–1.0 millimeters in length) as *S. globulus* (Fig. 35.2). It is known to cause death in coot and common moorhen in Europe, but the Wisconsin outbreaks are the first documentation of this parasite causing mortality in North America. The susceptibility of other North American birds, beside coot, remains unknown. At the Wisconsin location, various waterfowl species were dying from infections of *S. globulus* while the coot were dying from *L. polyoon* infections. None of the waterfowl were found to be infected with *L. polyoon*.

The life cycle for *L. polyoon* is not known, although the suspected first intermediate host is the snail *Bithynia tentaculata*. Investigations at the National Wildlife Health Center (NWHC) have disclosed that snails that were collected in Wisconsin were infected with a cercariae or a larval form of the parasite that fits the literature description of *L. polyoon*. Additional studies are required to confirm that these larval forms are *L. polyoon*. Also, the intermediate host for the infective larval form or metacacariae has not yet been found. *L. polyoon* infects primarily the upper and middle areas of the small intestine. No significant field signs are associated with infection by *L. polyoon*. Gross lesions seen at necropsy include severe enteritis characterized by thickening of the intestinal wall and a fibrous-to-caseous core of necrotic debris that blocks the lumen of the intestine (Fig. 35.6).

The location where these outbreaks occurred is a lake that is drained by a stream, that is underlain with sand, and that has substantial growths of water weeds. The shoreline has been extensively developed for home sites and other human use, and the lake is used for recreation. *L. polyoon* has not been reported to infect humans.

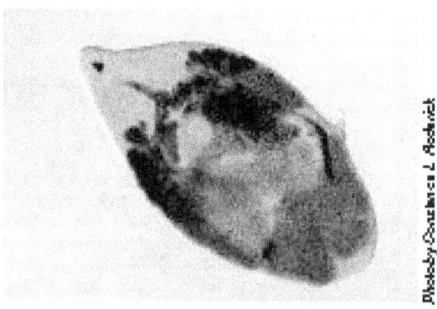

Fig. 35.4 *Leyogonimus polyoon* from the small intestine of an American coot.

Fig. 35.5 American coot and various species of waterfowl from die-off in Shawano Lake, Wisc. Coot mortality was due to *Leyogonimus polyoon*, but waterfowl mortality was due to *Sphaeridiotrema globulus*.

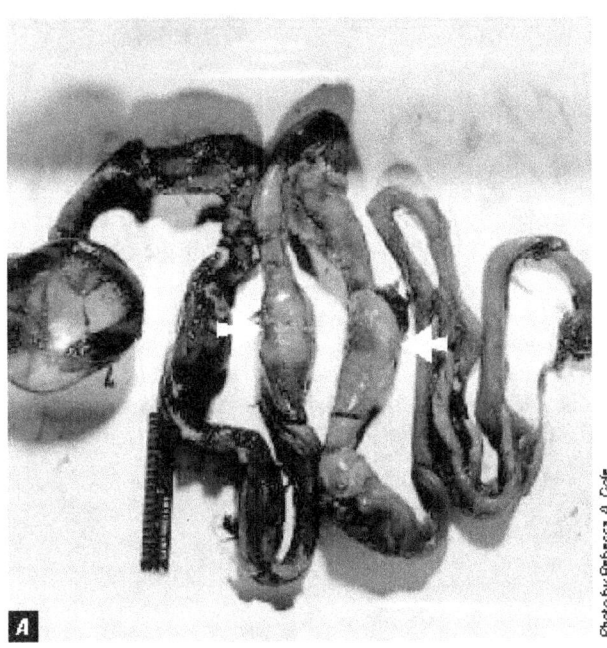

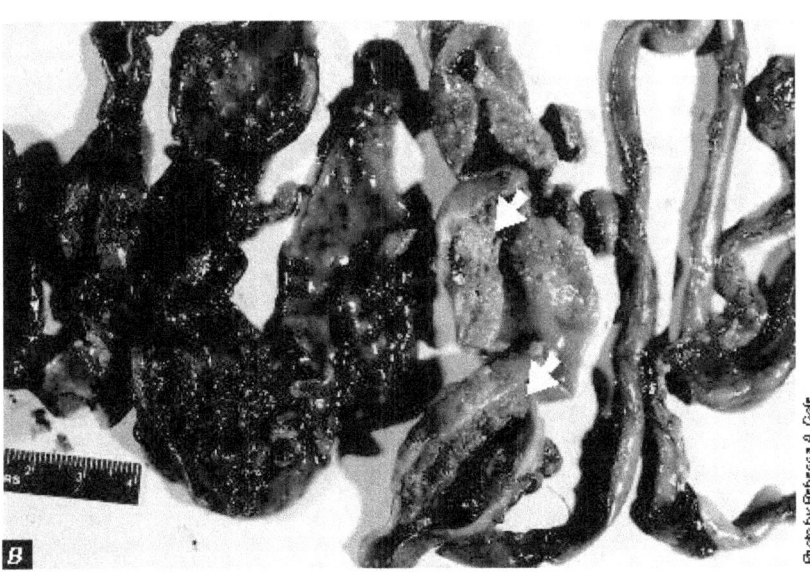

Fig. 35.6 (A) Gastrointestinal tract from an American coot that was infected with **Leyogonimus polyoon**. Note the enlarged or swollen areas (arrows). (B) Intestinal tract of an American coot. The intestinal tract has been incised to expose cheesy cores of dead tissue debris caused by **Leyogonimus polyoon** (arrows).

Cestodes

Tapeworms are common in wild birds, but they seldom cause death. Heavy burdens of these parasites may reduce the vigor of the bird and serve as a predisposing factor for other disease agents, or the parasites may occlude the intestine (Fig. 35.7). One genus, *Gastrotaenia* sp., lives in the gizzard and penetrates the keratohyalin lining or the horny covering of the gizzard pads, causing inflammation and necrosis. *Cloacotaenia* sp. inhabit the ureter or the tubular area that transports wastes from the kidneys to the cloaca in some waterfowl.

Nematodes

Trichostrongylidosis, the disease that is caused by *Trichostrongylus tenius*, is not currently a significant problem within the United States. However it is included because trichostrongylidosis, in natural populations of grouse in their native habitat in Scotland and elsewhere in the United Kingdom, demonstrates the impact that a parasite can have on the population dynamics of the bird host that it infects. *T. tenius* is a common nematode or roundworm that is found within the ceca of some types of wild birds, primarily grouse, geese, and poultry. This parasite has a direct life cycle that is closely associated with host food preferences for terrestrial vegeta-

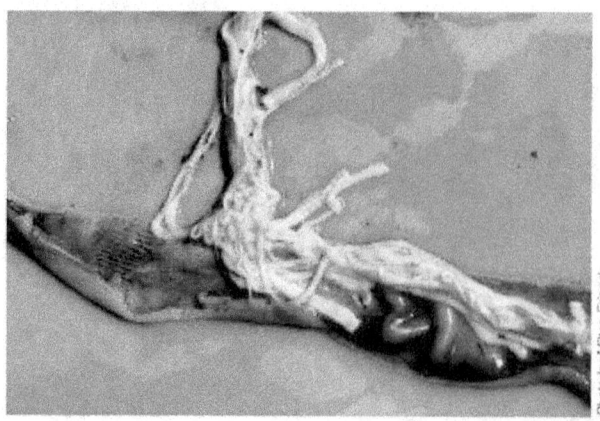

Figure 35.7 *Large numbers of tape worms may occlude the intestine.*

tion (Fig. 35.8). Trichostrongylidosis outbreaks can occur whenever birds are hatching because of the synchronous phase of the ecology of the parasite and the feeding habits of bird hosts.

Hatchability of the parasite eggs and survival of the free-living larval stages are moisture and temperature dependent. The parasite eggs do not develop under dry conditions and the free-living larvae are generally killed by freezing tem-

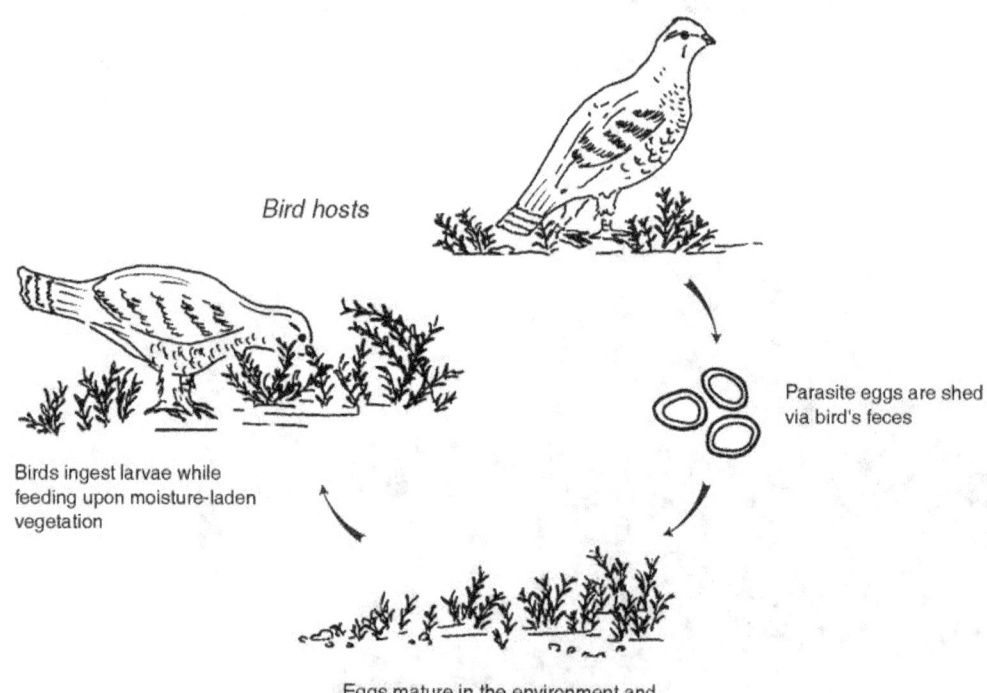

Figure 35.8 *Life cycle of the nematode Trichostrongylus tenius.*

peratures. When the feces that contain the parasite eggs are kept moist and when the ambient air temperature is suitable, first-stage larvae develop within the egg and hatch in about 2 days; free-living second-stage larvae develop within another 1-1/2-to-2 days; and development to infective third-stage larvae requires an additional 8–16 days. Studies in Scotland indicate that infective larvae crawl to the tips of moist heather and accumulate there in drops of water provided by the misty weather. Grouse that feed on the tips of the heather ingest the larvae along with their food. Infective larvae molt twice more within the ceca of the bird before they become sexually mature adults.

Infections have been reported in chicken, turkey, guinea fowl, pheasant, quail, pigeons, ducks, and geese; but infections are most notable for red grouse because this parasite has clearly been shown to regulate natural populations of this species. Because adult worms can survive in their bird host for more than 2 years, all adult birds evaluated in some populations have been found to be infected, thereby, providing a reservoir for infection of young. In addition, larvae can arrest development within the ceca, overwinter, and then resume development in the spring. The temperature and moisture requirements for hatchability of the parasite eggs and survival of larvae results in synchronized availability of parasites during the period of production of young grouse. This results in the primary occurrence of disease and mortality in grouse during the spring. The hatchlings are exposed to infective larvae soon after the diet for chicks changes from insects to vegetation.

The impacts of infection by *T. tenius* are greater than just chick mortality. These parasites also decrease available energy for egg laying by adult birds. The resting metabolic rate is increased and there is a decrease in food intake by the bird. The resulting impact is reduced fecundity within the population (fewer chicks) along with high chick mortality. This combination of impacts controls population levels.

Echinuria uncinata is a common nematode that infects the proventriculus of various waterfowl species. The life cycle is indirect, and the parasite uses zooplankton, especially *Daphnia* sp., as intermediate hosts (Fig. 35.9). Worms mature in a duck approximately 51 days after it eats infected zooplankton. Adult worms, which are approximately 5 millimeters long, burrow headfirst into the mucosa and submucosa of the proventriculus, causing tissue swelling and inflammation. Tumor-like nodules form and can be large enough to obstruct the lumen of the proventriculus. This parasite can be especially dangerous to waterfowl where zooplankton blooms coincide with the hatching of young. Often birds that are late to hatch and do not have fully developed immune systems can consume enough infected zooplankton in a very short period of time to become severely infected. In areas where water is shallow, where zooplankton populations are numerous, and where birds are crowded into the area, this roundworm can be transmitted to many birds during a short time.

This occurred within a population of the endangered Laysan duck on Laysan Island, Hawaii during the fall of 1993. A drought had struck the island and the brine flies that the

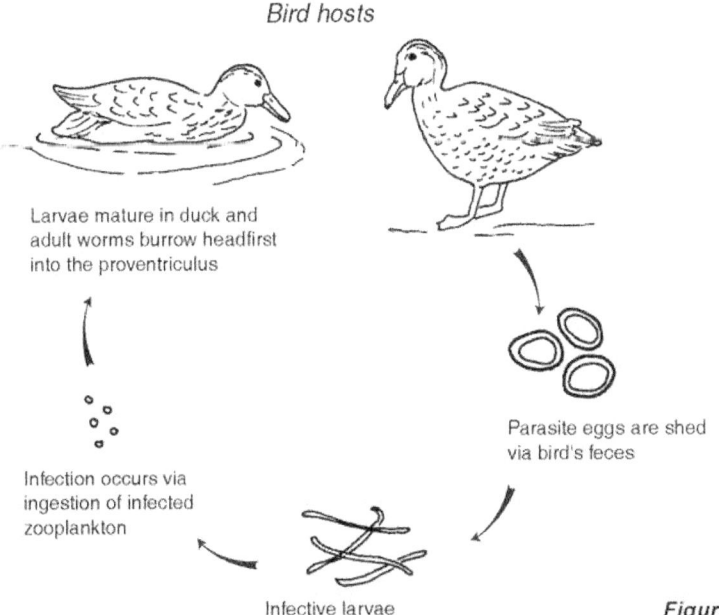

Figure 35.9 Life cycle of the nematode ***Echinuria uncinata***.

ducks feed heavily on were believed to be scarce. This depressed food base may have resulted in a reduced level of nutrition and may have compromised the ability of the birds to withstand infections by *E. uncinata*. Birds from this die-off were severely emaciated, had thickened proventriculi, and nodules along the proventriculi and intestines. The glands within the proventriculus were severely distorted, which suggests that the function of the proventriculus was compromised. Blood samples taken from sick birds suggested that they were emaciated and severely infected with parasitic worms. It was thought that the combination of the drought, aggregation of birds around freshwater seeps, and scarce food sources, combined with the severe parasitism, caused the ducks to die. Other examples exist where the combination of overcrowded waterfowl, a zooplankton population explosion, large numbers of infected zooplankton, and high retention of worms within waterfowl resulted in sufficient pathology by the worms to cause clinical disease and death. Maintaining fast water flow to prevent zooplankton explosions has been successfully employed for disease prevention in captive flocks such as those at waterfowl parks. *E. uncinata* is widely distributed geographically and has not been reported to infect humans.

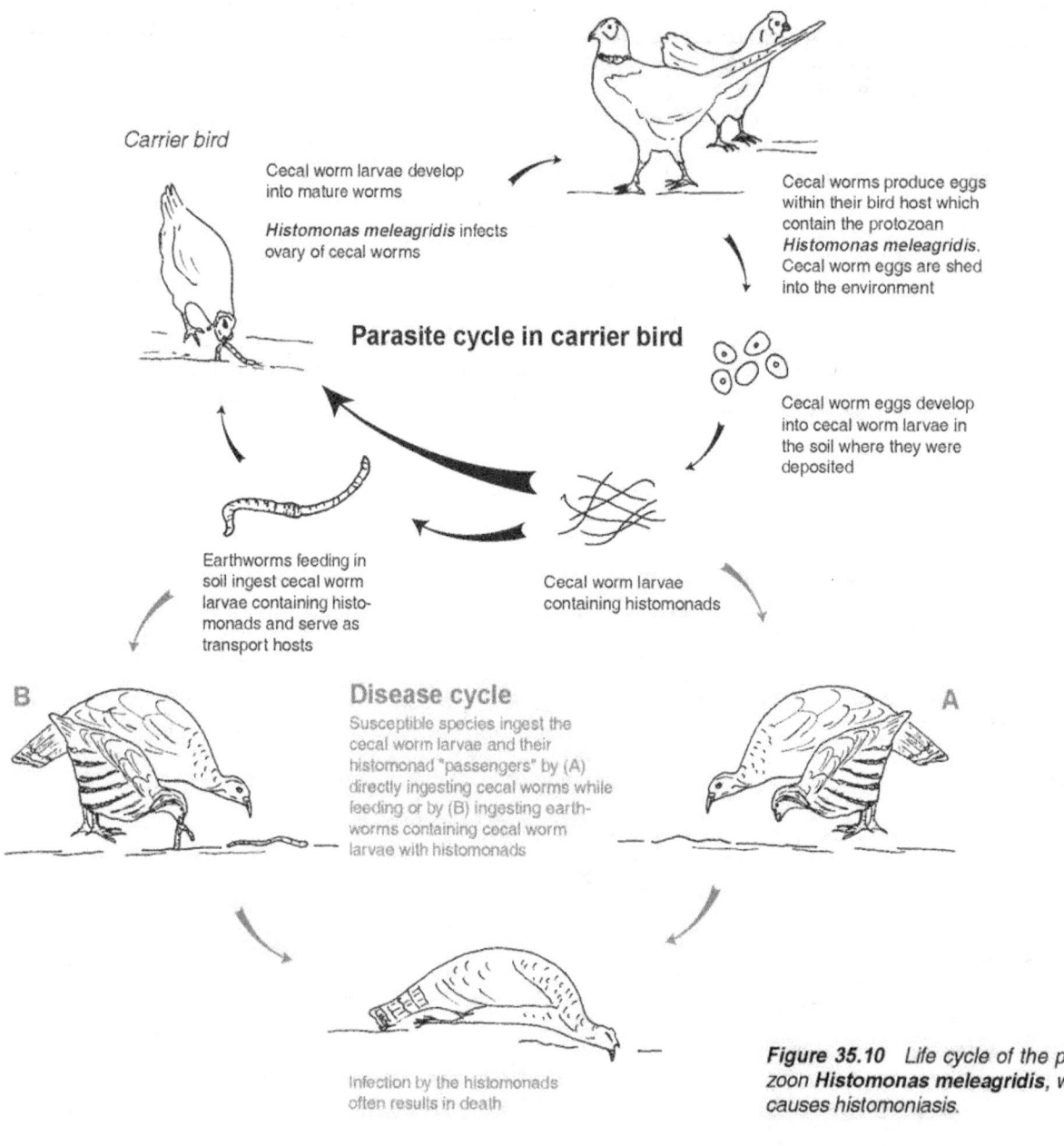

Figure 35.10 Life cycle of the protozoon **Histomonas meleagridis**, which causes histomoniasis.

Protozoa

Histomoniasis is capable of causing catastrophic losses in the wild turkey, a species whose restoration has become a major wildlife management success story. *Histomonas meleagridis*, the protozoan that causes histomoniasis, utilizes the cecal worm, *Heterakis gallinarum* (a nematode), as a vector for entry into the bird hosts (Fig. 35.10). The disease is commonly called blackhead because infections sometimes cause a bluish or blackish appearance of the skin of the head in some birds due to an excessive concentration of reduced hemoglobin in the blood or cyanosis.

Earthworms and other soil invertebrates can become part of the parasite's life cycle when they feed on fecal-contaminated soil that contains cecal worm eggs infected with histomonads. The cecal worm larvae and histomonads are stored in the body of the earthworm and are transmitted to birds when worms are fed upon. However, earthworms are not required for the life cycle; cecal worm larvae that contain histomonads may be ingested by birds when they feed in a contaminated environment.

Most, if not all, gallinaceous birds are susceptible hosts. Turkey, grouse, chicken, and partridge develop severe disease and suffer high mortality rates that can exceed 75 percent of those infected. Disease is less severe in Hungarian partridge and bobwhite quail. In contrast, pheasant and some other species often do not exhibit signs of disease, but they instead become carriers that maintain the disease cycle. Canada geese that were examined at the NWHC have also been found to have a histomoniasis-like disease. In North America, wild turkey and bobwhite quail are the species most commonly infected in the wild. The disease is found worldwide.

There are no clinical signs specific to histomoniasis. Wild turkeys affected with this disease often are listless, have an unthrifty appearance of ruffled feathers, and stand with drooped wings. The birds may appear depressed, and their feces are often sulfur-yellow in color. This fecal coloration generally occurs early in the disease and, combined with other field signs, it is highly suggestive of histomoniasis. The primary gross lesions seen upon necropsy of infected birds are numerous large, pale grey, discrete circular crater-like areas of necrosis or tissue death within the liver (Fig. 35.11) and thickened caecal walls that often also become ulcerated and hemorrhagic. The lumen of the ceca may also be obstructed by aggregations of yellowish necrotic debris referred to as cecal cores (Fig. 35.12).

Disease prevention should be the major focus for addressing histomoniasis. The introduction of gallinaceous bird species that are disease carriers, such as pheasants, into habitat occupied by highly susceptible species, such as wild turkey, is unwise and it can have catastrophic results. Similarly, because chickens are often carriers of *H. gallinarum* (cecal worms), and often shed histomonads, spreading uncomposted chicken manure onto fields can distribute cecal worm eggs to wild and susceptible species.

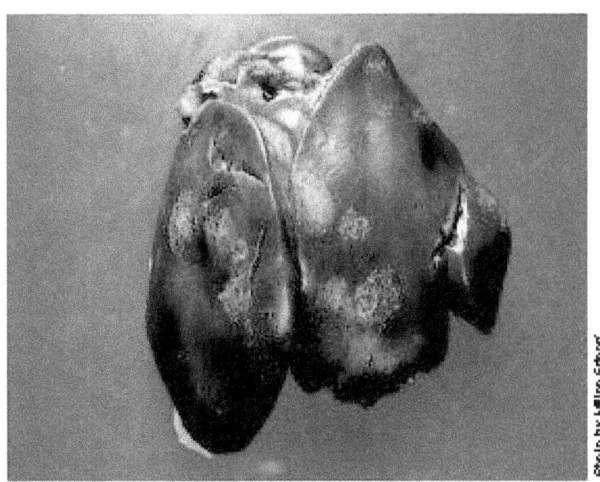

Figure 35.11 Large, pale areas in the liver of a bird infected with *Histomonas sp.*

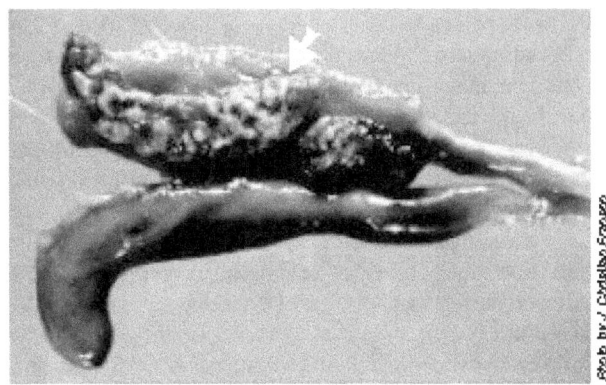

Figure 35.12 Cecal core (necrotic debris) in cecum of a bird infected with *Histomonas sp.*

Histomoniasis has caused the deaths of wild turkeys that were provided feed in barnyards frequented by chickens. Therefore, when attempting to reestablish wild turkey flocks in areas where they no longer exist and during periods of inclement weather that create food shortages for wild birds, placement of feed stations should be done with consideration of potential carriers of *H. meleagridis* or *H. gallinarum*.

Ectoparasites

In addition to being vectors that transmit disease to birds, ectoparasites can be direct causes of illness and death. Just a few adult ticks feeding on a small bird can cause anemia, reduced growth, weight loss, and contribute in other ways to a depressed state of health. The fowl tick, a soft-bodied tick of the family Argasidae, is the most important poultry ectoparasite in many countries and it is often a factor limiting raising chickens and turkeys. Chickens have also been reported to suffer tick paralysis, which is a motor paralysis or paralysis of the voluntary muscles, from bites of *Argas* sp. ticks. Tick paralysis in songbirds has been associated with the bite of the hard-bodied bird tick, *Ixodes brunneus*. Fatal paralysis from bites by this tick has been reported in numerous species of small birds. The engorged ticks in fatal cases are generally found on the bird's head and they may be attached to its eyelids. Death results from a powerful neurotoxin that is secreted by the tick while it feeds on the bird. Other species of *Ixodes* ticks have been associated with tick paralysis and mortality in marine birds, including albatross and petrels. An ascending motor paralysis that starts at the feet, progresses for 7–10 days, and ends in death has been reported.

Heavy infestations of lice, mites, fleas, flies, and other biting insects have also been responsible for causing illness and even death of wild birds, especially among nestlings. Conditions caused by these insects range from feather loss and skin damage from acariasis or mange, to myiases or infestation with fly maggots, and anemia. Mites of the genus *Knemidocoptes* are the primary cause of mange in birds, and the mites belong to the same family (Sarcoptidae) of mites that cause mange in mammals and humans. The *Knemidocoptes* sp. mites are specific to birds and they are not a human health hazard.

More knowledge is needed about the role of ectoparasites as causes of bird death. Proper identification of the species associated with bird mortality is an important component of such assessments; therefore, the presence of insects within bird nests in which freshly dead nestlings are found should be recorded, representative parasite specimens should be collected along with any visible parasites on the carcass, and the parasites submitted should be with the carcass. Ticks and any heavy infestations of fleas, lice, and other insects on live birds being handled for banding or other purposes should also be noted, and, when practical, samples should be collected and submitted for identification to a parasitologist or disease diagnostic laboratory. An abundance of such parasites may be indicators of other health problems for the birds.

Rebecca A. Cole and Milton Friend

Supplementary Reading

Anderson, R.C., 1992, Nematode parasites of vertebrates, their development and transmission: Wallingford, United Kingdom, C•A•B International, University Press, 578 p.

Arnall, L., and Keymer, I.F., 1975, Bird diseases, an introduction to the study of birds in health and disease: Bailliere Tindall, London, T.F.H. Publications, 528 p.

Calnek, B.W. and others, 1997, Diseases of Poultry (10th ed.): Ames, Iowa, Iowa State University Press, 1,080 p.

Delehay, T.M., Speakman, J.R., and Mass, R., 1995, The energetic consequences of parasitism: effects of a developing infection of *Trichostrongylus tenus* (Nematoda) on red grouse (*Lagopus lagopus scoticus*) energy balance, body weight and condition: Parasitology, v. 110, p. 473–482.

Davidson, W.R., and Nettles, V.F., 1997, Field manual of wildlife diseases in the Southeastern United States (2d ed.): Athens, Ga., University of Georgia, 417 p.

Gagnon, C., Scott, M.E., and McLaughlin, J.D., 1993, Gross lesions and hematological changes in domesticated mallard ducklings experimentally infected with *Cyathocotyle bushiensis* (Digenea): Journal of Parasitology, v. 79, p. 757–762.

Hove, J., and Scott, M.E., 1988, Ecological studies on *Cyathocotyle bushiensis* (Digenea) and *Sphaeridiotrema globulus* (Digenea), possible pathogens of dabbling ducks in southern Quebec: Journal of Wildlife Diseases, v. 24, p. 407–421.

Hudson, P.J., 1986, The effect of a parasitic nematode on the breeding production of red grouse: Journal of Animal Ecology, v. 55, p. 85–92.

Huffman, J.E., and Roscoe, D., 1989, Experimental infections of waterfowl with *Sphaeridiotrema globulus* (Digenea): Journal of Wildlife Diseases, v. 25, p. 143–146.

Mucha, K.H., and Huggman, J.E., 1991, Inflammatory cell stimulation and wound healing in *Sphearidiotrema globulus* experimentally infected mallard ducks (*Anas platyrhynchos*): Journal of Wildlife Diseases, v. 27, p. 428–434.

Section 6
Biotoxins

Algal Toxins

Mycotoxins

Avian Botulism

Aerial view of a large dinoflagellate bloom in near-shore ocean waters
Photo by Peter Franks, Scripps Institute of Oceanography

Introduction to Biotoxins

"Ecological toxicology is the study of all toxicants produced by living organisms and of the ecological relationships made possible by these poisons."
(Hayes)

"In all communities chemical interrelations are important aspects of the adaptation of species to one another; in some communities chemical relations seem to be the principal basis of species niche differentiation and community organization." (Whittaker and Feeny)

"Undoubtedly there is much to be learned from finding out how the battle [between toxicants produced by living organisms and host defenses developed in response to these toxicants] has been fought for the last several million years."
(Hayes)

All quotes from:

Hayes, Wayland, J., Jr., 1991, in Hayes, Wayland J., Jr., and Laws, Edward R., Jr., eds., Handbook of pesticide toxicology, v. 1, General principles: New York, Academic Press, p. 7–8.

Biotoxins are usually defined as poisons that are produced by and derived from the cells or secretions of living organisms. These natural poisons include some of the most toxic agents known and they are found within a wide variety of life forms. Organisms that produce such toxins are generally classified as being venomous or poisonous. The classification of venomous is usually associated with animal life forms such as poisonous reptiles and insects that have highly developed cellular mechanisms for toxin production and that deliver their toxins during a biting (rattlesnake) or stinging (black widow spider) act. Poisonous organisms are generally thought of as those that deliver toxins by being ingested or by their secretions being ingested by another organism. Therefore, these toxins are essentially forms of food poisoning. Readers should appreciate that virtually all venomous organisms are poisonous but many poisonous organisms are not venomous. This Section will address poisonous, but not venomous, organisms, and it includes the perspective of biotoxins as products of plants and lower life forms.

Birds become poisoned by a broad array of biotoxins. The chapter about avian botulism involves microbial toxins produced within replicating *Clostridium botulinum* bacteria. The potency of toxins that are produced by the disease-causing Clostridia are legendary, and the toxins include such human diseases as tetanus and lethal botulism food poisoning. Avian botulism is currently the most important disease of waterfowl and shorebirds, nationally and internationally, and outbreaks of this disease commonly kill tens of thousands of birds during a single event. Up to a million birds have recently been lost within a single location during the course of a protracted outbreak.

Because many avian botulism die-offs occur on the same wetlands year after year, one of the primary areas of research on this disease has focused on identifying and understanding the microenvironmental characteristics that contribute to a mortality event. The development of wetland-specific risk assessment tools will enable wildlife disease specialists and natural resource managers to more effectively manage avian botulism.

Fungi are an additional source of microbial biotoxins that cause the death of free-ranging wild birds. Mycotoxins, which are toxins produced by fungi, have received considerable study because of their effects on food animals and humans. In poultry, for example, many types of mycotoxins are known to cause problems that include mortality, decreased growth, impaired reproduction, immunosuppression, and pathologic effects on a variety of other organ systems. Although these toxins have received little study in wildlife, a growing body of literature documents similar effects of mycotoxins in a variety of free-ranging species. The chapter about mycotoxins illustrates the capabilities of aflatoxins and trichothecenes to cause large-scale bird losses as the result of bird ingestion of food contaminated by molds that produce these toxins. As more becomes known about the occurrence of mycotoxins in the natural environment, and as analytical techniques for the specific toxins become more commonly available, it is likely that more and more cases of mycotoxicosis will be reported in wildlife.

The range of living organisms that cause poisoning in wild birds is further illustrated by plant toxins in the chapter about algal toxins. Less is known about poisoning of birds from toxic plants than is known about poisoning from bacterial and fungal toxins. Plant toxins other than algal toxins that have caused bird mortality have rarely been reported. Choke cherry seeds contain chemical compounds that release cyanide upon digestion if the seed capsule is broken during digestion. Songbirds have been killed by cyanide poisoning from eating these seeds. Waterfowl mortality has been attributed to ingestion of castor beans, which results in intoxication from ricin, the active ingredient within the seed that causes poisoning. A small number of other reports of plant toxins causing wild bird mortality also exist.

The so-called algal toxins are produced by a variety of organisms, including true algae, dinoflagellates (aquatic protozoa), and blue-green algae, and are the least understood of the biotoxins covered in this Section. Algal blooms, especially red tides and blue-green blooms, wreak aesthetic and economic havoc in many freshwater and marine environments because of the potential for toxins to be present. Perhaps one of the most widely recognized toxins in this group is saxitoxin, the agent of paralytic shellfish poisoning, which causes occasional human deaths and renders many tons of shellfish inedible throughout the world.

Algal toxins are likely to become increasingly recognized as a cause of waterbird mortality. Eutrophication of inland waterbodies due to nutrient loads is causing more algal blooms within those waters, many of which are used by large numbers of water birds. Enhanced technology and increased study are needed to better understand the ecology of algal blooms and the production of toxic components that are hazardous to bird life. With the exception of avian botulism, biotoxins as a cause of disease in wild birds have received little study. However, there should be no debate regarding the need for study since disease caused by biotoxins extends beyond direct mortality. Impaired immune system function or immunosuppression and cancers caused by biotoxins have both been documented in animals and humans. Other effects on wildlife are also likely because of the diversity of disease impacts seen in humans and domestic animals.

Chapter 36
Algal Toxins

Synonyms
Red tide toxins, phycotoxins

Periodic blooms of algae, including true algae, dinoflagellates, and cyanobacteria or blue-green algae have been reported in marine and freshwater bodies throughout the world. Although many blooms are merely an aesthetic nuisance, some species of algae produce toxins that kill fish, shellfish, humans, livestock and wildlife. Pigmented blooms of toxin-producing marine algae are often referred to as "red tides" (Fig. 36.1). Proliferations of freshwater toxin-producing cyanobacteria are simply called "cyanobacterial blooms" or "toxic algal blooms." Cyanobacterial blooms initially appear green and may later turn blue, sometimes forming a "scum" in the water (Fig. 36.2).

Although algal blooms historically have been considered a natural phenomenon, the frequency of occurrence of harmful algae appears to have increased in recent years. Agricultural runoff and other pollutants of freshwater and marine wetlands and water bodies have resulted in increased nutrient loading of phosphorus and nitrogen, thus providing conditions favorable to the growth of potentially toxic algae. The detrimental impact of red tides and cyanobacterial blooms on wetland, shore, and pelagic species has long been suspected but not often been substantiated because information on the effects of these toxins in fish and wildlife species is lacking and diagnostic tools are limited.

Cause

Some dinoflagellates and cyanobacteria produce toxins that can affect domestic animals and humans. Some of these toxins such as domoic acid, saxitoxin (paralytic shellfish poisoning or PSP toxin), brevetoxin, and cyanobacterial toxins (including anatoxins, microcystins, and nodularins) have been suspected, but they have rarely been documented, as the cause of bird mortality (Table 36.1). Marine algal toxins such as domoic acid, saxitoxin, and brevetoxin that bioaccumulate or are magnified in the food chain by fish and shellfish, and anatoxins from freshwater cyanobacteria, affect the nervous system; cyanobacteria that contain microcystins or nodularin cause liver damage.

The effects of some harmful algae are not related to toxin production but rather are related to depleted dissolved-oxygen concentrations in water caused by algal proliferation, death, and decay, or night respiration. Other harmful effects include occlusion of sunlight by large numbers of algae and physical damage to the gills of fish caused by the structure of some algal organisms. All of these effects can

Figure 36.1 *Aerial view of a large dinoflagellate bloom in near-shore ocean waters. The organism responsible for this bloom is not a toxin producer; however, toxic blooms may have a similar appearance.*

Figure 36.2 *A cyanobacterial or blue-green algal bloom.*

Table 36.1 Documented instances of wild bird mortality caused by algal toxins.

Toxin	Algal species	Toxin type(s)	Migratory bird species affected	Route of exposure
Cyanobacterial	**Microcystis** sp., **Anabaena** sp., **Aphanizomenon** sp., **Nodularia** sp., and **Oscillatoria** sp.	Hepatotoxins (microcystins and nodularin) Neurotoxins (anatoxin-a and anatoxin-a(s))	Unidentified ducks, geese, and songbirds, Franklin's gull, American coot, mallard, American wigeon	Oral (water)
Domoic acid (amnesic shellfish poisoning)	**Pseudonitzschia** sp.	Neurotoxin	Brown pelican, Brandt's cormorant	Oral (food items)
Saxitoxin (paralytic shellfish poisoning)	**Alexandrium** sp.	Neurotoxin	Shag, northern fulmar, great cormorant, herring gull, common tern, common murre, Pacific loon, and sooty shearwater	Oral (food items)
Brevetoxin	**Gymnodinium** sp.	Neurotoxin	Lesser scaup	Oral (food items)

lead to mortality of aquatic invertebrates, aquatic plants, or fish and may produce an environment conducive to botulism. Other marine algal toxins (okadaic acid, neosaxitoxin, ciguatoxin, and *Pfiesteria* exotoxin) and cyanobacterial toxins (saxitoxin, neosaxitoxin, and cylindrospermopsin) have not yet been identified as causes of bird mortality events, but increased awareness and further research may establish a relationship.

Species Affected

Many bird and mammal species can be affected by algal toxins. Most reports of mortality in birds are of die-offs that occur in conjunction with a bloom. Sometimes algal toxins are found in potential food items; however, there have been very few instances in which the algal toxin has been isolated from the ingesta or tissues of affected birds. Domoic acid poisoning caused mortality in brown pelicans and Brandt's cormorants on the central California coast. Brevetoxin has been suspected as the cause of mortality in lesser scaup, and saxitoxin has been strongly suspected as the cause of mortality in sea birds (common terns, shags, great cormorants, northern fulmars, herring gulls, common murres, Pacific loons, sooty shearwaters, and others). Cyanobacterial toxicosis has been suspected in mortalities of free-ranging ducks, geese, eared grebes, gulls, and songbirds.

Distribution

Many of the organisms responsible for red tides are widely distributed and, in recent years, the organisms seem to be markedly spreading. Natural events such as hurricanes can disperse organisms, and it is suspected that some organisms may be transported long distances in ship ballast waters. Another factor that may encourage algal proliferation in both marine and freshwater systems is increased nutrient loading. Certain algae occur more commonly in some areas than others and it is useful to know which ones are problems in specific locations. Good sources of information about algal blooms are the State public health department or the State division of marine resources or marine fisheries.

Seasonality

There have not been enough confirmed instances of wild bird mortality caused by red tides and cyanobacterial blooms to establish seasonal patterns of occurrence.

Field Signs

Field signs reported are variable and they depend on the toxin involved. Domoic acid poisoning of brown pelicans caused neurologic signs that included muscle tremors, a characteristic side-to-side head movement, pouch scratching, awkward flight, toe clenching, twisting of the head over the back, vomiting, and loss of the righting reflex just before death. Brandt's cormorants that also were involved in this mortality event were easily approached and handled, but they did not exhibit the neurologic signs seen in the pelicans. Sea birds suspected of having been poisoned by saxitoxin exhibited paralysis and vomiting. Clinical signs observed in lesser scaup suspected of having been poisoned by brevetoxin included lethargy, weakness, reluctance or inability to fly, head droop, and excessive ocular, nasal, and oral discharge.

White Pekin ducklings that were experimentally exposed to brevetoxins exhibited lethargy, loss of muscle coordination or ataxia, spastic head movements, head droop to one side, and leg extension to the rear during rest. Clinical signs in muscovy ducks dosed with anatoxin-a(s) included excessive salivation, regurgitation of algae, diarrhea, tremors, reduced responsiveness and activity, incoordination, difficulty breathing, excessive thirst, congestion in foot webs, wing and leg weakness, and recumbency and intermittent seizures prior to death.

Gross Lesions

No characteristic or diagnostic gross lesions have been described for most types of algal toxin poisonings of wild birds. Many of the toxins, particularly the neurotoxins, have a chemical effect that does not produce a grossly observable lesion. Birds that ingest toxic blooms of *Microcystis* may have notable lesions of necrosis or tissue death and hemorrhage in the liver. These lesions have been reported in domestic mammals and birds, including ducks, that died as a result of exposure to a toxic *Microcystis* algal bloom or that were experimentally dosed with microcystin.

Diagnosis

Definitive diagnosis of algal toxicosis is difficult. Circumstantial evidence, such as the occurrence of a marine red tide or freshwater cyanobacterial bloom in conjunction with a die-off, and supportive clinical and pathologic findings, such as a lack of evidence of the presence of other types of toxins or infectious disease, are often used to reach a presumptive diagnosis. Analysis of the upper gastrointestinal tract contents or tissues of affected birds for algal toxins is possible but the tests are not yet widely available. In addition, there are no established toxic thresholds for wildlife species. Even when levels of particular toxins can be measured it may be difficult to assess their significance. Recently developed methods permit detection of microcystins in animal tissues and gastrointestinal contents by using enzyme linked immunosorbent assay (ELISA) technologies. Also, it is now possible to detect saxitoxin in urine and blood samples from affected animals by using highly sensitive neuroreceptor assays.

A sample of organisms from the bloom may be useful or necessary for diagnosis. Because of the ephemeral nature of blooms, collect algal samples during the die-off event as soon as possible after carcasses are found. Contact a diagnostic laboratory for advice on appropriate sample collection.

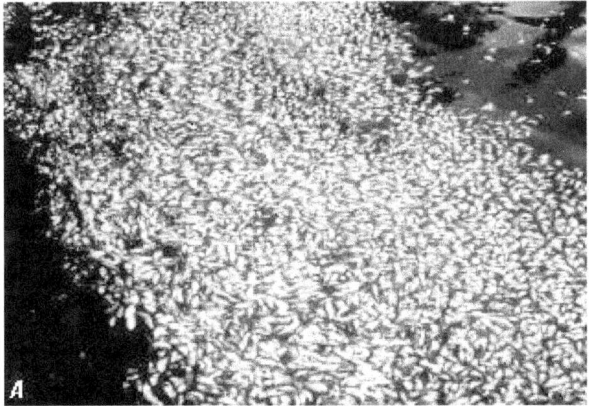 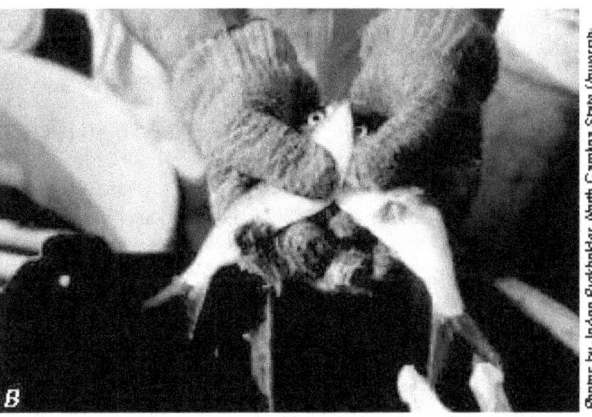

*Figure 36.3 (A and B) These fish were killed by **Pfiesteria** sp., an organism that has caused neurological problems, including prolonged amnesia, in people exposed by aerosols in a laboratory. This organism has only been fully described recently, and it has not been reported to cause mortality in birds; however, it may be encountered by biologists investigating concurrent bird and fish kills.*

Control

Because it is difficult to identify algal toxins as the cause of wildlife mortalities, there has been little opportunity to consider control measures. Currently, there is much interest in algal toxins and their threat to human water and food supplies. Identification of the conditions that trigger harmful algal blooms may aid in developing strategies to prevent red tides or freshwater cyanobacterial blooms and associated wildlife mortality. Controlling nutrient loading through reduced fertilizer use, improved animal waste control, and improved sewage treatment may reduce the number, or likely locations, of toxic algal blooms. Careful monitoring and early detection of potentially toxic algal blooms could allow time to initiate actions to prevent or reduce bird mortality.

Human Health Considerations

Most red tide and toxic freshwater cyanobacteria are not harmful unless they are ingested. However, some organisms irritate the skin and others release toxic compounds into the water and, if aerosolized by wave action, these compounds may cause problems when people inhale them (Fig. 36.3). When investigating wildlife mortality that is occurring in conjunction with a known red tide or cyanobacterial bloom, contact the local public health department or a diagnostic laboratory for information on precautions you may need to take. As in the investigation of all wildlife mortality events, wear rubber or latex gloves when handling carcasses.

Lynn H. Creekmore

Supplementary Reading

Anderson, D.A., 1994, Red tides: Scientific American, August, p. 62–68.

____, 1994, The toxins of cyanobacteria: Scientific American, January, p. 78–86.

Forrester, D.J., Gaskin, J.M., White, F.H., Thompson, N.P., Quick, J.A., Jr., Henderson, G.E., Woodard, J.C., and Robertson, W.D., 1977, An epizootic of waterfowl associated with a red tide episode in Florida: Journal of Wildlife Diseases, v. 13, p. 160–167.

Nisbet, I.C.T., 1983, Paralytic shellfish poisoning: Effects on breeding terns; Condor, v. 85, p. 338–345.

Work, T.M., Barr, B., Beale, A. M., Quilliam, M.A., and Wright, J.L.C., 1993, Epidemiology of domoic acid poisoning in brown pelicans (*Pelecanus occidentalis*) and Brandt's cormorants (*Phalacrocorax penicillatus*) in California: Journal of Zoo and Wildlife Medicine, v. 24, p. 54–62.

Chapter 37
Mycotoxins

Mycotoxins are toxins produced by molds (fungi) that, when they are ingested, can cause diseases called mycotoxicosis. These diseases are are not infectious. The effects on the animal are caused by fungal toxins in foods ingested, usually grains, and are not caused by infection with the fungus. Many different molds produce mycotoxins and many corresponding disease syndromes have been described for domestic animals. However, only two types of mycotoxin poisoning, aflatoxicosis and fusariotoxicosis, have been documented in free-ranging migratory birds.

Until recently, sickness or death caused by mycotoxins were rarely reported in migratory birds. Identification of mycotoxins as the cause of a mortality event can be difficult for a number of reasons. The effects may be subtle and difficult to detect or identify, or the effects may be delayed and the bird may have moved away from the contaminated food source before becoming sick or dying. Also, grain containing toxin-producing molds can be difficult or impossible to recognize because it may not appear overtly moldy.

Techniques to detect and quantify a variety of mycotoxins important to domestic animal and human health are available through many diagnostic laboratories that serve health needs for those species. These same techniques are applicable for wildlife. Further study and improved diagnostic technology is likely to result in identification of additional types of mycotoxins as causes of disease and death in waterfowl and other wildlife.

Aflatoxin Poisoning

Synonyms
Aflatoxicosis

Cause
Aflatoxins are a group of closely related toxic compounds produced by the fungi *Aspergillus flavus* or *A. parasiticus*. Four types of aflatoxins commonly are found in grains contaminated by these fungi: aflatoxin B1, aflatoxin B2, aflatoxin G1, and aflatoxin G2. These compounds become more toxic when they are metabolized after they are ingested. Aflatoxin B1 is the most commonly occurring and the most toxic. Aflatoxins present in very low concentrations, in parts per billion (ppb), can cause toxicosis.

Domestic ducklings are quite sensitive and the effects of aflatoxin exposure have been studied extensively in this species. In one study using 1-day-old ducklings, the LD_{50}, which is the dose of toxin required to produce death in 50 percent of the test animals via a single dose or single day's feeding, of aflatoxin B1 was 360 ppb aflatoxin.

Aflatoxins are often associated with groundnuts (peanuts) and corn, but they also have been found in other grains and nuts. *Aspergillus* sp. fungi can proliferate in improperly stored grain that has a moisture content of greater than 14 percent, relative humidity greater than 70 percent, and temperature greater than 70 °F. These fungi also can invade grains in the field, especially when there is drought stress, insect damage, or mechanical damage.

Species Affected
Aflatoxins can affect humans, many species of warm-blooded domestic and wild animals, and fish, most notably rainbow trout. Animals that consume grain are more likely to be affected than those that do not. Susceptibility depends on species, age, and diet. In general, birds are more susceptible than mammals, and young birds are more susceptible than adult birds. Mortality events caused by exposure to aflatoxins have been reported in free-ranging birds including a variety of duck species (mallard, black duck, lesser scaup, gadwall, and blue- and green-winged teal), Canada geese, snow geese, and sandhill cranes.

Distribution
Within the United States, the problem of aflatoxin-contaminated grain as a cause of disease in domestic animals and humans has been associated with the Southeastern and Gulf Coast States. Documented wildlife mortality events caused by aflatoxicosis are few; of those reported in wild birds, most occurred in Texas. However, the major fungi that produce aflatoxins, *A. flavus* and *A. parasiticus*, are widespread in temperate and tropical environments.

Seasonality
Most mortalities caused by exposure to acutely toxic levels of aflatoxins are reported in the fall and winter and coincide with times during migration and wintering when cranes and waterfowl are consuming waste grain in fields (Fig. 37.1). Mortality can occur at any time of the year when contaminated grain is provided at birdfeeding stations.

Field Signs
Field signs of aflatoxicosis reported in waterbirds vary from depression and lethargy to blindness, lack of awareness of surroundings, inability to fly, tremors, and wing flapping. Often the birds are simply found dead.

Figure 37.1 Canada geese feeding on standing waste corn in a winter field.

Figure 37.2 Pale, enlarged liver with multiple, focal hemorrhages from a bird with acute aflatoxicosis.

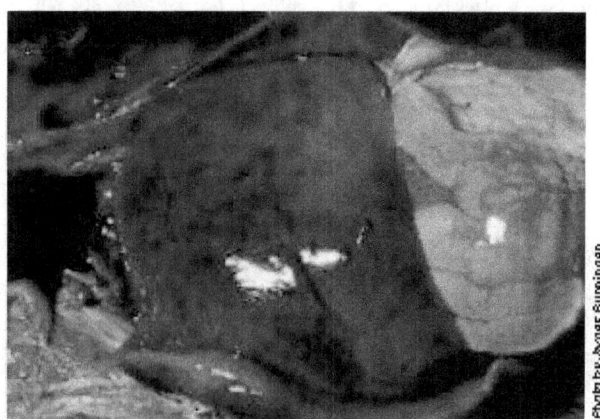

Figure 37.3 Hemorrhagic liver from a bird with acute aflatoxicosis.

Gross Lesions

Lesions of aflatoxicosis can be variable, depending on the amount of aflatoxin ingested and the length of time the animal is exposed. Birds exposed to high levels over a short period of time may have an enlarged, swollen pale liver (Fig. 37.2). Liver hemorrhages may be found in multiple focal areas, or may be diffuse and involve most of the liver tissue (Fig. 37.3). Hemorrhages and fluid may be observed in many organs in the chest and abdomen. Inflammation and bleeding of the gastrointestinal tract lining, which may cause the intestines to appear blackish-red throughout their entire length, may also be observed.

Birds that are exposed to small amounts of aflatoxin over a long period of time may not die suddenly, but rather may have chronic health problems. Chronic effects, which include appetite loss, weight loss, and general ill health, can be more insidious and difficult to definitively relate to aflatoxin exposure. Chronic exposure also may produce a shrunken, fibrous liver with regenerative nodules or tumors. In laboratory tests, aflatoxin B1 has been shown to cause genetic mutations, liver cancers, and, possibly, fetal defects. Chronic low level aflatoxin exposure also is known to suppress the immune system, which may predispose animals to infectious diseases.

Diagnosis

Diagnosis is made by examining tissues for gross lesions and typical microscopic lesions, which include liver tissue death or necrosis and proliferation of lesions in the bile duct. Measurement of aflatoxin levels in ingesta and tissues collected from affected birds and from the grain suspected of being contaminated is also crucial for confirming the diagnosis. The samples of choice include whole refrigerated carcasses for necropsy as well as grain that affected birds have been eating. Because mycotoxin occurrence can vary widely within an agricultural field, it is important to try to obtain a representative sample of the suspect grain. If possible, transport the grain frozen, and ensure that the sample remains frozen so that fungal growth and toxin production secondary to improper postcollection storage does not occur.

Control

Wildlife should not be fed grain that has levels of aflatoxins in excess of those allowed for use in human or domestic animal food (20 ppb for consumption by humans or young animals and dairy cattle; 100 ppb for mature poultry). In years when aflatoxins are a problem, grain from fields that are frequently used by wildlife should be checked for aflatoxin levels. If the fields are aflatoxin-contaminated, deep plowing of the contents can make the grain unavailable to wildlife. If the fields cannot be plowed, hazing wildlife from the area can lessen their exposure.

Human Health Considerations

Handling aflatoxin-poisoned sick or dead birds does not pose a human health risk. However, birds known to have died from acute aflatoxicosis should not be consumed.

Fusariotoxin Poisoning

Synonyms

Fusariomycotoxicosis, trichothecene mycotoxicosis, T-2 toxicosis, vomitoxicosis, zearalenone toxicosis

Cause

Fusariotoxin poisoning is caused by toxins produced by fungi of the genus *Fusarium*. There are two classes of toxins produced by these organisms: metabolites that have properties similar to the hormone estrogen such as zearalenone (F-2 toxin), and the trichothecenes. Zearalenone has been linked to hyperestrogenic or feminizing syndromes in domestic animals, but similar effects have not yet been observed in wild species. More than 50 trichothecene toxins have been identified from *Fusarium* cultures and field samples. However, the trichothecenes have rarely been documented as the cause of mortality in free-ranging birds.

The most frequently occurring trichothecene toxin in the United States is deoxynivalenol, commonly called vomitoxin. Others include T-2 toxin, diacetoxyscirpenol (DAS), neosolaniol and iso-neosolaniol. The feedstuffs involved in *Fusarium* sp. toxin production include corn, wheat, barley, oats, peanuts, and sometimes forages. *Fusarium* toxins differ from other mycotoxins in that they tend to be produced during the colder seasons of the year.

Species Affected

Poisoning caused by trichothecene toxins has been documented in domestic mammals, poultry, and waterfowl. However, reports of fusariotoxicosis in free-ranging waterfowl and other migratory birds are rare and poorly documented. Mortality attributed to trichothecenes has occurred on several occasions in free-ranging sandhill cranes. It is suspected that wild waterfowl could be affected by trichothecenes because these toxins tend to be produced in low temperature conditions when waterfowl make heavy use of waste grain as food.

Distribution

Fusarium sp. are widespread in the environment and they commonly occur as plant pathogens or contaminants in stored or waste grain and other plant parts. Toxin production from *Fusarium* sp. is most commonly a problem in the North-Central United States and Canada. However, the only documented locations of wild bird mortality caused by *Fusarium* toxins are in Texas and New Mexico, and these die-off events involved sandhill cranes that fed on *Fusarium*-contaminated peanuts.

Seasonality

Fusarium invasion often occurs during wet conditions in the summer and fall while crops are in the field. However, the organism also can grow in stored grain. The optimal temperatures for mold growth and toxin production differ. Temperatures that support growth of the vegetative form of the fungus are between 64 and 77 °F, but actual toxin production tends to occur at temperatures between 40 and 65 °F and, in some cases, has even been documented at near-freezing temperatures. Colder temperatures favoring toxin production coincide with times that cranes and waterfowl are using waste grain in fields during their fall and winter migration.

Field Signs

Different trichothecenes have different effects. In general, clinical signs in domestic poultry and geese include feed refusal, vomiting, and gastrointestinal bleeding. Some birds have been described as having neurologic abnormalities. Free-ranging sandhill cranes diagnosed with trichothecene toxicosis had difficulty keeping their balance and had flaccid paralysis or weakness of the neck and wing muscles. This created a stance characterized by a drooping head and wings (Fig. 37.4). The more toxic trichothecenes cause immune suppression and may predispose birds to secondary infections.

Gross Lesions

Inflammation and ulceration of the skin and mucosal surfaces of the oral cavity and upper gastrointestinal tract are the most commonly reported lesions in domestic animals and were observed in affected sandhill cranes (Fig. 37.5). Gross lesions described in sandhill cranes also included subcutaneous fluid over the head and neck (Fig. 37.6) and multiple hemorrhages and pale areas in skeletal muscle.

Figure 37.4 A sandhill crane suffering from fusariotoxicosis. Notice the wing and head droop.

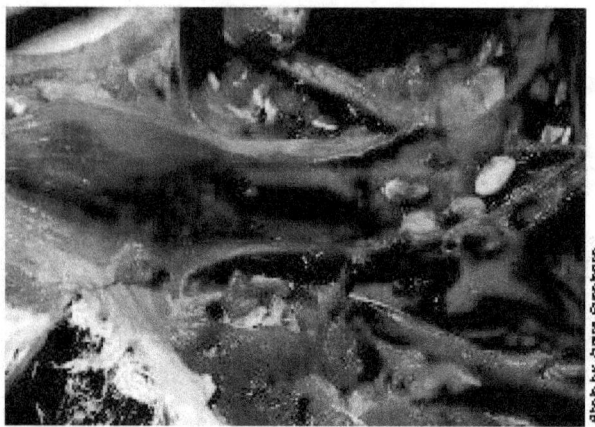

Figure 37.5 Inflammation and ulceration of the mucosal surface of the esophagus in a sandhill crane with fusariotoxicosis.

Figure 37.6 Fluid beneath the skin of the head and neck of a sandhill crane with fusariotoxicosis.

Figure 37.7 Plowing of this field made these peanuts unavailable to the cranes and ended a fusariotoxicosis mortality event.

Diagnosis

Diagnosis is made through observing the appropriate field signs, finding gross as well as microscopic tissue lesions, and detecting the suspected toxin in grains, forages, or the ingesta of affected animals. However, the tests required to detect these toxins are complex and few diagnostic laboratories offer tests for multiple trichothecenes. The samples of choice include both refrigerated and frozen carcasses for necropsy examination and a representative sample of the suspected contaminated grain source. Because the toxin is produced under cold conditions, the grain sample should be frozen rather than refrigerated for shipment to the diagnostic laboratory.

Control

Wildlife should not be fed grain with levels of fusariotoxins in excess of those recommended for use in domestic animal food. In years when fusariotoxins are a problem, grain from fields that are frequently used by wildlife should be checked for toxins. If the fields are significantly contaminated, deep plowing of the contents of the field can make the grain unavailable to wildlife (Fig. 37.7). If the fields cannot be plowed, hazing wildlife from the area can lessen their exposure.

Human Health Considerations

Handling fusariotoxin-poisoned sick or dead birds does not pose a human health risk, but birds known to have died from acute fusariotoxicosis should not be consumed. The more potent trichothecenes in grain or forages may present a health hazard when the fungal spores and contaminated plant parts are inhaled or when they contact the skin.

Lynn H. Creekmore

Supplementary Reading

Pier, A.C., 1981, Mycotoxins and animal health, *in* Cornelius, C.E., and Simpson, eds., Advances in veterinary science and comparative medicine: New York, Academic Press, p. 186–243.

Robinson, R.M., Ray, A.C., Reagor, J.C., and Holland, L.A., 1982, Waterfowl mortality caused by aflatoxicosis in Texas: Journal of Wildlife Diseases, v. 18, p. 311–313.

Roffe, T.J., Stroud, R.K., and Windingstad, R.M., 1989, Suspected fusariomycotoxicosis in sandhill cranes (Grus canadensis): clinical and pathological findings: Avian Diseases, v. 33, p. 451–457.

Windingstad, R.M., Cole, R.J., Nelson, P.E., Roffe, T.J., George, R.R., and Dorner, J.W., 1989, Fusarium mycotoxins from peanuts suspected as a cause of sandhill crane mortality: Journal of Wildlife Diseases, v. 25, p. 38–46.

Wobeser, G.A., 1997, Mycotoxins, *in* Diseases of wild waterfowl (2nd ed): New York, N.Y., Plenum Press, p. 189–193.

Chapter 38
Avian Botulism

Synonyms
Limberneck, Western duck sickness, duck disease, alkali poisoning

Cause

Avian botulism is a paralytic, often fatal, disease of birds that results when they ingest toxin produced by the bacterium, *Clostridium botulinum*. Seven distinct types of toxin designated by the letters A to G have been identified (Table 38.1). Waterfowl die-offs due to botulism are usually caused by type C toxin; sporadic die-offs among fish-eating birds, such as common loons and gulls, have been caused by type E toxin. Type A botulinum toxin has also caused disease in birds, most frequently in domestic chickens. Types B, D, F, and G are not known to cause avian botulism in North America.

Table 38.1 Botulinum toxins and primary species affected.

Toxin type	Animals affected	Risk for humans
A	Poultry, occasionally	High
B	Horses	High
C	Wild birds, cattle, horses, poultry	Low
D	Cattle	Low
E	Fish-eating birds	High
F	*	Unknown
G	*	Unknown

*Rarely detected in nature; too little information for species evaluations.

C. botulinum is an oxygen-intolerant or anaerobic bacterium that persists in the form of dormant spores when environmental conditions are adverse. The spores are resistant to heating and drying and can remain viable for years. Spores of type C botulism strains are widely distributed in wetland sediments; they can also be found in the tissues of most wetland inhabitants, including aquatic insects, mollusks, and crustacea and many vertebrates, including healthy birds. Botulinum toxin is produced only after the spores germinate, when the organism is actively growing and multiplying. Although the bacteria provide the mechanism for toxin production, the gene that encodes for the toxin protein is actually carried by a virus or phage that infects the bacteria. Unfortunately, little is known about the natural factors that control phage infection and replication within the bacteria, but several factors may play a role, including the bacterial host strain and environmental characteristics, such as temperature and salinity.

Because botulinum spores and the phages that carry the toxin gene are so prevalent in wetlands, they are not considered to be a limiting factor in the occurrence of outbreaks in waterbirds. Other factors are thought to be more critical in the timing and location of botulism outbreaks; these include optimal environmental conditions for spore germination and bacterial growth, suitable material or substrates that provide energy for bacterial replication, and a means of toxin transfer to birds. It is likely that toxin production, toxin availability to birds and, subsequently, botulism outbreaks in birds are largely controlled by these ecological factors.

As with other bacteria, temperature plays a critical role in the multiplication of *C. botulinum*, with optimal growth in the laboratory occurring between 25° and 40°C. Most botulism outbreaks take place during the summer and fall when ambient temperatures are high (Fig. 38.1). Winter botulism outbreaks have been documented in some locations, but these are generally thought to be due to residual toxin produced during the previous summer. Conditions that elevate wetland sediment temperatures and decrease dissolved oxygen, including the presence of decaying organic matter and shal-

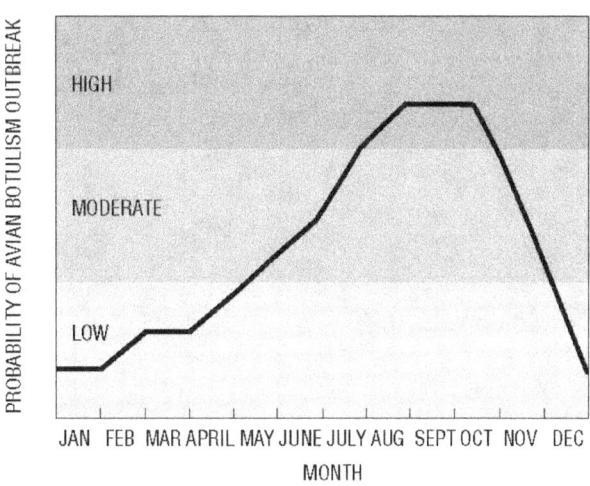

Figure 38.1 Relative seasonal probability of type C botulism in North American water birds.

low water, may increase the risk of botulism outbreaks (Fig. 38.2). However, these conditions are not prerequisite to an outbreak because botulism has occurred in large river systems and in deep, well-oxygenated wetlands, which suggests that other environmental conditions may be more critical. In studies conducted by the National Wildlife Health Center, several environmental factors, including pH, salinity (Fig. 38.3), temperature, and oxidation-reduction potential in the sediments and water column, appeared to significantly influence the likelihood of botulism outbreaks in wetlands.

In addition to permissive environmental conditions, *C. botulinum* also requires an energy source for growth and multiplication. Because it lacks the ability to synthesize certain essential amino acids, the bacterium requires a high protein substrate; it is essentially a "meat lover." The most important substrates for toxin production in natural wetlands have never been identified, but there are many possibilities, including decaying organic matter or any other protein particulates. Decomposing carcasses, both vertebrate and invertebrate, are well known to support toxin production. Human activities can also increase the available substrate for toxin production in wetlands (Table 38.2). For example, wetland flooding and draining, pesticides, and other agricultural pollutants may kill aquatic life, thereby providing more substrate for toxin production. Raw sewage and rotting vegetation are other potential sources of energy.

Although many substrates are suitable for botulinum toxin production, in order for a botulism outbreak to occur the toxin must be in a form that is available to birds. In some cases, decaying organic matter may be directly ingested, but in other cases there must be some means of toxin transfer from the substrate to the birds, presumably through zooplankton or invertebrate food items that inadvertently consumed toxin. Invertebrates are unaffected by the toxin and, because they feed on decaying matter, they can effectively act to concentrate toxin. Although most waterfowl will not directly consume a vertebrate carcass, they will readily ingest any maggots that fall off of it. In this way, botulism outbreaks often become self-perpetuating. This has become known as the carcass-maggot cycle of avian botulism (Fig. 38.4).

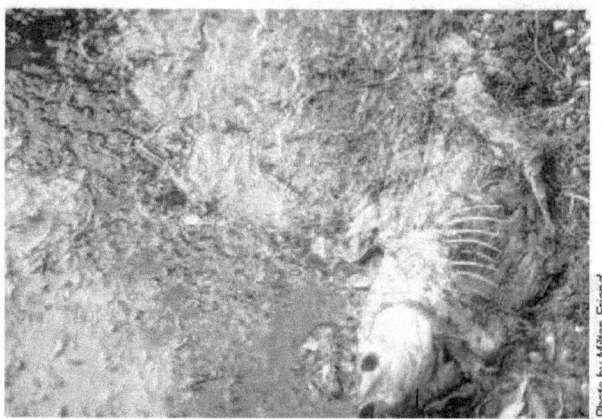

Figure 38.2 *Decaying organic matter may increase the risk of avian botulism outbreaks.*

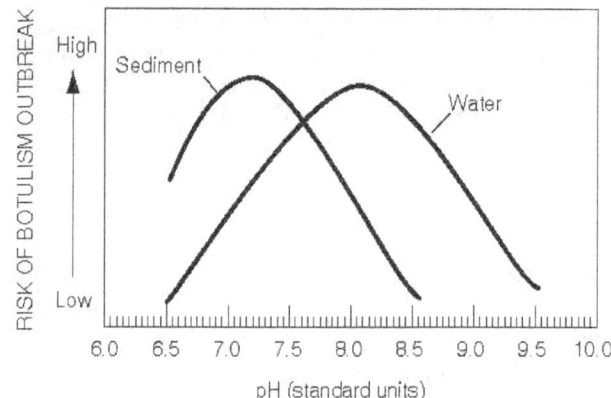

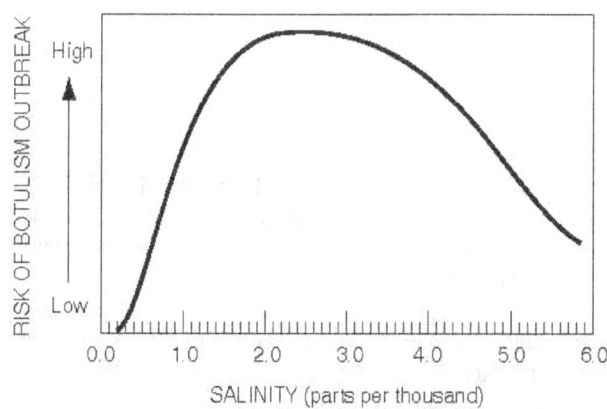

Figure 38.3 *Relationship of pH and salinity to avian botulism outbreaks.*

Table 38.2 Human activities speculated to contribute to avian botulism outbreaks in wetlands.

Action	Consequences of action
Fluctuating water levels for flooding and drying	Deaths of terrestrial and aquatic invertebrates and fish
Pesticides and other chemical inputs into wetlands from agriculture	Deaths of aquatic life
Raw sewage discharges into wetlands	Nutrient enhancement resulting in "boom and bust" invertebrate populations and oxygen depletion causing deaths of aquatic and plant life

Carcass-maggot cycle of avian botulism

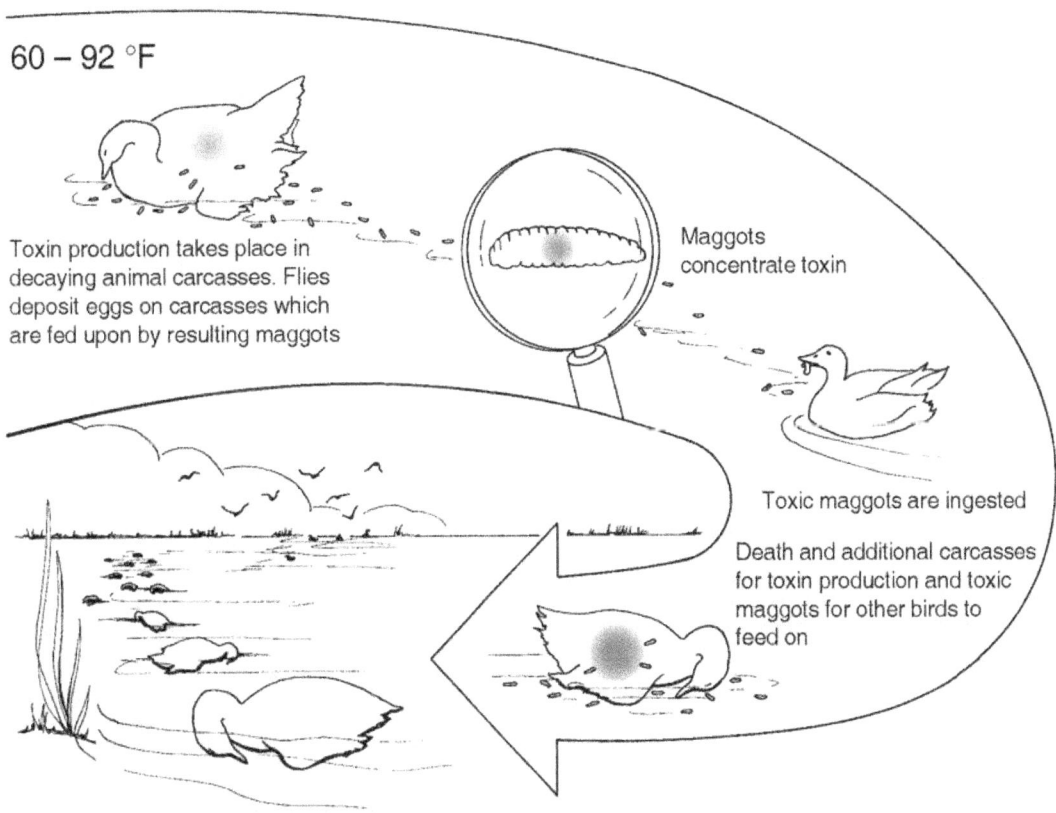

Figure 38.4 Carcass-maggot cycle of avian botulism.

Species Affected

Many species of birds and some mammals are affected by type C botulism. In the wild, waterbirds suffer the greatest losses, but almost all birds are susceptible to type C botulism. The exception is vultures, which are highly resistant to type C toxin. Foraging behavior is probably the most significant host determinant for botulism. Filter-feeding and dabbling waterfowl and probing shorebirds appear to be among the species at greatest risk (Fig. 38.5). Mortality of wild raptors from botulism has been associated with improper disposal of poultry carcasses. Among captive and domestic birds, pheasants, poultry, and waterfowl are the most frequently affected (Fig. 38.6). Cattle, horses, and ranch mink are also susceptible to type C botulism. Although dogs and cats are usually regarded as being resistant to type C toxin, a few cases have been reported in dogs, which is a factor to consider when dogs are used to retrieve carcasses during outbreaks. Also, type C botulism occurred in captive African lions which were fed toxin-laden chickens.

Losses vary a great deal from year to year at site-specific locations and from species to species. A few hundred birds may die one year and tens of thousands or more the following year. More than a million deaths from avian botulism have been reported in relatively localized outbreaks in a single year, and outbreaks with losses of 50,000 birds or more are relatively common (Table 38.3).

On a worldwide basis, avian botulism is probably the most important disease of migratory birds.

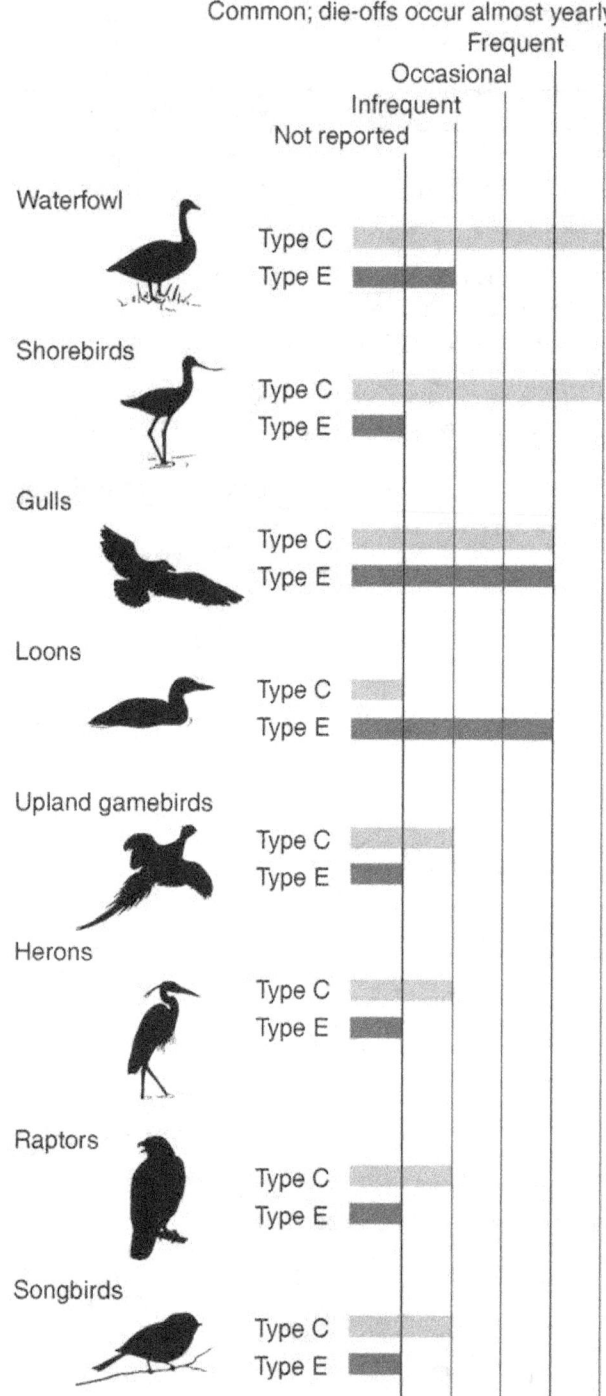

Figure 38.5 Frequency of botulism in major groups of wild birds.

274 Field Manual of Wildlife Diseases: Birds

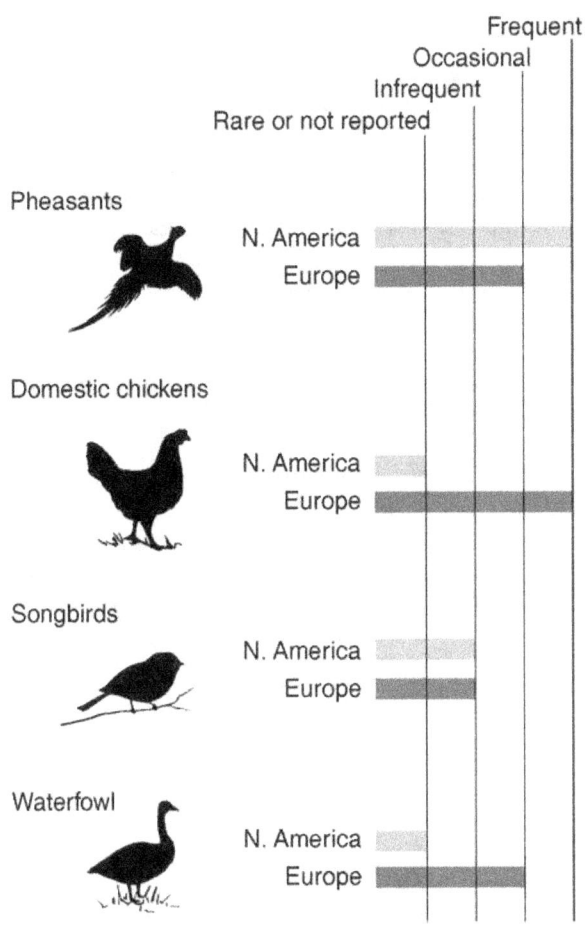

Figure 38.6 Frequency of botulism in captive birds.

Table 38.3 Major waterfowl botulism outbreaks in the United States and Canada.

Location	Year	Estimated loss
Utah and California	1910	"Millions"
Lake Malheur, Oregon	1925	100,000
Great Salt Lake, Utah	1929	100,000–300,000
Tulare Basin, California	1941	250,000
Western United States	1952	4–5 million
Montana	1978	50,000
Montana	1979	100,000
Great Salt Lake, Utah	1980	110,000
Canada (Alberta)	1995	100,000
Canada (Manitoba)	1996	117,000
Canada (Saskatchewan)	1997	1 million
Great Salt Lake, Utah	1997	514,000

Distribution

Outbreaks of avian botulism have occurred in the United States and Canada since the beginning of the century, if not earlier. Outbreaks have also been reported in many other countries; most of these reports are recent, usually within the past 30 years (Fig. 38.7). Most type C botulism outbreaks within the United States occur west of the Mississippi River; however, outbreaks have occurred from coast-to-coast and border-to-border, and the distribution of the disease has greatly expanded since the early 1900s (Fig. 38.8). Type E outbreaks in birds are much less frequent and, within the United States, have been confined to the Great Lakes region.

Seasonality

July through September are the primary months for type C avian botulism outbreaks in the United States and Canada. However, outbreaks occur as late as December and January and occasionally during early spring in southern regions of the United States and in California. Type E outbreaks have occurred during late fall and spring.

Field Signs

Lines of carcasses coinciding with receding water levels generally typify the appearance of major botulism die-offs, although outbreaks have also occurred in impoundments containing several feet of water, lakes with stable water levels, and in large rivers. When receding water conditions are involved, botulism is typically a disease of the water's edge and seldom are sick or dead birds found very far from the edge of vegetation bordering the water or the original water's edge (Fig. 38.9). In impoundments where water levels are relatively stable, affected birds are likely to be found in areas of flooded vegetation. Botulism-affected birds also tend to congregate along vegetated peninsulas and islands (Fig. 38.10).

Healthy birds, sick, and recently dead birds will commonly be found together during a botulism outbreak, along with carcasses in various stages of postmortem decay. Often, species representing two, three, or even more orders of birds suffer losses simultaneously.

Figure 38.7 Type C botulism outbreaks in wild birds.

Avian botulism affects the peripheral nerves and results in paralysis of voluntary muscles. Inability to sustain flight is seen early in botulism, but this sign is not useful for distinguishing botulism-intoxicated birds from those affected by other diseases. Because ducks suffering from botulism cannot fly and their legs become paralyzed, they often propel themselves across the water and mud flats with their wings (Fig. 38.11) (see also Fig. 1.2 in Chapter 1, Recording and Submitting Specimen Data). This sequence of signs contrasts with that of lead-poisoned birds, which retain their ability to walk and run although flight becomes difficult (see Chapter 43).

Paralysis of the inner eyelid or nictitating membrane (Fig. 38.12) and neck muscles follow, resulting in inability to hold the head erect (Fig. 38.13). These are the two most easily recognized signs of avian botulism. When birds reach this stage, they often drown before they might otherwise die from the respiratory failure caused by botulinum toxin.

Figure 38.8 Locations of avian botulism outbreaks in the United States.

Figure 38.9 Typical scene of avian botulism. Dead birds are often found along the shore in parallel rows that represent receding water levels.

Figure 38.10 Botulism-affected birds tend to congregate along vegetated peninsulas and islands. Both dead and sick birds are evident in the photograph.

Figure 38.11 Botulism-intoxicated birds that have lost the power of flight and use of their legs often attempt escape by propelling themselves across water or land using their wings.

Figure 38.12 Paralysis of the inner eyelid is a common sign in botulism-intoxicated birds.

Figure 38.13 Paralysis of the neck muscles in botulism-intoxicated birds results in inability to hold the head erect (limberneck). Death by drowning often results.

Gross Lesions

There are no characteristic or diagnostic gross lesions in waterfowl dying of either type C or type E botulism. Often, affected birds die by drowning, and lesions associated with drowning may be present.

Diagnosis

The most widely used test for avian botulism is the mouse-protection test, although an enzyme-linked immunosorbent assay (ELISA) for type C toxin has been developed recently. For the mouse test, blood is collected from a sick or freshly dead bird and the serum fraction is then inoculated into two groups of laboratory mice, one group of which has been given type-specific antitoxin. The mice receiving antitoxin will survive, and those that receive no antitoxin will become sick with characteristic signs or die if botulism toxin is present in the serum sample. The ELISA is an *in vitro* test that detects inactive as well as biologically active toxin.

A presumptive diagnosis is often based on a combination of signs observed in sick birds and the absence of obvious lesions of disease when the internal organs and tissues of sick and dead birds are examined. However, this initial diagnosis must be confirmed by the mouse-protection or ELISA test to separate avian botulism from algal poisoning, castorbean poisoning, and other toxic processes that cause similar signs of disease. Avian botulism should be suspected when maggots are found as part of the ingesta of gizzard contents of dead birds (Fig. 38.14), however, such findings are rare. After a bird ingests toxin, it takes several hours to days before the bird develops signs of the disease and dies. By this time, most food items ingested at the time of intoxication have been eliminated.

Prevention and Control

Prevention of avian botulism outbreaks in waterbirds will depend on a thorough understanding of the interactions between the agent, the host, and the environment. Because botulism spores are so ubiquitous in wetlands and are resilient, attempts to reduce or eliminate the agent are not currently feasible, but some actions can be taken to mitigate environmental conditions that increase the likelihood of outbreaks.

Management of Environment

Attempts should be made to reduce organic inputs into wetlands or to eliminate factors that introduce large amounts of decaying matter. For example, in areas that are managed primarily for migratory waterfowl (ducks, geese, swans), reflooding land that has been dry for a long time is not recommended during the summer. Similarly, avoid sharp water drawdowns in the summer because they could result in fishkills and die-offs of aquatic invertebrates whose carcasses could then become substrates for *C. botulinum* growth. In areas managed primarily for shorebirds, water drawdowns provide essential habitat; thus, botulism control must focus on cleaning up any vertebrate carcasses that may result from drawdowns.

Prompt removal and proper disposal of vertebrate carcasses by burial or burning, especially during outbreaks, are highly effective for removing substrates for toxin production. The importance of prompt and thorough carcass removal and proper disposal cannot be overemphasized. Several thousand toxic maggots can be produced from a single waterfowl carcass (Fig. 38.15). Consumption of as few as two to four of these toxic maggots can kill a duck, thereby perpetuating the botulism cycle. It is not uncommon to find three or

Figure 38.14 Waterfowl and other birds readily feed on maggots as exemplified by their presence in the gizzard of this duck that died from avian botulism.

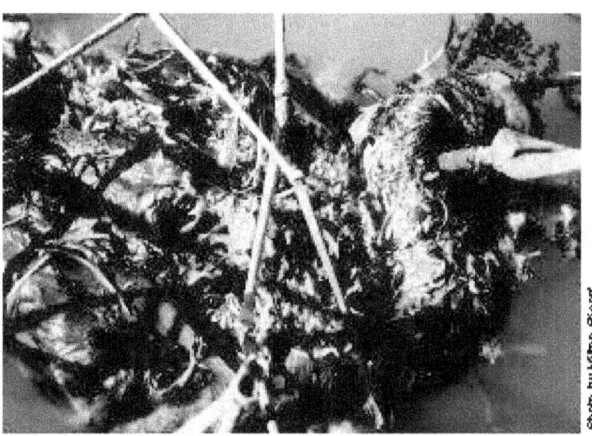

Figure 38.15 Several thousand toxic maggots can be produced from a single waterfowl carcass.

four freshly dead birds within a few feet of a maggot-laden carcass. Failure to carry out adequate carcass removal and disposal programs can cause a rapid build-up of highly toxic decaying matter and toxin-laden maggots, thereby accelerating losses in waterbirds, as well as seeding the environment with more botulism spores as the carcasses decompose.

Many botulism outbreaks occur on the same wetlands year after year and within a wetland there may be localized "hot spots." Also, outbreaks often follow a fairly consistent and predictable timeframe. These conditions have direct management implications that should be applied towards minimizing losses. Specific actions that should be taken include accurately documenting conditions and dates of outbreaks in problem areas, planning for and implementing intensified surveillance and carcass pickup and disposal, and modifying habitat to reduce the potential for botulism losses or deny bird use of major problem areas during the botulism "season" or both.

Management actions for minimizing losses from avian botulism

Document environmental conditions, specific impoundments or areas of outbreaks, and dates of occurrence and cessation.

Plan for and implement intensive surveillance and vertebrate carcass pickup and disposal starting 10–15 days before the earliest documented cases until 10–15 days after the end of the botulism "season."

Where possible, monitor and modify environmental conditions to prevent the pH and salinity of wetlands from reaching or being maintained within high hazard levels.

Avoid water drawdowns for rough fish and vegetation control during warm weather. Collect vertebrate carcasses (fish) and properly dispose of them if drawdowns are necessary during summer and warm fall months.

Construct wetland impoundments in botulism-prone areas in a manner that facilitates rapid and complete drainage thereby encouraging bird movement to alternative impoundments.

Because fish carcasses can also serve as sites for *C. botulinum* growth, they should be promptly removed during fish control programs in marshes, or fish control programs should be restricted to the cooler months of year (the nonfly season). Also, bird collisions with power lines that cross marshes have been the source of major botulism outbreaks because carcasses from these collisions served as initial substrates for toxin production within marshes. Therefore, if possible, power lines should not be placed across marshes used by large concentrations of waterbirds.

Numerous outbreaks of avian botulism have been associated with sewage and other wastewater discharges into marshes. This relationship is not presently understood, but outbreaks have occurred often enough that wetland managers should discourage the discharges of these effluents when many waterfowl or shorebirds are using the area or are likely to use an area during warm weather.

Treatment of Sick Birds

Botulism-intoxicated waterfowl can recover from the disease. If sick birds are provided with freshwater and shade, or injected with antitoxin, recovery rates of 75 to 90 percent and higher can result (Fig. 38.16). In contrast to waterfowl, very few coot, shorebirds, gulls, and grebes survive botulism intoxication, even after treatment. Experience to date with these species indicates that rehabilitation efforts may not be worthwhile.

Because avian botulism most often afflicts waterfowl in the seasons when they are flightless due to wing molt, biologists and rehabilitators must be careful to distinguish between birds in molt and birds with early stages of botulism, because the behavior of these birds may be similar. Molting birds are very difficult to catch, and birds that cannot be captured with a reasonable effort should not be pursued further. Birds that are suffering from botulism can easily be captured when they lose the ability to dive to escape pursuit. Birds at this level of intoxication still have a high probability of surviving if proper treatment is administered.

When botulism-intoxicated birds are treated, they should be maintained under conditions or holding pens that provide free access to freshwater, maximum provision for shade, the opportunity for recovered birds to fly out of the enclosure when they choose to, and minimum disturbance (including the presence of humans). It is also important to remove carcasses daily from holding pens to prevent the buildup of toxic maggots.

Costs associated with capturing and treating sick birds are high. Therefore, the emphasis for dealing with avian botulism should be on prevention and control of this disease rather than on treatment of intoxicated birds. However, antitoxin should be available for use in case endangered species are affected.

Human Health Considerations

Botulism in humans is usually the result of eating improperly home-canned foods and is most often caused by type A or type B botulinum toxin. There have been several human cases of type E botulism in North America from eating improperly smoked or cooked fish or marine products. Type C botulism has not been associated with disease in humans, although several outbreaks have been reported in captive primates. Thorough cooking destroys botulinum toxin in food.

Duck type	Amount of antitoxin (cc)	Percentage recovered	
		Moderate[1] clinical signs	Severe[1] clinical signs
Pintail			
	0.5	91.9	69.0
	1.0	93.5	73.3
Green-winged teal			
	0.5	93.5	57.4
	1.0	91.2	58.1
Mallard			
	0.5	81.6	66.3
	1.0	84.9	67.8
Shoveler			
	0.5	89.2	53.1
	1.0	95.1	57.6

[1] Condition of bird when treated with antitoxin

Figure 38.16   Recovery rates of ducks in response to antitoxin therapy.

Tonie E. Rocke and Milton Friend

Supplementary Reading

Eklund, M.W., and Dowell, V.R., eds., 1987, Avian Botulism: An International Perspective: Springfield, Ill., Charles C. Thomas, 405 p.

Hariharan, H., and Mitchell, W.R., 1977, Type C botulism: the agent, host spectrum and environment: The Veterinary Bulletin, v. 47, p. 95–103.

Reed, T.M., and Rocke, T.E., 1992, The role of avian carcasses in botulism epizootics: The Wildlife Society Bulletin, v. 20, p. 175–182

Rocke, T.E., 1993, *Clostridium botulinum*, *in* Gyles, C.L. and Thoen, C.O., eds., Pathogenisis of bacterial infections in animals (2d ed.): Ames, Iowa, Iowa State University Press, p. 86–96.

Rocke, T.E., Smith, S.R., and Nashold, S.W., 1998, Preliminary investigation of the *in vitro* test for the diagnosis of avian botulism in wild birds: The Journal of Wildlife Diseases, v. 34 p. 744–751.

Section 7
Chemical Toxins

Organophosphorus and Carbamate Pesticides

Chlorinated Hydrocarbon Insecticides

Polychlorinated Biphenyls

Oil

Lead

Selenium

Mercury

Cyanide

Salt

Barbiturates

Miscellaneous Chemical Toxins

Oiling disrupts normal feather structures and function
Photo by Nancy J. Thomas

Introduction to Chemical Toxins

"Dosage alone determines poisoning."
(Translation of Paracelsus)

Many kinds of potentially harmful chemicals are found in environments used by wildlife. Some chemicals, such as pesticides and polychlorinated biphenyls (PCBs), are synthetic compounds that may become environmental contaminants through their use and application. Other materials, such as selenium and salt, are natural components of some environments, but contaminants of others. Natural and synthetic materials may cause direct poisoning and death, but they also may have adverse effects on wildlife that impair certain biological systems, such as the reproductive and immune systems. This section provides information about some of the environmental contaminants and natural chemicals that commonly cause avian mortality; microbial and other biotoxins are addressed in the preceding section.

Direct poisoning and mortality of wildlife caused by exposure to chemical toxins are the focus of this section. However, the indirect effects of chemicals may have significantly greater impacts on wildlife populations than the direct effects. Behavioral changes that affect survival, reproductive success and the survival of young, and that impair the functioning of the immune system are examples of indirect chemical toxicity that are known to occur but that are beyond the scope of this publication. For additional information readers are directed to more comprehensive treatments of environmental toxicology and to publications that focus on specific chemicals and their effects on wildlife.

The diagnosis of chemical poisoning as the cause of wildlife mortality is a challenging task because of the vast array of chemicals that wildlife may be exposed to (Table 1), the variable biological responses following concurrent exposure to multiple chemicals, the absence of tissue residues for some chemical toxins, and the lack of specific pathological changes associated with most chemical toxins in tissues. The diagnostic process can be greatly facilitated by a thorough field observation record, comprehensive background information about the circumstances of a mortality event, and by properly collecting, handling, and preserving samples submitted to the diagnostic laboratory (see Section 1). Sources of assistance for the investigation of wildlife mortality, when toxins are suspected, are listed in Appendix B.

Areas Covered

The chapters that follow address chemical toxins that are recurrent causes of avian mortality. The chapters discuss chemicals that cause frequent and sometimes large-scale mortality events, as well as some chemicals that are less significant, because they are restricted to certain geographic areas or have been recently recognized as emerging problems.

Pesticides
    Organophosphorus and carbamate compounds
    Chlorinated hydrocarbons
Polychlorinated biphenyls
Oil
Lead
Selenium
Mercury
Cyanide
Salt
Barbiturates

Quote from:

Philipus Aureolus Paracelsus, a German-Swiss physician and alchemist who lived from 1493 to 1541.

Table 1 Examples of chemical toxins to which wildlife may be exposed.

Pesticides

This group includes chemicals that are used to kill or repel organisms that are unwanted in particular situations. Insecticides are generally the best known pesticides but others, their target organisms, and examples of compounds within those groups include the following:

Pesticide type	Target organisms	Compounds
Acaricides	Mites, ticks, spiders	Permethrin, Phosmet, Methiocarb, Bomyl®, Carbofuran, Demeton (Systox®)
Algacides	Algae	Copper sulfate, Potassium bromide, Chlorine
Antibiotics	Bacteria	Phenol, Nitrapyrin
Avicides	Birds	Avitrol®, Fenthion, Compound 1080, Starlicide®
Fungicides	Fungi	Thiram, Ziram, Captan, Hexaconazole
Herbicides	Plants	Diquat®, Alachlor (Lasso®), Atrazine
Molluscicides	Snails and slugs	Bayluscide®, Methiocarb, Zectran®
Nematocides	Nematodes (worms)	Terbufos (Counter®), Isazofos (Triumph®), Aldicarb (Temik®), Carbofuran, Diazinon
Piscicides	Fish	Rotenone, Antimycin
Repellents	Mammals Birds	Thiram Methiocarb
Rodenticides	Rodents	Warfarin, Diphacinone, Brodifacoum (Talon®), Chlorophacinone

Metals

Wildlife may be exposed to metals when they are components of pesticides, such as mercury and cadmium in fungicides, or through other routes, such as aquatic food chains with high mercury levels.

Metal	Source
Arsenic	Used as an insecticide and preservative; present in wastes from metal smelting and glass manufacturing.
Cadmium	Used as a fungicide; waste from electroplating and production of plastics and batteries.
Chromium	Industrial effluents from ore refinement, chemical processing.
Copper	Used as a fungicide, an algicide, and in agriculture.
Lead	Mine tailings, ingestion of particulate lead deposited during sporting activities.
Mercury	Used as a fungicide in paper mills and other industrial and agricultural uses; combustion of fossil fuels.
Selenium	Irrigation drain water from soils with high selenium concentrations; combustion of fossil fuel; sewage sludge.
Zinc	Found throughout the environment; higher levels in areas of industrial discharge.

Petroleum

Wildlife may be exposed to many forms of petroleum, ranging from crude oils to highly refined forms, such as fuel oil.

Others

Many manufactured compounds, such as antifreeze (ethylene glycol) and certain drugs (such as euthanasia agents), present hazards to exposed wildlife.

Chapter 39

Organophosphorus and Carbamate Pesticides

Synonyms

Organophosphates, OPs

The insecticidal properties of organophosphorus (OP) and carbamate compounds were first discovered in the 1930s, and the compounds were developed for pesticide use in the 1940s. They have been used increasingly since the 1970s when environmentally persistent organochlorine pesticides, such as DDT and dieldrin, were banned for use in the United States. Organophosphorus and carbamate pesticides are generally short-lived in the environment (usually lasting only days to months instead of years) and, generally, chemical breakdown is accelerated as temperatures or pH or both increase.

Cause

The toxicity of OP and carbamate pesticides is due to the disruption of the nervous system of an invertebrate or a vertebrate through the inhibition of cholinesterase (ChE) enzymes. These enzymes are involved in transmitting normal nerve impulses throughout the nervous system. An acute pesticide dose reduces the activity of ChEs, and nerve impulses cannot be transmitted normally. This can paralyze the nervous system, and it may lead to death, usually from respiratory failure.

Species Affected

It is possible for a wide variety of vertebrate species to be affected by OP or carbamate pesticides. However, birds appear to be more sensitive than other vertebrates to the toxic effects of OP and carbamate pesticides. More than 100 avian species have been poisoned by these pesticides. Waterfowl, passerines, and raptors are the species most commonly identified in reported OP- and carbamate-related mortalities in the United States (Fig. 39.1). Raptors and other bird species become victims of secondary poisoning when they scavenge dead animals poisoned by pesticides or when they feed on live animals or invertebrates that are unable to escape predation because of pesticide intoxication.

Age, sex, diet, and body condition all are factors that affect a bird's susceptibility to pesticide poisoning. Generally, embryos and young birds, particularly the dependant or altricial birds, appear to be more sensitive to OP or carbamate compounds than adults. Dietary deficiencies, low fat reserves, poor physiological condition, and high energy needs, such as migration or high metabolic rates, may increase vulnerability to these compounds. Behavioral traits may also increase the potential for exposure to OP or carbamate compounds. Species at increased risk are those that congregate in areas of treated habitats, gorge on a food source (like geese), forage in treated substrates, or feed on target organisms shortly after applications of these compounds.

> **Common routes of exposure of birds to OP and carbamate pesticides include:**
>
> Consumption of:
>   Treated seeds
>   Vegetation with pesticide residues
>   Dead or struggling poisoned insects
>   Granular formulations as grit, food, or coincidentally with other food items
>   Carrion killed by a pesticide
>   Food intentionally baited with pesticide
>   Live animals intoxicated with pesticide
>   Water contaminated with pesticide from runoff or irrigation
> Inhalation
> Absorption through the skin

Also, there can be considerable variability in the sensitivity of individual species to these pesticides (Table 39.1).

Table 39.1 Toxicity for birds of organophosphorus pesticides and carbamate pesticides.
[Modified from Hoffman and others, 1995. LD_{50} is the single oral dose of pesticide in milligrams per kilogram of body weight that is required to kill 50 percent of the experimental population]

		Species LD_{50}		
Compound	Class	Mallard duck	Ring-necked pheasant	Red-winged blackbird
Aldicarb	Carbamate	3.4	5.3	1.8
Carbaryl	Carbamate	>2,000	707	56
Carbofuran	Carbamate	0.5	4.1	0.4
Methiocarb	Carbamate	13	270	4.6
Mexacarbate	Carbamate	3.0	4.6	10
Azinphos-methyl	OP	136	75	8.5
Dimethoate	OP	42	20	6.6
Ethion	OP	>2,000	1,297	45
Phorate	OP	0.6	7.1	1.0
Temephos	OP	79	35	42

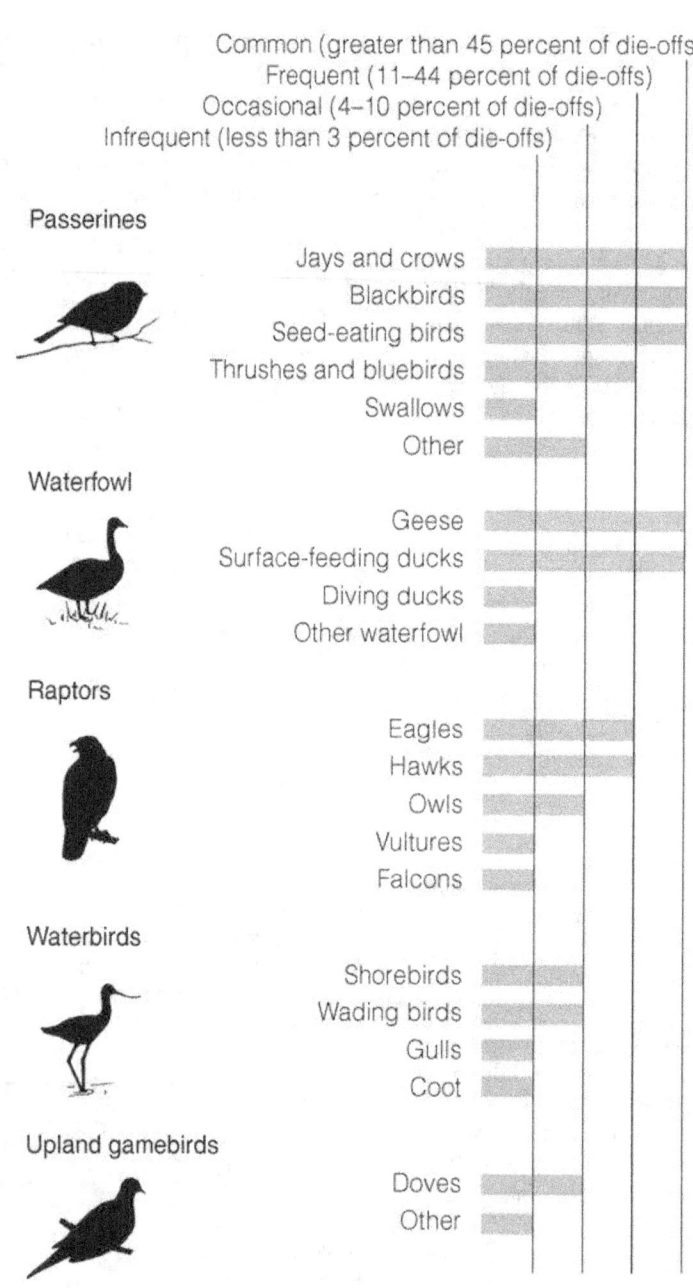

Figure 39.1 Frequency of occurrence of major groups of birds in documented organophosphorus and carbamate pesticide mortality events from 1986–95 (National Wildlife Health Center data base).

Distribution

Organophosphorus and carbamate compounds are used throughout the world as insecticides, herbicides, nematocides, acaricides, fungicides, rodenticides, avicides, and bird repellants. These compounds are applied in a wide variety of habitats including agricultural lands, forests, rangelands, wetlands, residential areas, and commercial sites. Wild bird deaths from OP and carbamate poisoning have been reported throughout the United States (Fig. 39.2). In more than half of these mortality incidents, the pesticide source is unknown (Fig. 39.3). Known applications of these compounds fall into five groups: approved applications in 1) agricultural land uses such as field and row crops, pastures, orchards, and forests; 2) residential and urban sites for turf in parks, golf courses, yards, and other urban pest control uses; 3) livestock uses such as pour-ons or feed products; 4) vertebrate pest control; and 5) malicious pesticide use, such as baiting to intentionally harm wildlife (Fig. 39.3).

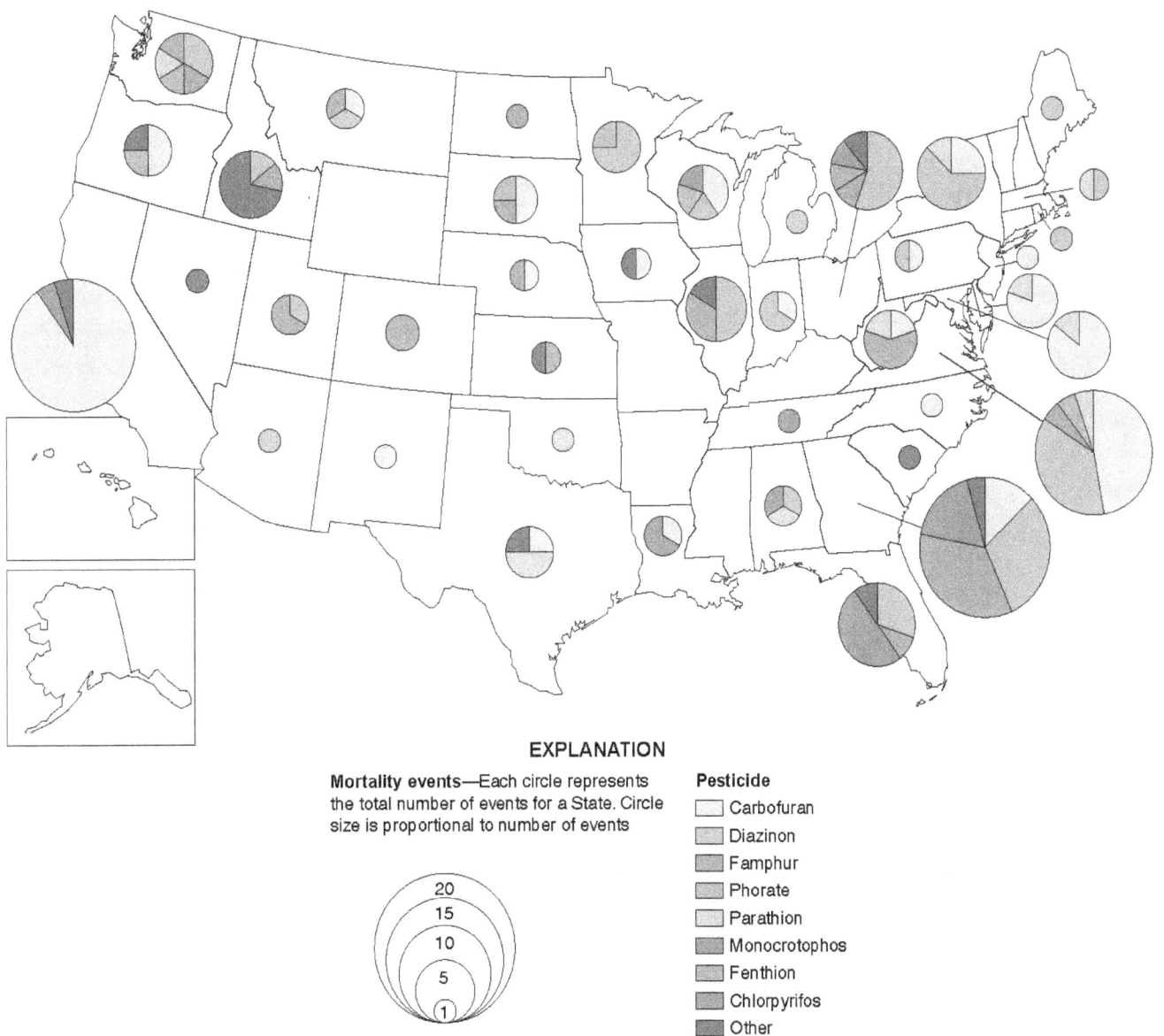

Figure 39.2 Distribution of 181 avian mortality events caused by organophosphorus and carbamate pesticides, 1986–1995 (National Wildlife Health Center data base).

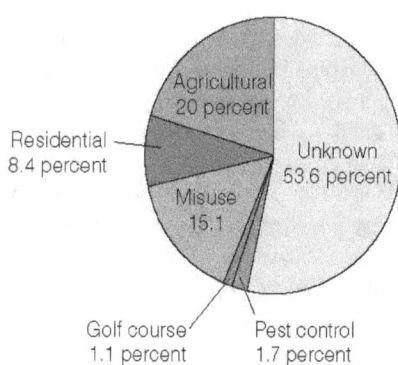

Figure 39.3 Applications associated with avian mortality caused by organophosphorus and carbamate pesticides from 1986–95 (National Wildlife Health Center data base).

Seasonality

Because OP and carbamate pesticides are typically short-lived in the environment, seasonality of avian mortality is generally associated with pesticide applications (Fig. 39.4). In documented mortality events in the United States, February was the peak month for the onset of bird die-offs, and most of these die-offs occurred in the southern United States, where the growing season starts early in the year.

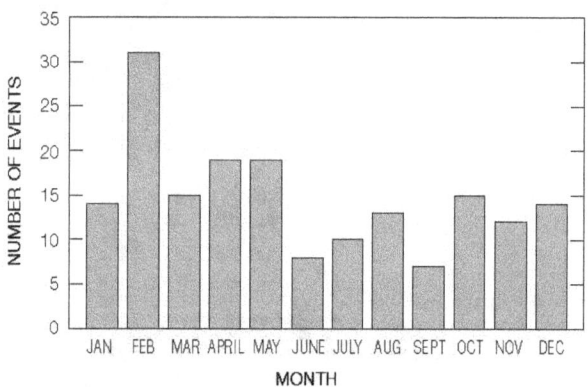

Figure 39.4 Avian mortality events due to organophosphorus and carbamate pesticides by date of onset, 1986–95 (National Wildlife Health Center data base).

Field Signs

Mortality can be the first sign noted in a pesticide poisoning, but the observer may find other clues at the scene of a mortality event. Live affected birds may exhibit convulsions, lethargy, paralysis, tremors, or other nonspecific neurological signs.

Clinical signs and bird behaviors that are commonly associated with acute exposure to cholinesterase-inhibiting pesticides
[Modified from Mineau, 1991]

Convulsions
Hyperexcitability
Incoordination of muscular action (ataxia)
Muscular weakness (myasthenia)
Difficult breathing (dyspnea)
Rapid breathing (tachypnea)
Vomiting
Defecation
Diarrhea
Spasmodic contraction of anal sphincter (tenesmus)
Lethargy
Induced tranquility
Head and limbs arched back (opisthotonos)
Slight paralysis (paresis)
Blindness
Contraction of pupils (miosis)
Dilation of pupils (mydriasis)
Drooping of eyelid (ptosis)
Protrusion of eyes (exopthalmia)
Excessive tear formation (lacrimation)
Excessive thirst (polydypsia)
Bleeding from nares (epistaxis)
Erection of contour feathers (piloerection)

Birds that die rapidly with pronounced neurological signs may leave evidence of their struggle even after death, such as vegetation clenched in their talons (Fig. 39.5) or vegetation that they disturbed during thrashing or convulsions. Animals may not have time to disperse before the toxin takes effect, and carcasses of multiple species, especially predators and granivorous or insectivorous wildlife, may be found within the same area following OP or carbamate exposure.

Birds can also be affected by a sublethal dose of an OP or carbamate pesticide. Sublethal exposure may contribute to other causes of mortality in birds, such as trauma. In some instances when birds have died due to trauma from a vehicle impact, a building strike, or predation, decreased brain ChE has been demonstrated, which indicates pesticide exposure. The sublethal dose of pesticide likely impaired the nervous system enough to alter behavior, thus making the animal more vulnerable to a traumatic cause of death. Special studies that evaluated sublethal OP or carbamate compound exposure in birds have found other effects to birds, including a reduced ability to regulate body temperature; impaired reproduction; and reduced tolerance to cold stress, which can cause reduced activity, leading to decreased feeding and weight loss. Altered behaviors such as reduced nest attentiveness and changes in singing by passerines have also been observed.

Figure 39.5 Vegetation clenched in the talons of a bald eagle, which is a finding consistent with organophosphorus or carbamate poisoning in raptors.

Gross Findings

Granular material or the presence of dye or both in the gastrointestinal tract are conspicuous findings that implicate pesticide ingestion. The necropsy finding of freshly ingested food in the upper gastrointestinal tract of a carcass is a good indicator of death by intoxication, especially when a large amount of a uniform food item is present. Feathers, flesh, hair, or other animal parts in the stomachs of raptors or of scavengers are common in secondary poisoning, whereas ingested grain is often found in waterfowl and passerines (Fig. 39.6). The food item may indicate the pesticide source, and the food can then be analyzed for specific chemical compounds.

The gross lesions that are associated with acute mortality from pesticide poisoning in birds are nonspecific and are usually minimal. Reddening of the intestinal wall, or even hemorrhage (Fig. 39.7), is observed occasionally with ingestion of certain pesticides. Redness and excess fluid in the lungs may be observed; these findings are consistent with respiratory failure. However, these changes are not unique to pesticide poisonings; they can be found in animals that died from other causes.

Diagnosis

A diagnostic evaluation is essential. A diagnosis of pesticide poisoning in birds is based on evidence of ChE inhibition in the brain or the blood and identification of pesticide residues in gastrointestinal contents. In many instances, depressed ChE activity will be the first indication that OP or carbamate pesticides caused a mortality event. A necropsy is necessary to rule out other causes of mortality or to identify contributing causes.

Brain ChE activity is a reliable indicator of OP and carbamate exposure in dead birds, but the absence of ChE depression does not reliably rule out poisoning. Brain ChE activity is measured and compared to normal brain ChE activity of the same species to determine the decrease in enzyme activity from normal levels (Appendix D). A decrease in brain ChE activity of 25 percent or more from normal indicates exposure to a cholinesterase-inhibiting compound (OP or carbamate pesticide); a decrease of 50 percent or more from normal is evidence of lethal exposure. Because of the variation in results between laboratories and the variability even between methods and procedures within a lab, it is important to compare results with controls from the same laboratory using the same method and not interpret analytical results from two or more laboratories or from two or more analytical methods.

Analyses can be carried one step further to differentiate the effect of OP from carbamate compounds by measuring the enzyme activity of a sample after incubation at 37–40 °C and comparing it to the initial measurement. Enzyme activity that returns toward a normal level after incubation, or that reactivates, indicates that carbamate poisoning is likely because carbamates tend to release their bond with ChE over time at increased temperatures or in aqueous environments. Because reactivation can occur with some pesticides, depressed brain ChE activity in a pesticide-poisoned bird may be difficult to document if the carcass has remained in a warm environment for an extended period of time. Another method that is used to differentiate an OP from a carbamate compound exposure is reactivation analysis, during which 2-PAM, a cholinesterase regenerating agent, is added to the sample and the change in brain ChE activity is then measured. Reactivation of ChE activity using 2-PAM occurs only when an OP compound is bound to the enzyme.

When a pesticide die-off is suspected, it is important to chill carcasses immediately. If diagnostic evaluation cannot be initiated within 24–48 hours, carcasses should be frozen as soon as possible to prevent further change in brain ChE activity. Also, when normal brain ChE activity values are not known for a particular species, control samples collected from normal birds of the same species are needed in order to compare ChE values.

In birds that recover from OP or carbamate poisoning, brain ChE activity will typically increase but it may remain below normal levels for up to 3 weeks, depending on the compound and on the dose received. Cholinesterase activity in blood from live birds may be used as an indicator of pesticide exposure; however, blood ChE activity is more variable than brain ChE activity. Cholinesterase enzymes in the blood are more sensitive than brain ChE to OP and carbamate pesticides; therefore, pesticide exposure quickly and dramatically depresses blood ChE activity, which then rapidly returns to normal levels.

One advantage of measuring blood ChE activity is that a nonlethal sample can be taken to provide evidence of OP or carbamate pesticide exposure in live birds. A disadvantage of measuring blood ChE activity is that interpretation is difficult because normal blood ChE activity varies among spe-

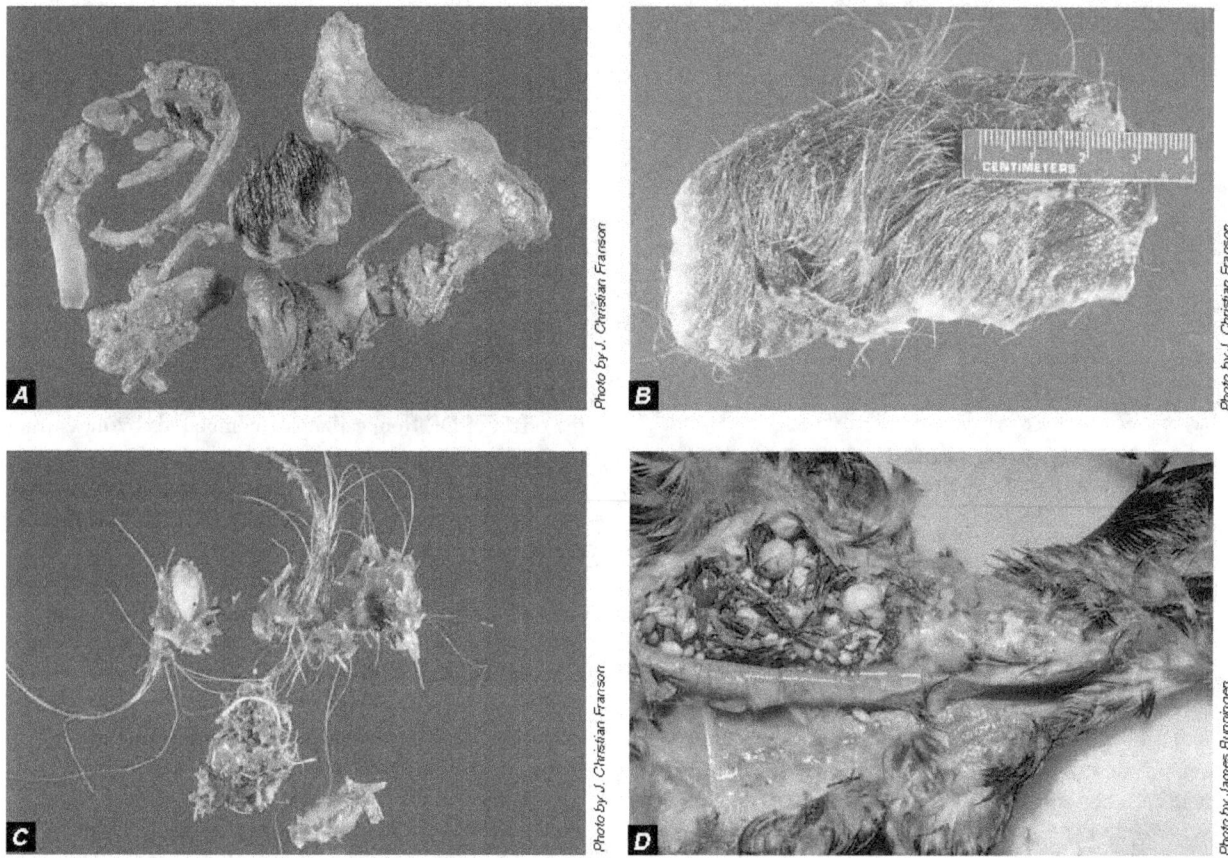

Figure 39.6 Examples of food items found in the gastrointestinal tracts of birds that died from organophosphorus or carbamate poisoning. **(A)** Pig remains from the crop of a bald eagle. **(B)** Bovine skin from the stomach of a bald eagle. **(C)** Bovine hair from the stomach of a magpie. **(D)** Corn in the esophagus of a mallard.

cies, age, sex, and body condition, and because a diurnal ChE variation may occur in some species. The reactivation analysis described above, which is used to differentiate an OP- from a carbamate-induced intoxication when measuring brain ChE activity, can also be used to evaluate blood ChE activity. In live animals, a presumptive diagnosis can also be made by reversing the neurological signs with proper medical treatment.

Specific compound residues may be identified in gastrointestinal contents. Mass spectrometry and gas chromatography are the usual analytical methods. Table 39.3 lists the compounds that were identified as the cause of mortality in the documented wild bird mortality events illustrated in Fig. 39.2.

Table 39.3 Specific organophosphorus and carbamate pesticides known to cause wild bird mortality events.

Carbamates	Organophosphorus compounds	
Carbofuran	Chlorpyrifos	Fenthion
Methiocarb	Diazinon	Fonofos
Oxamyl	Dicrotophos	Methamidophos
Aldicarb	Dimethoate	Monocrotophos
	Disulfoton	Parathion
	Famphur[1]	Phorate
	Fenamiphos	Phosphamidon
	Fensulfothion	Terbufos

[1] Famphur is regulated by the Food and Drug Administration as a drug.

Control

When a die-off with a confirmed pesticide poisoning diagnosis has occurred, birds should be denied use of the pesticide-affected area. Carcass pickup is necessary to prevent secondary toxicity to scavengers and prevent mortality from other causes related to decomposing carcasses, such as botulism. Any remaining pesticide in bags, on treated seed, bait, or grain must be removed to prevent further mortality.

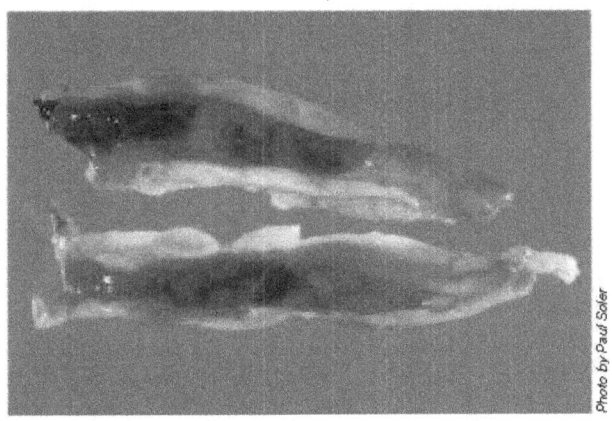

Figure 39.7 Hemorrhage in the intestine is an occasional finding in waterfowl that died of exposure to pesticides, particularly organophosphorus compounds.

Followup to wildlife mortality incidents due to pesticide poisoning is important for determining the source and the use of a chemical. Documented wildlife mortality from approved pesticide applications is considered by regulatory authorities for developing label use restrictions and for licensing pesticide formulations. Malicious use of pesticides to kill unwanted wildlife is against the law, and legal means can be employed to stop illegal use.

Persons who apply pesticides need to consider wildlife use and environmental conditions when they apply the chemicals. Migration patterns of the wildlife that use the area, the presence of nesting and breeding species, and weather conditions, such as the potential for aerial drift or runoff into wetlands or ponds, are among the factors that should be considered. Pesticides should be applied only as directed; the use of alternate chemicals or formulations that pose less risk to nontarget species should also be considered. Buffer zones at crop perimeters will provide more protection to areas used by wildlife. Agricultural land planted adjacent to wetlands should be plowed parallel to a wetland to minimize runoff.

Human Health Considerations

Human exposure to OP or carbamate pesticides can result in serious illness or even death. Exposure can occur through inhalation, absorption through the skin, or by ingestion. When pesticides that may be associated with wildlife mortality incidents are investigated, field procedures should be scrutinized to avoid inadvertent exposure of personnel to pesticides. Persons who collect carcasses or field samples must prevent their exposure by wearing nonpermeable gloves, rubber boots, or other appropriate clothing that will prevent skin absorption, and respirators should be used if chemical inhalation is possible.

Poisoning in humans should be treated as a serious medical emergency. When someone seeks medical attention for exposure to an OP or carbamate compound, the attending physician should be informed that the person may have been exposed to these chemicals. Patients can be monitored by blood sampling to evaluate their blood ChE levels. Aggressive treatment of acute intoxication does not protect against the possibility of delayed onset neurotoxicity or persistent neurological defects. Certain compounds have been documented to cause delayed effects in humans. An intermediate syndrome that occurs within 24–96 hours after exposure has recently been described with intoxications of fenthion, dimethoate, monocrotophos, and methamidophos. Muscles of the limbs and those innervated by cranial nerves are affected, causing palsies, respiratory depression, and distress. Another delayed neurotoxicity from some OP compounds can occur 1–2 weeks after exposure. Initially, incoordination develops, and it can progress to moderate to severe muscle weakness and paralysis. This delayed effect was documented with some OP compounds that were rarely used as pesticides, but the effect may be a potential risk with similar compounds that are more commonly used today if sufficient exposure to the compound occurs. These delayed effects could be a problem in wildlife, but they have not been recognized yet in any wildlife species.

Linda C. Glaser

Supplementary Reading

Amdur, M.O., Doull, J., and Klaassen, D.C., eds., 1991, Casarett and Doull's Toxicology, The basic science of poisons, (4th ed.): Elmsford, N.Y., Pergamon Press, 1,033 p.

Grue, C.E., Fleming, W.J., Busby, D.G., and Hill, E.F., 1983, Assessing hazards of organophosphate pesticides in wildlife, *in* Transactions of the 48th North American Wildlife & Natural Resources Conference: Washington, D.C., The Wildlife Management Institute, p 200–220.

Hill, E.F. and Fleming, W.J., 1982, Anticholinesterase poisoning of birds: field monitoring and diagnosis of acute poisoning: Environmental Toxicology and Chemistry 1:27–38.

Hill, E.F., 1995, Organophosphorus and carbamate pesticides, *in* Hoffman, D.H., Rattner, B.A., Burton, G.A., Jr., and Cairns, J., Jr., eds., Handbook of ecotoxicology: Boca Raton, Fla., Lewis Publishers, p 243–274.

Mineau, P., ed., 1991, Cholinesterase-inhibiting Insecticides, Their impact on wildlife and the environment, chemicals in agriculture v. 2.: Amsterdam, The Netherlands, Elsevier Science Publishing, 348 p.

Smith, G.J., 1987, Pesticide use and toxicology in relation to wildlife: Organophosphorus and carbamate compounds: Washington, D.C., U.S. Department of the Interior, Fish and Wildlife Service, Resource Publication 170, 171 p.

Chapter 40
Chlorinated Hydrocarbon Insecticides

Synonyms
Organochlorines, OCs

Chlorinated hydrocarbon insecticides (OCs) are diverse synthetic chemicals that belong to several groups, based on chemical structure. DDT is the best known of these insecticides. First synthesized in 1874, DDT remained obscure until its insecticidal properties became known in 1939, a discovery that earned a Nobel Prize in 1948. The means of synthesizing the cyclodiene group, the most toxic of the OCs, was discovered in 1928 and resulted in a Nobel Prize in 1950. The insecticidal properties of cyclodienes, which include aldrin, dieldrin, and endrin (Table 40.1), were discovered about 1945. OCs became widely used in the United States following World War II. Their primary uses included broad spectrum applications for agricultural crops and forestry and, to a lesser extent, human health protection by spraying to destroy mosquitoes and other potential disease carriers. These compounds also became widely used to combat insect carriers of domestic animal diseases.

Cause

Chlorinated hydrocarbon insecticides are stored in body fat reserves or are lipophilic, and they remain in the environment for long periods of time after application. They bioaccumulate or are readily accumulated by animals through many exposure routes or repeated exposure and they tend to biomagnify or accumulate in higher concentrations in animals that are higher in the food chain. This combination of bioaccumulation and biomagnification can harm or kill wildlife, especially some species of birds. The highly toxic cyclodiene compounds cause direct mortality of birds as well as secondary poisoning, which results when birds prey on organisms dying from insecticide applications. Reproductive impairment is the primary effect of the less acutely toxic DDT and its metabolites, DDD and DDE. The cumulative storage of OC residues within body fat reserves presents an additional hazard for birds. Rapid use and depletion or mobilization of fat reserves during migration, food shortages, and other stressful conditions release OC residues into the blood. The residues are then carried to the brain, where they can reach toxic levels resulting in acute poisoning.

Species Affected

Acute mortality from exposure to OCs has been documented in many bird species (Table 40.1). However, the toxicity for birds of different types of these insecticides varies greatly (Tables 40.2 and 40.3). In general, birds that are higher in the food chain are more likely to be affected by OCs present in the environment than birds that are lower in the food chain. This is especially true for fish-eating birds and raptors (Fig. 40.1). Environmental biomagnification of these contaminants can be seen in the mortality of robins and other birds from DDT. Leaves from trees that were sprayed with DDT to control Dutch elm disease had high residues of DDT (174–273 parts per million) shortly after spray applications. When the leaves dropped in the fall, they still contained 20–28 parts per million of DDT. This leaf litter, along with spray residue that reached the ground, produced high DDT residues in the top levels of soil. Earthworms that fed in those soils concentrated the residues to a level high enough to kill birds that fed on them. Another hazard is OC seed dressings, which are used to prevent insect damage to agricultural crops, that may be ingested by waterfowl and other seed or grain-eating birds.

Distribution

Exposure to chlorinated hydrocarbon insecticides is global, and residues of these compounds are found in nearly every environment, even in Antarctica and the Arctic. Avian mortalities from OCs have been reported from Europe, Asia, North America, and South America. Poisoning may occur anywhere that birds are exposed to point sources of these chemicals or through bioaccumulation and biomagnification. Because of their environmental persistence and global movement, residues of chlorinated hydrocarbon insecticides impact bird health long after they become environmental contaminants and at locations far from the original application sites. For example, DDT compounds, polychlorinated biphenyls (PCBs), and dioxin-like compounds were recently found in black-footed albatross adults, chicks, and eggs on Midway Atoll in the Pacific.

Seasonality

Exposure of birds to OCs is most likely during spring and summer in countries where these compounds are still used to control insect pests during the growing season, but exposure may occur any time that residues are present in food sources. For example, waterfowl and other birds that fed on

Table 40.1 Examples of avian mortality events caused by chlorinated hydrocarbon insecticides.

Insecticide	Purpose of application	Means of bird exposure	Bird group affected	Principal species affected	Event location and time period
DDT	Spray application to control Dutch elm disease.	Biomagnification in terrestrial food chain.	Passerines	Robin and other small birds.	New England, Midwest; late 1940s to 1950s.
DDD	Spray application to control gnats.	Biomagnification in aquatic food chain.	Grebes	Western grebe	Clear Lake, California, 1950s.
Aldrin	Treatment of rice seed to combat agricultural pests.	Consumption of treated seeds, use and depletion of stored fat reserves during migration or periods of stress.	Waterfowl	Fulvous whistling duck, snow goose, blue-winged teal.	Texas, 1970s
Heptachlor	Treatment of wheat seed to control agricultural pests.	Consumption of treated seeds.	Waterfowl	Canada goose	Oregon, 1970s
Toxaphene	Spray application to control agricultural pests.	Direct contact with and consumption of contaminated food.	Waterfowl	Ducks and coot	California, 1960s
	Spray application for fisheries management.	Ingestion of contaminated food.	Waterfowl	Blue-winged teal, shoveler, mallard.	Nebraska, 1960s
Dieldrin	Spray application to control agricultural pests.	Biomagnification in food chain.	Raptors	Bald eagle, peregrine falcon	Nationwide, 1960s and 1970s.
Endrin	Spray application to control orchard rodents.	Direct contact with spray; consumption of contaminated food; biomagnification; use and depletion of fat reserves.	Gallinaceous birds, raptors, geese	Quail, chukar partridge, goshawk, Cooper's hawk, barn owl, Canada goose.	Washington, 1960s 1970s, 1980s.
Chlordane	Dry formulations to control soil pests and termites.	Consumption of contaminated food; biomagnification.	Raptors, passerines	Great horned owl, American kestrel, Cooper's hawk, blue jay, robin, starling.	New York, Maryland, New Jersey, 1980s.

Table 40.2 Toxicity for the mallard duck of some chlorinated hydrocarbon insecticides.
[Modified from Heinz and others, 1979. LC_{50} is the insecticide concentration, in parts per million, in feed that is required to kill 50 percent of birds during a given period of time. LD_{50} is the insecticide amount, in milligrams per kilogram of body weight, in a single dose that is required to kill 50 percent of birds. ppm, parts per million; mg/kg, milligrams per kilogram; >, greater than; ≥, greater than or equal to. — no data available]

Insecticide	Subacute exposure LC_{50} (ppm)	Acute exposure LD_{50} (mg/kg)
Aldrin	155	520
Chlordane	858	1,200
DDT	1,869	>2,240
Dieldrin	169	381
Endosulfan	1,053	33
Endrin	22	5.6
Heptachlor	480	≥2,000
Lindane	—	>2,000
Mirex	>5,000	>2,400
Toxaphene	538	71

Table 40.3 Relative acute toxicity of chlorinated hydrocarbon insecticides for birds.
[Modified from Hudson and others, 1984. LD_{50} is the insecticide amount, in milligrams per kilogram of body weight, in a single dose that is required to kill 50 percent of birds. mg/kg, milligrams per kilogram; >, greater than; <, less than.]

Species	LD_{50} (mg/kg)								
	Aldrin	Chlordane	DDT	Dieldrin	Endosulfan	Endrin	Lindane	Mirex	Toxaphene
Canada goose				<141					
Mallard duck	520	1,200	>2,240	381	31–45	5.6	2,000	2,400	70.7
Fulvous whistling duck	29.2			100–200					99
Sandhill crane			>1,200						100–316
Pheasant	16.8	24–72	1,334	79	80 to >320	1.8		>2,000	40
Sharp-tailed grouse						1.1			19.9
Gray partridge				8.8					23.7
Chukar partridge				25.3					
Bobwhite quail	6.6								85.5
California quail		14.1	595	8.8		1.2			23.7
House sparrow				47.6					
Horned lark									581

endrin-treated winter wheat seed have died in the autumn, and raptors have died yearround. Reproductive effects are manifested during the breeding season, but the exposure that causes these effects can occur at any time of year.

Field Signs

Thin eggshells that often collapse under the weight of the nesting bird and eggs that break during incubation (Fig. 40.2) are classic signs of exposure to DDT and some other OCs. Clinically ill birds suffering from acute poisoning often exhibit signs of central nervous system disorders such as tremors, incoordination, and convulsions (Fig. 40.3). Other birds may be lethargic and exhibit additional behavioral changes (Table 40.4).

Gross Lesions

Birds dying of chronic exposure to OCs are often emaciated (Fig. 40.4). Those that die acutely usually exhibit no lesions. The pathological effects attributed to exposure to these compounds (Table 40.4) are not unique and, therefore, they cannot be used as the only basis for diagnosis.

Diagnosis

Residue analysis combined with necropsy findings, clinical signs, and an adequate field history are generally required for a diagnosis of chlorinated hydrocarbon insecticide poisoning. Brain is the tissue of choice for residue analysis because chemical concentrations that indicate poisoning in birds have been determined for several of these compounds. Take care not to contaminate tissues for residue analysis. Submit the entire carcass whenever possible, otherwise remove the head and send it intact to the laboratory. When it is necessary to remove the brain or other tissues for analysis, rinse the instruments with a solvent, such as acetone or hexane, to remove chlorinated hydrocarbon insecticide residues from them. Place the tissues in solvent-rinsed glass containers or wrap them in aluminum foil. The foil should not have been prepared by a manufacturer that uses oils made of animal fats. A "K" on the package label indicates that no animal fats were used in the manufacturing process.

Experimental studies have been done in an attempt to establish lethal brain levels for OCs in various species of birds (Fig. 40.5). DDE levels in the brains of bald eagles thought to have died from this contaminant have ranged from 212 to 385 parts per million (wet weight), and these levels are consistent with brain DDE levels of kestrels that died from experimental dosing studies (213–301 parts per million, wet weight). These findings are important for interpreting field data (Fig. 40.6). However, interpretation of residue values is complicated by the simultaneous occurrence of other contaminants that may combine with, interact with, or inhibit the toxic effects of any individual compound. Other factors, such as sex, age, and nutritional level also may affect toxicity.

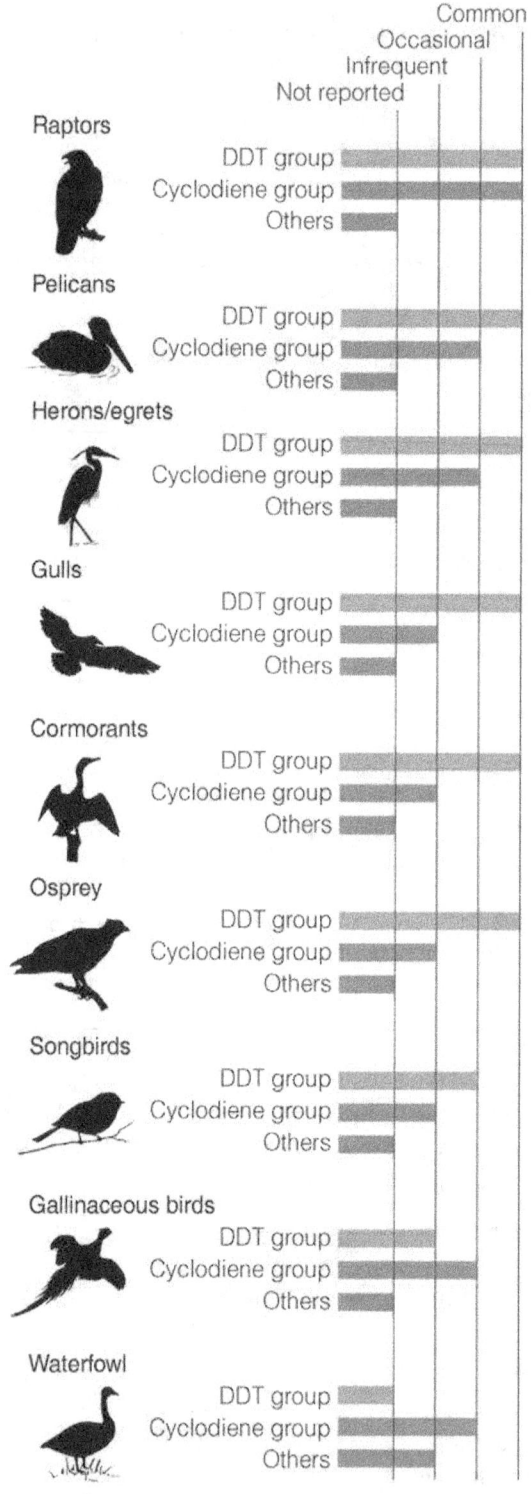

Figure 40.1 Relative importance of health effects caused by chlorinated hydrocarbon insecticides in selected free-living birds.

Figure 40.2 The flattened egg within this white-faced ibis clutch was caused by DDE.

Figure 40.3 Damage to the central nervous system of birds by chlorinated hydrocarbon insecticides results in the type of aberrant posture seen in **(A)** this hen pheasant and **(B)** this mallard duck.

Table 40.4 Most commonly reported effects from chlorinated hydrocarbon insecticide exposures of birds.

General effect	Specific effects
Behavioral	Lethargy, slowness, depression
	Locomotive and muscle incoordination (ataxia)
	Tremors and convulsions
	Reduced nest attentiveness and nest abandonment
	Violent wing beating
	Aberrant wing and body carriage
	Muscle spasms causing the body to bend backwards and become rigid (opisthotonos)
Reproductive	Embryo mortality
	Decreased egg hatchability
	Decreased egg production
	Eggshell thinning
	Egg breakage during incubation
Pathological	Emaciation; muscle wasting and absence of fat
	Congestion of the lungs, kidneys, and especially the liver have been reported in pheasants dying from dieldrin poisoning
	Increased liver weight
	Small spleens have been reported
	In general, pathological changes are not readily evident at the gross level, and microscopic changes are not diagnostic
Immunological	Increased susceptibility to infectious disease
Other	Disruption of salt gland function by DDE

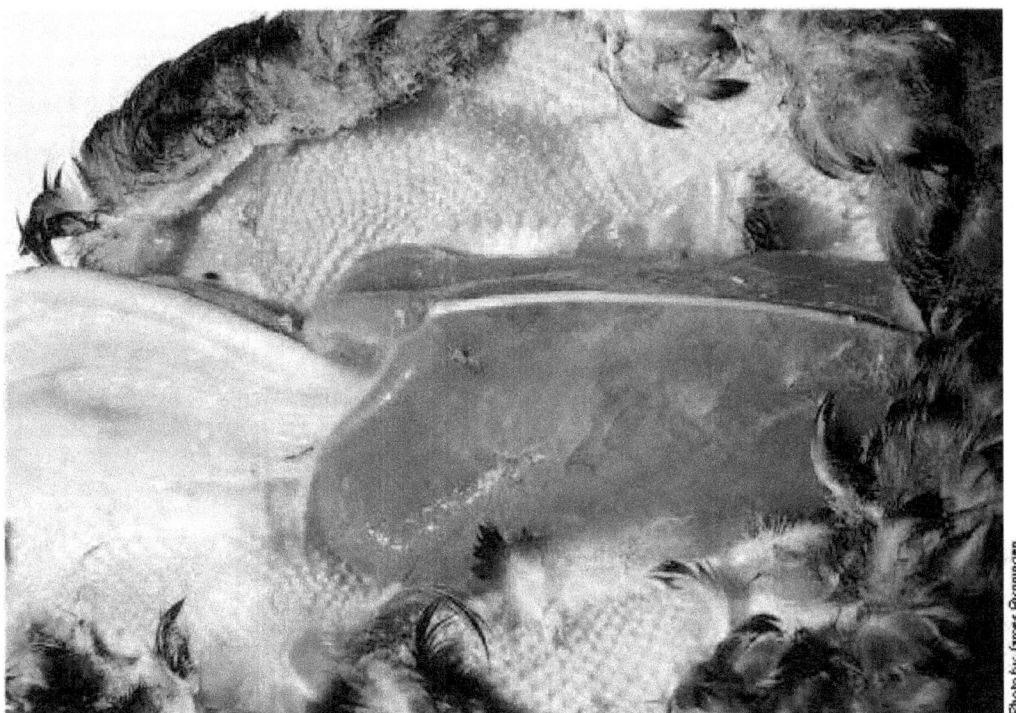

Figure 40.4 Chronic toxicity from chlorinated hydrocarbon insecticides can result in emaciation, demonstrated by the prominent keel and lack of subcutaneous fat in this black duck. In addition, emaciation caused by the rapid use and depletion of body fat stores due to stresses of migration, inadequate food supplies, and other causes can concentrate body residues of chlorinated hydrocarbons in the brain and cause acute toxicity.

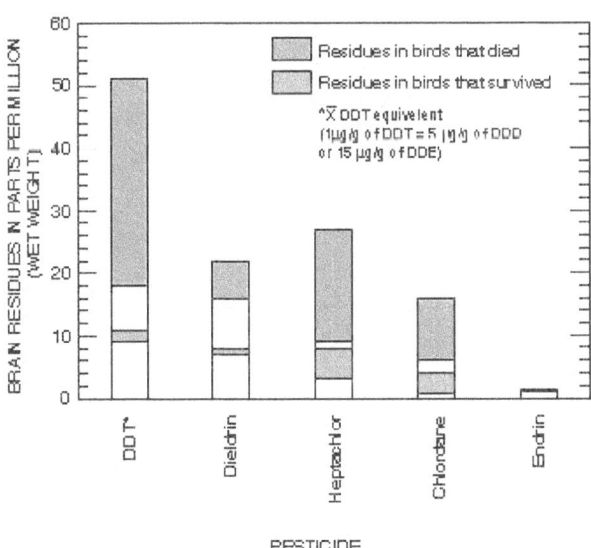

Figure 40.5 Mean chlorinated hydrocarbon insecticide residues in brains of experimentally dosed passerines.

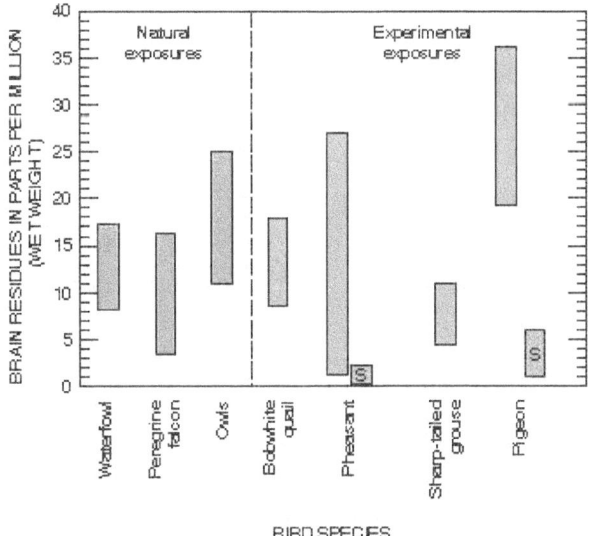

Figure 40.6 Range of dieldrin residues in brains of selected avian species. [S is the range of concentration within which some of the birds survived.]

Control

Because uses of most OCs have been banned or greatly curtailed in the United States, controlling wildlife exposure to these compounds depends largely on properly disposing of existing stores, preventing leakage into the environment, and preventing malicious use. The spreading of these compounds to environments where they are no longer used will continue until suitable alternative pest controls are found. Also, migratory wildlife that are exposed to these compounds in areas where they are still used may not exhibit effects until they reach other areas on their migratory route.

Human Health Considerations

As with many of the other toxins discussed in this section, residues of chlorinated hydrocarbons in birds are stored in tissues and are not transferred to humans through routine handling of carcasses. Exceptions include situations where a person could somehow come in contact with the pesticide, for example, in the stomach contents of a bird or on its skin or feathers. It is always wise to handle carcasses with disposable gloves, and to treat unknown mortalities as possible sources of infectious agents transferrable to humans.

Milton Friend and J. Christian Franson

Supplementary Reading

Blus, L.J., 1995, Organochlorine pesticides, *in* Hoffman, D.J., and others, eds., Handbook of ecotoxicology: Boca Raton, Fla., Lewis Publishers, p. 275–300.

Blus, L.J., Wiemeyer, S. N., and Henny, C.J., 1996, Organochlorine pesticides, *in* Fairbrother, A., and others, eds., Noninfectious diseases of wildlife (2nd ed.): Ames, Iowa, Iowa State University Press, p. 61–70.

Heinz, G.H., Hill, E.F., Stickel, W.H., and Stickel, L.F., 1979, Environmental Contaminant studies by the Patuxent Wildlife Research Center, *in* Kenaga, ed., Avian and mammalian wildlife toxicology: Philadelphia, Penn., American Society for Testing and Materials, Special Technical Publication 693, p. 9–35

Hudson, R.H., Tucker, R.K., and Haegele, M.A., 1984, Handbook of toxicity of pesticides to wildlife, (2d ed.): Washington, D.C., U.S. Department of the Interior, Fish and Wildlife Service, Resources Publication 153, 90 p.

Peakall, D.B., 1996, Dieldrin and other cyclodiene pesticides in wildlife, *in* Beyer, W.N., and others, eds., Environmental contaminants in wildlife: interpreting tissue concentrations: Boca Raton, Fla., Lewis Publishers, p. 73–97.

Chapter 41
Polychlorinated Biphenyls

Synonyms
PCBs, aroclors, chlorinated biphenyls

Polychlorinated biphenyls (PCBs) are industrial compounds with multiple industrial and commercial uses (Table 41.1). PCBs are chemically inert and stable when heated. These properties contribute greatly to PCBs having become environmental contaminants. The chemical inertness and heat stability properties that make PCBs desirable for industry also protect them from destruction when the products in which they are used are discarded. These same properties also enable PCB residues to persist in the environment for long periods of time and to be transported worldwide when contaminated particulate matter travels through waters, precipitation, wind, and other physical forces.

PCBs have a physical structure similar to DDT, and, like DDT, they are classified as aromatic hydrocarbons which contain one or more benzene rings. The presence of chlorine results in DDT, PCBs, and other compounds with similar structures commonly being referred to as chlorinated hydrocarbons. The toxicity of these compounds is associated with the amount of chlorine they contain. The trade name of Aroclor® for PCBs that were produced by a manufacturer in the United States contains a numerical designation that specifies the amount of chlorine present in a particular formulation. For example, Aroclor® 1221 contains 21 percent chlorine while Aroclor® 1254 contains 54 percent chlorine. The first two digits designate the number of carbons in the formulation. The chemical structure of PCBs results in the possibility of many different forms or isomers, (more commonly called congeners) of these compounds. PCBs in other countries have different trade names than Aroclor® (Table 41.2).

Cause
Like other chlorinated hydrocarbons, PCBs accumulate in the fat of animals or are lipophilic, and they tend to become concentrated at higher levels of the food chain. In general, persistence increases for PCBs that are made with higher amounts of chlorine. Birds are most susceptible to PCB compounds of the mid-chlorination range (42–54 percent).

Species Affected
Mammals, especially mink, are more susceptible than birds and invertebrates to direct toxicity from PCBs. The highest tissue concentrations of these compounds are found among birds, especially marine species that are at the top of complex oceanic food webs and among fish-eating birds, such as cormorants, that use large inland water bodies. For example a 12.9-fold increase has been reported from plankton to fish in a Lake Michigan food web. Although direct toxicity for birds is generally low (Table 41.3), PCBs are powerful inducers of liver enzyme systems that increase the metabolism of hormones. PCBs may have caused thin eggshells in double-crested cormorants and white pelicans, and under experimental conditions, in ring-doves and (perhaps) in Coturnix quail and mallard ducks. Unfortunately, there is insufficient knowledge to clearly define the impacts of PCBs on bird reproduction, especially in field situations, because tissue residues are often highly correlated with other lipophilic compounds, such as organochlorines. Findings have generally been inconclusive, but the greatest effects have been seen in gallinaceous birds such as pheasants, chickens, and doves.

Distribution
PCBs were first identified in the tissues of wildlife in Sweden, and they are now known to occur in a wide variety of wildlife and other species, including humans, throughout the world. PCBs are clearly global contaminants, and they are the most abundant of the chlorinated hydrocarbon pollutants in the global ecosystem with the possible exception of petroleum products. Industrial wastes released into aquatic systems, point sources of contamination from manufacturing facilities, landfills receiving waste from such facilities, and combustion and other disposal of products containing PCBs are generally recognized sources of contamination. Another less well-known source of PCB contamination of the environment was the use of PCBs during the 1950s and 1960s as additives to extend the residual life and effectiveness of expensive chlorinated insecticides such as chlordane, aldrin, dieldrin, and benzene hexachloride. It is estimated that more than 1.5 metric tons of PCBs have been produced worldwide. PCB manufacturing in the United States was discontinued in 1978.

The variable environmental distribution of PCBs results from their physical and chemical properties, which influence their rates of distribution, retention, and degradation in different environments. This results in great differences in the relative concentrations of the different forms of PCBs found in wildlife samples from different geographic areas and is also a reflection of the magnitude of local and regional con-

Table 41.1 Uses of polychlorinated biphenyls (PCBs) in industry and products for society.

Properties
- Heat stability
- Chemical stability
- Ability to be mixed with organic compounds
- Slow degradation

Industrial uses
- Lubricants, hydraulic fluids, grinding fluids
- Heat transfer agents, insulators
- Plasticizers
- Dielectric sealants
- Dedusting agents
- Protective coatings

Common products that have contained PCB additives
- Wire and cable coating
- Impregnants for braided cotton-asbestos insulation
- Printing inks and mimeograph inks
- Preparation of imitation gold leaf
- Pigment vehicle for decoration of glass and ceramics
- Essential components of coating for flameproofing cotton drill for outer garments and for rendering olive-drab canvas fire retardant, water-repellant, and rot-proof (tents, tarpaulins)
- Moistureproof coating for wood, paper, concrete, and brick
- Asphalt, roof coatings
- High quality precision casting wax; waxes used in making dental castings and costume jewelry
- Sealers for masonry, wood, fiberboard, and paper
- Window envelopes
- Polystyrene, polyethylene, neoprene, polybutene, silicone rubber, crepe rubber
- Plasticizers in paints
- Life extenders and sometimes toxicity synergists for pesticides containing DDT, dieldrin, lindane, chlordane, aldrin, and benzene

tamination patterns, environmental transport processes, and the composition of PCB residues in the food chain.

Seasonality

Exposure to PCBs is not seasonally dependent; except that in warm weather, PCB residues may vaporize or evaporate with liquid from contaminated areas, and thus, increase the risk of airborne exposure.

Field Signs

Direct mortality of wild birds from exposure to PCBs rarely occurs. We are only aware of one such event having been documented. The number of different PCBs present in the environment further complicates evaluations because of different impacts and lethality associated with these different compounds. Nonspecific signs associated with acute exposure of birds to toxic levels of PCBs include lethargy, lack of locomotive and muscle coordination or ataxia, tremors, and other observations. Behavioral modifications and impaired reproductive performance may also occur and would be more readily detected at the population rather than individual level (Table 41.4).

Gross Lesions

There are no diagnostic lesions associated with exposure to PCBs. Enlarged liver and kidneys, atrophy of the spleen and the bursa of Fabricius, emaciation, and excess fluids around the heart have been associated with chronic exposure.

Excess fluid or edema in tissues has been found in some cases of acute PCB exposure, and this suggests that PCBs may interfere with tissue permeability or cardiac function or both. PCBs have been shown to cause physical defects in embryos, or be teratogenic, in chickens, and they also cause

Table 41.2 Trade names for polychlorinated biphenyls (PCBs).

Trade name	Country of manufacturer	Manufacturer
Aroclor®	United States of America	Monsanto
Clophens®	Germany	Bayer
Fenclors®	Italy	Caffaro
Phenoclors®; Pyralenes®	France	Prodelec
Kanechlors®	Japan	Kanegafuchi
Others have been produced in Czechoslovakia and the former USSR		

Table 41.3 Relative toxicity of polychlorinated biphenyls (PCBs) for birds. [Adapted from Eisler, 1986. LC_{50} is the contaminant concentration in the diet that is required to kill 50 percent of the test animals in a given period of time; by comparison, the LC_{50} for mink to Aroclors® 1242 and 1254 is 8.6 and 6.7, respectively. mg/kg, milligrams per kilogram. >, greater than. —, no data available.]

Species	LD_{50} (mg/kg of Aroclor®)			
	1221	1242	1254	1260
Bobwhite quail	>6,000	2,098	604	747
Mallard duck	—	3,182	2,699	1,975
Ring-necked pheasant	>4,000	2,078	1,091	1,260
Japanese quail	>6,000	>6,000	2,898	2,186
European starling, red-winged blackbird, brown-headed cowbird	—	—	1,500	—

Table 41.4 Reported effects of polychlorinated biphenyls (PCBs) in birds.

Type of impact	Examples
Behavioral	Lethargy
	Locomotive and muscle incoordination or ataxia
	Tremors and convulsions
	Reduced nest attentiveness and protection of eggs
Reproductive	Embryo mortality resulting in decreased hatchability of eggs
	Decreased egg production
	Egg shell thinning
Pathological	Accumulation of fluid within the pericardial sac or hydropericardium
	Excess fluid or edema in body tissues and organs
	Atrophy of bursa of Fabricius, spleen, and other lymphoid tissues
	Enlarged livers that are firm and light colored
	Bill and foot deformities (from embryonic exposure)
Immunological	Increased susceptibility to infectious disease
Other	Weight loss
	Debilitation

a condition analogous to chick edema disease. This condition results in the leakage of body fluids into various organs and tissues. However, the presence of dioxins as contaminants within the PCB formulations may be the actual cause of these lesions.

Diagnosis

Diagnosis of acute poisoning is based on PCB residues in tissues, and as for most other chlorinated hydrocarbons, mortality is best diagnosed from residues found in brain tissue. However, the concentrations of PCBs that indicate poisoning vary greatly with the specific formulation of PCBs, the species of bird, and, often, the presence of other environmental contaminants. Detection of subacute effects, such as poor reproductive performance and immunosuppression, is also confounded by these same factors. Comparison of residues in the tissues of birds suspected of being poisoned with residues in tissues of normal birds of the same species in nearby or regional sites can be diagnostically useful along with knowledge of PCB deposition and discharges in the area. Comparisons are sometimes difficult because of the varying effects of different PCB mixtures and the interactions that occur between PCBs, other pollutants, and other disease agents. Many toxic and biochemical responses from PCB exposure occur in multiple species and body organ systems.

Residue levels alone will generally not be sufficient data for making a diagnosis. Necropsy findings combined with laboratory analyses, including residue evaluations, knowledge of environmental conditions and events at the field site, and response of different species to PCB exposure are all needed for sound judgements to be reached.

Control

Prevention of the entry of PCBs into the environment and containment or removal of PCB contamination that is already present are necessary to reduce exposure of wildlife. PCB sales in the United States were stopped in the 1970s, but large amounts are still present in the environment due to environmental persistence and to global transport by winds and other means from locations where PCBs are still used. Improper disposal of products that contain PCBs through landfills and incineration at temperatures that are too low (below 1,600 °C) to destroy PCBs can cause further environmental contamination. However, more stringent air-quality standards in the United States and other nations have diminished the potential that PCBs in incinerated materials will be added to the environment through combustion.

Bird use of heavily contaminated sites should be prevented to the extent feasible by habitat manipulation, physical barriers, scaring devices, and other appropriate means. Knowledge of PCB levels in specific environments should be gained prior to developing those areas for wildlife, including the use of dredge material to create artificial islands for bird nesting habitat. PCB and heavy metal loads in sediments should also be considered in decisions regarding dumping dredge materials.

Human Health Considerations

PCBs are known to accumulate in humans, and health advisories are often issued about consuming wildlife from heavily contaminated environments. Residues in wildlife can only be transferred to humans by consuming contaminated tissues. As with most chlorinated hydrocarbons, the greatest concentrations of residues are in fat tissue, and removing fatty parts of the carcass prior to cooking can significantly reduce potential human exposure. Although PCB residues cannot be transferred to humans from wildlife by means other than consumption, the cause of death is seldom known when dead wildlife are encountered and the risk of exposure to disease agents that can be transmitted by contact should not be taken. Always wear gloves or use other physical barriers to prevent personal contact with the carcass.

Milton Friend and J. Christian Franson

Supplementary Reading

Eisler, R., 1986, Polychlorinated biphenyl hazards to fish, wildlife, and invertebrates: a synoptic review: Fish and Wildlife Service Biological Report 85(1.7), 72 p.

Hoffman, D.J., Rice, C.P., and Kubiak, T.J., 1996, PCBs and dioxins in birds, *in* Beyer, W.N., and others, eds., Environmental contaminants in wildlife: interpreting tissue concentrations: Boca Raton, Fla., Lewis Publishers, p. 165–207.

O'Hara, T.M., and Rice, C.D., 1996, Polychlorinated biphenyls, *in* Fairbrother, A., and others, eds., Noninfectious diseases of wildlife (2nd ed.): Ames, Iowa, Iowa State University Press, p. 71–86.

Rice, C.P., and O'Keefe, P., 1995, Sources, pathways, and effects of PCBs, dioxins, and dibenzofurans, *in* Hoffman, D.J., and others, eds., Handbook of ecotoxicology: Boca Raton, Fla., Lewis Publishers, p. 424–468.

Chapter 42
Oil

Synonyms
Petroleum

Each year, an average of 14 million gallons of oil from more than 10,000 accidental spills flow into fresh and saltwater environments in and around the United States. Most accidental oil spills occur when oil is transported by tankers or barges, but oil is also spilled during highway, rail, and pipeline transport, and by nontransportation-related facilities, such as refinery, bulk storage, and marine and land facilities (Fig. 42.1). Accidental releases, however, account for only a small percentage of all oil entering the environment; in heavily used urban estuaries, the total petroleum hydrocarbon contributions due to transportation activities may be 10 percent or less. Most oil is introduced to the environment by intentional discharges from normal transport and refining operations, industrial and municipal discharges, used lubricant and other waste oil disposal, urban runoff, river runoff, atmospheric deposition, and natural seeps. Oil-laden wastewater is often released into settling ponds and wetlands (Fig. 42.2). Discharges of oil field brines are a major source of the petroleum crude oil that enters estuaries in Texas.

Figure 42.2 *Wastewater laden with petroleum being discharged into a settling pond.*

Cause

Birds that are exposed to spilled or waste petroleum can be affected both externally and internally. Oil contamination of feathers (Fig. 42.3) disrupts their normal structure and function, and it results in the loss of insulation for warmth and waterproofing. Oiled birds lose the ability to fly, and they frequently die from hypothermia, starvation, exhaustion, or drowning. Birds that are exposed to oil during their reproductive season can also transfer lethal doses of the contaminant to their eggs during incubation. Even small quantities of oil (5–20 microliters) externally applied to eggs can kill embryos. Birds can also ingest, inhale, or absorb oil when exposed to a spill or while preening contaminated plumage. The toxic effects of ingested oil vary, depending on the type of oil and on the species of birds affected. These effects include gastrointestinal irritation and hemorrhaging, anemia, reproductive impairment, depressed growth, and osmoregulatory dysfunction (Table 42.1). Polycyclic aromatic hydrocarbons (PAH) contribute to the toxicity of crude petroleum and refined petroleum products, but the amounts of PAH in petroleum products vary greatly.

Unfortunately, the effects of petroleum pollution can persist long after the visible spill is cleaned or dispersed. Petroleum persistence in the water column is usually less than 6 months, but it can be much longer (more than 10 years) in other components of the environment. Chronic losses may result when birds ingest oil in contaminated food items. For example, oil from the 1989 Exxon Valdez spill is still se-

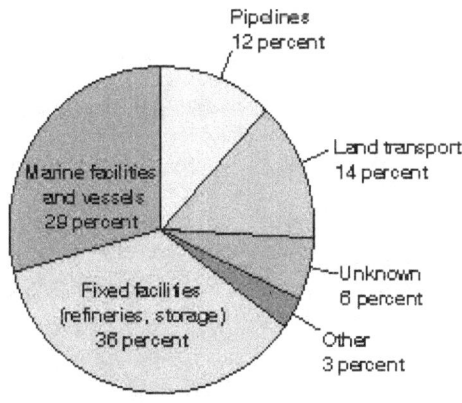

Figure 42.1 *Origin of petroleum spills, 1987–94.*

questered in bivalve communities within the areas of contamination and, thus, is still available to birds and other wildlife that feed on bivalves. Subtle effects on reproduction, such as decreased egg production, reduced fertility and hatchability, and decreased sperm production, as well as reduced immunologic function and impaired disease resistance, may occur as a result of ingesting oil-contaminated food (Table 42.1).

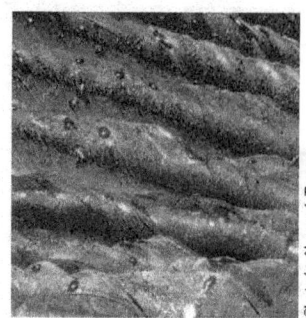

Figure 42.3 Oiling disrupts normal feather structure and function.

Table 42.1 Commonly reported effects of oil toxicosis in birds.

Impact	Consequences
Mechanical	
Loss of waterproofing and insulation value of feathers	Wetting, chilling, and hypothermia leading to death
	Exhaustion due to depletion of body stores, inability to feed, and greater expenditure of energy to maintain body heat and stay afloat
	Altered behavior
	Drowning
Toxicological	
Pathological changes in tissue	Inflammation of gastrointestinal lining
	Malformations
	Embryotoxicity
Physiologic disruptions	Altered endocrine function
	Liver and kidney disorders
	Altered blood chemistry
	Blood disorders including anemia
	Impaired salt (nasal) gland function resulting in disruption of osmoregulation
Reproductive	
	Embryotoxicity
	Impaired reproduction
Other	
	Reduced growth and development
	Reduced immunologic function
	Impaired disease resistance

Species Affected

A wide variety of birds and other wildlife have been affected by oil. The bird species affected depend on the location of the oil and the behavior of the birds. Species that suffer the greatest losses are gregarious, spend most of their time on the water, often near shipping lanes, and dive into the water to find food or to avoid disturbance. Seabirds, such as auks, guillemots, murres, puffins, sea ducks, and penguins, are particularly susceptible to contamination from oil spills (Fig. 42.4). In addition, annual losses of marine birds occur due to natural oil seeps along the Santa Barbara Channel of the California coast.

Seasonality

Species with high reproductive rates may quickly recover from a spill, but for species with low reproductive rates, such as brown pelicans, oil pollution can cause catastrophic losses and it may take decades for populations to return to prespill numbers. Even oiled brown pelicans that have been successfully rehabilitated have reduced reproductive success.

Winter storms increase the likelihood of transport spills, making January, February, and March the peak spill season. This is also the time of year when seabirds and waterfowl congregate in wintering areas, resulting in an increased potential for significant bird losses.

Sea and bay ducks (scoters, scaups, oldsquaws, canvasback) that tend to concentrate on wintering grounds and diving birds (grebes, loons, and mergansers) that overwinter in marine environments or on large water bodies with commercial shipping are quite vulnerable to oil pollution, especially during winter months. Eiders are vulnerable most of the year.

Distribution and Extent of Mortality

The oiling of migratory birds is not limited to specific geographic areas. Accidental oil spills have occurred in all 50 States including inland waters, such as rivers and nonnavigable waters, and in open coastal waters, ports and harbors. Although it is not possible to accurately estimate the number of birds lost to oil pollution, in many cases the mortality has been substantial (Table 42.2). Bird losses of 5,000 or more are common for larger oil spills. Reports are usually of the numbers of oiled birds found dead or moribund on the shore, but these estimates may be inaccurate because of search biases, accessibility of the shore, losses of birds that have sunk to the bottom, and other factors. An important source of error in estimating losses in marine environments is the unknown proportion of oiled birds that die at sea but that do not reach the coast.

In addition to accidental spills, other opportunities for animal exposure to oil occur in association with oil production, petroleum refining, and highly industrialized locations throughout the United States. Persistent oil pollution is a chronic problem around marinas and ports due to discharges from shipping and boating activities and storage tank clean-

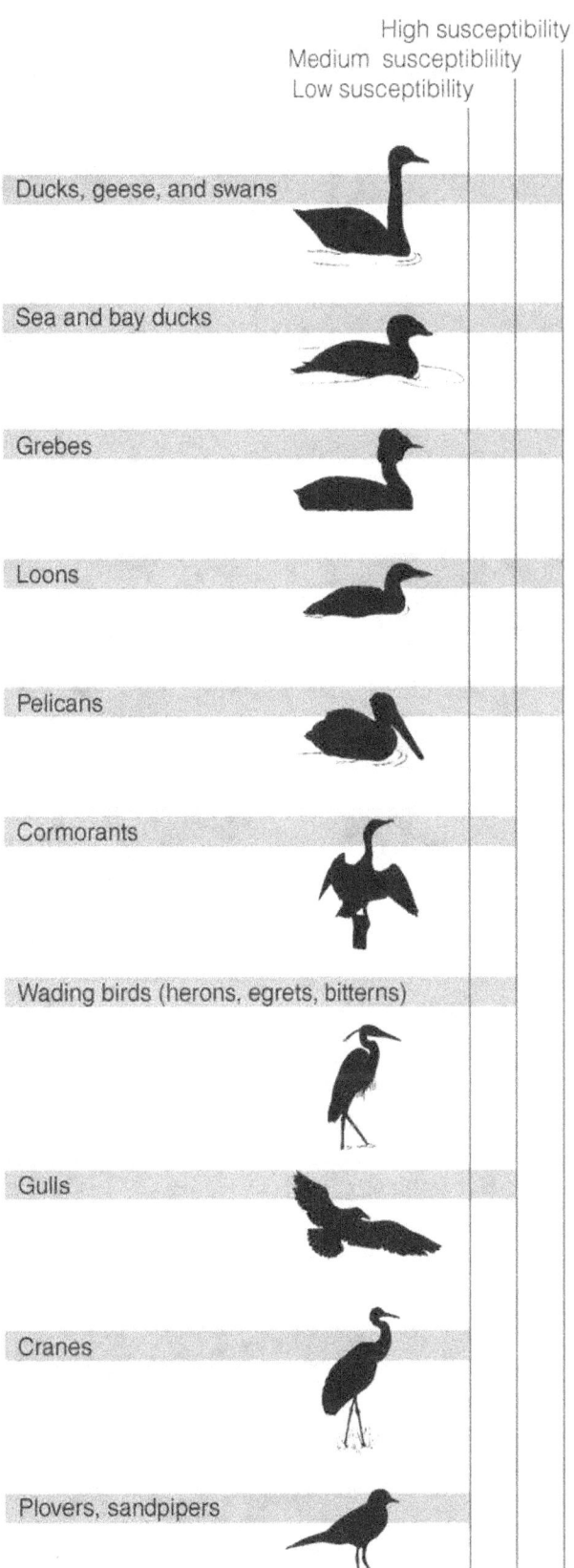

Figure 42.4 Relative susceptibility of birds to oiling.

Table 42.2 *Examples of bird mortality from oil spills.*

Vessel or source	Year	Site	Estimated bird mortalities
Exxon Valdez	1989	Prince William Sound, Alaska	350,000–390,000
Nestucca	1988	Grays Harbor, Washington	50,000
Amoco Cadiz	1978	Brittany, France	20,000
Barge STC-101	1976	Chesapeake Bay, Virginia	20,000–50,000
Torrey Canyon	1967	English Channel	30,000
Gerda Maersk	1955	Elbe River, Germany	500,000

Figure 42.5 *These oiled birds were recovered from oil-field wastewater pits in the southwestern United States.*

Figure 42.6 *Oiled birds become wet and chilled because oil damages feather waterproofing and insulating properties.*

ing, but, unfortunately, the numbers of birds affected by oil pollution in these areas are unknown. In the Playa Lakes regions of eastern New Mexico, northwestern Texas, and western Oklahoma, open pits and tanks containing oil and oil-field wastes have been reported to claim the lives of approximately 100,000 birds each year (Fig. 42.5).

Field Signs

Major oil spills are frequently accompanied by intensive media coverage, and they may be well publicized before slicks or affected birds appear. However, small spills, especially those of unknown origin, often go unnoticed except for the appearance of a few contaminated birds. Oiled birds are frequently wet and chilled because the oil damages feather waterproofing and insulating properties (Fig. 42.6); birds may ride lower in the water than normal because they have lost feather buoyancy. Oiling is suggested when water birds leave the water for islands, rocks, pilings, and other surfaces because they are chilled (Fig. 42.7). Birds that survive for 48 hours or more after oiling are often thin, and even close to starvation, because they have stopped feeding and are using first body fat and then muscle tissue to produce heat in response to chilling.

Matting of the feathers occurs from external oiling. Oil can usually be seen or smelled on the feathers, but some light, transparent oils may be difficult to detect. One useful technique for detecting oiling is to place a few feathers from the bird in a pan of water and watch for an oil sheen to appear (Fig. 42.8). An enzyme-linked immunosorbent assay (ELISA), which detects PAH in oil, can provide quick confirmation of the presence of petroleum products on fur or feathers.

Gross Lesions

Necropsy findings of birds that die from oil exposure are highly variable. Birds are often emaciated, and oil may be present in their trachea, lungs (Fig. 42.9), digestive tract, and around the vent. The lining of the intestine may be reddened, or the intestine may contain blood. The salt glands, which are located over the eyes, may appear swollen (Fig. 42.10), and the adrenal glands may be enlarged. A variety of other changes in the normal appearance of tissues and organs may also be present, but no specific or consistent lesion is typical in animals that are exposed to oil.

Diagnosis

Diagnosis of oiling is seldom a problem; visible oil on the bird or in the environment usually suffices (Fig. 42.11). However, proving that oil has caused mortality is more complex. For damage assessments and cause-of-death determinations, it must be determined that oiling did not occur after the death of the animals in question.

Chemical analyses of tissues or eggs are difficult to use for diagnosis because the chemical composition of petroleum products is complex. Therefore, good background information and field observations are an integral part of specimen submission to diagnostic laboratories (see Chapter 1, Recording and Submitting Specimen History Data). Submit whole carcasses whenever possible.

Control

Treatment of oil spills within the States, territorial possessions, and territorial waters of the United States is legislatively mandated by the Oil Pollution Act of 1990. The Act mandates the inclusion of a fish and wildlife response plan within the National Contingency Plan and the creation of Area Contingency Plans. These plans provide for an integrated response to a spill with assigned agency responsibilities for protecting fish and wildlife and environmental cleanup.

In the event of a spill, contact the National Response Center at the 24-hour, toll free number 1-(800)-424-8802. The National Response Center will advise the responsible agencies (Coast Guard, Environmental Protection Agency,

Figure 42.7 Common murre out of water due to oiling.

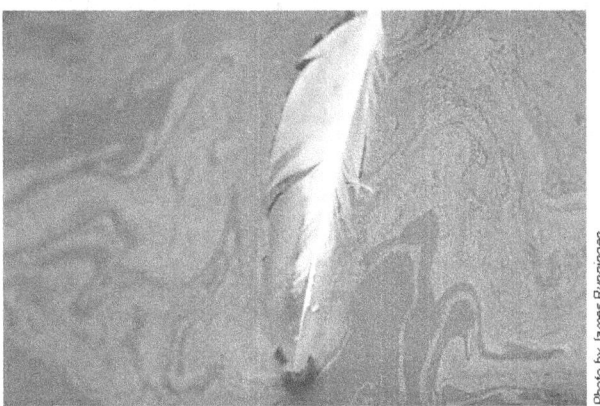

Figure 42.8 If external oiling is suspected, place feathers on water and watch for oil sheen.

Figure 42.9 In severe cases, oil may be inhaled and may discolor the lungs, such as in this Canada goose.

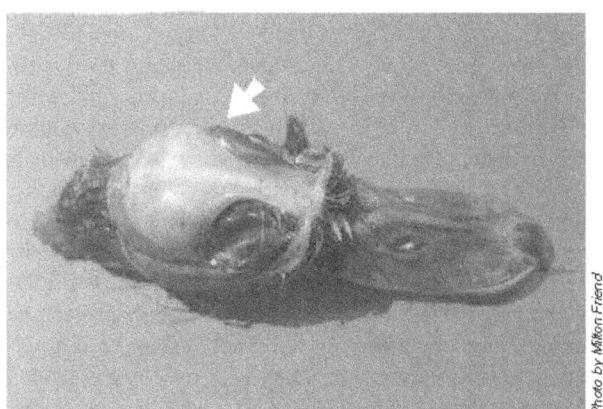

Figure 42.10 Swollen salt glands.

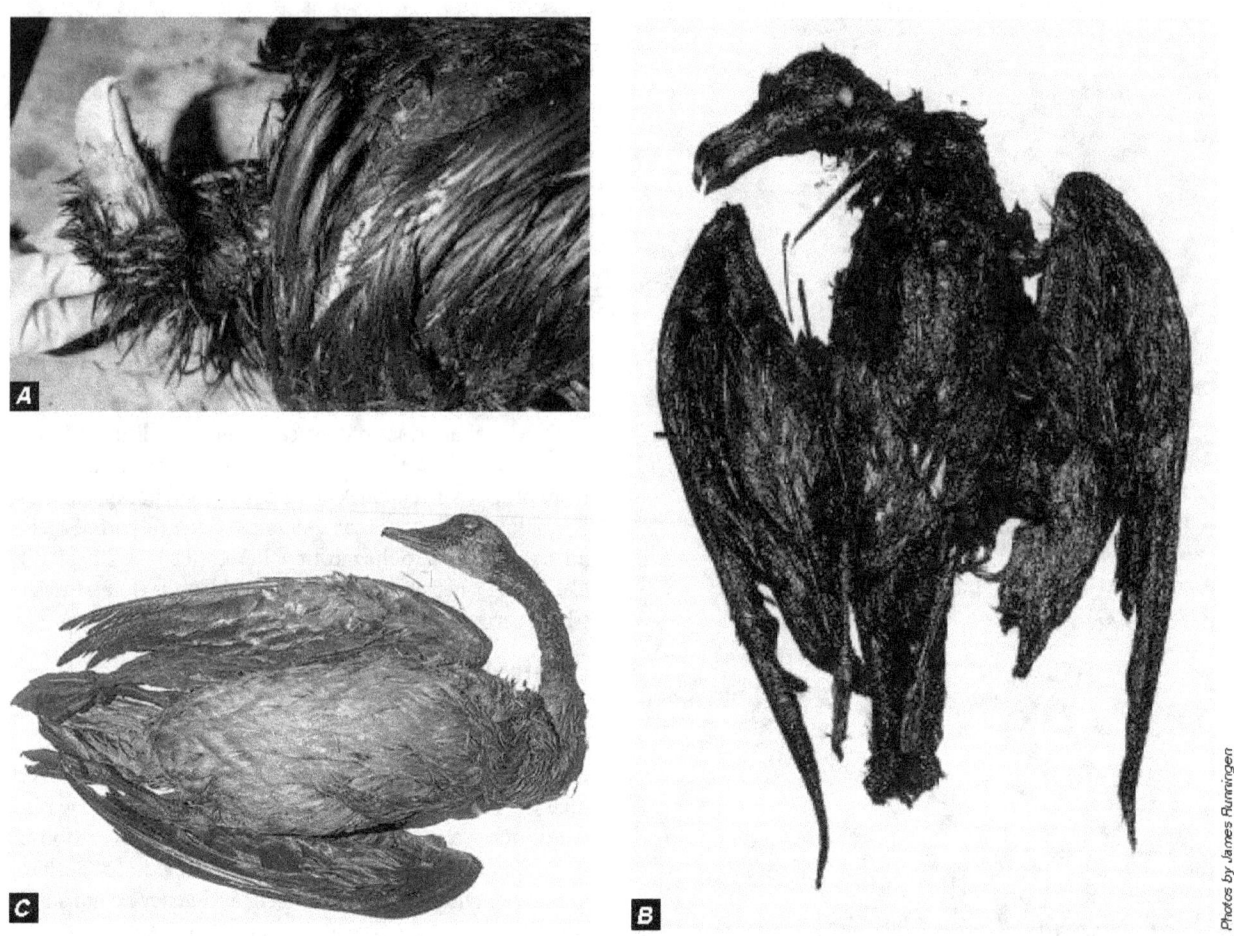

Figure 42.11 Diagnosis of oiling is facilitated when oil is plainly visible on the carcass, such as on this bald eagle (A), herring gull (B), and Canada goose (C).

and natural resource trustees) who will then respond to the event. In some States, notably California, State agencies may have lead responsibility for oil spills.

Cleaning oiled birds may not be justified on a "population" basis, but it is desired by the public, required by both State and Federal laws, and warranted when rare, threatened, or endangered species are involved. Contingency plans that were developed under the Oil Pollution Act address wildlife rehabilitation. Do not attempt to rehabilitate oiled animals without knowledge of cleaning techniques. For situations that do not require a response mandated by the Act, obtain advice from State wildlife resource agencies and the private sector (Table 42.3).

Scaring devices and other forms of disturbance can be used to discourage bird use of oil-polluted areas. If a polluted area is being used or is likely to be used by endangered species, it may be helpful to initiate actions that will attract the birds to other locations while the spill is contained and cleaned. All actions taken, including wildlife rehabilitation, should be in concert with those mandated to address oil spills.

Human Health Considerations

Direct contact with petroleum, handling oiled wildlife, and activities associated with the cleanup are all potentially hazardous to humans. Health impacts due to the toxic effects of petroleum include contact dermatitis, increased skin cancer risk, eye irritation, and problems associated with inhaling volatile components of petroleum products. These products may be contaminated with other chemicals including polychlorinated biphenyls (PCBs) and organophosphates. Wear protective clothing to prevent direct exposure of oil to skin surfaces. Preventing injuries during spill containment and cleanup requires a cool head, advice from experts, and close supervision of workers — especially volunteers. Two major concerns are drowning and hypothermia.

Workers should not enter the water, climb slippery cliffs, or put themselves in hazardous situations to rescue birds. Also, the birds themselves present a hazard. Many sea birds have sharp, "spearing" beaks and often aim for the eyes of their predators — and their caretakers. Always wear goggles when handling these birds.

Table 42.3 Sources of information for rehabilitation of oiled birds.

Many individuals and groups have expertise in the rehabilitation of oiled birds and other wildlife. The following are major programs that conduct this type of activity.

Program and address	Telephone
Tri-State Bird Rescue and Research, Inc. 110 Possum Hollow Rd., Newark, DE 19711	302-737-9543
California Department of Fish and Game Office of Oil Spill Prevention and Response Oiled Wildlife Care Network Wildlife Health Center University of California, Davis, CA 95616	530-752-4167
International Bird Rescue Research Center 699 Potter St., Berkeley, CA 94710	510-841-9086

Tonie E. Rocke

Supplementary Reading

Albers, P.H., 1995, Oil, biological communities and contingency planning, *in* Fink, L., and others, eds., Wildlife and oil spills: response, research, and contingency planning: Hanover, Pa., The Sheridan Press, p. 1–10.

Albers, P.H., 1995, Petroleum and individual polycyclic aromatic hydrocarbons, *in* Hoffman, D.J., and other, eds., Handbook of Ecotoxicology: Boca Raton, Fla., Lewis Publishers, p. 330–355.

Bourne, W.R.P., 1976, Seabirds and pollution *in* Johnson, R., ed., Marine Pollution: London, Academic Press, p. 403–502.

Burger, A.E., 1993, Estimating the mortality of seabirds following oil spills: effects of spill volume. Marine Pollution Bulletin, v. 26, p. 140–143.

Flickinger, E.L., 1981, Wildlife mortality at petroleum pits in Texas: Journal of Wildlife Management, v. 45, p. 560–564.

Hoffman, D.J., 1990, Embryotoxicity and teratogenicity of environmental contaminants to bird eggs: Reviews of Environmental Contamination and Toxicology, v. 115, p. 39–89.

Chapter 43
Lead

Synonym
Plumbism

Lead poisoning of waterfowl is neither a new disease nor a subject without controversy. The use of lead shot for waterfowl hunting within the United States has been prohibited and efforts are underway to ban the use of lead fishing sinkers and prohibit the use of lead shot for nonwaterfowl hunting. The first documented reports within the United States of lead-poisoned waterfowl were from Texas in 1874. Numerous other reports and studies added to those findings during the years and decades that followed. However, strong opposition to nontoxic shot requirements prevented full implementation of them until 1991. A full transition to nontoxic shot shells for all hunting and to nontoxic fishing sinkers and jig heads for fishing within the United States will not happen easily. The continued use of lead shot and lead fishing weights and the large amounts of these materials previously deposited in environments where birds feed assure that lead poisoning will remain a common bird disease for some time.

Cause
Lead poisoning is an intoxication resulting from absorption of hazardous levels of lead into body tissues. Lead pellets from shot shells, when ingested, are the most common source of lead poisoning in birds. Other far less common sources include lead fishing sinkers, mine wastes, paint pigments, bullets, and other lead objects that are swallowed.

Species Affected
Lead poisoning has affected every major species of waterfowl in North America and has also been reported in a wide variety of other birds. The annual magnitude of lead poisoning losses for individual species cannot be precisely determined. However, reasonable estimates of lead-poisoning losses in different waterfowl species can be made on the basis of mortality reports and gizzard analyses. Within the United States, annual losses from lead poisoning prior to the 1991 ban on the use of lead shot for waterfowl hunting were estimated at between 1.6 and 2.4 million waterfowl, based on a fall flight of 100 million birds. Followup studies have not been conducted since the ban on lead shot to determine current losses from lead poisoning. This disease still affects waterfowl and other species due to decades of residual lead shot in marsh sediments, continued deposition from allowable use of lead shot during harvest of other species, noncompliance with nontoxic shot regulations, target shooting over areas where birds may feed, and from other sources of lead.

Lead poisoning is common in mallard, northern pintail, redhead, and scaup ducks; Canada and snow geese; and tundra swan. The frequency of this disease decreases with increasing specialization of food habits and higher percentages of fish in the diet. Therefore, goldeneye and merganser ducks are seldom affected (Fig. 43.1). A surprising recent finding has been lead poisoning in spectacled and common eiders on their Alaskan breeding range, where the intensity of hunting is far less than in the contiguous 48 States. These findings demonstrate that lead poisoning can afflict birds even without heavy hunting pressure. Among land birds, eagles are most frequently reported dying from lead poisoning (Fig. 43.2). Lead poisoning in eagles and other raptors generally is a result of swallowing lead shot embedded in the flesh of their prey. With the exception of waterfowl and raptors, lead poisoning from ingesting lead shot is generally a minor finding for other species (Table 43.1). However, lead poisoning has been reported in partridge, grouse, and pheasants subjected to intensive shooting in uplands of Europe. Lead poisoning in pheasants in Great Britain was reported as early as 1875.

Lead poisoning due to ingesting lead fishing weights has been reported in numerous species. The greatest number of reports are from swans as a group, common loon, brown pelican, Canada goose, and mallard duck (Fig. 43.3). Laysan albatross chicks on Midway Atoll suffer high lead exposures and mortality from ingesting lead-laden paint chips flecking off of vacant military buildings (Fig. 43.4).

Distribution
Losses occur coast-to-coast and border-to-border within the United States. Documented lead poisoning in birds varies widely between States and does not necessarily reflect true geographic differences in the frequency of occurrence of this condition. For example, although the geographic distribution of lead poisoning in bald eagles is closely associated with their wintering areas, the number of lead poisoning cases from Wisconsin and Minnesota is disproportionately high. Because submission of bald and golden eagles for examination from different areas is highly variable, no direct comparison can be made between States regarding the number of lead-poisoned eagles (Fig. 43.5A). The reported distribution of lead poisoning in eagles and waterfowl depends

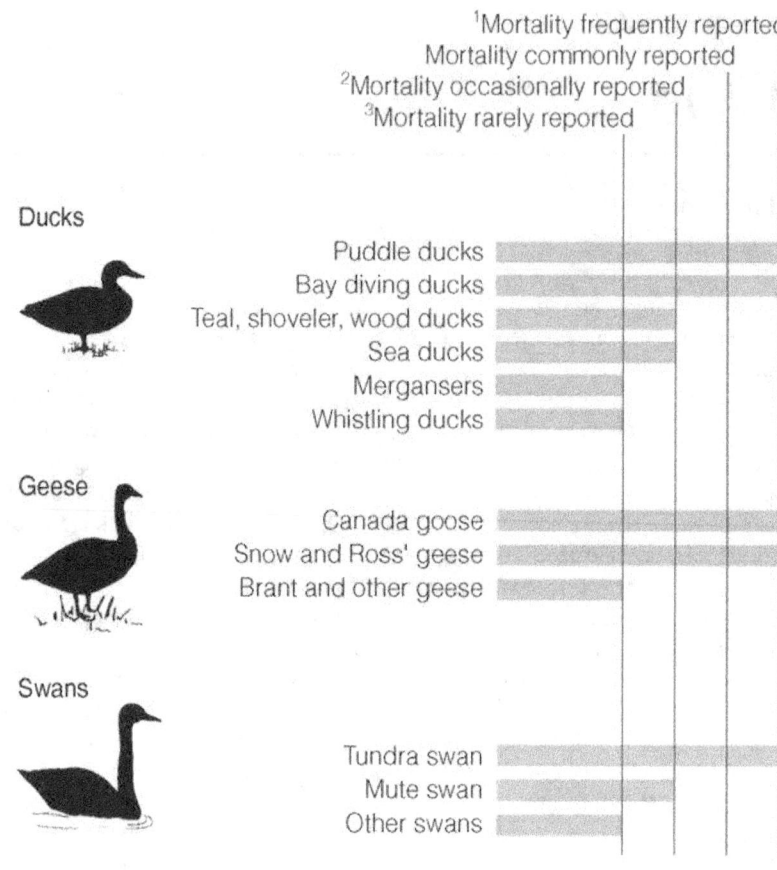

Figure 43.1 Relative occurrences of lead shot poisoning in North American waterfowl.

on the numbers of birds submitted for complete disease diagnostic evaluations. In areas where few birds are examined, the frequency of lead poisoning and other diseases will be underestimated. Even where many bird carcasses are adequately evaluated, the number of diagnoses made reflects minimum numbers of lead-poisoning cases. The general distribution of this disease in waterfowl on the basis of lead shot-ingestion surveys and documented mortality prior to nontoxic shot requirements is shown in Fig. 43.5B.

Lead poisoning has also been reported as a cause of migratory bird mortality in other countries (Fig. 43.6). Several of these countries have implemented nontoxic shot requirements and several others are beginning to address this issue.

Seasonality

Birds can can die from lead poisoning throughout the year, although birds are most often poisoned by lead after the waterfowl hunting season has been completed in northern areas and during the later part of the season in southern areas of the United States. January and February are peak months for cases in tundra swans, Canada geese, and puddle ducks. Spring losses are more commonly reported for diving ducks. Tundra swans are also frequently lead poisoned during spring migration.

Field Signs

Lead-poisoned waterfowl are often mistaken for hunting season cripples. Special attention should be given to waterfowl that do not take flight when the flock is disturbed and to small groups of waterfowl that remain after most other birds of that species have migrated from the area. Lead-poisoned birds become reluctant to fly when approached and those that can still fly are often noticeably weak flyers — unable to sustain flight for any distance or flying erratically

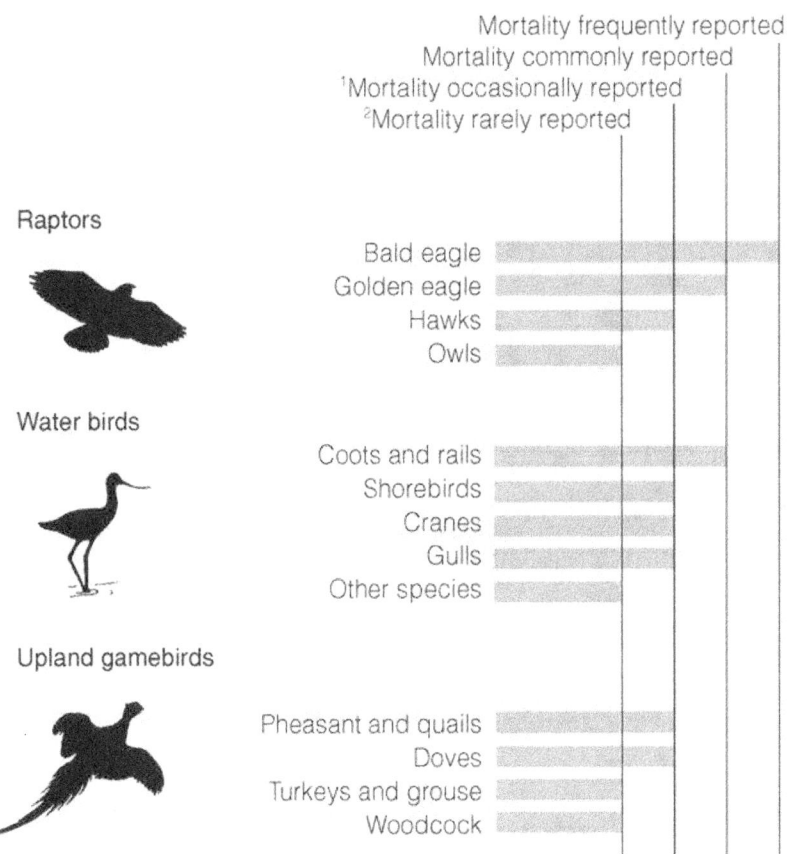

[1]Lead shot ingestion studies generally indicate low levels of exposure to lead shot
[2]Lead shot ingestion studies genreally indicate little or no lead shot ingestion

Figure 43.2 *Relative occurrence of lead shot poisoning in groups of birds other than waterfowl.*

and landing poorly. Birds that attempt to escape pursuit by running may exhibit an unsteady gait. In lead-poisoned Canada geese, the head and neck position may appear "crooked" or bent during flight; a marked change in the tone of call is also sometimes evident in this species. As the disease progresses and waterfowl become flightless, the wings are held in a characteristic "roof shaped" position (Fig. 43.7), which is followed by wing droop as the birds become increasingly moribund (Fig. 43.8). Fluid may discharge from the bill and often a bird may not attempt to escape in the presence of humans.

Lead-poisoned waterfowl are easily captured during advanced stages of intoxication (Fig. 43.9). Because severely affected birds generally seek isolation and protective cover, well-trained retrieving dogs can help greatly to locate and collect these birds. An abundance of bile-stained feces on an area used by waterfowl (Fig. 43.10) is suggestive of lead poisoning and warrants ground searches even if other field signs have not been observed. Green-colored feces can also result from feeding on green wheat and other plants, but the coloration is somewhat different.

Gross Lesions

Lead-poisoned waterfowl are often emaciated because of the prolonged course of the illness and its impact on essential body processes. Therefore, many affected birds appear to be starving; they are light in weight, have a "hatchet-breast" appearance (Fig. 43.11), and the undersurface of their skin is devoid of fat (Fig. 43.12). The vent area of these birds is often stained with a bright green diarrhea (Fig. 43.13). The heads of Canada geese may appear puffy or swollen because serum-like fluids accumulate in the tissues of the face (Fig. 43.14).

Lesions observed at necropsy of lead-poisoned birds that

Table 43.1 Documented North American cases of lead poisoning in free-ranging nonwaterfowl species.

Nonendangered species			
Upland gamebirds			
Ring-necked pheasant	Hungarian partridge	Bobwhite quail	Scaled quail
Wild turkey	Mourning dove		
Raptors			
Golden eagle	Northern harrier	Rough-legged hawk	
Red-tailed hawk	Prairie falcon	Turkey vulture	
Wetland birds			
Common loon	Double-crested cormorant	Greater sandhill crane	Lesser sandhill crane
White pelican	American coot	Royal tern	Flamingo
Great blue heron	White ibis	Great egret	Snowy egret
Sora rail	American avocet	Black-necked stilt	Marbled godwit
Pectoral sandpiper	Western sandpiper	Long-billed dowitcher	Laughing gull
Herring gull	Glaucous-winged gull	California gull	Laysan albatross[1]
Endangered species			
California condor	Brown pelican	Whooping crane[2]	
Bald eagle	Mississippi sandhill crane	Peregrine falcon	

[1] The cause of poisoning was ingestion of paint chips rather than lead shot, bullets, or fishing tackle.
[2] The cause of poisoning was particulate lead of unknown origin but not lead shot or fishing tackle.

have died after a prolonged illness generally consist of the following:

1. Severe wasting of the breast muscles (Fig. 43.11).
2. Absent or reduced amounts of visceral fat (Fig. 43.12).
3. Impactions of the esophagus or proventriculus in approximately 20–30 percent of affected waterfowl. These impactions may contain food items, or combinations of food, sand, and mud. The extent of impaction may be restricted to the gizzard and proventriculus, extend to the mouth, or lie somewhere in between (Fig. 43.15).
4. A prominent gallbladder that is distended, filled with bile, and dark or bright green (Fig. 43.16).
5. The normally yellow gizzard lining is discolored a dark or bright green (Fig. 43.17). Gizzard contents are also often bile-stained.
6. Lead pellets or small particles of lead are often present among gizzard and proventricular contents. Pellets that have been present for a long time are well worn, reduced in size, and disk-like rather than spherical (Fig. 43.18). Careful washing of contents is required to find smaller lead fragments. X-ray examination is often used to detect radiopaque objects in gizzards, but recovery of the objects is necessary to separate lead from other metals. Flushing contents through a series of progressively smaller sieves is one method of pellet recovery.

Less obvious pathological changes include wasting of internal organs such as the liver, kidneys, and spleen; areas of paleness in the heart muscle; a flabby-looking heart; and paler-than-normal-looking internal organs and muscle tissue.

The above field signs and gross lesions provide a basis for a presumptive diagnosis of lead poisoning. However, none of these signs or lesions is diagnostic by itself and all can result from other causes. Also, many of the above signs and lesions are absent in birds that die acutely following an overwhelming lead exposure.

Diagnosis

A definitive diagnosis of lead poisoning as a cause of death is based on pathological and toxicological findings supplemented by clinical signs and field observations. The presence or absence of lead shot or lead particles in the gizzard contents is useful information and should be recorded, but it is not diagnostic. The liver or kidneys are the tissues of choice for toxicology analysis, with liver tissue being more commonly used. If you suspect lead poisoning and cannot submit whole birds to the diagnostic laboratory, remove the liver or kidney tissue, wrap the specimens separately in aluminum foil, and freeze them until they are submitted for analysis. Collect the entire liver or one entire kidney. However,

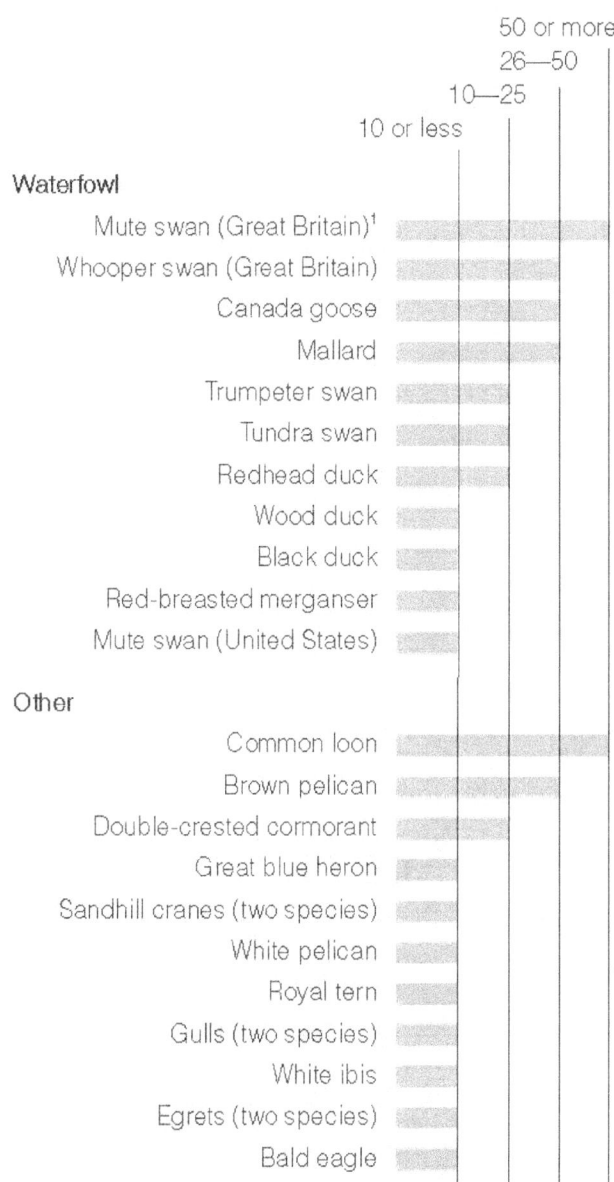

[1] Great Britain banned the use of lead sinkers in 1987.

Figure 43.3 Number of reported lead poisoning occurrences following ingestion of lead sinkers and jigs through 1994.

Figure 43.4 The droopy wings and unthrifty appearance of this Laysan albatross chick are the result of lead poisoning caused by ingestion of lead-laden chips that flecked off abandoned buildings. The paint had high concentrations of lead.

Lead 321

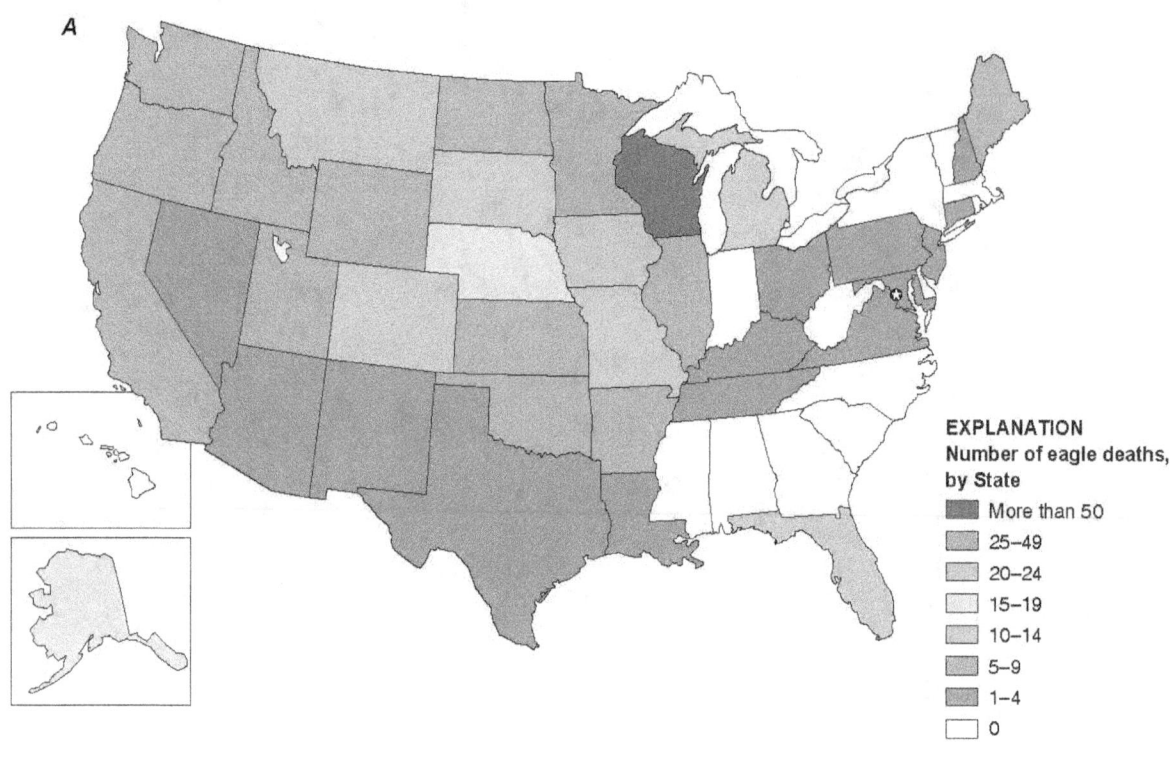

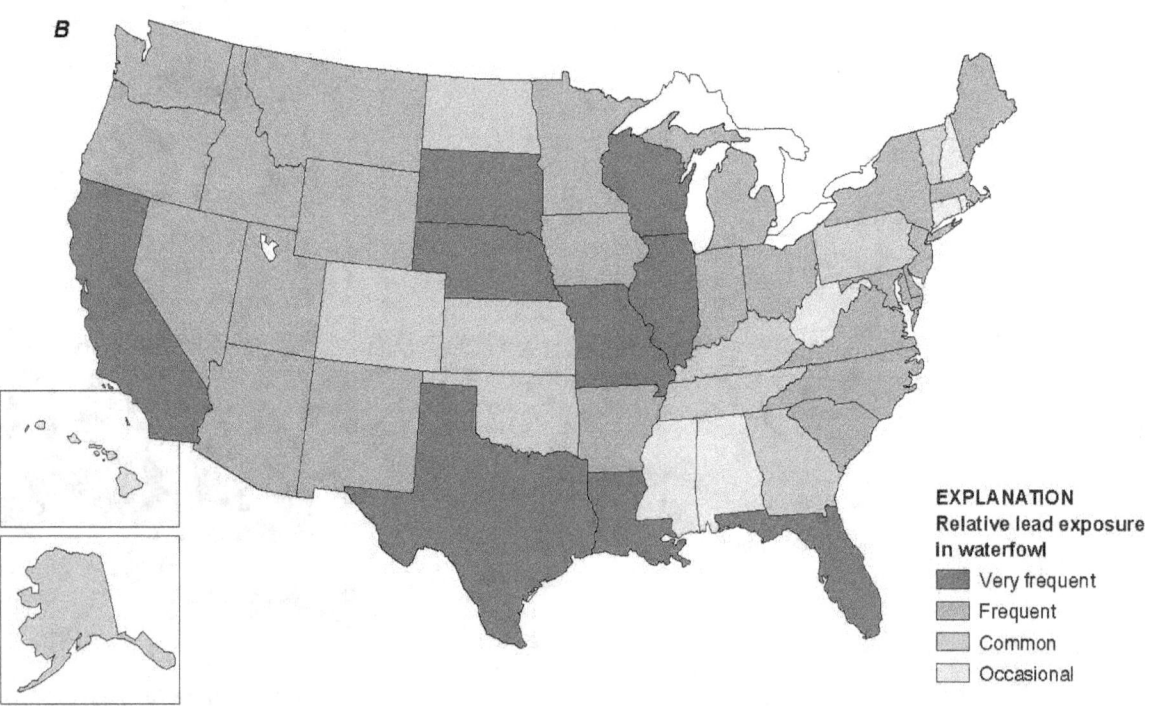

Figure 43.5 (A) Diagnosed cases of lead poisoning in bald eagles though mid-April, 1996. (B) Relative occurrence of lead exposure in waterfowl prior to the 1991 ban on use of lead shot for waterfowl hunting. Evaluation is based on gizzard analysis and reported mortality.

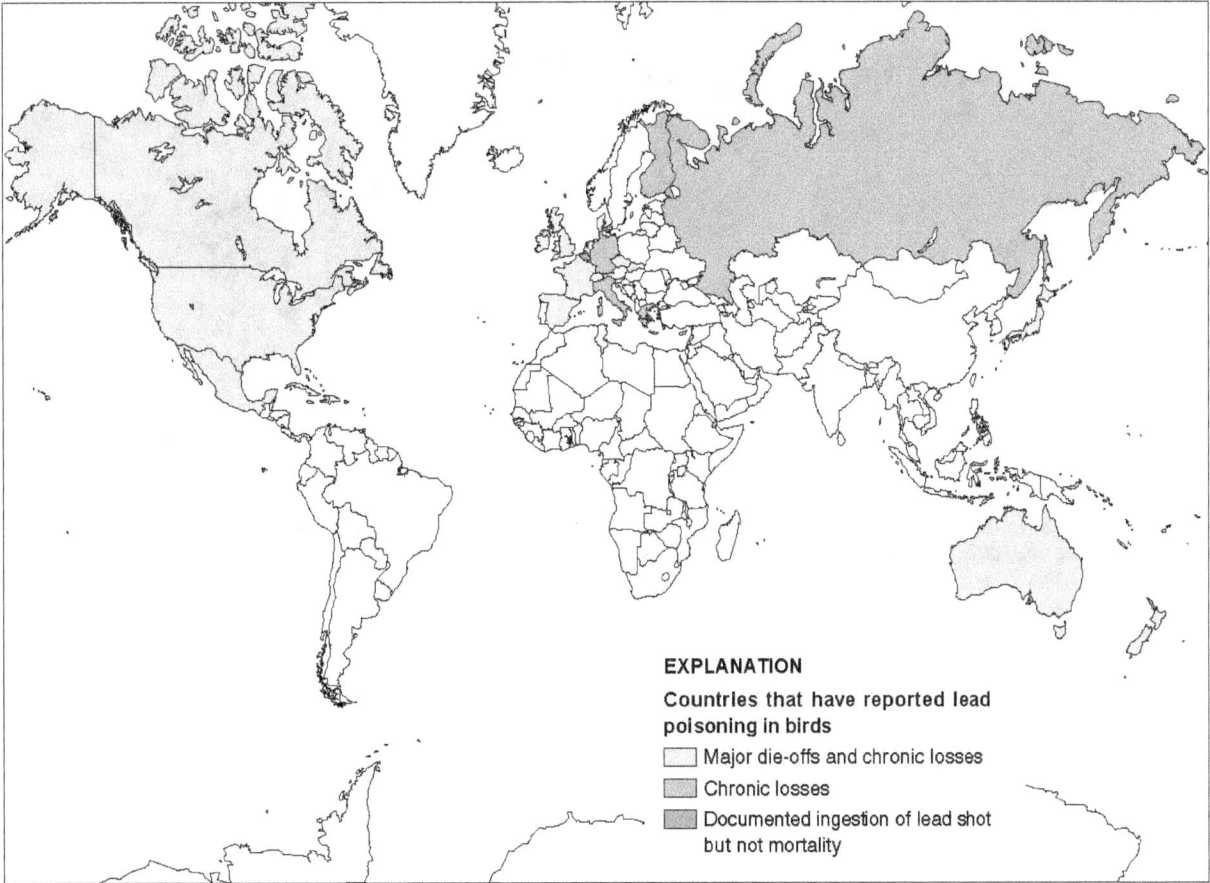

Figure 43.6 Countries that have reported lead poisoning in birds.

because toxicology is but one aspect of reaching a lead-poisoning diagnosis, make every effort to submit whole birds for analysis. Liver lead values of 6–8 parts per million or higher on a wet-weight basis or 20–30 parts per million on a dry-weight basis are suggestive of lead poisoning when other signs of lead poisoning are present.

Lead levels in populations of live birds can be evaluated by using whole blood. Collect a minimum of 2–5 milliliters of blood in lead-free tubes containing an anticoagulant such as sodium citrate or EDTA. Evidence of lead exposure can also be obtained through indirect measurements involving blood enzymes. Measurement of protoporphyrin IX in red blood cells is the most popular assay because only a few drops of blood are needed and testing is inexpensive once appropriate instrumentation is obtained. Elevated blood protoporphyrin levels are correlated with lead exposure and serve as a sensitive screening assay, but they do not provide direct measurement of the amount of lead in blood. This technique has its greatest value in identifying populations from which more direct measurements should be taken and for screening blood samples to determine which should be tested for blood lead concentrations. Confirm correct procedures for collecting blood samples for lead analysis with the diagnostic laboratory before collecting the samples. Keep blood samples chilled until submitting them for analysis, regardless of the assay that will be used. Write the date and time of collection on the tube along with the specimen number and other information identifying the sample and its origin.

The diagnosis of lead poisoning as a disease or poisoning syndrome, but not as a cause of death, can be made from tissue residues alone when there are sufficient residue data for the species in question or closely related species. The amount of tissue residue variability that exists between species can be considerable and it is also influenced by the route of lead exposure such as ingestion vs. inhalation (Fig. 43.19). For example, rock doves (pigeon) are highly resistant to high concentrations of lead when they are compared with other birds, but most lead exposure in rock doves is from automobile emissions in cities. Rock doves that have ingested lead shot have greatly increased tissue lead levels, can exhibit behavioral changes consistent with lead toxicity in other species, and can die from the toxic effects of lead.

Figure 43.7 Characteristic "roof-shaped" position of the wings in (A) a lead-poisoned mallard (leading bird) and (B) a snow goose.

Figure 43.8 Wing droop in a tundra swan in advanced stages of lead intoxication.

Figure 43.9 Inability of these lead-poisoned Canada geese to escape capture by humans illustrates their great vulnerability to predation.

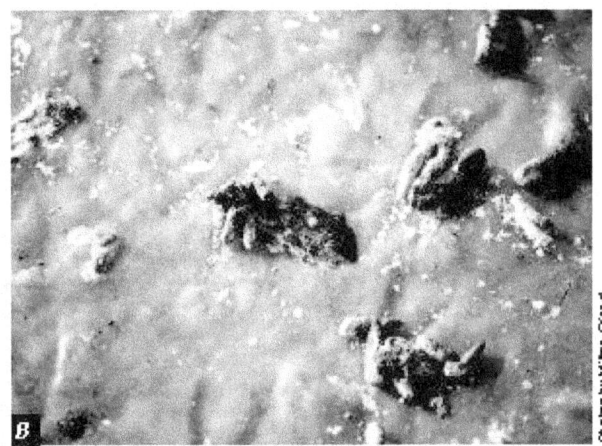

Figure 43.10 Waterfowl feces provide presumptive evidence of lead poisoning. Examination of **(A)** feces where waterfowl are concentrating and **(B)** observations of an abundance of bright green-colored feces should be reason to search for sick birds and carcasses.

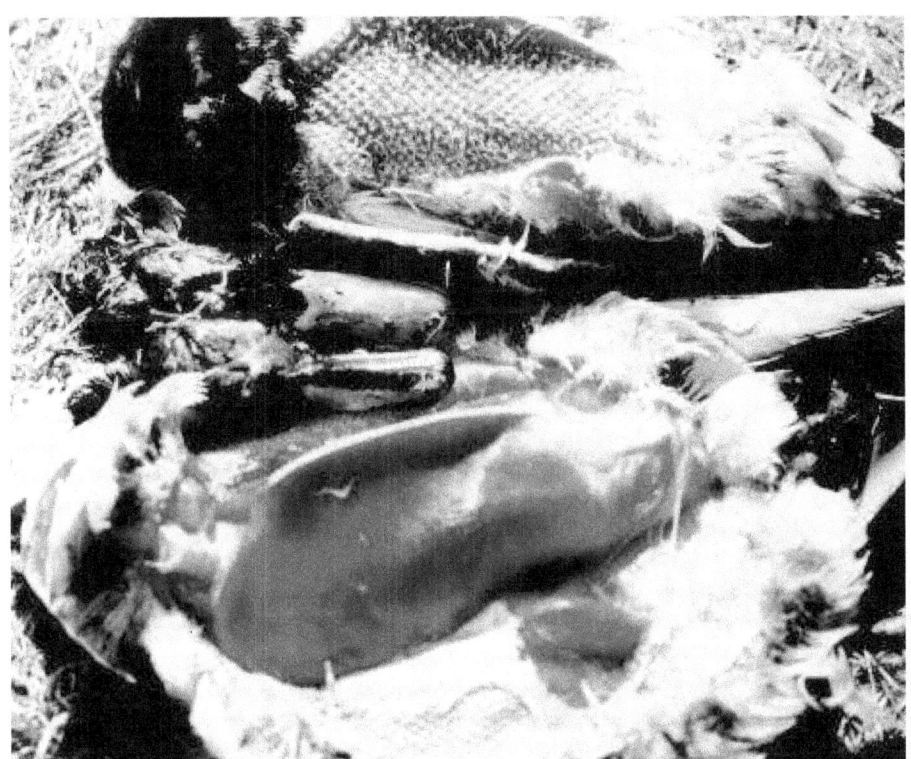

Figure 43.11 "Hatchet-breast" appearance of a lead-poisoned mallard (top bird) and northern pintail. The skin has been removed from the breast of the pintail to further illustrate the severe loss of muscle tissue.

Figure 43.12 Loss of subcutaneous fat is often extreme in lead-poisoned birds. **(A)** The undersurface of the skin of this pintail is totally devoid of fat, in contrast with **(B)** the abundance of yellow fat present in the mallard (bottom bird) that had died of avian cholera. Note also the absence of fat in the visceral area and along the knees of the northern pintail (top bird) in comparison with the mallard.

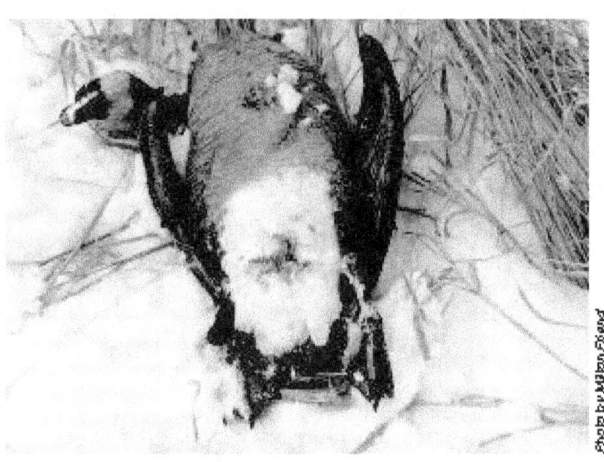

Figure 43.13 Bright green staining of the vent area is often indicative of lead poisoning.

Figure 43.14 The heads of lead-poisoned Canada geese often appear puffy or swollen.

Figure 43.15 Examples of impactions in lead-poisoned birds. **(A)** Impaction of corn in digestive tract of a hen mallard, extending from the gizzard to the mouth; **(B)** snow goose with an impaction of grasses. **(C)** Tundra swan with impaction of grasses and some seeds, extending from the mouth to the gizzard; and **(D)** a more limited impaction in a drake mallard.

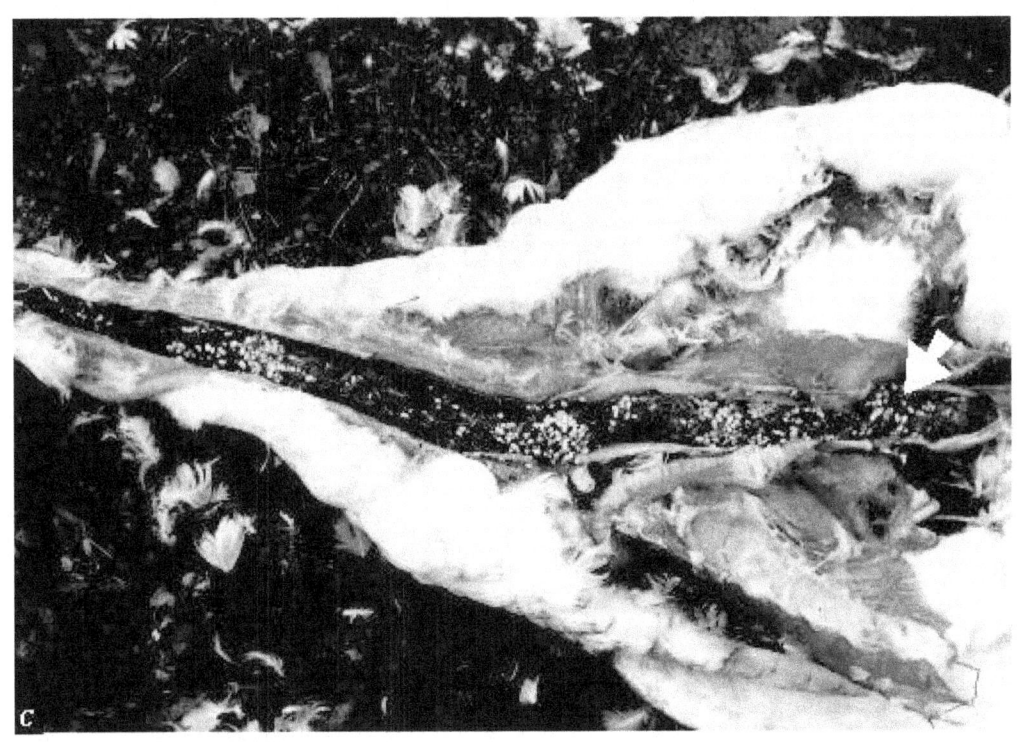

Figure 43.16 The gallbladder (top arrow) of lead-poisoned birds is often distended and filled with bright green bile. Note also the lead shot present in the gizzard (bottom arrow) of this bird.

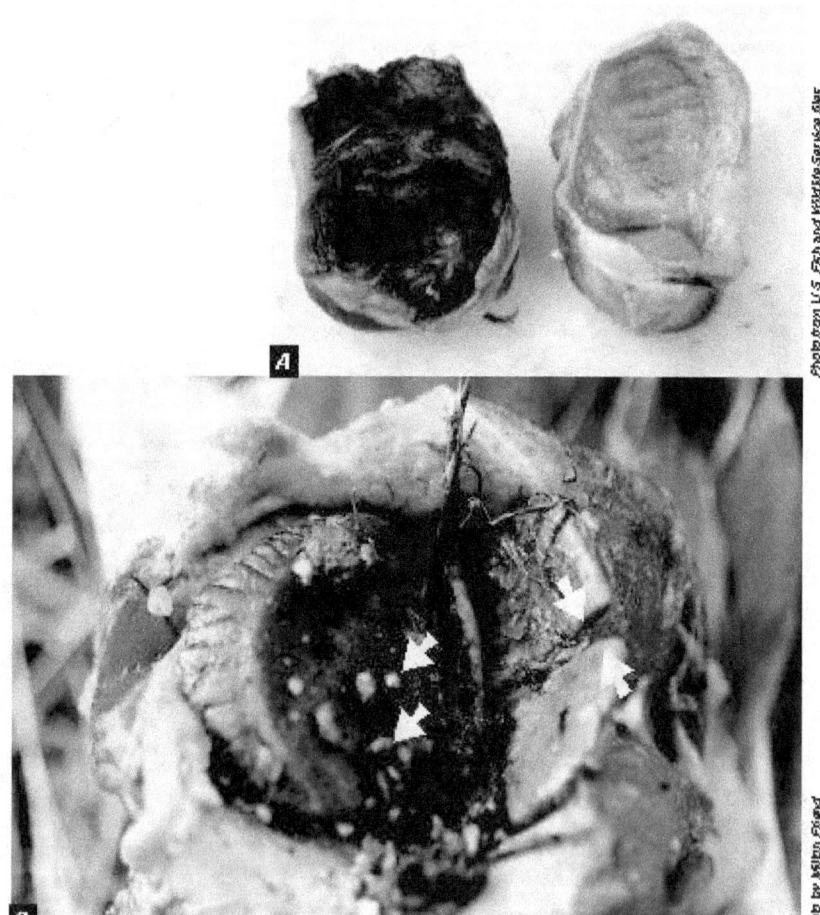

Figure 43.17 **(A)** Comparison between the appearance of the gizzard lining of a lead-poisoned mallard (left) and a normal mallard (right). **(B)** Pathological changes in the gizzard of a lead-poisoned bird. Note green-stained coloration and hard appearance of tissue. The gizzard lining has split (arrow) because the tissue has become so brittle. Note also the presence of lead shot among the grit in the center of the pad.

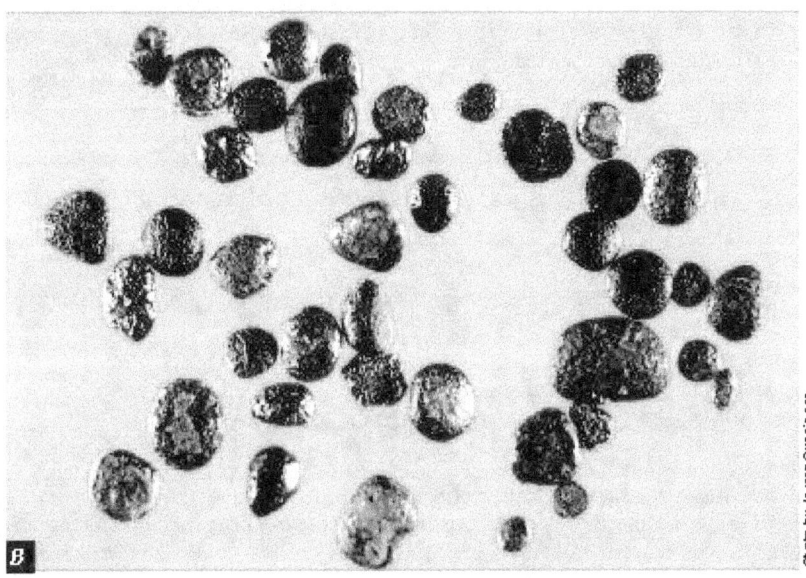

Figure 43.18 (A) Lead shot in Canada goose gizzard. Note the presence of corn. Corn and other cereal grains intensify the toxicity of lead. (B) Lead shot, originally spherical, that have been worn down in the waterfowl gizzard. Note the flattened, disk-like shape of many of these pellets.

Control

Two actions can often be taken to reduce the magnitude of mortality from lead poisoning when die-offs occur: denying birds use of problem areas, and rigorous pickup and proper disposal of dead and moribund birds.

Denying birds use of problem areas requires knowing where the birds are picking up the lead. This is complicated by the fact that signs of intoxication may not appear until 1 week after lead ingestion, and birds may not start dying until 2–3 weeks after lead ingestion. Habitat modification of contaminated areas is also useful in some instances, but differences in feeding habits must be considered. For example, placing additional water on an area may protect puddle ducks from reaching lead shot on the bottom of wetlands, but this may create an attractive feeding area for diving ducks. Similarly, draining an area may prevent waterfowl from using an area and ingesting shot, but it may create an attractive feeding area for shorebirds or pheasants. Therefore, control plans must consider the broad spectrum of wildlife likely to use the area at the time action will be taken. Rigorous pickup and proper disposal of lead-contaminated waterfowl carcasses is required to prevent raptors and other scavenger species from ingesting them. The high percentage of waterfowl with embedded body shot provides a continual opportunity for lead exposure in raptors that far exceeds the opportunity for ingestion of shot present in waterfowl gizzards.

Other management practices that have been used to reduce losses from lead poisoning on site-specific areas include tillage programs to turn lead shot below the surface of soil so that shot is not readily available to birds, planting food crops other than corn and other grains that aggravate the effects of lead ingestion, and requiring the use of nontoxic shot in hunting areas. The potential contributions of the first two practices toward reducing lead-poisoning losses among birds are, at best, limited and temporary. Supplemental grit has also been placed in wetlands in the belief that

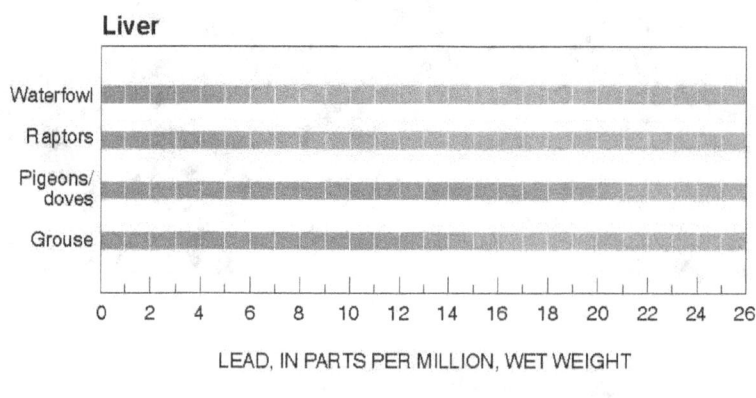

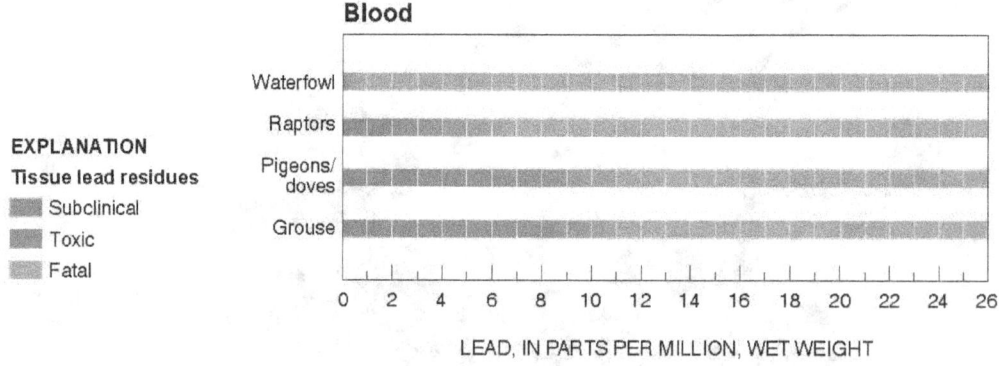

Figure 43.19 Lead residues in the liver and blood associated with subclinical, toxic, and mortality findings in several groups of birds. Variability within groups occurs because of differences in species and a variety of factors affecting toxicity within species.

birds intentionally ingest shot because grit is not available; as with tillage and food crops, any benefits are limited and temporary. The use of nontoxic shot is the only long-term solution for significantly reducing migratory bird losses from lead poisoning.

The strong correlation between exposure of waterfowl to lead and the use of lead shot for hunting waterfowl was vividly demonstrated by National Wildlife Health Center sponsored studies that compared tissue lead levels and gizzard analyses in a subpopulation of Canada geese as they migrated from their breeding grounds to their wintering grounds. Nontoxic shot requirements were in place at some sampling sites but not at others. Lead exposure was significantly less where nontoxic shot requirements existed.

Since lead shot has been banned for hunting waterfowl in the United States, attention has turned to regulating the use of lead fishing sinkers and lead jig heads. The Environmental Protection Agency has been petitioned to address the problem of bird mortality from these sources (Fig. 43.20). Prohibitions against using lead fishing weights below certain sizes have already been initiated on some Federal lands and other areas. The number of cases of lead poisoning in swans in the Thames Valley of England was reduced by 70 percent in 2 years following enactment of the 1987 ban on use of split lead shot and other fishing sinkers up to 1 ounce in size. Sizes larger than those that can be ingested by birds have not yet become a focus for concern.

The use of lead shot for target shooting and hunting on uplands is also receiving increased attention. In general, ingestion rates for lead shot in upland species are far less than those for waterfowl, even for doves (Table 43.2). The harvest of doves is somewhat analogous to waterfowl hunting in that large numbers of shells are often fired over the same location year after year (Fig. 43.21.). However, the duration of intense shooting on specific sites tends to be much less for doves than for waterfowl and the hunting area is generally tilled annually for agricultural purposes.

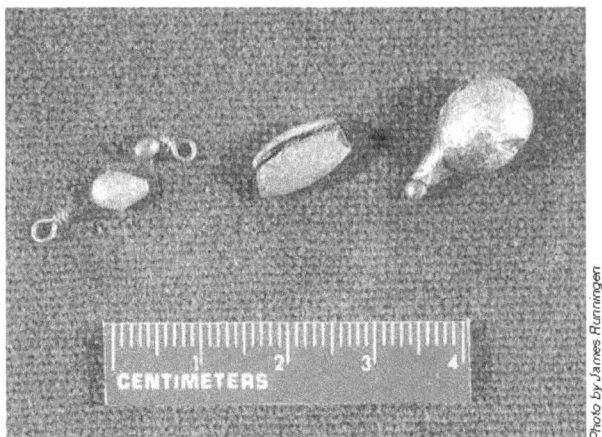

Figure 43.20 Fishing weights found in the stomachs and gizzards of birds that died from lead poisoning.

Veterinary treatment of lead-poisoned birds is generally not a reasonable approach. However, endangered species or other birds of high individual value that are lead poisoned may warrant treatment. In those instances, treatment should be done only by qualified persons familiar with and skilled in the proper use of lead-chelating chemicals. Under the best of circumstances, the results of treatment are unpredictable and the success rate low.

Human Health Considerations

People do inadvertently consume lead-poisoned birds. Although this is not desirable, no appreciable risks to human health exist. Most lead present in the body of a lead-poisoned bird is in organs such as the liver and kidneys rather than in the flesh. The dose relation (milligrams of lead per kilogram of body weight) and lead excretion processes are such that a great number of lead-poisoned birds would need to be consumed in a relatively short time before toxic levels of lead could build up in the human body. Persons who eat

Table 43.2 Percentage of upland gamebirds reported with ingested lead shot, by State.

Species	State	Percentage with ingested lead shot
Mourning doves	Alabama	1.0
	Eastern seaboard (Maryland to South Carolina)	2.4
	Indiana	2.3
	Maryland	1.0–6.5
Scaled quail	New Mexico	0.4
Bobwhite quail	New Mexico	1.8

Figure 43.21 *High bag limits and the large number of shells generally expended to reach a bag limit on swift-flying mourning doves results in large amounts of lead shot being deposited in uplands. Because most of the doves are harvested over agricultural fields, tillage helps to reduce the potential for that shot being ingested.*

the liver, kidneys, and other soft tissues from lead-poisoned birds would consume more lead than those who eat only the muscle tissue of these birds. Persons who consume waterfowl bones would be additionally exposed to lead, because lead is stored long-term in bone.

There are a few documented cases of humans developing lead poisoning after having accidentally ingested lead shot embedded in the meat they ate. This type of lead poisoning is rare, perhaps due to caution exercised when eating hunter-killed wildlife so as to avoid potential damage to teeth from biting into shot. Lead shot that is ingested can also become lodged in the appendix, resulting in appendicitis. This does not happen often, and it happens most in people who hunt waterfowl for subsistence. It is also possible that humans may ingest tiny fragments of lead that may be present in tissues of wildlife killed with lead shot.

Milton Friend

Supplementary Reading

Franson, J.C., 1996, Interpretation of tissue lead residues in birds other than waterfowl, *in* Beyer, W. N., and others, eds., Environmental contaminants in wildlife, interpreting tissue concentrations: Boca Raton, Fla., Lewis Publishers, p. 265–279.

Franson, J.C., Petersen, M.R., Meteyer, C.U., and Smith, M.R., 1995, Lead poisoning of spectacled eiders (*Somateria fischeri*) and of a common eider (*Somateria mollissima*) in Alaska: Journal of Wildlife Diseases, v. 31, no. 2, p. 268–271.

Sanderson, G. C., and Bellrose, F. C., 1986, A review of the problem of lead poisoning in waterfowl: Illinois Natural History Survey, 172, Special Publication 4, 34 p.

Scheuhammer, A. M., and Norris, S. L., 1996, The ecotoxicology of lead shot and lead fishing weights: Ecotoxicology, v. 5, p. 279–295.

Scheuhammer, A. M., Perrault, J.A., Routhier, E., Braune, B.M., and Campbell, G.D., 1998. Elevated lead concentrations in edible portions of game birds harvested with lead shot. Environmental Pollution, v. 102, p. 251–257.

Chapter 44
Selenium

Synonyms
Selenosis

Cause
Selenium is a naturally occurring element that is present in some soils. Unlike mercury and lead, which also are natural environmental components, selenium is an essential nutrient in living systems. The amount of dietary selenium required by animals depends upon many factors, including the availability of certain other metals such as zinc and copper, as well as vitamin E and other nutrients. Muscle damage results if dietary selenium is deficient, but dietary excess can be toxic.

Species Affected
Selenium poisoning or toxicosis has been documented in many avian species as well as in mammals and humans. The vulnerability of animals to selenium poisoning is primarily associated with the use of heavily contaminated habitats. Plants and invertebrates in contaminated aquatic systems may accumulate selenium in concentrations that are toxic to birds that consume them. In an experimental study with mallard ducklings, it was demonstrated that exposure to selenium in contaminated food items enhanced the birds' susceptibility to infectious diseases.

Distribution
The potential for selenium poisoning exists wherever bird habitat is created over sites with high soil concentrations of selenium and where point-source releases of selenium, for example from smelter emissions and sewage sludge, contaminate the environment.

Kesterson Reservoir in California is a classic example of bioaccumulation of selenium in wetlands created in an area with selenium-rich soils. The reservoir became a sump for wastewater return flows from irrigated soils that were rich in selenium. The continual addition of selenium-laden return wastewater leads to toxic concentrations of selenium in food items of birds. The result is reproductive failure caused by embryonic deformities and death, as well as mortality of adult birds.

Seasonality
The seasonality of selenium poisoning depends on when birds use habitats that have high selenium concentrations.

Field Signs
There are no unique clinical signs of selenium poisoning. The primary field indications that selenium poisoning may be occurring in an area are poor avian reproductive performance, embryonic deaths and deformities, and occasional mortality of adults.

Gross Lesions
Deformities caused by selenium poisoning may include missing or abnormal body parts, especially wings, legs, eyes, and beaks, as well fluid accumulation in the skull (Fig. 44.1). Affected adults often are emaciated, but other gross lesions generally are absent.

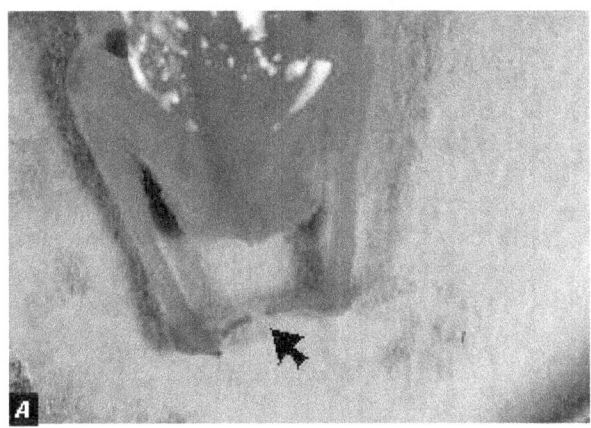

Figure 44.1 Embryonic deformities may be seen in the offspring of birds exposed to high levels of selenium. (A) A cleared and stained preparation of a coot embryo with fusion of the digits of both feet (arrow). (B) This mallard embryo has fluid accumulation over the back of the skull, and the mandible is only a remnant of normal (arrows).

Diagnosis

Diagnosis of selenium poisoning is complicated by its biological interactions with other elements, particularly mercury. These two elements often lessen or prevent the toxicity of each other when both are present. The diagnosis of selenium poisoning depends upon a history of potential exposure; gross developmental defects; microscopic lesions, primarily evidence of chronic liver damage, that are consistent with selenium toxicosis; and selenium levels in tissues and environmental samples such as food items, water, and sediment. In birds found dead at Kesterson Reservoir, mean selenium concentrations in livers and kidneys were about 95 parts per million dry weight, which is about 10 times higher than levels found in birds from a control area.

Control

The construction of artificial wetlands that are likely to attract bird use in areas of selenium-rich soils should be carefully evaluated for the potential for bioaccumulation of selenium in food items. It is preferable not to create wetlands where toxic concentrations of selenium can be expected. For existing wetlands, control measures should be directed at providing sources of clean water and at preventing environmental contamination by selenium through carefully disposing of selenium-containing wastes, including irrigation drainwater and sewage. The use of scare devices and other methods to prevent birds from using heavily contaminated areas should be considered.

Human Health Considerations

The ingestion of high levels of selenium can result in poisoning in humans. One should wear gloves when handling carcasses, but birds suspected of having died of selenium poisoning present no special hazard, because residues are biologically bound within tissues.

J. Christian Franson

Supplementary Reading

Eisler, R., 1985, Selenium hazards to fish, wildlife, and invertebrates: a synoptic review: Fish and Wildlife Service Biological Report 85(1.5), 57 p.

Heinz, G.H., 1996, Selenium in birds, *in* Beyer, W.N., and others, eds., Environmental contaminants in wildlife: interpreting tissue concentrations: Boca Raton, Fla., Lewis Publishers, p. 447–458.

Ohlendorf, H.M., and Hothem, R.L., 1995, Agricultural drainwater effects on wildlife in central California, *in* Hoffman, D.J., and others, eds., Handbook of ecotoxicology: Boca Raton, Fla., Lewis Publishers, p. 577–595.

Ohlendorf, H.M., 1996, Selenium, *in* Fairbrother, A., and others, eds., Noninfectious diseases of wildlife (2nd ed.): Ames, Iowa, Iowa State University Press, p. 128–140.

Chapter 45
Mercury

Synonyms

Minamata disease

Mercury has been used by humans for over 2,000 years and was associated with premature deaths of cinnabar (mercuric sulfide) miners as early as 700 B.C. More recent human poisonings have been related to agricultural and industrial uses of mercury. One of the best documented of these cases occurred in the 1950s in Minamata Bay, Japan, when mercury was discharged into the environment and accumulated in fish and shellfish used as human food. In addition to human poisonings, mercury poisoning or toxicosis has been identified in many other species.

Mercury is sometimes used to recover gold from stream sediments, and it may pose hazards to wildlife if it is released to the environment during ore recovery. Fungicidal treatment of seeds with mercury was common in the 1950s and 1960s, but this agricultural practice has been largely halted in the Northern Hemisphere.

Cause

Mercury is a heavy metal that is nonessential and toxic to vertebrates, and it occurs in both organic and inorganic forms. The organic forms, such as methylmercury, are generally the most toxic. However, inorganic mercury can be transformed into organic forms through a variety of biological processes. Mercury occurs naturally in soils and sediments, but it is also introduced into the environment by human activities (Fig. 45.1).

Species Affected

Birds affected by mercury include species that are exposed to high levels of the metal because of their feeding behavior (Fig. 45.2). Exposure may occur through accumulation of mercury in the aquatic food chain, agricultural uses of mercury as a fungicidal seed treatment, and from point-source industrial and mining discharge to the environment.

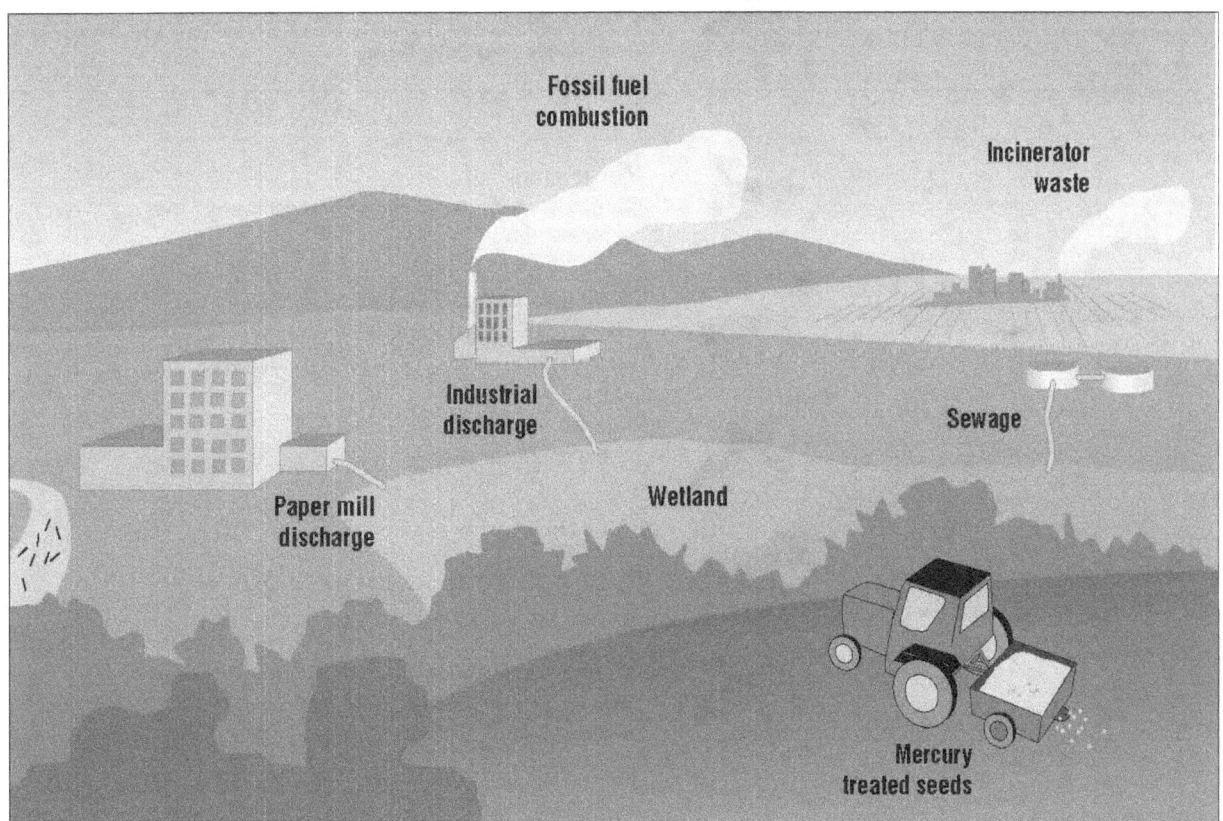

Figure 45.1   Sources of mercury contamination.

Species group and representative species

Loons
Common loon, red-throated loon

Wading birds
Common egret, great blue heron, black-crowned night heron

Pelicans
Brown pelican, white pelican, gannets

Cormorants
Double-crested cormorant

Mergansers
Common merganser, red-breasted merganser

Gulls and terns
Herring gull, common tern

Pelagic seabirds
Fulmars, shearwaters

Raptors
Bald eagle, osprey, golden eagle, owls

Gallinaceous birds
Ring-necked pheasant, chukar partridge, grouse, quail

Figure 45.2 Avian species most likely to be exposed to mercury.

Sources of mercury exposure for birds that live, nest, or feed in or near aquatic systems include industrial discharge, acid precipitation, and high mercury levels in fish and sediments. Gallinaceous birds, such as turkey and pheasant, may be exposed when they consume mercury-treated grain. Raptors, such as golden eagle and owls, may be secondarily exposed when they consume birds or small mammals that died from eating treated grain.

Major bird die-offs from mercury poisoning are rarely reported. Mortality from mercury poisoning is more of an insidious problem involving scattered mortalities. Some instances where mercury has been associated with mortality or sublethal effects are listed in Table 45.1.

Distribution

Mercury is present in fossil fuels and in some soils and sediments. The release of mercury into the atmosphere from burning of fossil fuels, the conversion of inorganic mercury to organic methylmercury and its cycling in aquatic systems, and accumulation through the food chain can expose wildlife to mercury and potential toxicity. Problems with mercury poisoning in birds traditionally have occurred in northern latitudes in areas affected by acid precipitation, at point-sources of industrial discharge, and in agricultural areas where mercury-based seed treatments have been used.

Seasonality

Seasonality is dependent only on the movement and foraging of birds that may put them at risk of mercury exposure while they feed in contaminated habitats.

Field Signs

Clinical signs of mercury poisoning in birds have been documented primarily from experimental feeding studies, and they include incoordination, tremors, weakness, ruffled feathers, and drooping eyelids. Experimental exposure of birds to high levels of mercury has caused acute death in less than 1 hour with few signs of intoxication. In free-ranging birds, most cases of mercury poisoning are probably more insidious, resulting in an emaciation syndrome and a variety of sublethal effects that may act together to cause eventual death (Table 45.2).

Gross Lesions

Birds suspected of having died of mercury poisoning often are emaciated, but no other gross lesions are noted.

Diagnosis

A diagnosis of mercury poisoning as cause of death can seldom be made on the basis of mercury concentrations in tissues alone. A complete necropsy examination with appropriate laboratory evaluations should be done by a qualified diagnostic laboratory. A diagnosis is generally based on total mercury concentrations of 20 parts per million wet weight or more in the liver or the kidneys and by the presence of microscopic lesions in tissues consistent with mercury poisoning. A definitive diagnosis is difficult, however, because the amounts of residues that would indicate mercury poisoning have not been determined for most bird species. Also, seabirds may naturally accumulate and tolerate higher levels of mercury than nonmarine birds. Another confounding factor is that selenium, which is an element that is essential to health, has been found to reduce the toxicity of mercury, and residues of both of these elements are often found in birds. A thorough history of field observations and background information about potential agricultural and industrial uses of mercury is an invaluable supplement to the specimens submitted.

Table 45.1 Reports of mercury exposure associated with mortality and sublethal effects in free-ranging birds.

Location	Species	Effect
Sweden	Pheasants, partridge, pigeon, magpie, passerines	Mortality
Sweden	Goshawk, Eurasian sparrowhawk, white-tailed eagle, peregrine falcon	Mortality
The Netherlands	Various raptors	Morbidity and mortality
Canada	Loons, turkey vulture	Mortality
Canada	Common tern	Poor reproduction
Scotland	Golden eagle	Poor reproduction
United States	Bald eagle	Poor reproduction
Canada	Loons	Poor reproduction

Table 45.2 Sublethal effects of mercury exposure from experimental studies.

Species	Effect(s)
Pheasants	Decreased egg weight, fertility, and hatchability
Starling	Microscopic kidney lesions
Mallard duck	Microscopic brain lesions, skeletal deformities; reduced clutch size, hatchability, embryonic growth; behavioral changes
Black duck	Reduced clutch size and hatchability
Red-tailed hawk	Neurologic signs of weakness and incoordination

Control

Prevention of exposure is required to control the lethal and sublethal effects of mercury poisoning in avian populations. Elimination of mercury discharge in industrial, mining, and sewage wastes, reduction of fossil fuel (especially coal) combustion, reduced inputs to (and thus releases from) municipal incinerators, and elimination of agricultural uses will reduce the amount of mercury entering the environment as a result of human activities. One factor to consider in the development of new wetlands is that the accumulation of mercury in aquatic biota is enhanced when terrestrial habitats are flooded. Little control is possible over low-level exposure to naturally occurring sources of mercury from soils and sediment.

Human Health Considerations

Mercury is a well-documented human health hazard. Avoid exposure to elemental mercury, which is volatile and can be inhaled in significant amounts in enclosed areas, mercury-based seed treatments, and mercury-contaminated food. One should wear gloves when handling carcasses, but birds thought to have died of mercury poisoning present no special hazard because the mercury is biologically bound to tissues within the carcass.

J. Christian Franson

Supplementary Reading

Eisler, R., 1987, Mercury hazards to fish, wildlife, and invertebrates: a synoptic review: U.S. Fish and Wildlife Service Biological Report 85(1.10), 90 p.

Hecky, R.E., Ramsey, D.J., Bodaly, R.A., and Strange, N.E., 1991, Increased methylmercury contamination in fish in newly formed freshwater reservoirs, *in* Suzuki, T., and others, Advances in mercury toxicology: New York, N.Y., Plenum Press, p. 33–52.

Heinz, G.H., 1996, Mercury poisoning in wildlife, *in* Fairbrother, A., and others, eds., Noninfectious diseases of wildlife (2nd ed.): Ames, Iowa, Iowa State University Press, p. 118–127.

Thompson, D.R., 1996, Mercury in birds and terrestrial mammals, *in* Beyer, W.N., and others, eds., Environmental contaminants in wildlife: interpreting tissue concentrations: Boca Raton, Fla., Lewis Publishers, p. 341–356.

Wren, C.D., Harris, S., and Harttrup, N., 1995, Ecotoxicology of mercury and cadmium, *in* Hoffman, D.J., and others, eds., Handbook of ecotoxicology: Boca Raton, Fla., Lewis Publishers, p. 392–423.

Chapter 46
Cyanide

Synonyms

Hydrocyanic acid poisoning, Prussic acid poisoning

Cause

Cyanide poisoning of birds is caused by exposure to cyanide in two forms: inorganic salts and hydrogen cyanide gas (HCN). Two sources of cyanide have been associated with bird mortalities: gold and silver mines that use cyanide in the extraction process and a predator control device called the M-44 sodium cyanide ejector, which uses cyanide as the toxic agent.

Most of the cyanide mortality documented in birds is a result of exposure to cyanide used in heap leach and carbon-in-pulp mill gold or silver mining processes. At these mines, the animals are exposed when they ingest water that contains cyanide salts used in mining processes or, possibly, when they inhale HCN gas. In heap leach mining operations, the ore is placed on an impermeable pad over which a cyanide solution is sprayed or dripped. The cyanide solution dissolves and attaches to or "leaches out" the gold. The cyanide and gold solution is then drained to a plastic-lined pond, which is commonly called the pregnant pond. The gold is extracted, and the remaining solution is moved into another lined pond, which is commonly called the barren pond. The cyanide concentration in this pond is increased so that the solution is again suitable for use in the leaching process, and the solution is used again on the ore heap (Fig. 46.1). Bird use of the HCN-contaminated water in the ponds (Fig. 46.2) or contaminated water on or at the base of the heap leach pads (Fig. 46.3) can result in mortality.

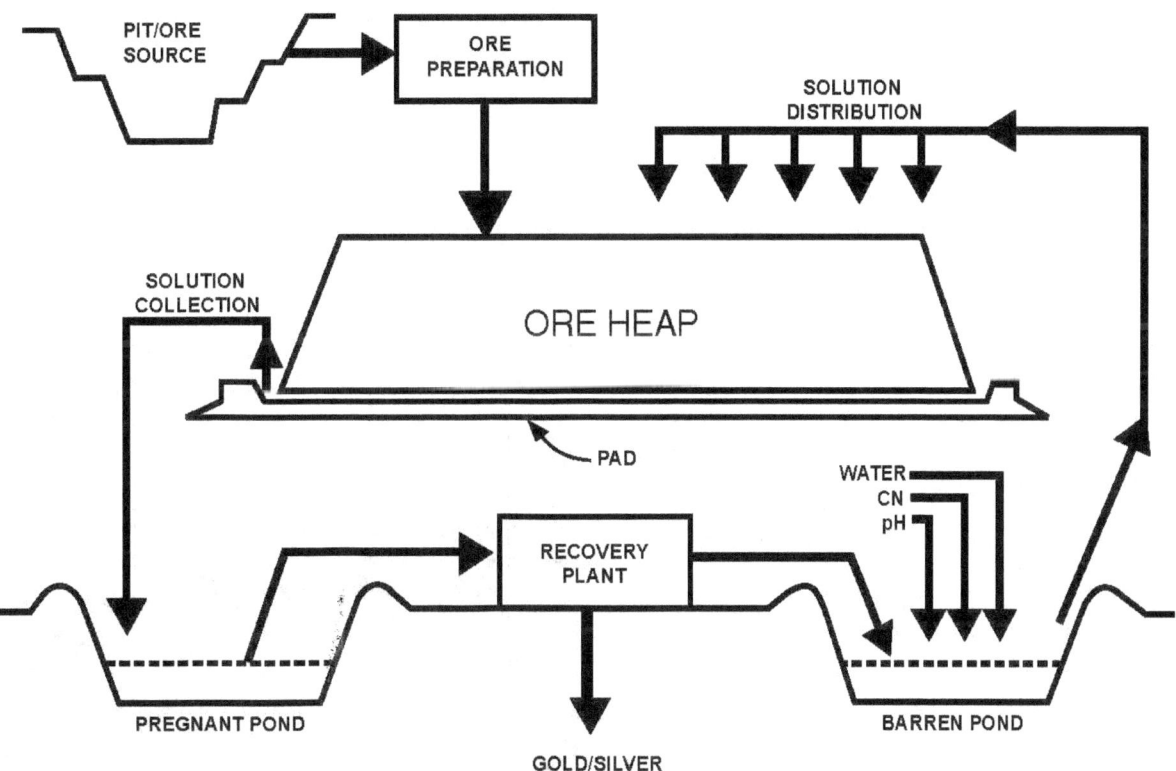

Figure 46.1 Schematic of a typical heap leach system (Graphic by Robert Hallock, U.S. Fish and Wildlife Service).

Mill tailings ponds produced by mines using the carbon-in-pulp mill process have also been responsible for migratory bird mortalities. In this process, crushed ore, cyanide solution, and carbon are placed together in a large vat. The cyanide solution extracts the gold from the ore, and the gold then adheres to the surface of the carbon. After the gold is extracted from the ore, the spent ore and the cyanide solution slurry are discharged to a mill tailings pond. The cyanide solution from the pond is drained, recharged, and reused in the extraction process. Tailings ponds range from 10 to several hundred surface acres and, in addition to open water, frequently have "mud flats" that are attractive to a wide variety of migratory birds. Cyanide concentrations are typically greatest near the spigots where mill slurry is discharged into the pond and are lowest in the solution reclamation areas.

The M-44 is a mechanical device designed to kill mammalian predators, specifically coyotes, by ejecting sodium cyanide into the animal's mouth (Fig. 46.4). Cyanide from M-44s has occasionally been documented as the cause of mortality in nontarget bird species, such as eagles and other scavengers, that are attracted by the bait and trigger the M-44 device.

Species Affected

Both birds and mammals can be killed by cyanide. From 1986–95, more than 3,000 cyanide-related mortalities involving about 75 species of birds representing 23 families were reported to the National Wildlife Health Center (NWHC). Waterbirds and passerines represented the greatest number of species affected (Fig. 46.5). Exposure to cyanide used in gold mining accounted for almost all of the mortalities; only one bird in these submissions, a bald eagle, was killed by an M-44.

Distribution

Mines that use cyanide in the gold- or silver-extraction process are located in many areas of the United States. However, most mines are concentrated in western States, particularly in arid areas (Fig. 46.6). Because water is limited in these areas, birds are often attracted to the water sources created by the mining operations. Bird mortality associated with mining operations in six States has been reported to the NWHC (Fig. 46.7).

The M-44 is used more commonly in the Western states, and its use is restricted by the Environmental Protection Agency and individual State regulations.

Seasonality

Cyanide toxicosis can occur at any time of the year. However, most mortalities associated with exposure to cyanide at mines are reported in the spring and fall months when birds are migrating through areas where mines are located.

Figure 46.2 Aerial view of a heap leach mine. Note the open ponds of water (arrows).

Figure 46.3 Heap leach pads at a mine that uses cyanide in the gold-extraction process. The water puddling at the base of the pad in the foreground contains cyanide.

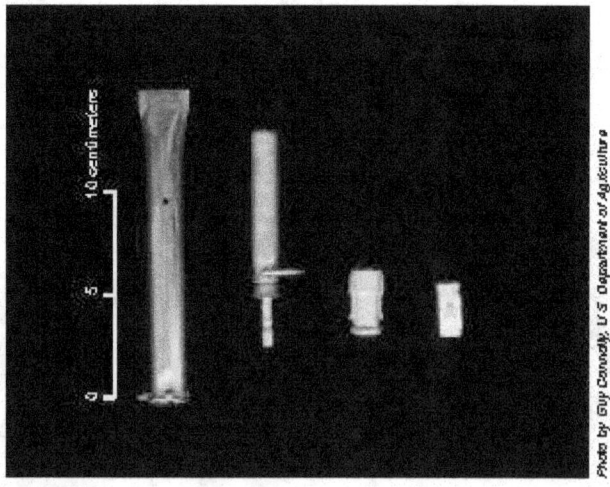

Figure 46.4 The M-44 device consists of a stake (left), an ejector, a top, and a capsule containing cyanide.

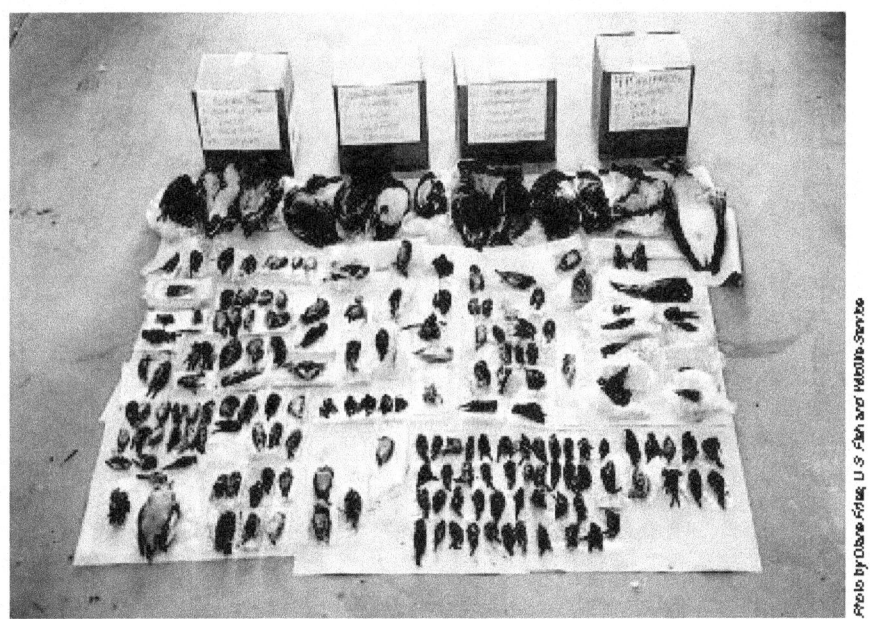

Figure 46.5 All of these birds were killed by cyanide-contaminated water at a heap leach gold mine. Note the diversity of the species present.

EXPLANATION

☐ States containing mines that use cyanide in mining operations. (Nevada has the highest concentration of these mines)

Figure 46.6 States containing mines that use cyanide in leaching operations.

Field Signs

Cyanide acts rapidly, and affected birds are most often found dead. Cyanide interferes with the body's ability to utilize oxygen in the blood. Although the blood is well oxygenated, this oxygen cannot be released to the tissues and the animal dies from lack of oxygen or anoxia.

Gross Lesions

Animals that die from cyanide toxicosis have bright red, oxygenated blood, and their tissues or organs, particularly the lungs, may appear congested with blood. The lungs of affected animals may also be hemorrhagic and edematous (Fig. 46.8). A yellow Day-Glo® fluorescent particle marker is used in the M-44 chemical mixture and animals exposed to cyanide through the M-44 device may have fluorescent yellow staining in the mouth or on the feathers or fur around the face. Visualization of this staining can be enhanced with ultraviolet light.

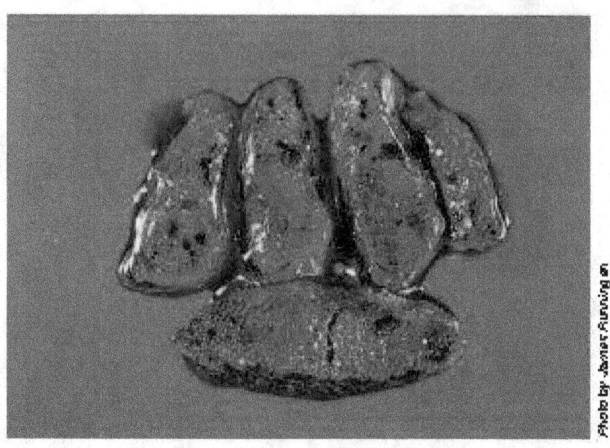

Figure 46.8 Lungs from a cyanide-poisoned bird. Note the congestion and edema.

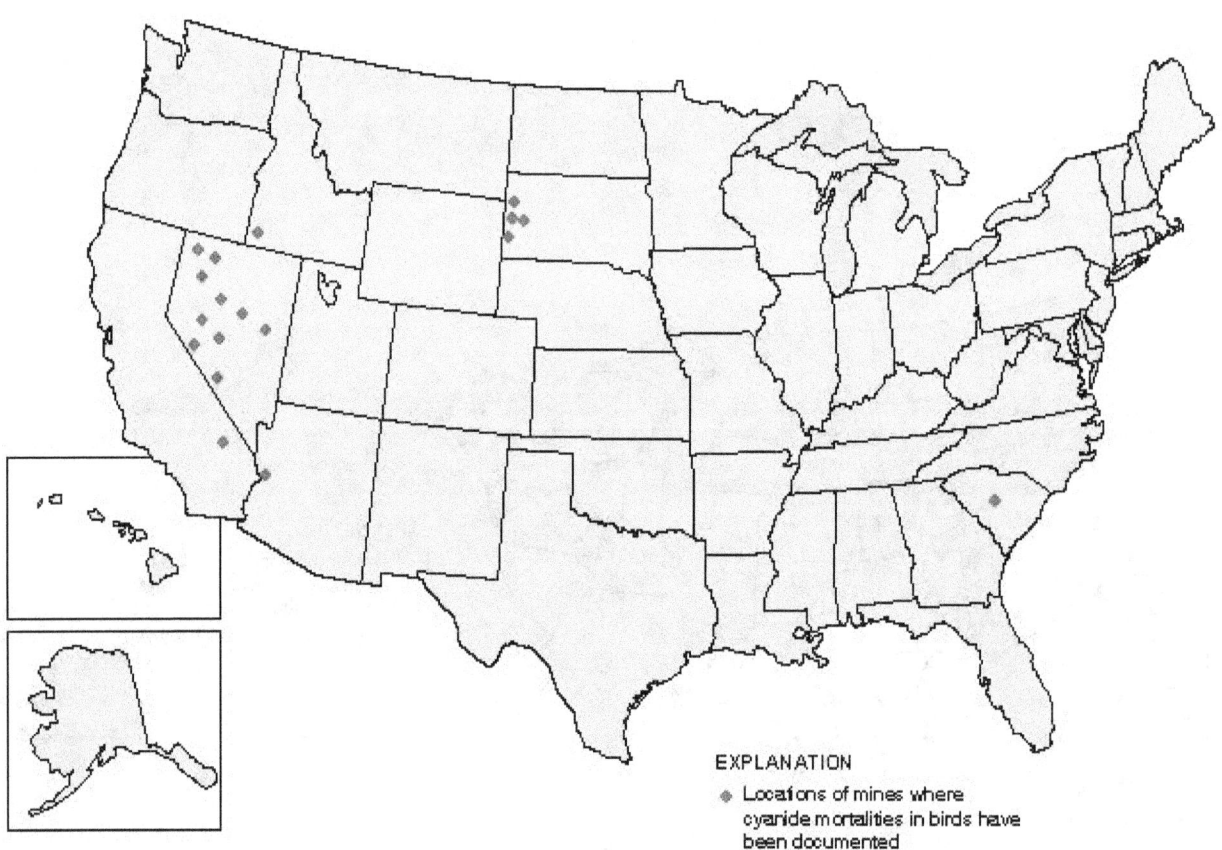

Figure 46.7 Locations of mines where cyanide mortalities in birds have been documented.

Diagnosis

Diagnosis is based on the field history, by the lack of gross lesions other than those described above, and by chemical analysis of tissues, such as the blood, heart, liver, and brain, to detect cyanide. Proper carcass handling is very important for meaningful chemical analysis results because cyanide levels in tissues can diminish rapidly after death unless the carcass or tissues are frozen. Consequently, the best sample to submit to the diagnostic laboratory is the whole carcass of a bird found freshly dead and frozen immediately after retrieval. The carcass should remain frozen during shipment to the diagnostic laboratory; this is one instance in which dry ice is recommended. Contact a diagnostic laboratory for advice on carcass handling and shipment.

Control

The primary method for preventing cyanide toxicosis at heap leach and carbon-in-pulp mill mining sites is to deny birds access to cyanide-contaminated water. This may or may not be difficult (or even possible) depending upon the size and configuration of a particular site. Successful methods used include netting over the solution ponds, covering heap leach collection channels, and designing mines that have no exposed solution ponds. Prevention of puddling in association with the heaps or netting over problem areas where puddling occurs are also beneficial. Detoxification or dilution have been the only successful means of preventing wildlife mortality at mill tailings ponds due to their large size and changing shapes. Hazing has not been very successful in preventing bird mortality at heap leach pads or heap leach and mill tailings ponds.

M-44s should be placed and baited to target only the intended species. Proper use of the M-44 lessens potential exposure of nontarget birds and mammals (Fig. 46.9).

Human Health Considerations

Cyanide gas can cause death in humans; therefore, care should be taken when visiting mining sites. Alkaline cyanide solutions that are allowed to become acidic release cyanide gas. Abandoned sites where the cyanide solutions are no longer monitored and the proper pH maintained pose the greatest risk. In some instances, protective equipment may be necessary for site inspection or carcass pick-up. Untrained persons should not handle the M-44 sodium cyanide ejector. An antidote is provided with the device, and the people authorized to handle the device should be trained to administer the antidote quickly in the case of an accident.

Lynn H. Creekmore

Figure 46.9 *(A) Closeup of a set M-44 device and (B) a completed M-44 set with a cow chip cover (arrow). Notice the warning sign. These signs are required at main entrances to areas in which M-44 devices are set and within 25 feet of each device.*

Supplementary Reading

Connolly, G., 1988, M-44 sodium cyanide ejectors in the animal damage control program, 1976–1986, *in* Crabb, A.A., and March, R.E., eds., Proceedings of the vertebrate pest conference (v. 13): Davis, Calif., University of California, p. 220–225.

Eisler, Ronald, 1991, Cyanide hazards to fish, wildlife, and invertebrates: A synoptic review: U.S. Fish and Wildlife Service Contaminant Hazard Reviews Report 23, Biological Report 85(1.23). 55 p.

Henny, C.J., Hallock, R.J., and Hill, E.F., 1994, Cyanide and migratory birds at gold mines in Nevada, USA: Ecotoxicology, v. 3, p. 45–58.

Proceedings of the Nevada wildlife/mining workshop, Reno, Nevada, March 27–29, 1990: Reno, Nev., Nevada Mining Association, 233 p.

Wiemeyer, S.N., Hill, E.F., Carpenter, J.W., and Krynitsky, J.A., 1986, Acute oral toxicity of sodium cyanide in birds: Journal of Wildlife Diseases, v. 22, no. 4, p. 538–546.

Chapter 47
Salt

Synonyms
Water deprivation, salt encrustation

Cause

Animals become victims of salt poisoning or toxicosis when toxic levels of sodium and chloride accumulate in the blood after they ingest large amounts of salt or, in some species, are deprived of water. For birds, salt sources may include saline water and road salt.

Normally, the salt glands of birds (Fig. 47.1) excrete sodium and chloride to maintain the proper physiologic chemical balance. However, when there has been insufficient time for acclimation of the salt gland to the saline environment, or when salt gland function is compromised by exposure to certain pesticides or oil, the electrolyte balance of the blood may be upset by the excess sodium and chloride, resulting in toxicosis. Salt accumulation on the outside of the body, or salt encrustation, is a greater problem for waterbirds that use very saline waters than is salt toxicosis. Salt encrustation can lead to exertion, acute muscle degeneration, and eventual drowning during the struggle to escape entrapment.

Species Affected

This infrequently reported toxicosis has affected gallinaceous birds, such as pheasants, and rock doves that consumed road salt and migratory waterbirds forced to use highly saline water. Mortality from salt encrustation most often involves diving ducks.

Distribution

Salt poisoning and salt encrustation can occur anywhere that birds use saline environments. However, salt poisoning may be more likely in northern latitudes where saline lakes remain open while nearby freshwater habitats freeze over and where salt is used for removing ice from roadways.

Seasonality

Salt poisoning and salt encrustation may affect birds at any time of the year. In winter or early spring, terrestrial birds may consume road salt for grit and mineral content. Migratory waterbirds are more likely to be poisoned during late autumn migration after they have spent several months on freshwater nesting grounds. Cold snaps that freeze freshwater areas along the migratory route may force birds to use more saline waters that remain open because of the high salt content. High winds can contribute to salt encrustation by continually covering birds with salt-laden water.

Field Signs

Clinical signs of salt poisoning may include muscle weakness, partial paralysis, and difficult breathing, all of which can be caused by a variety of other toxicoses. Carcasses may or may not be covered with salt (Fig. 47.2).

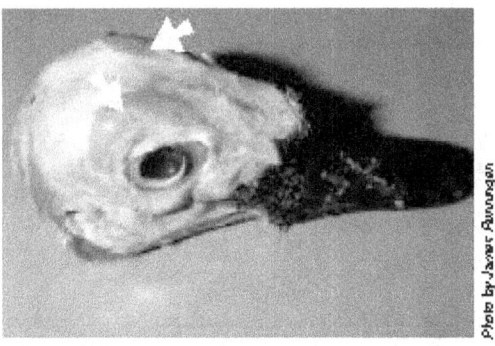

Figure 47.1 *The salt glands of birds are located just above the eyes (arrows).*

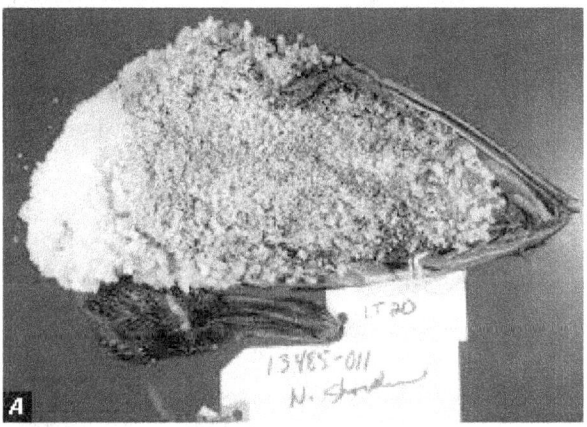

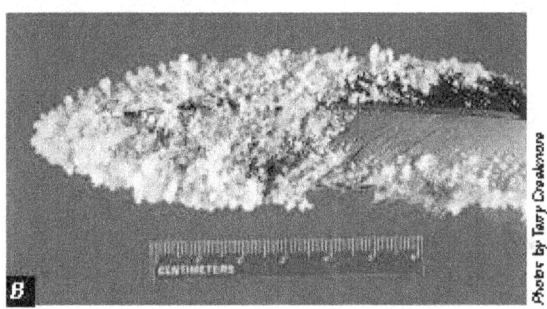

Figure 47.2 *Salt encrustation may completely cover the bird with salt (A), or salt may accumulate on margins of feathers (B).*

Salt 347

Gross Lesions

Gross lesions are nonspecific, and they may include reddening of the brain surface (Fig. 47.3), visceral gout (Fig. 47.4), fluid accumulation in the lungs, small hemorrhages on the viscera, and erosions on the surfaces of the eyes.

Diagnosis

Finding a source of salt exposure lends support to a sometimes difficult diagnosis of salt poisoning. Salt on the feathers provides further evidence, but is not in itself diagnostic. Refrigerated blood and frozen as well as formalin-fixed brain are the best tissues to collect for laboratory analysis. Because the body maintains a constant internal environment or homeostasis, sodium concentrations in these tissues normally deviate very little. Therefore, a comparison of sodium concentrations between suspect and reference specimens can be used to support a diagnosis of salt poisoning. Microscopic examination of formalin-fixed brain tissue is also useful when salt poisoning is suspected.

Control

Birds that are on highly saline lakes can be hazed to freshwater areas, if such areas exist nearby. Road salt should be used sparingly and should be stored out of reach of wildlife. Management practices that may expose birds to compounds that interfere with salt gland function, such as applications of organophosphorus and carbamate pesticides, should be done only when necessary and should be scheduled to allow arriving birds maximum time to adapt to saline environments.

Human Health Considerations

None.

J. Christian Franson and Milton Friend

Supplemental Reading

Friend, M., and Abel, J.H., Jr., 1976, Inhibition of mallard salt gland function by DDE and organophosphates, *in* Page, L.A., ed., Wildlife Disease: New York, N.Y., Plenum Press, p. 261–269.

Trainer, D.O., and Karstad, L., 1960, Salt poisoning in Wisconsin wildlife: Journal of the American Veterinary Association v. 136, p. 14–17.

Windingstad, R.M., Kartch, F.X., Stroud, R.K., and Smith, M.R., 1987, Salt toxicosis in waterfowl in North Dakota: Journal of Wildlife Diseases, v. 23, p. 443–446.

Wobeser, G.A., 1997, Salt and saline water, *in* Diseases of wild waterfowl (2nd ed): New York, N.Y., Plenum Press, p. 204–207.

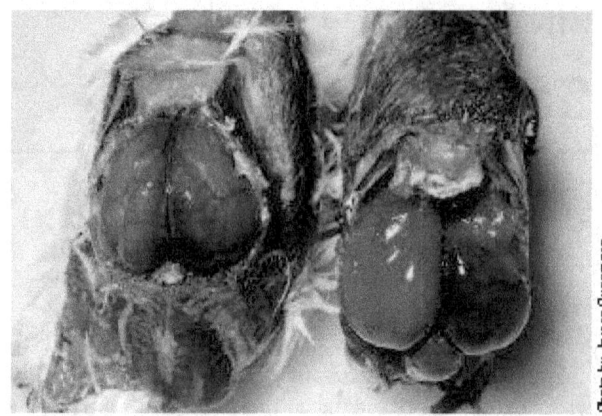

Figure 47.3 *The brains of salt-poisoned birds are sometimes very red and congested.*

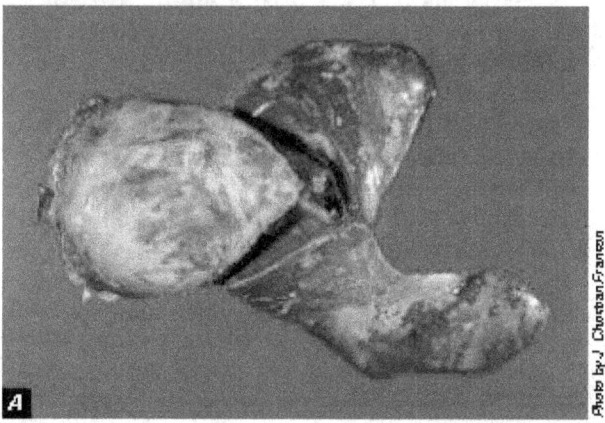

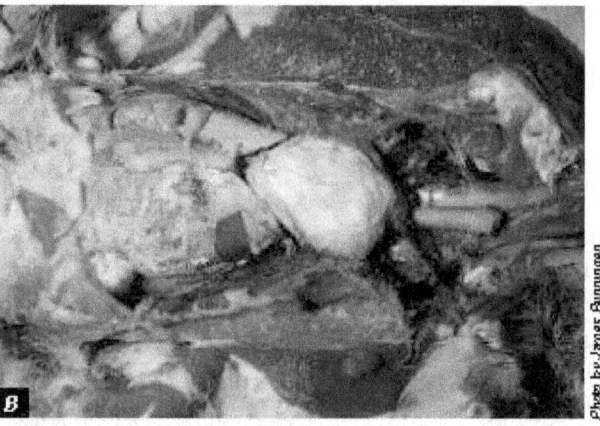

Figure 47.4 *(A and B) Visceral gout, or accumulation of gritty uric acid deposits on visceral surfaces, is a nonspecific lesion that is sometimes associated with salt poisoning.*

Chapter 48
Barbiturates

Synonyms
Pentobarbital poisoning, sodium pentobarbital poisoning

Cause
Barbiturate products are commonly used to euthanize domestic animals. The primary active component in euthanasia solutions is sodium pentobarbital, but some products also contain other minor ingredients (Fig. 48.1).

Euthanasia solutions are generally injected intravenously in domestic animals; therefore, after death, the solutions will be most concentrated in the blood and the highly vascularized organs, such as the liver or spleen, of the euthanized animal.

Euthanized carcasses that are available as carrion pose a hazard to scavenging birds and mammals. Large domestic animal carcasses, such as horses, that are not used for food or rendering but that are sufficiently valuable (monetarily or psychologically) to warrant veterinary services and euthanasia drugs are the most common sources of barbiturate poisoning in scavengers. In one instance in British Columbia, a single cow carcass was responsible for poisoning 29 bald eagles.

Circumstances that interfere with burial, such as frozen winter soil or bulky carcasses, result in euthanized carcasses being available for scavenger species. This problem could increase in the future if more stringent air-quality standards restrict carcass incineration.

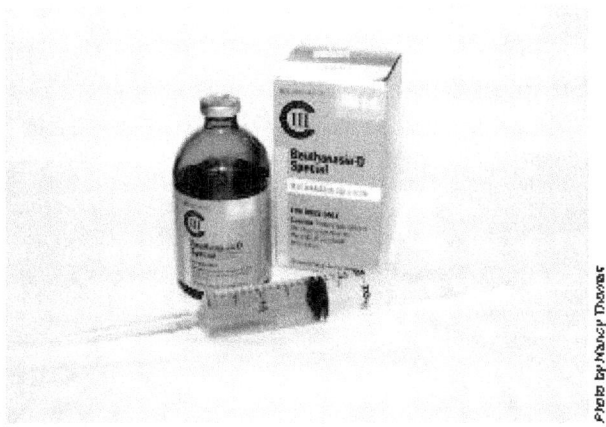

Figure 48.1 The active agent in most injectable euthanasia solutions is sodium pentobarbital.

Species Affected
Bald and golden eagles are the only free-living wildlife species that have been reported to have died of barbiturate poisoning. Raptors generally have a narrow tolerance for barbiturate compounds; therefore, an anesthetic dose is often close to a fatal dose in these species.

Distribution
As of 1997, the National Wildlife Health Center database contained records of 17 cases of barbiturate poisoning in eagles from six States (Fig. 48.2). Additional cases have been reported by other investigators.

Seasonality
Cases of barbiturate poisoning have been more frequent in late winter and early spring, but they are not confined to that period. Cases of barbiturate poisoning may be correlated with the spring thaw in northern climates, when carcasses thaw, and the internal organs become more readily available to scavengers. Residues in those carcasses become available to scavenger species at that time. Food supplies are often limited at this time, so scavenging is more common.

Field Signs
The most useful and specific field sign is the proximity of dead or moribund birds to a euthanized animal carcass that shows evidence of scavenging. In lieu of that, the proximity of dead or moribund birds to a domestic animal carcass of unknown origin is a less specific sign, but under that circumstance, barbiturates should be considered along with other poisons, such as pesticides.

Barbiturate-poisoned birds have been found near landfills in which euthanized animal carcasses were discarded. Landfills are legal disposal sites for carcasses in some States or locales.

Barbiturate poisoning may take hours to develop; therefore, poisoned birds can be found distant from the poison source. Eagles have been found beneath their roost trees without evident sources of poisoning.

Barbiturate-intoxicated birds are sedated, drowsy, sluggish, or comatose; have varying degrees of consciousness; and have slow heart and respiration rates. Although they may struggle to right themselves if they fall from a perch as toxicity progresses, signs of prolonged or violent struggling are unlikely. They are more likely to be found on undisturbed substrate. If more than one bird is exposed, the dose ingested and susceptibility to the poison may vary with each bird;

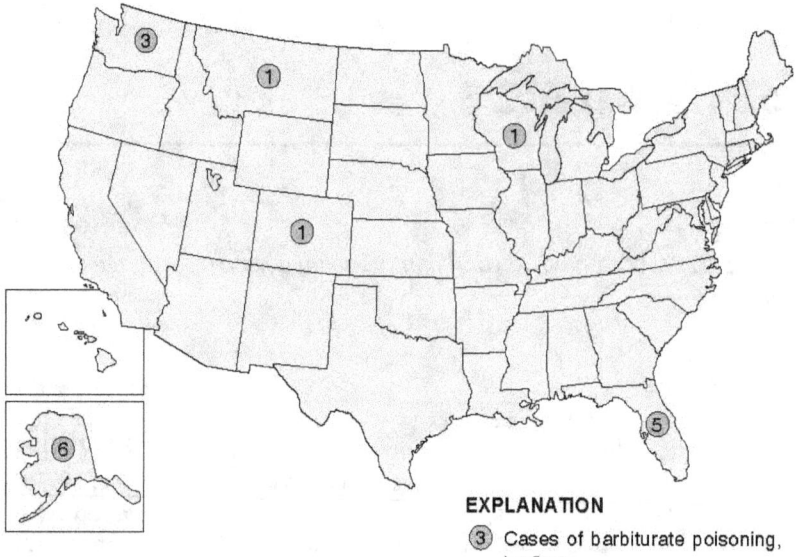

Figure 48.2 Seventeen cases of barbiturate poisoning in bald and golden eagles have been reported by the National Wildlife Health Center from six States as of 1997.

therefore, a range of signs from sublethal sedation to coma to death may be observed. Birds that are sedated or even comatose can recover if they are given supportive care until they metabolize the drug. Recovery may take several days.

Gross Lesions

There are no specific lesions. Ingesta may be present in the upper gastrointestinal tract as in other acute poisonings. The ingesta may be recognizable as domestic animal in origin. Barbiturate-poisoned birds are often in good body condition, thus reflecting the acute nature of this toxicosis.

Diagnosis

Analysis of liver or upper gastrointestinal contents detects pentobarbital and, sometimes, other components of euthanasia drugs. Liver analysis is more definitive for determining that a bird absorbed drug from the ingesta. Blood samples from live birds can be analyzed, but the clinical and field signs and the bird's recovery may be sufficient for a presumptive diagnosis.

Samples of blood-engorged organs, blood clots, or other tissue from scavenged sites in the suspect domestic animal carcass can assist in tracing the source of the poison.

Control

Treatment

Birds found alive in the field are often hypothermic (exhibiting low body temperature); warming of less affected birds, in itself, may result in recovery. A veterinarian can provide supportive care, administer cardiac and respiratory stimulants, and remove the undigested crop contents so that no further drug is absorbed.

Management

State agricultural departments in the United States generally regulate carcass disposal to assure that carcasses are not available to scavengers. Circumstances such as frozen ground that prevents burial, poor compliance with regulations, or shallow burial may circumvent these regulations. Landfill regulations or policy can guarantee that carcasses are covered before scavenging is likely.

Cases of barbiturate poisoning are generally inadvertent. Prevention can be greatly enhanced by increasing awareness of the hazard among the public and veterinary community.

Human Health Considerations

None. Euthanized carcasses are not for human consumption. Barbiturates are not absorbed through the skin.

Nancy J. Thomas

Supplemental Reading

Langlier, K.M., 1993, Barbiturate poisoning in twenty-nine bald eagles, *in* Redig, P.T., and others, eds., Raptor Biomedicine: Minneapolis, Minn., University of Minnesota Press, p. 231–232.

Chapter 49
Miscellaneous Chemical Toxins

The previous chapters provide information about some of the chemical toxins that have lethal effects on wild birds. The material presented in Section 7, Chemical Toxins, is far from comprehensive because wild birds are poisoned by a wide variety of toxic substances. Also, monitoring of wild bird mortality is not yet organized so that diagnostic findings can be extended to reflect the relative impacts among the types of toxins, within populations, or among species, geographic areas, and time. The data that are available are not collectively based on random sampling, nor do specimen collection and submission follow methodical assessment methods. Instead, most data simply document individual bird poisoning events. The inherent biases in this information include the species of birds observed dead (large birds in open areas are more likely to be observed dead than small forest birds); the species of birds likely to be submitted for analysis (bald eagles are more likely to be submitted than house sparrows); collection sites (agricultural fields are more likely to be observed than urban environments); geographic area of the country; season; reasons for submissions; and other variables. Nevertheless, findings from individual events reflect the causes of mortality associated with those events and collectively identify chemical toxins that repeatedly cause bird mortalities which result in carcass collection and submission for diagnostic assessment.

The tables that follow illustrate the relative occurrence of poisoning by different types of toxic substances for wild bird carcasses evaluated at the National Wildlife Health Center during the period of 1984 through 1995. This information was compiled to reflect the relative frequency of poisoning in different groups of birds as a function of the number of years that mortality occurred, the number of multiple-death events, and the number of years that had multiple-species deaths.

As noted above, biases in collecting and submitting carcasses prevent extrapolating these data to population impacts. The specimens that were evaluated depend on submissions from field personnel who had detected avian mortality events, and, for various reasons, had sought a diagnosis of the causes of mortality. Therefore, the tables simply reflect a relative accounting of what types of toxins were found most commonly to be the cause of death of the species that were submitted for evaluation. These data are not without meaning, because they clearly identify specific causes of poisoning in various groups of wild birds.

Carbofuran stands out as a frequent cause of mortality of a variety of bird species (Table 49.1). Diazinon was the most frequently diagnosed pesticide-induced cause of mortality in waterfowl, and famphur and carbofuran had similar prominence for eagles (Tables 49.1 and 49.2). As should be expected, chlorinated hydrocarbon pesticides were not frequently determined to be the cause of wild bird mortality (Table 49.3) now that these pesticides have been replaced by organophosphates, carbamates, and other compounds. Strychnine was a frequent cause of eagle mortality among compounds used as rodenticides and repellents (Table 49.4).

More than 30 different toxic substances were diagnosed as the cause of bird mortalities in specimens submitted (Tables 49.1 through 49.5). The substances included naturally occurring materials such as selenium and sodium as well as synthetic products such as insecticides, and data in the tables are limited to those substances that caused direct lethal effects. As previously noted, there are many possible impacts of chemical toxins in addition to immediate toxicity that cause illness and death; some of these impacts involve interactions with other chemical or biological agents.

Residue analyses by themselves are often insufficient determinants of cause of mortality from chemical toxins be-

Table 49.1 Relative occurrence of carbamate-caused mortality in free-ranging birds, 1984–95.
[Frequency of occurence: ● frequent, ◉ common, ◌ occasional, ○ infrequent or not reported]

Compound	Species						
	Eagles	Hawks	Waterfowl	Gulls/terns	Crows[1]	Songbirds	Doves
Aldicarb	◉	○	○	○	◌	○	○
Carbofuran	●	◉	◉	◌	◉	◉	◌
Methiocarb	○	○	○	○	◌	○	○
Unspecified	◌	○	◉	○	○	◌	○

[1] Includes vultures, ravens, magpies, and crows.

Table 49.2 Relative occurrence of organophosphorus-caused mortality in free-ranging birds, 1984–95.

[Frequency of occurence; ● frequent, ● common, ● occasional, ○ infrequent or not reported]

Compound	Species								
	Eagles	Hawks	Owls	Waterfowl	Cranes	Shorebirds	Crows[1]	Songbirds	Doves
Chlorpyrifos	○	○	○	○	○	○	●	○	○
Coumaphos	●	○	○	○	○	○	○	○	○
Diazinon	○	○	○	●	○	○	●	●	○
Dimethoate	○	○	○	●	○	○	○	○	○
Disulfoton	○	○	○	○	○	○	●	●	○
Famphur	●	●	●	○	○	○	●	●	●
Fenthion	●	●	●	○	○	○	○	●	○
Fonofos	○	○	○	●	○	○	○	○	○
Monocrotophos	○	○	○	●	○	○	○	●	●
Parathion	●	●	○	●	○	●	○	●	○
Phorate	●	○	○	●	○	○	○	○	○
Terbufos	●	○	○	○	○	○	○	●	○
Unspecified	●	●	●	○	●	○	○	○	○

[1] Includes vultures, ravens, magpies, and crows.

Table 49.3 Relative occurrence of chlorinated-hydrocarbon-caused mortality in free-ranging birds, 1984–95.

[Frequency of occurence: ● frequent, ● common, ● occasional, ○ infrequent or not reported]

Compound	Species		
	Eagles	Owls	Songbirds
Dieldrin	●	●	●
Heptachlor	●	○	○

Table 49.4 Relative occurrence of rodenticides and repellents as causes of mortality in free-ranging birds, 1984–95.

[Frequency of occurence: ● frequent, ● common, ● occasional, ○ infrequent or not reported]

Compound	Species				
	Eagles	Hawks	Waterfowl	Crows[1]	Songbirds
Avitrol®	○	○	○	○	●
Brodifacoum	●	○	●	○	○
1080	●	○	○	○	○
Strychnine	●	●	●	●	●
Thallium	●	○	○	○	○
Zinc phosphide	○	○	●	○	○

[1] Includes vultures, ravens, magpies, and crows.

Table 49.5 Relative occurrence of miscellaneous toxicants as causes of mortality in free-ranging birds, 1984–95.
[Frequency of occurence: ● frequent, ◉ common, ◎ occasional, ○ infrequent or not reported]

Species	Compound							
	Chloride	Cyanide	Ethylene glycol	Fluorine	Hydrogen sulfide	Penta-barbitol	Selenium	Sodium
Eagles	○	◎	○	○	○	●	○	○
Hawks	○	◎	○	○	○	○	○	○
Owls	○	○	○	○	◎	○	○	○
Waterfowl	◎	◉	○	◎	○	○	◉	◉
Cranes	○	◎	○	○	○	○	○	○
Grebes	○	◎	○	○	○	○	◉	◎
Pelicans	○	○	○	○	○	○	○	◎
Gulls/terns	○	◎	○	○	○	○	○	○
Shorebirds	○	◉	○	○	○	○	○	○
Egrets[1]	○	○	○	○	○	○	◉	○
Crows[2]	○	○	◎	○	○	○	○	○
Songbirds	○	◉	○	○	○	○	○	○
Doves	○	◎	○	○	○	○	○	◎
Swallows	○	◎	○	○	○	○	○	○
Quail	○	○	○	○	◎	○	○	○

[1] Includes long-legged wading birds such as herons and egrets.
[2] Includes vultures, ravens, magpies, and crows.

cause of species variations, lack of residue for some types of compounds, and other variables. Similarly, the often-quoted 16th Century statement that, "Dosage Alone Determines Poisoning" is modified by such factors as route of exposure and other important factors.

Chemical toxins are, and will continue to be, important causes of wildlife mortality. Documentation of mortality from chemical toxins requires rigorous diagnostic work. Determination of wildlife impacts will best be accomplished through methodical monitoring programs that allow sound evaluations of changes in the status and trends of specific compounds and their impacts on wild bird populations by geographic area.

Milton Friend

Section 8
Miscellaneous

Electrocution

Miscellaneous Diseases

Vertebral column deformity (scoliosis) in a bald eagle
Photo by James Runningen

Introduction to Miscellaneous Diseases

"Nature is far from benign; at least it has no special sentiment for the welfare of the human versus other species." (Lederberg)

The fact that "Nature is far from benign" is clearly evident from the preceding chapters of this Manual. The diseases and other conditions described are the proverbial "tip of the iceberg" relative to the number of specific causes of ill health and death for free-ranging wild birds, but the wild bird health problems described account for most major wild bird disease conditions seen within the United States. However, the full toll from disease involves many other causes of illness and death that individually may cause substantial die-offs. Two examples of these other causes of die-offs are the deaths of Canada geese that ingest dry soybeans, which then expand and cause lethal impactions within the moist environment of digestive tract, and the poisoning of ducks from rictin, a naturally occurring toxic component of castor beans. Some of these lesser-known causes of disease and mortality may become increasingly important in the future because landscape and other changes could result in environmental conditions that may enhance the interface between specific disease agents and susceptible bird species.

This final Section of the Manual includes some of the lesser-known causes of avian mortality. The first chapter provides an overview of electrocution in birds, with a special emphasis on eagles. The second chapter is a miscellaneous chapter that highlights a significant disease of domestic ducklings not yet known to exist in wild birds, disease caused by stress due to improper handling of birds, and several other conditions that might be encountered by biologists who work with birds. These other conditions include tumors, traumatic injuries, weather, nutritional factors, and drowning as causes of avian illness and death. These two chapters expand the scope of disease presented in the previous chapters and provide additional perspectives of the diverse causes of avian mortality. It is our hope that the collective information provided in this Manual will stimulate those interested in the conservation and well-being of avian species to give greater consideration to disease in the management strategies employed for the conservation of these species.

Quote from:

Lederberg, J., 1993, Viruses and humankind: intracellular symbiosis and evolutionary competition, *in* Morse, S.S., ed., Emerging viruses: Oxford, England, Oxford University Press, p. 3.

Chapter 50
Electrocution

Cause

Power lines and power poles present a potential electrocution hazard to wild birds. Many birds, especially raptors, select power poles for perching, and, sometimes, for nesting (Figs. 50.1–3). If a bird's appendages bridge the gap between two energized parts or between an energized and a grounded metal part, electricity flows through the "bridge" that is filling the gap and the bird is electrocuted.

Most commonly, birds are electrocuted where conducting wires (conductors) are placed closer together than the wingspan of birds that frequent the poles (Fig. 50.2). Feathers are poor electrical conductors, but if contact is made between points on the skin, talons, or beak, or if the feathers are wet, conduction can occur. Common anatomical sites of contact include conduction between the wrists of each wing or between the skin of one wing and a foot or leg. The resulting shock causes severe, usually fatal, cardiovascular injury.

Because conductors on distribution lines are placed closer together than high voltage transmission lines, birds are more frequently electrocuted on distribution lines despite their lower voltage.

In addition to one to three conductors, power poles may also carry ground wires, transformers, or grounded metal crossarm braces. Complicated wiring configurations that put multiple energized and grounded metal parts near attractive perching or nesting sites are the most hazardous configurations (Fig. 50.3).

Species Affected

Electrocution is primarily a problem of large raptors in open habitat, particularly treeless areas. Golden eagles are by far at greatest risk, but other eagles, large buteos, falcons, and the largest owls, such as the great horned owl, are also susceptible. The large wingspan of these birds appears to be the single most important factor in their susceptibility.

In addition to their size, the perching behavior of these bird species puts them at greater risk. Species that prefer exposed high perches are more likely to be attracted to power poles, as are the species that use a "still hunting" technique in which they perch and visually search the landscape for prey rather than hunting in flight.

Immature and subadult raptors are more commonly electrocuted. This predisposition is presumably related to their inexperience and awkwardness in taking off and landing.

Figure 50.1 A bald eagle using a power pole as a perch.

Figure 50.2 This is a hazardous situation because the eagle's wings can contact two conductors at once.

Figure 50.3 An eagle nest on the top of a power pole.

Distribution

Bird electrocutions are most common in the western plains of the United States where open shrub and grassland habitats are common, and are less prevalent in forested habitat (Fig. 50.4). However, birds may be electrocuted wherever electrical lines are above ground.

Generally, electrocutions are more prevalent in sites where a susceptible species' prey base is present and where suitable perches, other than power structures, are lacking. In the western plains, elevated perches are at a premium, and the more susceptible raptor species are abundant. The combination of golden eagles, jackrabbits, grassland habitat, and dangerous power pole configurations can be expected to be lethal. Similar conditions exist on the Russian steppes. Electrocution is a major cause of mortality for the Russian steppe eagle and for other raptors that nest on power poles and use them for perches in this largely treeless area (Fig. 50.5).

Figure 50.5 Power lines that are not designed to prevent electrocution and that cross largely treeless areas, such as this line on the Russian steppes, pose a significant hazard for large raptors that use the poles as perches for hunting and as nesting platforms.

Seasonality

Birds can be electrocuted during any season, but there can be seasonal fluctuations in electrocution frequency that are related to weather conditions or bird behavior. Electrocutions are more frequent during periods of rain and snow because of the increased conductivity of wet feathers. Inclement wet weather may also combine with windy conditions so that birds are less stable while landing and taking off. Where distribution lines are oriented with crossarms perpendicular or diagonal to the prevailing wind, more electrocutions occur.

Golden eagles may make greater use of power poles as night roosts during migration and wintering. This habit may make them more prone to electrocution as they stretch out to dry their wings in the morning sun.

Inattentiveness during seasonal mating behaviors or territorial conflicts have also been reported to predispose birds to electrocution.

Field Signs

Electrocuted birds often die immediately, so they are found near a power pole or beneath a power line.

The electrical hazard may be apparent in the configuration of the nearby pole. The conductors and other electrical hardware on the pole may be close together. The greatest hazards may be at corner poles where extra wires (jumpers) are required to provide a change in direction, or at poles with transformers or grounded metal equipment near the conductors (Fig. 50.6).

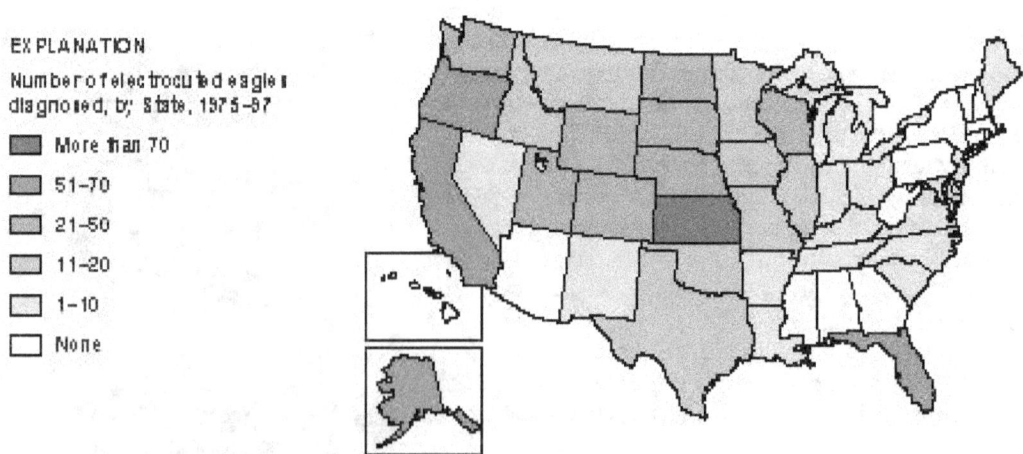

Figure 50.4 Number of electrocuted eagles diagnosed per State from 1975-95. (From unpublished data from the National Wildlife Health Center.)

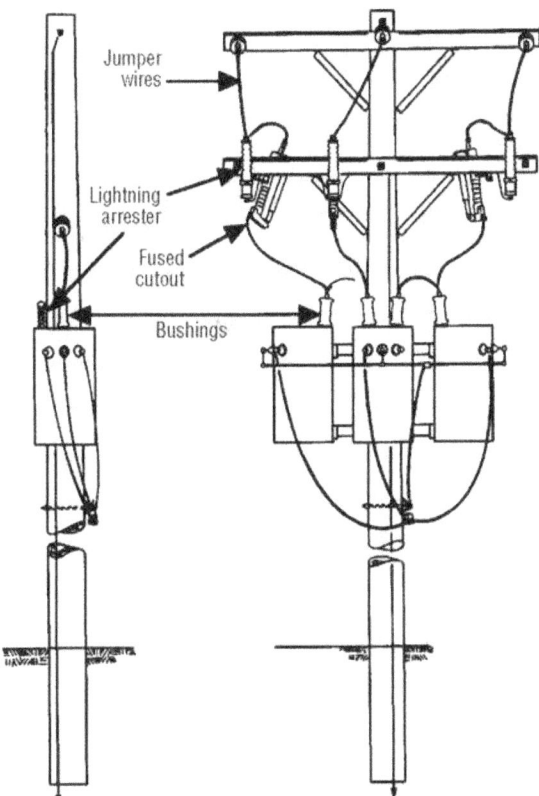

Figure 50.6 Complicated wiring that is configured with transformers, jumpers, and additional hardware is often responsible for raptor electrocutions. (Graphic provided by Monte Garrett, PacifiCorp)

Bird electrocutions can cause power outages; therefore, a history of electrical power disruption can help confirm the diagnosis and fix the location and time of electrocution.

Electrocuted birds may catch on fire and ignite vegetation beneath the power structures.

If a bird is electrocuted because the prey item or wet nest material it is carrying comes in contact with an energized part, then these items may be found with the carcass or clutched in its talons.

Gross Lesions

The hallmark of electrocution is burn marks. Burns are generally confined to the sites of body contact with the electrical source; however, if the feathers are ignited then the entire carcass may be charred (Fig. 50.7). Burn marks from fatal electrocutions can have a remarkable range in appearance from very subtle feather disruption to limb amputation. Burns cause the feather edges to curl or twist (Fig. 50.8), and light-colored feathers may be discolored brown or charred. Burns on avian skin appear as dry blisters, particularly on the scales of the feet or legs (Fig. 50.9A and B). The margins of these blisters may be brown or charred. Severe, deep burns can extend through the skin, cauterize muscles and tendons, liquefy fat, and even fracture bones.

Sublethal bird electrocutions are uncommon. In these cases, a single limb is usually affected. Initially, burns may be seen on the skin or the feathers at the contact site. Later, the only evidence may be the loss of blood supply to a wing or foot and eventual gangrene. If the damage can be removed by surgical amputation, some electrocuted birds can recover and be kept permanently in captivity.

Diagnosis

A diagnosis of electrocution is based on the presence of burns and an absence of evidence of other causes of death. Hemorrhages in the subcutaneous tissue and internal organs suggest cardiovascular injury and can support the diagnosis.

A field history that includes proximity to an electrical line is helpful but not sufficient in itself. Birds may collide with

Figure 50.7 An electrocuted bald eagle that is charred over most of its body.

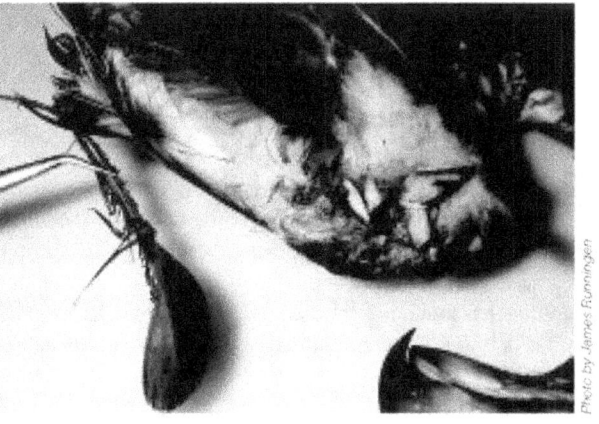

Figure 50.8 Electrical burns on the wing feathers of a bald eagle. Note also the fracture and charring of nearby bones.

electrocution. Power companies benefit by reducing costly power outages, by avoiding liability for migratory bird mortalities, and by the positive public image that is generated by control projects.

When new electrical installations are planned, the design can take into consideration the likelihood of raptor electrocution. The risk can be evaluated in advance by considering raptor concentrations and behavior along the installation route. Structures in raptor migratory corridors, as well as nesting and wintering ranges, may pose a risk.

Human Health Considerations

Under normal circumstances, there is no exposure.

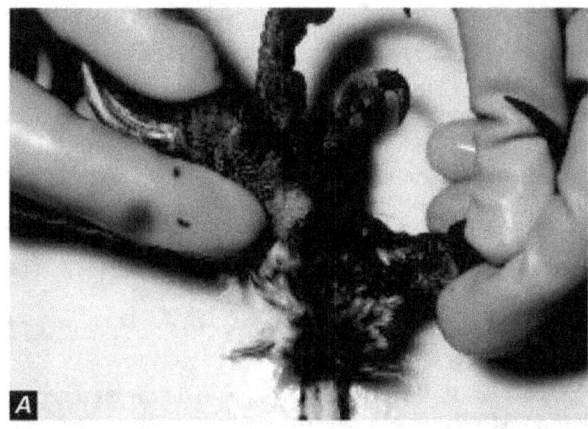

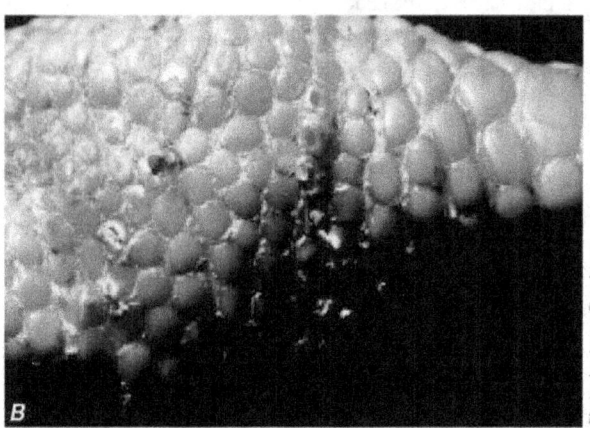

Figure 50.9 (A) A large burn on a golden eagle's foot. (B) Multiple small, subtle burns in the scales on a bald eagle's foot.

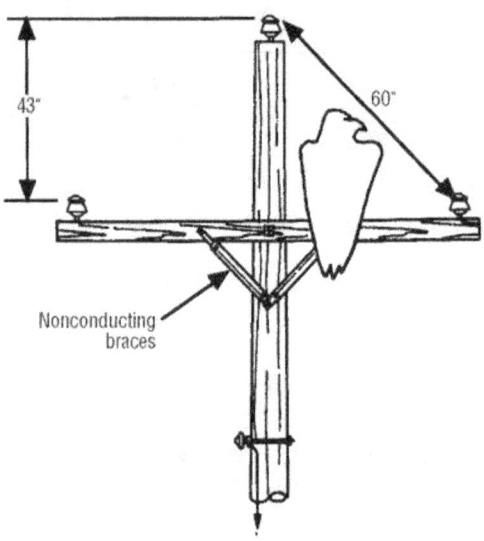

Figure 50.10 A safe wiring configuration separates the conductors and other energized hardware so that large raptors are unable to touch two pieces of hardware simultaneously. (Graphic provided by Monte Garrett, PacifiCorp)

power lines, be shot while perching, or fall from perches after poisoning or illness; therefore, location is not definitive for electrocution.

Control

Raptor electrocutions generally can be reduced by adopting safe electrical pole and line configurations or managing raptor perching. Safe wiring configurations separate the wires and the grounded metal parts so that raptors cannot simultaneously touch two of them at once (Fig. 50.10). Existing installations that contain hazardous configurations can be modified by insulating or reconfiguring the wiring. Rather than comprehensive modifications, an economical but effective approach is to modify selected poles based on field observations of bird use and mortality. If reconfiguring or insulating the wires is not feasible, then access to the hazardous perch can be blocked and safer, alternate perches can be provided. Despite the inherent equipment costs of modification, electrical power companies are often proactive in preventing bird

Nancy J. Thomas

Supplementary Reading

Avian Power Line Interaction Committee, 1996, Suggested practices for raptor protection on power lines: The state of the art in 1996: Edison Electric Institute/Raptor Research Foundation, Washington, D.C., 125 p.

Bevanger, K., 1994, Bird interactions with utility structures: Ibis, v. 136, p. 412–425.

Chapter 51
Miscellaneous Diseases

This concluding chapter is intended to further inform the reader of the broad spectrum of causes affecting the health of wild birds by illustrating a variety of disease conditions that are not described elsewhere in this Manual. The information in this chapter is not intended to represent a comprehensive description of other causes for ill-health and death in wild birds. Instead, examples are provided of some less commonly reported conditions that, in some instances, illustrate larger health issues. Too little is known about these conditions to currently assess their biological significance as mortality factors in wild birds.

Disease in Hatchlings and Young

Much of what is known about disease in free-ranging wild birds is the result of observations and investigations of fully grown birds. Nevertheless, the knowledge gained from domestic poultry and captive-reared wild birds has often demonstrated great disease impacts for young birds. Loss of young can have significant impacts on population levels (see Trichostrongylidosis in Chapter 35); therefore, special vigilance is needed to prevent the introduction of disease into free-ranging populations that have the potential for high mortality of young.

Duck hepatitis is an example of a disease of domestic ducks that could cause mortality of young free-ranging birds if it were to spread to free-ranging populations (Figs. 51.1–3). This highly fatal, rapidly spreading viral disease is found worldwide and is economically important to all duck-raising operations because of the high potential of mortality if it is not controlled. Young pheasants, goslings, and young guinea fowl have all suffered high mortality following experimental infection with duck hepatitis virus, thereby illustrating a greater host range than waterfowl. Mallard ducklings are also killed by this virus, and adult mallards have been reported to serve as mechanical or noninfected transport hosts for the movement of duck hepatitis virus between commercial duck-raising operations. Clinical signs and mortality in mallards have been confined to ducklings less than 3-weeks old. However, birds that recovered from infection have been reported to shed the virus in their feces for up to 8 weeks postinfection.

Plastic Debris

Improper disposal of several types of products made from plastic causes problems for birds. Some of these problems can result in mortality. They can frequently be reduced by educating people about the problems and by other means (Figs. 51.4–6).

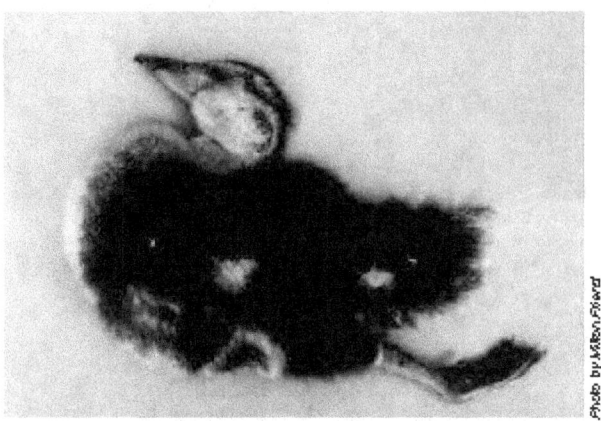

Figure 51.1 Typical terminal position of ducklings that die from duck hepatitis. This posture is referred to as opisthotonos, and it is characterized by the body being somewhat bowed forward with the head and bottom of the feet bent backward.

Figure 51.2. Typical liver lesions of mallard ducklings that died from duck hepatitis. Note the color change and enlargement of the two infected livers (A and B) compared with the liver from an uninfected duckling of the same age (C). The principal lesions, in addition to the greatly enlarged liver, are hemorrhages over varying amounts of the surface area. The more discrete areas of hemorrhage are referred to as petechia (for the very small isolated areas) or punctate (dotlike), and the broader areas of hemorrhage as ecchymotic.

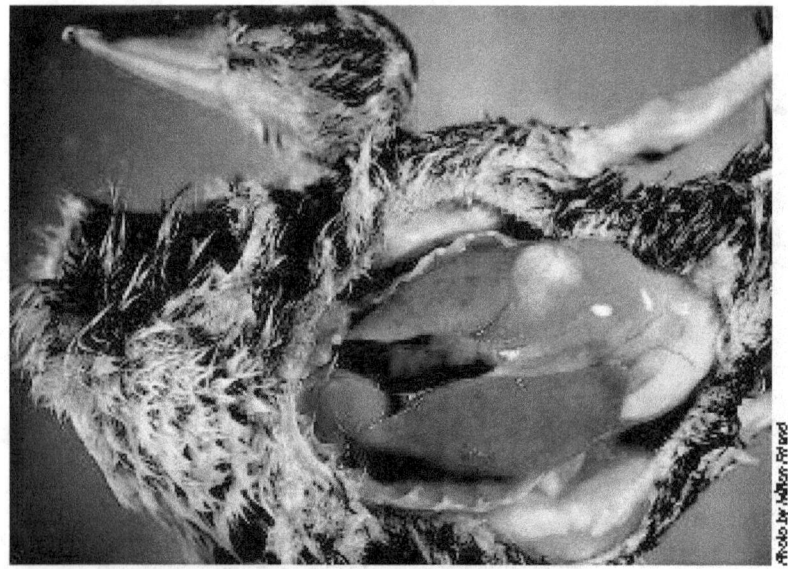

Figure 51.3 Mallard duckling infected with duck hepatitis. The livers of infected birds generally become so swollen that they fill much of the bird's abdominal cavity.

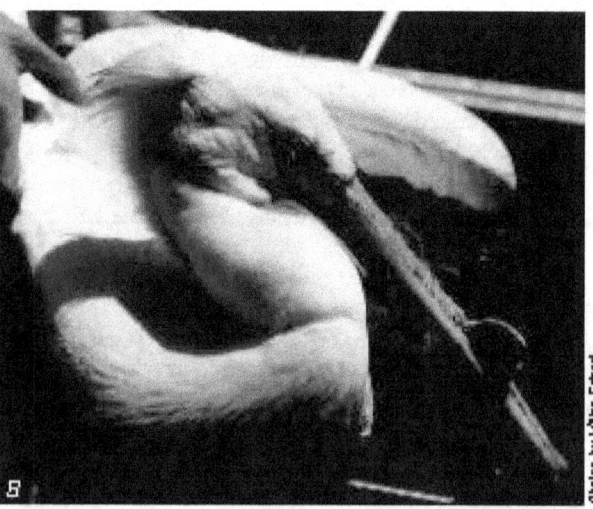

Figure 51.4 (A) Improperly discarded fishing line carried to the top of this tree by a double-crested cormorant became a "hangman's noose" and strangled the bird in this photograph. The line tangled around the tree top and it also looped around the bird's neck when it attempted to fly from its perch above a small urban lake. (B) Discarded fishing line wrapped around the bill of this white pelican would have resulted in death by starvation had the bird not been captured and the line removed. Note also the constricted areas of the pouch caused by the line.

Figure 51.5 A Canada goose with a plastic 6-pack ring entangled around its neck (arrow). Birds accidentally acquire these rings when they place their heads through them as they feed on the ground.

Figure 51.6 These discarded plastic materials were found in the stomach of an albatross chick. Items such as these are ingested as food by adult birds when they feed at sea and reach the chick when the adult regurgitates food to feed its young. Fortunately, most debris of this type is voided by the chicks without causing them harm. However, birds can suffer intestinal blockages and other ill effects.

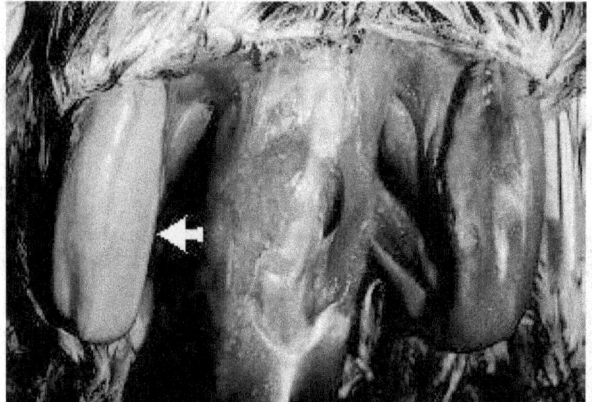

Figure 51.7 Light colored muscle of leg (arrow) represents capture myopathy in a sandhill crane.

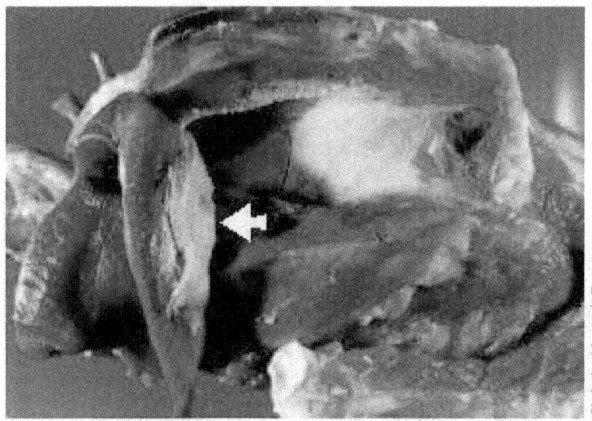

Figure 51.8 Light colored area in breast muscle (arrow) of a peregrine falcon with capture myopathy.

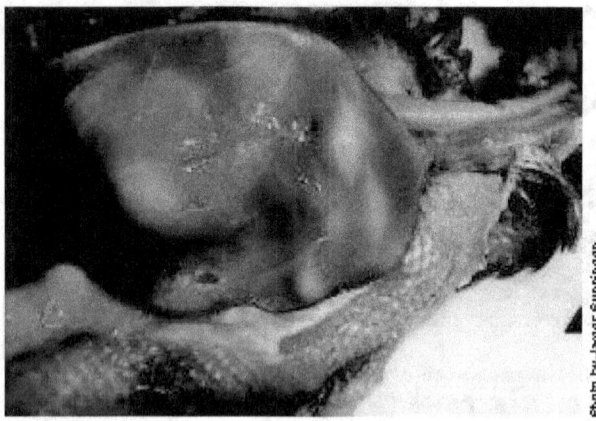

Figure 51.9 Tumors in breast muscle of a Canada goose.

Disease Due To Handling

Improper judgements and procedures by humans while they pursue, handle, and transport wild animals, including birds, during wildlife management activities can induce capture myopathy (Figs. 51.7–8). More descriptive names include over-straining disease, transport myopathy, exertional myopathy, muscle necrosis, white muscle disease, and stress myopathy. These names convey that improper handling or stress can cause a bird to overexert and result in stress-related injury to its muscles. Tissue damage is a result of complex physiological processes, not physical trauma such as bruising from impact. Mortality has been reported in a wide variety of bird species including flamingos, cranes, waterfowl, raptors, gulls, wild turkey, and other species. This disease of overexertion results in severe damage to striated muscles including the heart. Birds may die hours or even several days after they have been released, thereby leaving their human captors and handlers unaware of the damage that was done. The potential for inducing this disease should be carefully considered during the planning phases of wildlife capture, handling, and transportation, and measures should be taken to minimize risks. Warm environmental temperatures are often a risk factor as are the duration of pursuit, the method and duration of restraint, placement of birds in unfamiliar surroundings, and noise associated with human activities. Situations that have induced capture myopathy in birds include trapping and handling operations involving drop nets and rocket nets; drive-trapping, handling, and translocation of flightless birds; and handling birds so that marking devices, including radio transmitters, can be placed on them. All of these needed activities can be done safely if proper consideration is given to capture myopathy and the steps that can be taken to avoid inducing this disease.

Tumors

Neoplasms or tumors are infrequent findings in free-ranging wild birds, but they are found (Figs. 51.9–12). Tumors are formed by the abnormal progressive multiplication of cells into uncontrolled (by the body) new tissue that appears as various growths within tissues and organs. These growths may be noninvasive or benign, or they may spread to other tissues and parts of the body and be malignant. Tumors result from multiple causes. Virus-induced tumors, such as the herpesvirus that causes an important infectious poultry disease known as Marek's disease, are transmissible. Tumors formed due to other than infectious agents have been reported from all major body systems of birds, the reproductive, digestive, respiratory, nervous, and endocrine systems, in addition to the skin surfaces.

Less than 1 percent of the wild birds for which postmortem examinations were done at the National Wildlife Health Center (NWHC) over a span of more than 20 years (1975–1998) had tumors. These findings are consistent with those of other disease diagnostic laboratories that process large numbers of free-ranging wildlife. A notable exception at the NWHC has been a high prevalence of tumors in Mississippi sandhill cranes received from the wild (Figs. 51.11, 12). The cause(s) of the tumors in this endangered species remains undetermined.

Trauma

Many wild birds are injured and killed each year from impacts with buildings, wires, and other products of the human environment (Figs. 51.13, 14). Birds that have large wing spans, such as cranes and eagles, are among those commonly found with fractured wings and other injuries from collisions with power lines and wire fences. Road kills of raptors that feed on carrion are common. Whenever it is feasible, bird flight patterns and bird use of local habitat should be considered in the routing of power transmission lines, wind power generation units, and roads. Protective measures against bird strikes should be employed when they are warranted if less hazardous alternative routings cannot be accomplished. Monitoring for road kills of birds and observations of birds feeding on carcasses can indicate food shortages for species such as eagles and can be mitigated by establishing short-term feeding stations that move the birds from the roadways to safer locations during the period of food scarcity.

Other

Wild birds are subject to major direct losses from weather. Waterfowl and other species have been frozen to the ice by their feet and feathers (Fig. 51.15), and strong winds associated with hurricanes have filled coastal beaches with large numbers of birds with fractured wings. Heavy snows and storms that coat vegetation with a thick layer of ice deprive

Figure 51.10   Tumor on the leg of a ruffed grouse.

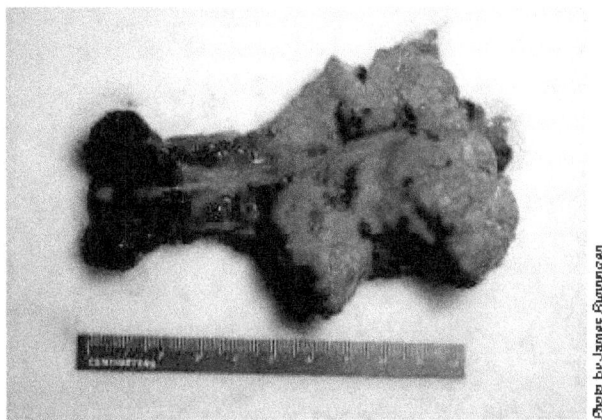

Figure 51.11   Tumor attached to the kidneys of a Mississippi sandhill crane.

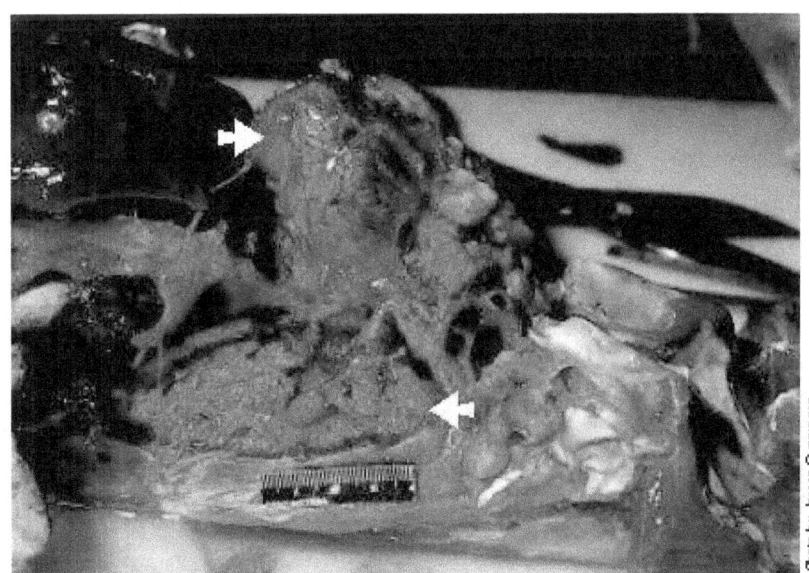

Figure 51.12   A malignant tumor covering the heart (top arrow) and lungs (bottom arrow) of a Mississippi sandhill crane.

Figure 51.13 Collision with fences, power lines, and other structures is a significant mortality factor for birds. This whooping crane died after striking a fence.

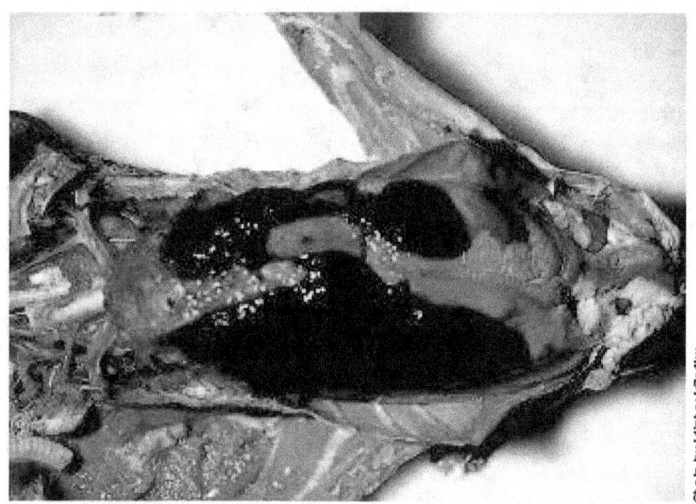

Figure 51.14 Trauma from collision often results in massive internal hemorrhage.

Figure 51.15 Severe weather can cause large losses of wildlife. These Canada geese became entrapped by ice when high winds accompanied by temperatures that rapidly dropped below freezing during a spring storm quickly turned this shallow wetland into a frozen body of water. The high winds prevented flight, and the water splashing over the birds froze them in place. Severe traumatic injuries resulted as the birds tried to free themselves from the ice.

wild birds and other wildlife of access to food and can result in starvation (Fig. 51.16). Numerous other weather-related situations also affect bird health.

Malnutrition resulting in starvation is but one aspect of nutritional diseases that may affect birds. Nutritional diseases are a complex subject area that is beyond the scope of this Manual, and they are mentioned only to make the reader aware of them. Nutritional diseases involve excess intake as well as deficiencies. Changes in bird diets associated with landscape changes due to agriculture can contribute to nutritional diseases. For example, excesses of dietary protein and vitamin deficiency may occur due to extensive feeding on agricultural grains rather than natural food sources. Visceral gout may result (Fig. 51.17). Under experimental conditions, substances that are toxic to the kidneys (nephrotoxic agents) and diets deficient in Vitamin A and high in calcium have caused avian gout.

Wild birds also drown. Drowning may be an outcome of extreme weather conditions that aquatic birds are sometimes subject to; exhaustion of passerines during migration, which causes them to drop into water bodies that they may be traversing at the time; and as a result of other factors, such as the feathers of aquatic birds becoming waterlogged from oil contamination or nonfunctioning preen glands that prevent birds from "waterproofing" their feathers.

Various deformities due to a variety of causes are also seen in wild birds (Fig. 51.18). Some deformities result from exposure to excess levels of selenium; others may result from exposure to synthetic compounds, nutritional disorders, or injury to tissues during early developmental stages of the bird; they may be of genetic origin; or result from other causes. Deformities are not commonly observed because birds that are afflicted with such conditions are likely to be more vulnerable to factors that reduce their chance for survival. Therefore, clusters of observations of deformities should be viewed as an indication of a larger problem and warrant investigation to determine the underlying cause.

Milton Friend and Nancy J. Thomas

Supplementary Reading

Fairbrother, A., Locke, L.N., and Hoff, G.L., 1996, Noninfectious diseases of wildlife, (2d ed.): Ames, Iowa, Iowa State University Press, 219 p.

Wallach, J.D., and Cooper, J.E., 1982, Nutritional diseases of wild birds, in Hoff, G.L., and others, eds., Noninfectious diseases of wildlife: Ames, Iowa, Iowa State University Press, p. 113–126.

Figure 51.16 Ice that coats vegetation may prevent access to food, resulting in starvation.

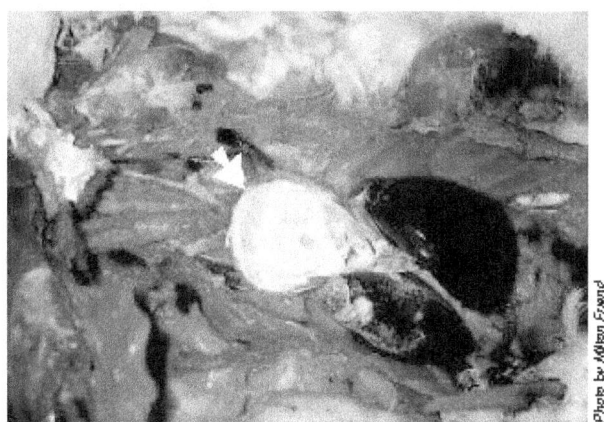

Figure 51.17 Dietary protein imbalances can cause visceral gout, exhibited by an accumulation of white, gritty deposits on surfaces of organs, such as the heart (arrow).

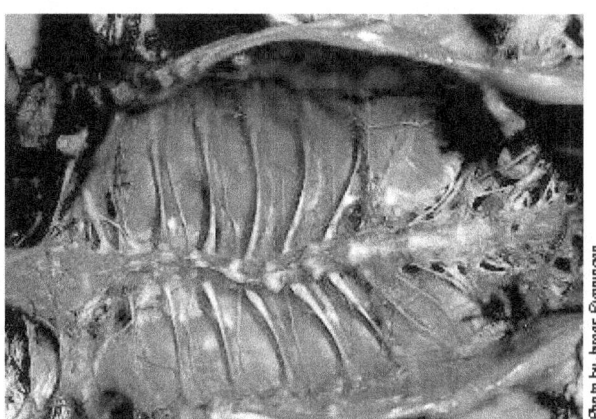

Figure 51.18 Vertebral column deformity (scoliosis) in a bald eagle.

Appendix A

Sample specimen history form

Submitter's name: Affiliation:

Address: Telephone: E-mail:

Date collected:

Method of collection: [found dead, euthanized (describe method) etc.]

Collector's name:

Specific die-off location:

State: County: Latitude/longitude:

Environmental factors: (Record conditions such as storms, precipitation, temperature changes, or other changes that may contribute to stress.)

Disease onset: (The best estimate of when the outbreak started.)

Species affected: (The diversity of species affected may provide clues to the disease involved.)

Age/sex: (Any selective mortality related to age and sex.)

Morbidity/mortality: (Ratio of sick animals to dead animals.)

Known dead: (Actual pickup figures.)

Estimated dead: (Consider removal by scavengers or other means.)

Clinical signs: (Any unusual behavior and physical appearance.)

Population at risk: (Number of animals in the area that could be exposed to the disease.)

Population movement: (Recent changes in the number of animals on the area and their source or destination, if known.)

Problem area description: (Land use, habitat types, and other distinctive features.)

Comments: (Additional information/observations that may be of value such as past occurrences of disease in area.)

Appendix B

Sources of wildlife diagnostic assistance in the United States

Assistance in obtaining a diagnosis of wildlife illness or death is available from a variety of sources. However, it is advisable to make inquiries before the need arises about available services, the estimated response time for completing work, and who to contact when assistance is required.

The following wildlife disease programs can offer information, assistance, and services.

Wildlife Disease Programs

1. U.S. Department of Interior,
   U.S. Geological Survey,
   Biological Resources Division
   National Wildlife Health Center
   6006 Schroeder Road
   Madison, WI 53711
   Telephone (608) 270-2400
   Web site: http://www.emtc.usgs.gov/nwhchome.html

2. State fish and game agencies. Several States have wildlife disease programs. Among those are Alaska, California, Colorado, Florida, Idaho, Michigan, New Jersey, New York, Wisconsin, and Wyoming. Contact the State fish and game agency headquarters to inquire about assistance.

3. Regional wildlife disease programs. Two regional programs are presently affiliated with universities:

   Southeastern Cooperative Wildlife Disease Study
   College of Veterinary Medicine
   University of Georgia
   Athens, GA 30602
   Telephone (706) 542-1741

   Northeastern Research Center for Wildlife Diseases
   University of Connecticut
   Department of Pathobiology
   Storrs, CT 06269-3089
   Telephone (860) 486-4000

4. University programs. Several other universities, for example, the University of Florida—Gainesville and Virginia Polytechnical Institute and State University—Blacksburg, are involved in wildlife disease activities. Inquiries at schools of veterinary medicine and departments of veterinary or animal science at universities throughout the United States will reveal additional sources of wildlife disease diagnostic assistance.

5. Private sector. Some private consultants also deal with wildlife disease problems.

Domestic Animal Disease Programs

1. U.S. Department of Agriculture,
   Animal and Plant Health Inspection Service
   National Veterinary Services Laboratories
   P.O. Box 844, Ames, Iowa 50010
   Telephone (515) 239-8600.

   This facility accepts diagnostic specimens that have been referred to it through appropriate State or Federal channels.

2. State departments of agriculture. Animal disease diagnostic laboratories exist to serve domestic animal needs, but will often accept wildlife specimens.

3. Private sector. Veterinarians in private practice often have both interest and expertise in wildlife diseases and may become involved with these problems.

Additional sources of assistance for investigating wildlife mortality events when chemical toxins are suspected.

Federal Government

U.S. Fish and Wildlife Service (FWS)

The Division of Environmental Contaminants (DEC) is the FWS focal point for issues associated with chemical toxins. Information on contaminants can be obtained from the Central Office in Washington, D.C.

Telephone: (703) 358-2148
Web site: http://www.fws.gov/~r9dec/ecprog.html and from DEC staff in the seven FWS Regional Offices.

DEC biologists are assigned to field offices throughout the FWS Regions. They work on specific contaminant issues in each Region and are available to provide information and assistance regarding mortality event investigations.

Toxic spill coordinators are located in each Regional Office, providing a focal point for response actions.

FWS Regional Offices are located in:
Portland, Ore.; Albuquerque, N. Mex.; Fort Snelling, Minn.; Atlanta, Ga.; Hadley, Mass.; Denver, Colo.; and Anchorage, Alaska.

U.S. Geological Survey, Biological Resources Division (BRD)

The National Wildlife Health Center, which is the BRD Science Center in Madison, Wis., provides information about and assists in investigating wildlife mortality events.

Telephone: (608) 270-2400
Web site: http://www.emtc.usgs.gov/nwhchome.html

Research on chemical toxins is carried out at several BRD Science Centers. Those Centers maintain in-depth technical knowledge regarding the fate and impacts of chemicals in the environment. The Patuxent Wildlife Research Center in Laurel, Md. is an internationally recognized source of information on the effects of contaminants, particularly on avian species.

Telephone: (301) 497-5500
Web site: http://www.pwrc.usgs.gov/

U.S. Environmental Protection Agency (EPA)

Office of Solid Waste and Emergency Response

Telephone: (703) 308-8413
Web site: http://www.epa.gov/epaoswer/

The Office of Pollution Prevention and Toxics (OPPT) assesses the hazards and risks posed by industrial chemicals to human health and the environment. The Environmental Effects Branch in OPPT can provide information on the toxicity of chemicals to aquatic and terrestrial organisms.

Telephone: (202) 260-1268

Office of Pesticide Programs maintains the Ecological Incident Information System, which is a data base on mortality of non-target organisms caused by pesticides.

Telephone: (703) 305-5392

State Government

Many State natural resource agencies have environmental contaminant programs that provide a mechanism to report suspected chemical toxin problems. Some States have groups that investigate mortality events associated with chemicals, and that may be able to provide field assistance and chemical analysis.

Natural resource agencies in several States maintain wildlife health programs, which respond to wildlife mortality events.

State veterinary diagnostic laboratories often have toxicologists on staff who have specific knowledge of toxic problems within the region.

Poison Control Centers

The National Animal Poison Control Center at the University of Illinois College of Veterinary Medicine provides a fee-based service directed to prevention and treatment of adverse effects of chemical exposures in animals. This service is staffed by veterinary health professionals who have access to a wide range of information specific to animal poisoning.

Telephone (800) 548-2423, (900) 680-0000
Web site: http://www.cvm.uiuc.edu/NAPCC/NAPCC.html

The American Association of Poison Control Centers (AAPCC) has certified about 40 regional poison information centers throughout the U.S. that focus on human exposure to chemical toxins. These centers function to provide poison information, telephone management and consultation, collect pertinent data, and deliver professional and public education. The national AAPCC office is in Washington, D.C.
Telephone: (202) 362-7217. A directory of the regional centers is available at Web site: http://www.pitt.edu/~martint/pages/rpiclist.htm

Colleges of Veterinary Medicine

Most colleges of veterinary medicine have toxicology departments staffed with experts in the area of animal toxicology.

Analytical Laboratories

Choosing an analytical laboratory requires attention to methods used, quality assurance/quality control (QA/QC), and cost. Laboratories should be using methods that are appropriate to the analysis required in the matrix (material being analyzed) that is submitted. Minimum quality control data provided by the laboratory should include:

(1) The results of analysis of spiked samples, or recovery. A known amount of the compound being analyzed for is added to the appropriate matrix. The recovery is the amount of the compound that was recovered in the analysis, and it is expressed as a percentage of the amount of compound added.
(2) A replication of results, or an agreement of analyses of duplicate samples.
(3) The results of blank samples, or an absence of the compound being analyzed for in a "clean" sample of the appropriate matrix.
(4) The results of analysis of standard reference samples. A sample with a known quantity of the compound is prepared by an independent laboratory, and this sample is then analyzed by the laboratory being evaluated. Although good QA/QC adds to the expense of analytical work, the alternative may be an incorrect diagnosis.

Some of the analytical laboratories that have been used by the FWS DEC and others include:

For inorganic analyses:
  Environmental Trace Substance Laboratory
  University of Missouri - Rolla
  101 USBM Bldg., 1300 North Bishop Ave.
  Rolla, MO 65409-0530
  Telephone: (314) 341-6607

  Research Triangle Institute
  3040 Cornwallis Road, Bldg. 6
  Research Triangle Park, NC 27709-2194
  Telephone: (919) 541-6896

  Geochemical & Environmental Research Group
  833 Graham Road
  College Station, TX 77845
  Telephone: (409) 690-0095

For organic analyses:
  Geochemical & Environmental Research Group (see above)

  Mississippi State Chemical Laboratory
  Mississippi State University
  Hand Chemical Lab, Rm 201, Morrill Road
  Mississippi State, MS 39762
  Telephone: (601) 325-3251

The above listing is not intended to be comprehensive, nor does it constitute endorsement by the Federal government. Rather, it illustrates the diversity of possible sources of assistance. Individual circumstances and events dictate which of these sources will be most useful in specific situations.

Appendix C

Sources of supplies used for collecting, preserving, and shipping specimens

Company	Address and telephone	Item
Scientific Products	319 West Ontario Chicago, IL 60610 Tel: (800)323-4515	Whirl-Pak® bags, formalin, wide-mouth, plastic jars, indelible markers
Curtis Matheson Scientific, Inc.	P.O. Box 1546 Houston, TX 77251 Tel: (713)820-9898	Whirl-Pak® bags, wide-mouth, plastic jars, indelible markers
Fisher Scientific Co.	711 Forbes Ave. Pittsburgh, PA 15219 Tel: (800)766-7000	Whirl-Pak® bags, formalin, indelible markers
Thomas Scientific	P.O. Box 99 Swedesboro, NJ 08095 Tel: (609)467-2000	Whirl-Pak® bags, wide-mouth, plastic jars, indelible markers
VWR Scientific	P.O. Box 66929 O'Hare Amp Chicago, IL 60666 Tel: (800)932-5000	Whirl-Pak® bags, indelible markers
Local hospital and medical supply businesses		Whirl-Pak® bags, wide-mouth, plastic jars
Some pharmacies		formalin
Freund Can Co.	167 W. 84th Street Chicago, IL 60620 Tel: (312)224-4230	metal paint cans with lids
U.S. General Services (gov't agencies only)	GSA Customer Supply Center Administration (GAS) 5619 W. 115th St. Worth, IL 60482 Tel: (800)262-0570	filament strapping tape, plastic bags, indelible markers
GSA Federal Supply Service	1500 E. Bannister Rd. Kansas City, MD 64131 Tel: (816)926-7315	filament strapping tape, plastic bags, indelible markers

Company	Address and telephone	Item
Local hardware, sports, and discount stores		Styrofoam®, shipping coolers, indelible markers, ice packs, filament strapping tape
Polyfoam Packers Corp.	2320-T Foster Avenue Wheeling, IL 60090 Tel: (800)225-7443	biomedical shippers and mailers

Protective clothing, gloves, and disinfectants can also be obtained from many of the sources listed above. The above list is not intended to be comprehensive, nor does it constitute endorsement by the Federal government.

Appendix D

Normal brain cholinesterase activity values

Species	Mean[1]	Standard deviation	Sample size
Avocet, American	19.4	2.9	5
Blackbird, Red-winged	24.5	1.2	5
Bobwhite, Northern	16.3	2.1	7
Brant, Black	14.4	0.8	5
Coot, American	20.5	5.0	12
Cormorant, Double-crested	29.3	2.4	5
Cowbird, Brown-headed	19.7	2.6	5
Crane, Mississippi Sandhill	16.6	2.3	15
Crane, Sandhill	17.5	1.4	8
Crane, Whooping	15.1	1.5	9
Dove, Mourning	22.8	3.3	11
Duck, American Wigeon	10.5	1.0	12
Duck, Blue-winged Teal	19.5	3.6	6
Duck, Green-winged Teal	13.5	2.0	17
Duck, Kola	11.0	1.2	9
Duck, Mallard	11.0	1.6	75
Duck, Muscovy	11.6	3.5	8
Duck, Northern Shoveler	14.7	1.2	5
Duck, Pintail	11.6	1.3	24
Duck, Ruddy	13.4	1.4	8
Duck, Wood	10.2	1.5	9
Eagle, Bald	16.0	2.6	156
Eagle, Golden	16.0	2.2	57
Egret, Common	17.4	1.9	5
Egret, Snowy	25.0	1.3	5
Falcon, Peregrine	18.6	3.2	27
Goldfinch, American	17.8	1.5	10
Goose, Canada	11.9	2.0	36
Goose, Canada (Aleutian)	14.0	3.6	8
Goose, Ross	14.2	2.3	9
Goose, Snow	13.6	2.6	42
Goose, White-fronted	12.1	1.2	9
Grebe, Eared	14.7	1.4	17
Grosbeak, Evening	20.3	3.1	5
Gull, Ring-billed	23.9	6.5	8
Hawk, Red-tailed	17.5	1.4	7
Hawk, Sharp-shinned	21.5	2.0	6
Heron, Black-crowned Night	15.6	3.0	5
Heron, Great Blue	13.3	2.1	8
Loon, Common	17.3	4.4	5
Owl, Great-horned	15.5	2.0	7
Owl, Screech	18.7	1.4	5
Owl, Spotted	14.6	2.0	9

Species	Mean[1]	Standard deviation	Sample size
Parrot, Puerto Rican	19.4	2.0	6
Pelican, American White	13.0	1.3	13
Pelican, Brown	11.2	1.2	16
Sandpiper, Semipalmated	14.1	1.1	5
Siskin, Pine	16.9	2.3	16
Stork, Wood	18.7	4.2	8
Swan, Trumpeter	11.3	1.3	10
Swan, Tundra	11.6	1.7	14
Tern, Least California	44.1	9.1	5
Woodcock, American	16.8	1.1	5
Woodpecker, Red-cockaded	38.4	4.8	5

[1] Cholinesterase activity is expressed as micromoles acetylthiocholine hydrolyzed per minute per gram of wet weight brain tissue (Hill, E.F. and Fleming, W.J., 1982, Environmental toxicology and chemistry 1:27–38).

Milton Smith

Appendix E

Common and scientific names of birds in text

Albatrosses, shearwaters, and petrels (Order Procellariiformes)
    Albatrosses (Family Diomedeidae)
        Black-footed albatross (*Diomedia nigripes*)
        Laysan albatross (*Diomedia immutabilis*)
    Shearwaters and petrels (Family Procellariidae)
        Northern fulmar (*Fulmarus glacialis*)
        Petrels (*Pterodroma* sp, *Oceanodroma* sp, *Oceanites* sp)
        Sooty shearwater (*Puffinus griseus*)
Cormorants, pelicans, and tropicbirds (Order Pelecaniformes)
    Cormorants and Shags (Family Phalacrocoracidae)
        Brandt's cormorant (*Phalacrocorax penicillatus*)
        Cape cormorant (*Phalacrocorax capensis*)
        Double-crested cormorant (*Phalacrocorax auritus*)
        Great cormorant (*Phalacrocorax carbo*)
        Shags (*Phalacrocorax* sp)
    Gannets and boobies (Family Sulidae)
        Gannet (*Morus bassanus*)
    Pelicans (Family Pelecanidae)
        Brown pelican (*Pelecanus occidentalis*)
        White pelican (*Pelecanus erythrorhynchos*)
    Tropicbirds (Family Phaethontidae)
        Red-tailed tropicbird (*Phaethon rubricauda*)
        White-tailed tropicbird (*Phaethon lepturus*)
Cranes, coots, and rails (Order Gruiformes)
    Cranes (Family Gruidae)
        Brolga crane (*Grus rubicunda*)
        Common crane (*Grus grus*)
        Demoiselle crane (*Anthropoides virgo*)
        East African crowned crane (*Balearica regulorum gibbericeps*)
        Greater sandhill crane (*Grus canadensis tabida*)
        Hooded crane (*Grus monacha*)
        Lesser sandhill crane (*Grus canadensis canadensis*)
        Manchurian crane (red-crowned crane) (*Grus japonensis*)
        Mississippi sandhill crane (*Grus canadensis pulla*)
        Sarus crane (*Grus antigone*)
        Stanley crane (blue crane) (*Anthropoides paradisea*)
        White-naped crane (*Grus vipio*)
        Whooping crane (*Grus americana*)
    Rails, coots, and gallinules (Family Rallidae)
        American coot (*Fulica americana*)
        Common moorhen (*Gallinula chloropos*)
        Sora (*Porzana carolina*)
Emus and ostriches (Order Struthioniformes)
    Emu (Family Casuariidae, *Dromaius novaehollandiae*)
    Ostrich (Family Struthionidae, *Struthio camelus*)
Grebes (Order Podicipediformes, Family Podicipedidae)
    Eared grebe (*Podiceps nigricollis*)
    Western grebe (*Aechmophorus occidentalis*)

Grouse, quail, and partridges (Order Galliformes, Family Phasianidae)
    Grouse (Subfamily Tetraoninae)
        Blue grouse (*Dendragapus obscurus*)
        Ptarmigan (*Lagopus* sp)
        Red grouse (*Lagopus lagopus scoticus*)
        Ruffed grouse (*Bonasa umbellus*)
        Sage grouse (*Centrocercus urophasianus*)
        Sharp-tailed grouse (*Tympanuchus phasianellus*)
    Quail (Subfamily Odontophorninae)
        Bobwhite quail (Northern) (*Colinus virginianus*)
        California quail (*Callipepla californica*)
        Coturnix quail (*Coturnix coturnix*)
        Japanese quail (*Coturnix japonica*)
        Scaled quail (*Callipepla squamata*)
    Partridges and pheasants (Subfamily Phasianinae)
        Chukar partridge (*Alectoris chukar*)
        Gray partridge (Hungarian partridge) (*Perdix perdix*)
        Common peafowl (*Pavo cristatus*)
        Ring-necked pheasant (*Phasianus colchicus*)
    Turkeys (Subfamily Meleagridinae)
        Wild turkey (*Meleagris gallopavo*)
    Guinea fowl (Subfamily Numidinae, *Numida* sp)
Hawks, falcons, and vultures (Order Falconiformes)
    Hawks, eagles, and kites (Family Accipitridae)
        Eurasian sparrowhawk (*Accipiter nisus*)
        Bald eagle (*Haliaeetus leucocephalus*)
        Booted eagle (*Hieraaetus pennatus*)
        Cooper's hawk (*Accipiter cooperii*)
        Common buzzard (*Buteo buteo*)
        Ferruginous hawk (*Buteo regalis*)
        Golden eagle (*Aquila chrysaetos*)
        Northern goshawk (*Accipiter gentilis*)
        Imperial eagle (*Aquila heliaca*)
        Northern harrier (marsh hawk) (*Circus cyaneus*)
        Red-shouldered hawk (*Buteo lineatus*)
        Red-tailed hawk (*Buteo jamaicensis*)
        Rough-legged hawk (*Buteo lagopus*)
        Sharp-shinned hawk (*Accipiter striatus*)
        Steppe eagle (*Aquila nipalensis*)
        White-tailed eagle (*Haliaeetus albicilla*)
        Osprey (*Pandion haliaetus*)
    Falcons and caracaras (Family Falconidae)
        American kestrel (sparrowhawk) (*Falco sparverius*)
        Gyrfalcon (*Falco rusticolus*)
        Peregrine falcon (*Falco peregrinus*)
        Prairie falcon (*Falco mexicanus*)
        Red-headed falcon (*Falco chicquera*)
        Saker falcon (*Falco cherrug*)
    Vultures (Family Cathartidae)
        California condor (*Gymnogyps californianus*)
        King vulture (*Sarcoramphus papa*)
        Turkey vulture (*Cathartes aura*)
Herons, ibises, and storks (Order Ciconiiformes)
    Flamingos (Family Phoenicopteridae)

                Greater flamingo (*Phoenicopterus ruber*)
            Herons, egrets, and bitterns (Family Ardeidae)
                Bitterns (*Botaurus* sp, *Ixobrychus* sp)
                Black-crowned night heron (*Nycticorax nycticorax*)
                Great blue heron (*Ardea herodias*)
                Great egret (common or american egret) (*Casmerodius albus*)
                Snowy egret (*Egretta thula*)
            Ibises and spoonbills (Family Threskiornithidae)
                Glossy ibis (*Plegadis falcinellus*)
                White ibis (*Eudocimus albus*)
                White-faced ibis (*Plegadis chihi*)
            Storks (Family Ciconiidae)
                Wood stork (*Mycteria americana*)
    Loons (Order Gaviiformes, Family Gaviidae)
                Common loon (*Gavia immer*)
                Pacific loon (*Gavia pacifica*)
                Red-throated loon (*Gavia stellata*)
    Owls (Order Strigiformes)
        Typical owls (Family Strigidae)
                Barred owl (*Strix varia*)
                Eagle owl (*Bubo bubo*)
                Eastern screech owl (*Otus asio*)
                Great-horned owl (*Bubo virginianus*)
                Long-eared owl (*Asio otus*)
                Short-eared owl (*Asio flammeus*)
                Snowy owl (*Nyctea scandiaca*)
                Spotted owl (*Strix occidentalis*)
        Barn-Owls (Family Tytonidae)
                Barn owl (*Tyto alba*)
    Parrots, parakeets, and macaws (Order Psittaciformes)
        Parrots (Family Psittacidae)
                Budgerigar (*Melopsittacus undulatus*)
                Cockatiel (*Nymphicus hollandicus*)
                Hawk-headed parrot *(Deroptyus accipitrinus)*
                Lories (*Lorius* sp)
                Lorikeets (*Trichoglossus* sp)
                Lovebirds (*Agapornis* sp)
                Macaws (*Ara* sp)
                Puerto Rican parrot (*Amazona vittata*)
                Rosellas (*Platycercus* sp)
                Yellow-naped parrot (*Amazona auropalliata*)
    Perching birds (Order Passeriformes)
        Finches (Family Fringillidae)
                American goldfinch (*Carduelis tristis*)
                Apapane (*Himatione sanguinea*)
                Cassin's finch (*Carpodacus cassinii*)
                Eurasian bullfinch (*Pyrrhula pyrrhula*)
                Evening grosbeak (*Coccothraustes vespertinus*)
                Goldfinches (*Carduelis* sp)
                House finch (*Carpodacus mexicanus*)
                Pine siskin (*Carduelis pinus*)
                Purple finch (*Carpodacus purpureus*)
        Wood-warblers, tangers, grosbeaks, sparrows, and blackbirds (Family Emberizidae)
                Brown-headed cowbird (*Molothrus ater*)

            Chipping sparrow (*Spizella passerina*)
            Common grackle (*Quiscalus quiscula*)
            Dusky seaside sparrow (*Ammospiza nigrescens*)
            Northern cardinal (*Cardinalis cardinalis*)
            Orioles (*Icterus* sp)
            Red-winged blackbird (*Agelaius phoeniceus*)
            Rufous-sided towhee (*Pipilo erythrophthalmus*)
        Chickadees (Family Paridae)
            Chestnut-backed chickadee (*Parus rufescens*)
        Crows, jays, and magpies (Family Corvidae)
            American Crow (*Corvus brachyrhychos*)
            Blue jay (*Cyanocitta cristata*)
            Eurasian jay (*Garrulus glandarius*)
            Jackdaw (*Corvus monedula*)
            Magpie (*Pica* sp)
            Raven (*Corvus corax*)
            Rook (*Corvus frugilegus*)
            Steller's jay (*Cyanocitta stelleri*)
        Waxwings (Family Bombycillidae)
            Cedar waxwing (*Bombycilla cedrorum*)
        Weavers (Family Ploceidae, *Ploceus* sp)
        Old world sparrows (Passeridae)
            House sparrow (English sparrow) (*Passer domesticus*)
        Larks (Family Alaudidae)
            Horned lark (*Eremophila alpestris*)
        Mockingbirds and thrashers (Family Mimidae)
            Northern mockingbird (*Mimus polyglottos*)
            Thrashers (*Toxostoma* sp)
        Nuthatches (Family Sittidae, *Sitta* sp)
        Shrikes (Family Laniidae, *Lanius* sp)
        Starlings (Family Sturnidae)
            European starling (*Sturnus vulgaris*)
            Mynas (*Acridotheres* sp, *Gracula* sp)
        Swallows (Family Hirundinidae, *Hirundo* sp, *Tachycineta* sp, *Riparia* sp, *Stelgidopteryx* sp)
            Martins (*Progne* sp)
        Thrushes, solitaires, and bluebirds (Family Muscicapidae)
            American robin *(Turdus migratorius*)
            Bluebirds (*Sialia* sp)
            Eurasian blackbird (*Turdus merula*)
            Thrushes (*Turdus* sp, *Ixoreus* sp, *Hylocichla* sp, *Catharus* sp)
        Estrildid finches (Family Estrildidae)
            Java finch (*Padda oryzivora*)
Penguins (Order Sphenisciformes, Family Spheniscidae)
        Blackfooted penguin (*Spheniscus demersus*)
Pigeons and doves (Order Columbiformes, Family Columbidae)
        Band-tailed pigeon (*Columba fasciata*)
        Mourning dove (*Zenaida macroura*)
        Ringed turtle dove (*Streptopelia risoria*)
        Rock dove (common pigeon) (*Columba livia*)
        White-winged dove (*Zenaida asiatica*)
        Wood pigeon (*Columba palumbus*)
Shorebirds (Order Charadriiformes)
    Gulls, terns, skuas, and skimmers (Family Laridae)
        California gull (*Larus californicus*)

        Common tern (*Sterna hirundo*)
        Franklin's gull (*Larus pipixcan*)
        Glaucous-winged gull (*Larus glaucescens*)
        Herring gull (*Larus argentatus*)
        Kittiwakes (*Rissa* sp)
        Laughing gull (*Larus atricilla*)
        Least California tern (*Sterna antillarum*)
        Ring-billed gull (*Larus delawarensis*)
        Royal tern (*Sterna maxima*)
        Skua (*Catharacta skua*)
    Auks, murres, and puffins (Family Alcidae)
        Common murre (*Uria aalge*)
        Guillemots (*Cepphus* sp)
        Murres (*Uria* sp)
        Puffins (*Fratercula* sp)
        Razorbill (*Alca torda*)
    Plovers (Family Charadriidae, *Charadrius* sp, *Pluvialis* sp)
    Sandpipers, turnstones, surfbirds, and phalaropes (Family Scolopacidae)
        American woodcock (*Scolopax minor*)
        Curlews (*Numenius* sp)
        Long-billed dowitcher (*Limnodromus scolopaceus*)
        Marbled godwit (*Limosa fedoa*)
        Pectoral sandpiper (*Calidris melanotos*)
        Ruddy turnstone (*Arenaria interpres*)
        Semipalmated sandpiper (*Calidris pusilla*)
        Spotted sandpiper (*Actitis macularia*)
        Western sandpiper (*Calidris mauri*)
    Stilts and avocets (Family Recurvirostridae)
        American avocet (*Recurvirostra americana*)
        Black stilt (*Himantopus novaezelandiae*)
        Black-necked stilt (*Himantopus mexicanus*)
Waterfowl (Order Anseriformes, Family Anatidae)
    Dabbling ducks
        American wigeon (*Anas americana*)
        American black duck (*Anas rubripes*)
        Blue-winged teal (*Anas discors*)
        Gadwall (*Anas strepera*)
        Green-winged teal (*Anas crecca*)
        Koloa duck (*Anas wyvilliana*)
        Laysan duck (*Anas laysanensis*)
        Mallard (*Anas platyrhynchos*)
        Mottled duck (*Anas fulvigula*)
        Muscovy duck (*Cairina moschata*)
        Northern pintail (*Anas acuta*)
        Northern shoveler (*Anas clypeata*)
        White Pekin duck (*Anas platyrhynchos*)
        Wood duck (*Aix sponsa*)
    Diving ducks
        Bufflehead (*Bucephala albeola*)
        Canvasback (*Aythya valisineria*)
        Common goldeneye (American goldeneye) (*Bucephala clangula*)
        Greater scaup (*Aythya marila*)
        Harlequin duck (*Histrionicus histrionicus*)
        Lesser scaup (*Aythya affinis*)

    Oldsquaw (*Clangula hyemalis*)
    Redhead (*Aythya americana*)
    Ring-necked duck (*Aythya collaris*)
    Ruddy duck (*Oxyura jamaicensis*)
  Sea ducks
    Black scoter (common scoter) (*Melanitta nigra*)
    Common eider (*Somateria mollissima*)
    Common merganser (*Mergus merganser*)
    Mergansers (*Mergus* sp, *Lophodytes* sp)
    Red-breasted merganser (*Mergus serrator*)
    Spectacled eider (*Somateria fischeri*)
    Surf scoter (*Melanitta perspicillata*)
    White-winged scoter (*Melanitta fusca*)
  Geese (Tribe Anserini)
    Aleutian Canada goose (*Branta canadensis leucopareia*)
    Bean goose (*Anser fabalis*)
    Black brant (*Branta bernicla nigricans*)
    Brant (*Branta bernicla*)
    Canada goose (*Branta canadensis*)
    Hawiian goose (nene goose) (*Nesochen sandvicensis*)
    Ross' goose (*Chen rossii*)
    Snow goose (*Chen caerulescens*)
    White-fronted goose (*Anser albifrons*)
  Swans (Tribe Cygnini)
    Bewick's swan (*Cygnus columbianus bewickii*)
    Black swan (*Cygnus atratus*)
    Mute swan (*Cygnus olor*)
    Trumpeter swan (*Cygnus buccinator*)
    Tundra swan (whistling swan) (*Cygnus columbianus*)
    Whooper swan (*Cygnus cygnus*)
  Whistling ducks (Tribe Dendrocygnini)
    Fulvous whistling duck (*Dendrocygna bicolor*)
Woodpeckers (Order Piciformes, Family Picidae)
    Red-cockaded woodpecker (*Picoides borealis*)

Sources

American Ornithologists' Union. 1983. Check-list of North American birds, (6th ed.): Lawrence, Kansas, Allen Press, 877 p.

Howard, R. and Moore, A. 1991. A complete checklist of the birds of the world, (2nd ed.): San Diego, California, Academic Press, 622 p.

Monroe, B. L. Jr., and Sibley, C. G. 1993. A world checklist of birds: New Haven, Connecticutt, Yale University Press, 393 p.

Appendix F

Common and scientific names other than birds

Mammals
    African lion (*Panthera leo*)
    Bears (*Ursus americanus*; *U. arctos*; *U. maritimus*)
    Beaver (*Castor canadensis*)
    Bighorn sheep (*Ovis canadensis*)
    Bison (*Bison bison*)
    Bobcat (*Felis rufus*)
    Caribou (New World and Siberia) (*Rangifer tarandus*)
    Chipmunks (*Tamias striatus* and *Eutamias* sp.)
    Cottontail rabbit (*Sylvilagus floridanus*)
    Coyote (*Canis latrans*)
    Deer (*Odocoileus* sp.)
    Elk (*Cervus elaphus*)
    Foxes (*Vulpes* sp., *Urocyon* sp., and *Alopex lagopus*)
    Fur seal (*Callorhinus ursinus*)
    Jackrabbits (*Lepus* sp.)
    Lynx (*Felis lynx*)
    Mink (*Mustela vison*)
    Muskrat (*Ondatra zibethicus*)
    Nutria (*Myocaster coypus*)
    Pronghorn antelope (*Antilocapra americana*)
    Puma (Mountain lion) (*Felis concolor*)
    Raccoon (*Procyon lotor*)
    Rats (*Rattus* sp.)
    Reindeer (Old World and Greenland) (*Rangifer tarandus*)
    Sea lions (*Zalophus californianus* and *Eumetopias jubata*)
    Voles - small, mouse-like mammals primarily of the genera *Clethrionomys* and *Microtus*.
    Weasels (*Mustela* sp.)

Invertebrates
    Asian tiger mosquito (*Aedes albopictus*)
    Black widow spider (*Latrodectus mactans*)
    Caecal worm (*Heterakis gallinarum*)

Fish
    Rainbow trout (*Salmo gairdneri*)

Plants
    Choke cherry (*Prunus virginiana*)

Appendix G

Chemical names

Common Names	Chemical Names
1080	Sodium monofluoracetate
2-PAM	Pralidoxime chloride
Alachlor (Lasso®)	2-Chloro-N-(2,6-diethylphenyl)-N-(methoxymethyl)acetamide
Aldicarb (Temik®)	2-Methyl-2-(methylthio)propanal O-[(methylamino)carbonyl]oxime
Aldrin	1,2,3,4,10,10-Hexachloro-1,4,4a,5,8,8a-hexahydro-1,4:5,8-dimethanonaphthalene
Antimycin	3-Methylbutanoic acid 3-[3-(formylamino)-2-hydroxybenzoyl]amino-8-hexyl-2,6-dimethyl-4,9-dioxo-1,5-dioxonan-7-yl ester
Aroclor®	Group of polychlorinated biphenyls
Atrazine	6-Chloro-N-ethyl-N'-(1-methylethyl)-1,3,5-triazine-2,4-diamine
Avitrol®	4-Aminopyridine
Bayluscide®	5-Chloro-N-(2-chloro-4-nitrophenyl)-2-hydroxybenzamide compound respectively with 2-aminoethanol(1:1)
Beuthanasia-D-Special®	Sodium pentobarbital
Bomyl®	3-[(Dimethoxyphosphinyl)oxy]-2-pentenedioic acid dimethyl ester
Brodifacoum (Talon®)	3-[3-(4'-Bromo[1-1'-biphenyl]-4-yl)-1,2,3,4-tetrahydro-1-napthalenyl]-4-hydroxy-2H-1-benzopyran-2-one
Captan	3a,4,7,7a-Tetrahydro-2-[(trichloromethyl)thio]-1H-isoindole-1,3(2H)-dione
Carbofuran	2,3-Dihydro-2,2-dimethyl-7-benzofuranol methylcarbamate
Chlorophacinone	2-[(4-Chlorophenyl)phenylacetyl]-1H-indene-1,3(2H)-dione
Chlorpyrifos	Phosphorothioic acid O,O-diethyl O-(3,5,6-trichloro-2-pyridinyl) ester
Chlordane	1,2,4,5,6,7,8,8-Octachloro-2,3,3a,4,7,7a-hexahydro-4,7-methano-1H-indene
Clophens®	Group of polychlorinated biphenyls
Coumaphos	Phosphorothioic acid O-(3-chloro-4-methyl-2-oxo-2H-1-benzopyran-7-yl) O,O-diethyl ester
DDD	Dichlorodiphenyl dichloroethane
DDE	Dichlorodiphenyldichloroethylene
DDT	Dichloro diphenyl trichloroethane
Demeton (Systox®)	Phosphorothioic acid O,O-diethyl O-[2-(ethylthio)ethyl] ester mixture with O,O-diethyl S-[2-(ethylthio)ethyl]phosphorothioate
Diazinon	Phosphorothioic acid O,O-diethyl O-[6-methyl-2-(1-methylethyl)-4-pyrimidinyl] ester
Dicrotophos	Phosphoric acid 3-(dimethylamino)-1-methyl-3-oxo-1-propenyl dimethyl ester
Dieldrin	(1aα,2β,2aα,3β,6β,6aα,7β,7aα)-3,4,5,6,9,9-Hexachloro-1a,2,2a,3,6,6a,7,7a-octahydro-2,7:3,6-dimethanonaphth[2,3-b]oxirene

Common Names	Chemical Names
Dimethoate	Phosphorodithioic acid O,O-dimethyl S-[2-(methylamino)-2-oxoethyl] ester
Diphacinone	1,3-Inandione, 2-diphenylacetyl
Diquat®	Dipyrido(1,2-a:2',1'-c)pyrazinediium, 6,7-dihydro
Disulfoton	Phosphorodithioic acid O,O-diethyl S-[2-(ethylthio)ethyl] ester
EDTA	Ethylenediaminetetraacetic acid
Endosulfan	6,7,8,9,10,10-Hexachloro-1,5,5a,6,9,9a-hexahydro-6,9-methano-2,4,3-benzodioxathiepin 3-oxide
Endrin	(1α,2β,2aβ,3α,6α,6aβ,7β,7aα)-3,4,5,6,9,9-Hexachloro-1a,2,2a,3,6,6a,7,7a-octahydro-2,7:3,6-dimethanonaphth[2,3-b]oxirene
Enflurane	2-Chloro-1-(difluoromethoxy)-1,1,2- trifluoroethane
Famphur	Phosphorothioic acid O-[4-[(dimethylamino)sulfonyl]phenyl] O,O-dimethyl ester
Fenamiphos	(1-Methylethyl)phosphoramidic acid ethyl 3-methyl-4-(methylthio)phenyl ester
Fenbendazole	[5-(Phenylthio)-1H-benzimidazol-2-yl]carbamic acid methyl ester
Fenclors®	Group of polychlorinated biphenyls
Fensulfothion	Phosphorothioic acid O,O-diethyl O-[4-methylsulfinyl)phenyl] ester
Fenthion	Phosphorothioic acid O,O-dimethyl O-[3-methyl-4-(methylthio)phenyl ester
Fonofos	Ethylphosphonodithioic acid O-ethyl S-phenyl ester
Formalin	Formaldehyde solution
Halothane	2-bromo-2-chloro-1,1,1-trifluoroethane
HCN	hydrogen cyanide gas
Heptachlor	1H-1,4,5,6,7,8,8-Heptachloro-3a,4,7,7a-tetrahydro-4,7-methanoindene
Hexaconazole	(RS)-2-(2,4-dichlorophenyl)-1-(IH-1,2,4,-triazol-1-yl)-hexan-2-ol
Hexane	C_6H_{14}
Isazophos (Triumph®)	Phosphorothioic acid O-[5-chloro-1-(1-methylethyl)-1H-1,2,4-triazol-3-yl] O,O-diethyl ester
Isoflurane	2-Chloro-2-(difluoromethoxy)-1,1,1-trifluoroethane
Ivermectin	22,23-Dihydroabamectin
Kanechlors®	Group of polychlorinated biphenyls
Lindane	(1α,2α,3β,4α,5α,5β)-1,2,3,4,5,6-Hexachlorocyclohexane
Methamidophos	Phosphoramidothioic acid O,S-dimethyl ester
Methiocarb	3,5-Dimethyl-4-(methylthio)phenyl methylcarbamate
Methoxyflurane	2,2-Dichloro-1,1-difluoro-1-methoxyethane
Mirex	1,1a,2,2,3,3a,4,5,5,5a,5b,6-Dodecachlorooctahydro-1,3,4-metheno-1H-cyclobuta[cd]pentalene
Monocrotophos	(E)-Phosphoric acid dimethyl [1-methyl-3-(methylamino)-3-oxo-1-propenyl] ester
Nitrapyrin	2-Chloro-6-(trichloromethyl)pyridine
OC	Organochlorine
OP	Organophosphate

Common Names	Chemical Names
Oxamyl	2-(Dimethylamino)-N-[[(methylamino)-carbonyl]oxy]-2-oxoethanimidothioic acid methyl ester
PAH	Polycyclic aromatic hydrocarbon
Parathion	Phosphorothioic acid O,O-diethyl O-(4-nitrophenyl) ester
PCB	Polychlorinated biphenyl
Permethrin	3-(2,2-Dichloroethenyl)-2,2-dimethylcyclopropanecarboxylic acid (3-phenoxyphenyl)methyl ester
Phenoclors®	Group of polychlorinated biphenyls
Phenol	C_6H_6O
Phorate	Phosphorodithioic acid O,O-diethyl S- [(ethylthio)methyl] ester
Phosmet	Phosphorodithioic acid S-[(1,3-dihydro-1,3- dioxo-2H-isoindol-2-yl)methyl] O,O-dimethyl ester
Phosphamidon	Phosphoric acid 2-chloro-3-(diethylamino)-1- methyl-3-oxo-1-propenyl dimethyl ester
Pyralenes®	Group of polychlorinated biphenyls
Rotenone	[2R-(2α,6aα,12aα)]-1,2,12,12a-Tetrahydro-8,9-dimethoxy-2-(1-methylethenyl)[1]benzopyrano[3,4-b]furo[2,3-h]benzopyran-6(6aH)-one
Sleepaway®	Sodium pentobarbital
Starlicide®	3-Chloro-p-toluidine hydrochloride
Strychnine	$C_{21}H_{22}N_2O_2$
Terbufos (Counter®)	Phosphorodithioic acid S-[[(1,1-dimethylethyl)thio]methyl] O,O-diethyl ester
Thiabendazole	2-(4-Thiazoyl)-1H-benzimidazole
Thiram	Tetramethylthioperoxydicarbonic diamide
Toxaphene	Chlorinated camphene
Warfarin	4-Hydroxy-3-(3-oxo-1-phenylbutyl)-2H-1-benzopyran-2-one
Zectran®	4-(Dimethylamino)-3,5-dimethylphenol methylcarbamate (ester)
Ziram	(T-4)-Bis(dimethylcarbamodithioato-S,-S¹)zinc

Sources

Howard, P.H., and Neal, M., 1992, Distionary of chemical names and synonyms: Boca Raton, Florida, Lewis Publishers, 1,394 p.

Budavari, S., O'neil, M.J., Smith, A., and others, (eds.),1996, The Merck index, 12th ed.): Whitehouse Station, New Jersey, Merck and Co., 1,741 p.

Meister, R.T. (ed.), 1996, Farm chemicals handbook '96: Willoughby, Ohio, Meister Publishing Co., 446 p.

Appendix H

Conversion table

The units of measurement that are used by laboratories for recording the results of scientific or diagnostic tests are reported in this Manual in the metric or SI units (from the French "Le System International d'Unites") that are customary for those results. Other units of measurement, such as those that are used in the field for reporting area or length, are reported in the inch-pound units that are common in the United States. The lists of conversion factors and abbreviations below are for those who are interested in converting the basic units of measurement that are used in this Manual to the system of choice. For temperature, use the equations provided.

To convert SI to inch-pound

	Multiply	*By*	*To obtain*
Length			
	millimeters (mm)	0.039	inch (in.)
	centimeters (cm)	0.39	inch (in.)
	meter (m)	3.28	foot (ft)
	meter (m)	1.09	yard (yd)
	kilometer (km)	0.62	mile (mi)
Weight			
	kilogram (kg)	2.21	pound (lb)
	metric ton	1.10	short ton
Volume or capacity			
	milliliters (mL)	0.03	fluid ounce (fl. oz)
	microliters (μL)	0.003	fluid ounce (fl. oz)

Temperature
  °C, degrees Celsius, to °F, degrees Fahrenheit $9/5 \times °C + 32 = °F$

To convert inch-pound to SI

	Multiply	*By*	*To obtain*
Length			
	inch (in.)	25.4	millimeters (mm)
	foot (ft)	2.54	centimeters (cm)
	mile (mi)	0.3048	kilometer (km)
Area			
	acre	4,047	square meters (m^2)
Weight			
	pound (lb)	0.4356	kilogram (kg)
	short ton	0.907	metric ton

Temperature
  °F, degrees Fahrenheit, to °C, degrees Celsius
    $(°F - 32) \times 5/9 = °C$

Other abbreviated units of measurement used in this Manual:

  cc   cubic centimeter, a unit of measurement that indicates the dosage of a drug or substance that is administered intravenously or by injection.

  μg/g micrograms per gram

  mg/kg milligrams per kilogram

Conversion table for units of measurement that are frequently used with the toxicology of pesticides:

Concentration of compound in tissue, food, or water, in SI units	Concentration of compound in tissue, food, or water, in parts per million (ppm)
1 nanogram per gram (1 ng/g)	0.001
1 microgram per 100 grams (1 µg/100 g)	0.01
1 milligram per kilogram (1 mg/kg)	1
1 microgram per gram (1 µg/g)	1
1 microgram per 100 milligrams (1 µg/100 mg)	10
1 milligram per 100 grams (1 mg/100 g)	10
1 milligram per gram (1 mg/g)	1,000

Concentration of compound in air, in SI units	Concentration of compound in air, in parts per million (ppm)
1 microgram per liter (1 µg/L)	1

Glossary

Abdomen — the portion of the body that lies between the thorax and the pelvis.

Abdominal cavity — the space that contains the abdominal viscera (the liver, spleen, intestines, etc.).

Abdominal wall — the layers of muscles lying between the skin and the abdominal cavity.

Absorption — to take in a substance through the pores or cells of a tissue. The substance must pass through the tissue to be absorbed.

Acanthocephalans — cylindrical, unsegmented worms that attach to the host by a retractable proboscis with sharp hooks.

Acaracides – substances, such as pesticides, that kill mites.

Acariasis — infestation of the body by mites.

Accipiters — short winged, long-tailed hawks; North American species are goshawk, Cooper's hawk, and sharp-shinned hawk.

Acute — sharp or severe, such as an illness with a sudden onset and a relatively short course.

Air sacs — thin-walled sacs that communicate with the lungs and are part of the avian respiratory system.

Airsacculitis — inflammation of the air sacs in birds.

Alcids — typically, pelagic colonial nesting seabirds, including species such as auklets, guillemots, murres, murrelets, and puffins.

Algae — a special form of plant life that lacks true roots, stems, or leaves, and that ranges in size from microscopic single cells to multicellular structures, such as seaweeds.

Alimentary canal — the digestive tract.

Allergic disease — development of a hypersensitivity of the host to substances foreign to the body, primarily antigens and other proteins.

Altricial — refers to newly hatched birds that require care in the nest for some period of time.

Ambient temperature — room or environmental temperature.

Amino acids — organic compounds of specific composition from which proteins are synthesized.

Amphibians — coldblooded animals characterized by moist, smooth skin that live both on land and in water at various life stages and that have gills at some stage of development, that is, frogs, toads, salamanders.

Amplification host — a host in which disease agents, such as viruses, increase in number.

Amyloid deposit — a complex protein material that grossly resembles starch and that in certain abnormal conditions accumulates in various body tissues causing cellular damage and injury to the affected organ.

Anaerobic — absence of oxygen; often refers to an organism that grows, lives, or is found in an environment devoid of oxygen, such as the cellular form of `Clostridium botulinum`, which causes avian botulism.

Analgesia — the absence of normal sensitivity to pain, typically, being in a semiconscious state induced through an anesthetic.

Anemia — a reduction in the normal number of red blood cells, or erythrocytes, in the body.

Anesthetic — a drug used to temporarily deaden pain.

Anesthetic induction time — the time between administering an anesthetic chemical and the actual time when target nerves are deadened.

Animal pathogens — organisms such as viruses, bacteria, and parasites, that are capable of invading and infecting animal hosts and causing disease.

Annelids — a group of invertebrates characterized by the segmented worms, including those in marine and freshwater and earthworms in addition to leeches.

Anorexia — lack of appetite.

Anoxia — a total lack of oxygen caused by several mechanisms that prevent oxygen from reaching the mitochondria of cells. Anoxia indicates a level of oxygen in animal tissues that is below normal in the presence of an adequate blood supply.

Antibody — a specialized serum protein produced by the immune system in response to an antigen in an attempt to counteract the effects of the antigen; antibodies in the blood indicate exposure to specific antigens or disease agents.

Antidote — substances that counteract or prevent the action of a poison.

Antigen — any foreign substance (generally proteins) to which the body reacts by producing antibodies. Antigens may be soluble substances such as toxins, particulate matter such as pollen, or microorganisms such as bacteria and viruses.

Antiserum — a serum containing antibodies to specific antigens; can be used to test biological samples for the presence of specific antigens.

Antitoxin — any substance that counteracts the action of a toxin or poison; generally, a specific type of antibody produced in experimental animals as a result of exposure to a specific toxin. Botulism antitoxin, for example, can be produced by exposing an animal to low levels of botulism toxin over a long period of time and then harvesting serum from that animal to treat other animals.

Arbovirus — a virus that is transmitted by invertebrates of the phylum Arthropoda [insects, arachnids (spiders, mites, ticks, etc.) and crustaceans].

Arthropod — members of the phylum Arthropoda (insects, arachnids, and crustaceans).

Ascites — accumulation of fluid in the abdominal cavity.

Aseptic — free from infection; sterile.

Asexual reproduction — the formation of new individuals without the union with cells of the opposite sex and usually by an individual.

Asymptomatic — without visible signs of illness; an asymptomatic carrier is an organism that harbors a disease agent, but that shows no outward signs.

Ataxia — incoordination.

Avicides — chemical substances used to kill or repel birds.

Avirulent — not virulent, does not cause disease.

Bacterin — a vaccine consisting of killed bacteria that is used for protection against infection by a specific bacterial disease.

Bacteriophage — a virus that infects a bacterium.

Bacterium — singular for bacteria. Any of a group of microscopic, unicellular organisms that have distinct cell membranes and that lack a distinct nucleus surrounded by a nuclear membrane.

Barbiturate — a type of sedative or anesthetic that is chemically derived from barbituric acid.

Bay diving ducks — typically, ducks that feed in deep bodies of water, usually in coastal bays and deep lakes. Species include canvasback, goldeneyes, redhead, and scaup.

Benign — noninvasive, that is, tumors that do not spread to other parts of the body; not malignant.

Big game — hunted species of large mammals (from deer to elephants).

Bile — yellow-brown to greenish liquid secreted by the liver and stored in the gallbladder before excretion by way of the intestine. Bile is composed of metabolic breakdown products derived from hemoglobin and other metabolic waste products.

Bioaccumulation — the accumulation of long-lived toxins, such as chlorinated hydrocarbons, as a result of repeated exposure or of exposure from a variety of sources.

Biomagnification — an increase in concentrations of long-lived contaminants in animals at higher positions in the food chain.

Biota — the plant and animal life of an area.

Biotoxins — poisons produced by and derived from the cells or secretions of a living organism, either plant or animal.

Birds of prey — synonymous with raptors; includes eagles, hawks, falcons, kites, and owls.

Biting louse — see Hippobascid flies.

Black flies — small, bloodsucking, biting flies of the genus Simulium; vectors for Leucocytozoon infections.

Blood flukes — trematode parasites that are found in the blood cells of the host.

Brine flies — species of flies whose larvae live in brine.

Brooding — care of young birds by the adult.

Buffered formalin — a 3.7 percent solution of formaldehyde (equal to 10 percent formalin) to which sodium phosphate buffers have been added. Buffered formalin is the best overall fixative for tissue for later microscopic study.

Bumblefoot — an inflammation and, often, swelling of the foot of birds as the result of a bacterial infection.

Bursa of Fabricious — a saclike outgrowth of the cloaca of birds that is part of the avian immune system.

Buteos — a subfamily of the hawks characterized by soaring behavior, broad, rounded wings, and a broad, fanned tail, such as the red-tailed hawk.

Caecum (British; plural caeca) or cecum (American; plural ceca) — a large, blind pouch or sac (often a pair) at the junction of the small intestine and the large intestine.

Calcification — the process by which tissues become hardened by the deposition of calcium salts.

Canidia — fungal spores.

Canker — synonymous with trichomoniasis in doves and pigeons.

Capture myopathy — a state of immobility resulting from damage to skeletal and cardiac muscles caused by extreme physical exertion, struggle, or stress; may occur in wildlife as they are chased in capture attempts; may appear later when captured wildlife are under physical restraint; or may appear after they have been released.

Cardiac muscle — heart muscle.

Cardiovascular system — the heart and blood vessels by

which blood is pumped and circulated through the body.

Carnivores — refers to flesheating mammals in the Order Carnivora and includes dogs, skunks, weasels, cats, raccoons, etc.

Carrion — dead and decaying flesh.

Caseous — resembling cheese or curd.

Central nervous system — the brain and spinal cord.

Ceratopogonid flies — very small, bloodsucking gnats commonly known as punkies, no-see-ums, or sand flies.

Cercaria — the final free-swimming larval stage of a trematode parasite.

Cestodes — flattened, usually segmented, parasitic worms; tapeworms.

Chelating chemical — a chemical that combines with a metal ion in a firm, ringlike band and that prevents the metallic ion from having any further biochemical effect.

Chlorinated hydrocarbons — organic compounds characterized by the presence of chlorine; commonly refers to persistent chemicals with insecticidal properties; DDT and dieldrin are common examples.

Choana — one of the paired openings on the inner side of the maxilla (upper beak), near the back of the oral cavity, that opens into the nasal cavity.

Cholinesterase enzymes — enzymes that are particularly important in the transmission of nerve impulses; the activity of these enzymes is inhibited by exposure to organophosphorus and carbamate compounds, and death results when activity is greatly reduced.

Chronic — persisting for a relatively long time.

Chronic losses — mortality of attrition; small numbers of continual losses over extended periods of time.

Clinical sign — an abnormal physiological change or behavior pattern that is indicative of illness. Signs are externally observable, as contrasted with symptoms, which are subjective.

Cloaca — a common passage for the fecal, urinary, and reproductive discharges of most lower vertebrates (birds, reptiles, and amphibians).

Coalescence — the fusion or growing together of tissue damage from a disease agent.

Coldblooded vertebrates — species such as fishes and reptiles, which have blood that varies in temperature to approximately that of the surrounding environment.

Colibacillosis — infection with the bacterium Escherichia coli.

Colon — the large intestine.

Colonial nesters — birds that nest in large groups.

Comatose — in a coma or comalike state; an abnormal state of continuous deep unconsciousness.

Congener — a member of the same taxonomic grouping, such as polychlorinated biphenyls, that possess similar chemical structures.

Congenital abnormality — usually an anatomical malformation that results from incomplete growth during embryonic development. Also refers to an abnormal biochemical pathway caused by a genetic factor.

Congestion — the abnormal accumulation of blood in a tissue or organ; often causes a reddening of the affected area.

Contagious — capable of being transmitted from animal to animal, such as a contagious disease.

Coccidiasis — the presence of coccidia, protozoa of the subphylum Sporozoa.

Coccidiosis — a disease caused by coccidia, protozoa of the subphylum Sporozoa.

Cornea — the transparent tissue on the front of the eyeball that covers the iris and pupil, through which light passes to the interior.

Coronary band — a fatty band encircling the heart; in hooved animals, the germinal layer beneath skin at the junction of the skin and hoof.

Cracker shell — a shotgun shell that is loaded to produce a visible burst and loud sound in order to frighten animals.

Crop — a dilation of the esophagus at the base of the neck of some birds.

Crustacea — a specialized group of invertebrates that includes such diverse species as lobster, shrimp, barnacles, wood lice, and water fleas.

Cygnet — a young swan.

Cyanobacteria — a genus of bacteria composed of the blue-green algae; like the dinoflagellates, cyanobacteria are important sources of environmental toxins that can cause illness and death in humans and wildlife.

Cyanosis — a bluish discoloration of the skin and mucous membranes due to an excessive concentration of deoxygenated hemoglobin in the blood.

Cystocanth — an infective juvenile stage of thorny-headed worms (acanthocephalan parasites).

Cytoplasm — the aqueous part of the cell that is outside of the nucleus but that is contained within the cell wall. The cytoplasm is the site of most of the chemical activities of the cell.

Dabbling ducks — Ducks that feed on the surface or in shallow water, including mallard, American black duck, gadwall, American wigeon, northern pintail, northern shoveler, and teal. Also referred to as puddle ducks.

Definitive host — an organism in which sexually mature stages of a parasite occur.

Dehydration — a condition that results from excessive loss of body fluids.

Depopulation — the destruction of an exposed or infected group of animals.

Dermatophytosis — a fungal infection of the skin.

Dessication — the act or process of drying a substance.

Digestive tract (alimentary canal) — the organs associated with the ingestion, digestion, and absorption of food, such as the esophagus, stomach, and intestines.

Dinoflagellates — aquatic protozoa that are an important component of plankton. These single-celled organisms may be present in vast numbers, causing discoloration of the water referred to as "red tide." Some species secrete powerful neurotoxins.

Dioxins — a chemical component of defoliants, such as agent orange, that are considered to be carcinogenic (cause cancer), teratogenic (cause fetal abnormalities), and mutagenic (cause abnormal mutation rate).

Direct life cycle — a parasitic life cycle that requires only a single host for its completion.

Diurnal — active during the day.

Diving ducks — synonymous with bay diving ducks.

Domestic duck — ducks typically raised for market, such as the white Pekin.

Drive-trapping — capture of flightless birds during the molt and of other animals by herding them into a netted or fenced containment area.

Drop nets — suspended nets used to capture animals by remote release of the nets or triggering mechanisms at the net site.

Dyspnea — labored breathing.

Ecchymotic — a hemorrhagic, irregular-shaped area in tissues that is bruise-like in appearance and, often, in color.

Ecology — the study of the interrelationships between living organisms and their environment.

Ectoparasite — a parasite that lives on the external surface, or in the integument, of its host.

Ectotherms — species that rely on sources of heat outside themselves (i.e., coldblooded species).

Edematous — swelling of tissues due to abnormal accumulation of fluid in the intercellular tissue spaces; seepage of these fluids may result in accumulations within the body cavity.

EDTA — ethylenediamine tetra-acetic acid; a chelating agent that binds with lead and that is used in the treatment of lead poisoning.

EEE — eastern equine encephalomyelitis; a viral disease.

ELISA — a molecular-based enzyme-linked immonosorbent assay; a type of test used to detect either antigen or antibody.

Emaciation — a wasted condition of the body; excessive leanness.

Emasculatome — a veterinary instrument designed for bloodless castration of cattle or sheep; has been used for euthanasia of birds by cervical dislocation.

Encrustation — forming a crust or a covering; for example, salt encrustation.

Endemic — a disease that commonly is present within a population or a geographical area.

Endogenous phase — developmental phase of the life cycle of a parasite that occurs within the host.

Endoparasite — a parasite that lives within the body of its host.

Endotherms — warmblooded vertebrates; species able to internally regulate their body temperatures.

Enteritis — inflammation of the intestine.

Enzootic — an animal disease that commonly is present within a population or geographical area.

Epicardium — the outer covering of the heart.

Epidemic — the presence of a disease in a population or in an area in a higher than expected prevalence, or rate.

Epithelial cells — cells that cover the external and internal surfaces of the body.

Epizootic — a disease affecting a greater number of animals than normal; typically, occurrences involving many animals in the same region at the same time.

Epizootiology — the study of the natural history of disease in animal populations.

Erosion — wearing away; gradual disintegration.

Erythrocytes — red blood cells; serve to transport oxygen throughout the body.

Esophagus — the passage extending from the mouth to the stomach.

Estrogenic — possessing characteristics of the hormone estrogen; estrogenic compounds may elicit the development of feminine characteristics in male animals.

Etiologic agent — any living or nonliving thing, power, or substance capable of causing a disease.

Eutrophication — the excessive growth, caused by an oversupply of nutrients, of plants and algae in bodies of water.

Exotic disease — a disease that normally does not occur within a particular area.

Exotoxin — a toxin formed and excreted by bacterial cells.

Exsanguination (bleeding out) — the draining of blood from an animal.

Fastidious — refers to the very specific requirements for the culture of some bacteria.

Fauna — the animals of an area.

"Feather edge" — a long, shallow edge of a body of water that gradually deepens offshore.

Femur — the thigh bone of humans; the upper legbone in hooved mammals and birds. The bone between the pelvis and the knee.

Feral pigeon — rock dove.

Fibrin — an insoluble protein that forms a network of fibers during clotting of the blood.

Fibrinoperitonitis — fibrin-coated inflammation of the surfaces of the peritoneal cavity.

Fibrinous — a pathologic term referring to a threadlike sheet of material that may occur on surfaces of organs in some disease conditions; clotting factors in blood contribute to the structure of this material.

Flatworms — the common name for parasites of the phylum Platyhelminthes, flukes or trematodes.

Flukes — parasitic flatworms; also referred to as trematodes.

Flylarvae — maggots.

Fomite — an object that is not in itself harmful, such as a wooden object or article of clothing, but that may harbor pathogenic microorganisms and serve to transmit an infection to a living organism.

Food chain — ascending trophic levels within an ecosystem in which species at the lower level are the primary food base for the species at the next highest level.

Formalin — a liquid solution of formaldehyde that is used as a tissue fixative, usually to prepare tissues for microscopic examination.

Fossorial — refers to digging animals that live in burrows.

Frounce — synonymous with trichomoniasis in raptors.

Fungicides — chemicals that kill fungi.

Gallinaceous birds — heavy-bodied, chickenlike land birds. Includes ring-necked pheasant, quails, grouse, and wild turkey.

Gamete — one of two cells produced by a gametocyte; the union of male and female gametes initiates the development of a new individual during sexual reproduction

Gametocyte — an undifferentiated cell that develops into a gamete.

Gangrene — tissue death due to a failure of the blood supply to that tissue area followed by bacterial invasion and putrefication.

Gapes — see gapeworm.

Gapeworm — parasites of the trachea of birds; synonym for tracheal worms.

Gastrointestinal tract — the tubular organs that form a digestive pathway from the mouth to the vent, including the stomach and intestines.

Geographic information system — a specialized computer system for storage, manipulation, and presentation of layers of geographical information.

Gizzard — the enlarged muscular ventriculus (stomach) of many birds.

Granuloma — refers to a tumorlike mass or nodule; often associated with a response to an infection.

Haemoproteus — blood parasites transmitted by louse flies of the family Hippoboscidae and midges of the family Ceratopogonidae.

Hatchet-breast — a common term to describe the prominent, protruding breast keel seen as the result of the atrophy of the breast muscles. "The keel appears as sharp and as prominent as the back of a hatchet."

Hawaiian forest birds — native and introduced avifauna of the forested areas of the Hawaiian Islands. Includes such species as sparrows, finches, cardinals, honeycreepers, and thrushes.

Helminths — parasitic worms

Hemosporidia — protozoan blood parasites.

Hemoglobin — the oxygen-carrying pigment of red blood cells.

Hemozoin — a dark pigment produced from the hemoglobin in the host's red blood cells by malarial parasites that collect in tissues, such as the spleen and liver, causing those organs to appear grayish to dark brown or black.

Hepatitis — inflammation of the liver.

Hepatomegaly — enlargement of the liver.

Herbicides — chemicals used to kill unwanted vegetation.

Hermaphroditic — organisms that possess both male and female functional reproductive organs.

Herpesvirus — one of the major groups of related viruses that have DNA nucleic acids and that are further characterized by similar size, shape, and physiochemical reactions.

Herpetologists — those who study the natural history and biology of reptiles.

Heterogenous organism — one that is derived from a combination of different types of parent organisms.

Hippoboscid flies — a group of wingless and winged parasitic flies found on birds and mammals.

Histoplasmosis — a disease of humans caused by inhalation of the fungus *Histoplasma capsulatum*.

History — as it refers to wildlife disease investigations, a record of background information and chronological events associated with a die-off.

Homeostasis — the tendency toward equilibrium; refers to the capacity of living organisms to maintain internal body environmental conditions necessary for survival.

Husbandry practice — the care and maintenance of animals.

Hydropericardium — an excessive amount of fluid within the sac surrounding the heart.

Hypersensitivity — greater than normal sensitivity to stimuli or to biological agents.

Hypothermia — greatly reduced body temperature.

Hypovalemic shock — shock resulting from insufficient blood volume to maintain adequate cardiac output and blood pressure; caused by acute hemorrhage or excessive fluid loss.

Icthyologists — those who study the natural history and biology of fishes.

Immune — being resistant to a disease.

Immunosuppressive therapy — a medical treatment that suppresses the normal immune response.

Impaction — an abnormal accumulation of food or other ingested materials that become lodged in a section of the digestive tract.

Immune system — the combination of host body defenses that guard against infectious disease.

Inapparent — an infection in which the infectious agent exists within the host but that causes no recognizable signs of illness; the infectious agent may or may not be shed at irregular times.

Incidence — the number of new cases of a disease occurring in a population within a certain time period.

Inclusion body — a structure within the cytoplasm or nucleus of a cell; a characteristic of some viral diseases, inclussion bodies occur in only a few species.

Incubation period — the time interval required for the development of disease; the time between the invasion of the body by a disease agent and the appearance of the first clinical signs.

Indigenous — native to a particular area.

Indirect life cycle — a life cycle that requires more than one host for its completion.

Infection — the invasion and multiplication of an infectious agent in host body tissues.

Infectious agent — a living organism capable of invading another.

Infective — capable of producing infection.

Infestation — parasitic invasion of external surfaces of a host.

Insecticides — pesticides used to kill insects.

Intermediate host — an organism in which a parasite undergoes a stage of asexual development.

Intracellular parasite — a parasitic organism, usually microscopic, that lives within the cells of the host animal.

Involuntary muscle — muscle that is not under the control of the individual.

Isolate — refers to microorganisms; the separation of a population of organisms that occur in a particular sample (verb); for example, to isolate a bacterial or viral organism from a sample. As a noun, refers to the organism that was isolated; for example, a bacterial isolate was obtained from a sample.

Isopods — crustaceans with flattened bodies, such as sowbugs, pillbugs, and wood lice.

Joint capsule — the thick, fibrous capsule surrounding a joint, as around the knee.

Keel — the narrow middle portion of a bird's sternum.

Kites — hawk-like birds.

Lacrimal discharge — a discharge from the tear glands near the eye.

Laparotomy — a surgical procedure in which an incision is made into the abdominal cavity, often to determine the sex of birds for which plumage and other characteristics cannot be used for that purpose.

Larva — an immature parasitic life cycle stage; typically, the form of the parasite is unlike the mature stage.

Larynx — the musculocartilaginous structure at the upper part of the trachea; it guards the entrance to the trachea and secondarily serves as the organ of voice.

Latent — dormant or concealed; a latent infection refers to the situation in which a disease condition is not apparent.

Lentogenic — refers to a form of Newcastle disease virus that is mildly virulent as measured in chickens.

Leucocytozoon — blood parasites transmitted by black flies of the family Simulidae.

Lesion — an abnormal change in tissue or an organ due to disease or injury.

Lethargy — abnormal drowsiness or stupor.

Lousefly — see Hippoboscid flies.

Lyme disease — an infectious disease that is caused by the spirochete Borrelia burgdorferi and transmitted by ticks.

Lipophilic — having an affinity for fat; such as chemicals that accumulate in fat and fatty tissues.

Livestock — domestic animals raised for food and fiber commonly refers to animals such as hogs, sheep, cattle, and horses.

M-44 — a predator-control device that uses cyanide as the toxic component.

Macrocyst — a large cyst; a large spore case (fungi); an encapsulated reproductive cell of some slime molds.

Macrogamete — the female sexual form of the malaria parasite that is found in the gut of the mosquito vector.

Maggot — a soft-bodied larva of an insect, especially a form that lives in decaying flesh.

Malarias — infectious diseases caused by protozoan parasites that attack the red blood cells.

Malignant — spread from the location of origin to other areas; that is, tumors that are invasive and that spread throughout the body.

Marek's disease — an important infectious disease of poultry, that is caused by infection with a herpesvirus.

Marine birds — birds of the open ocean, typically pelagic, and often colonial nesters, such as alcids, shearwaters, storm petrels, gannets, boobies, and frigatebirds.

Meningoencephalitis — inflammation of the transparent covering (meninges) of the brain.

Mergansers — a group of waterfowl that are commonly referred to as "fish ducks" due to their food habits.

Meront — an asexual stage in the development of some protozoan parasites that gives rise to merozoites.

Merozoite — a stage in the life cycle of some protozoan parasites.

Mesogenic — refers to a form of Newcastle disease virus that is moderately virulent as measured in chickens.

Metabolic rate — an expression of the rate at which oxygen is used by cells of the body.

Metacercaria — the encysted resting or maturing stage of a trematode (fluke) parasite in the tissues of an intermediate host.

Microgamete — the male sexual form of the malaria parasite found in the gut of the mosquito vector.

Migratory birds — all birds listed under the provisions of the Migratory Bird Treaty Act.

Minamata disease — mercury poisoning of humans; named after an incident resulting from contamination within Minamata Bay, Japan.

Miracidium — the first larval stage of a trematode parasite, which undergoes further development in the body of a snail.

Mobilization — refers to the tendency of lipophilic chemicals [environmental contaminants, such as chlorinated hydrocarbons, that have an affinity for storage in adipose (fat) tissue] to be released into the bloodstream as fat stores are depleted.

Mollusks — species of the phylum Mollusca; includes snails, slugs, mussels, oysters, clams, octopuses, nautiluses, squids, and similar species.

Molt — the normal shedding of hair, horns, feathers, and external skin before replacement by new growth.

Moribund — a visible, debilitated state resulting from disease; appearing to be suffering from disease and close to death.

Motility/motile/nonmotile — these terms refer to whether or not a bacterial organism moves on a particular culture medium; such movement reflects the presence of flagellae. Thus, the absence or presence of motility is a useful characteristic for identifying bacteria.

Motor paralysis — paralysis of the voluntary muscles.

Mucosa — a mucous membrane.

Mucous membrane — the layer of tissue that lines a cavity or the intestinal tract and that secretes a mixture of salts, sloughed cells, white blood cells, and proteins.

Mucosal surface — a layer of cells lining the inside of the intestinal tract or other body part that secretes mucus.

Myocarditis — inflammation of heart muscle.

Myocardium — the middle and thickest layer of the heart wall; composed of cardiac muscle.

Mycosis — fungal infection.

Mycotoxin — a poison produced by various species of molds (fungi).

Myiasis — infestation of the body by fly maggots.

Nares — the external openings on the top of the bill of birds; the external orifices of the nose; the nostrils.

Nasal gland — a specialized gland of birds and some other species that serves to concentrate salt and secrete it from the body.

Nasal cavity — the forward (proximal) portion of the passages of the respiratory system, extending from the nares to the pharynx and separated from the oral cavity by the roof of the mouth.

NDV — Newcastle disease virus.

Necropsy — the methodical examination of the internal organs and tissues of an animal after death to determine the cause of death or to observe and record pathological changes.

Necrosis — the death of cells in an organ or tissue.

Necrotic — dead; exhibiting morphological changes indicative of cell death; in this Manual, necrotic lesions refer to areas of dead tissue.

Nematocides — chemicals used to kill nematode worms.

Nematodes — unsegmented, cylindrical parasitic worms; roundworm.

Neoplasm — see tumor.

Nervous system — specialized components of vertebrates, and, to a lesser extent invertebrates, that control body actions and reactions to stimuli and the surrounding environment.

Neurotoxin — toxins that cause damage to or destroy nerve tissue.

Nictitating membrane — the so-called third eyelid, a fold of tissue connected to the medial (side closest to the midline) side of the eye, which moves across the eye to moisten and protect it.

Nocturnal — species that are active during evening (nondaylight) hours.

Nodule — a small mass of tissue that is firm, discrete, and detectable by touch.

Nontoxic shot — shotshells with shotpellets that are not made of lead or other toxic metals; typically, soft iron is used, and is referred to as steel shot.

No-see-ums — see Ceratopognid flies.

Occlusion — a blockage or obstruction; the closure of teeth.

Oligochaetes — the earthworms and aquatic forms of the class Oligochaeta.

Oocyst — the encysted or encapsulated zygote in the stage of some protozoan parasites; often highly resistant to environmental conditions.

Opisthotonos — abnormal spasm of the neck and back muscles resulting in a body position in which the head and heels are involuntarily thrown back and the body is arched forward.

Osmoregulation — adjustment of osmotic pressure in relation to the surrounding environment.

Osteoporosis — loss of bone structure.

PAH — an acronym for polycyclic aromatic hydrocarbons.

Panzootic — a disease involving animals within a wide geographic area such as a region, continent, or globally.

Parasitism — an association between two species in which one (the parasite) benefits from the other (the host), often by obtaining nutrients.

Paratenic host — a host that has been invaded by a parasite, but within which no morphological or reproductive development of the parasite takes place; a "transport" host.

Paresis — partial paralysis.

Passerines — small- to medium-sized perching birds.

Pathogenic — the ability to cause disease.

Pathological — an adjective used to describe structural or functional changes that have occurred as the result of a disease.

PCB — acronym for polychlorinated biphenyls, a group of chlorinated aromatic hydrocarbons used in a variety of commercial applications. These compounds have long environmental persistence and have been a source for various toxic effects in a wide variety of fauna.

Pelagic — refers to living in or near large bodies of water, such as oceans or seas; typically, this term refers to avian species that only come to land areas during the breeding season.

Pericardium — the fibrous sac surrounding the heart.

Pericarditis — inflammation and thickening of the sac surrounding the heart.

Peritoneal cavity — the abdominal cavity, which contains the visceral organs.

Phage — a virus that has been isolated from a prokaryote (an organism without a defined nucleus, having a single double-standard DNA molecule, a true cell wall, and other characteristics). Most phages are bacterial viruses.

Pharynx — the musculomembranous passage between the mouth and the larynx and esophagus.

Pigeon milk — the regurgitated liquid that an adult pigeon feeds its young.

Pinnipeds — aquatic mammals that include the sea lions, fur seals, walruses, and earless seals.

Plaque — a patch or a flat area, often on the surface of an organ.

Plasmodium — blood parasites transmitted by mosquitos of the family Culicidae.

Plumage — the feather covering of birds.

Postmortem — examination and dissection of animal carcasses performed after the death of the animal. Also, changes that occur in tissues after death.

Poultry — domestic avian species, such as chickens and domestic ducks, geese, and turkeys.

Prefledglings — birds of the current hatch year that have not become feathered enough to fly.

Prevalence — the number of cases of a disease occurring at a particular time in a designated or defined area; rate.

Proboscis — a tubular process or structure of the head or snout of an animal, usually used in feeding; in this Manual, the tubular process of Acanthocephalan parasites is used for attachment to the host and feeding from it.

Protoporphyrin — a component of hemoglobin; useful in the diagnosis of exposure to lead.

Protozoan — a one-celled animal with a recognizable nucleus, cytoplasm, and cytoplasmic structures.

Psittacines — parrots, parakeets, and other species within the family Psittacidae.

Puddle ducks — see dabbling ducks.

Proventriculus — the first, or "glandular," stomach of a bird.

Puddle ducks — synonymous with dabbling or surface-feeding ducks.

Punkies — small, biting midges of the genus `Culicoides`; vectors for `Haemoproteus` infections. See ceratopognid flies.

Purulent — containing pus, as in a purulent discharge.

Range — the geographic distribution of a population or the area within which an individual animal moves (as in home range).

Raptors — synonymous with birds of prey. Birds, including hawks, owls, falcons, and eagles, that feed on flesh.

Reactivation — refers to the process by which cholinesterase enzyme activity returns to normal after carbamate exposure.

Rendering — a process by which animal carcasses are converted into fats and fertilizer.

Reptiles — coldblooded vertebrates that belong to the class Reptiles; such as., snakes, turtles, lizards.

Reservoir host — the host that maintains the disease agent in nature and that provides a source of infection to susceptible hosts.

Respiratory system — the collection of organs that provide oxygen to the organism and result in the release of carbon dioxide; typically, the trachia and lungs.

Rice breast disease — synonym for sarcocystis.

Rocket nets — remotely triggered, weighted firing devices that are propelled through the air by an explosive force carrying the netting to which they are attached over the birds or other animals being captured.

Rodenticides — toxic substances used to kill rodents.

Rodents — mammals that have chisel-like, ever growing incisor teeth that are used for gnawing; i.e., mice to beavers.

Rookery — a nesting area for some colonial birds, such as herons and egrets.

Roost sites — typically, locations where birds congregate at night in trees and other locations.

Rough fish — a term given to bottom-feeding freshwater fish with large scales, such as carp, buffalo, and similar species.

Roundworms — see nematodes.

Ruminants — hooved mammals possessing a rumen or first stomach, from which food or a cud is regurgitated for further chewing. Includes deer, elk, sheep, cattle, etc.

Salivary glands — the glands of the mouth that produce saliva.

Salt gland — see nasal gland.

Sandflies — see punkies.

Scavengers — animals that feed on dead, sick or injured prey. Includes crows, vultures, gulls, eagles, hawks, etc.

Schizogony — a type of asexual reproduction in some protozoan parasites in which daughter cells are produced by multiple nuclear divisions of the parasite (schizont).

Schizonts — the multinucleate, intermediate parasite stage that develops into merozoites within a host cell.

Scoliosis — an abnormal lateral curvature of the spine.

Sea ducks — ducks that frequent open ocean, although some species may be found on coastal bays or inland waters. Includes oldsquaw, eiders, scoters, and harlequin duck.

Secondary poisoning — intoxication of an animal as a result of eating a poisoned animal; for example, the poisoning of an eagle after it has fed on a duck that was poisoned by a chemical in treated grain. This differs from biomagnification, which involves increasing concentrations of toxic compounds within the body of organisms at increasing higher levels of a food chain.

Section 7 consultations — the Endangered Species Act requires discussion and evaluation of any proposed Federal activity, program, or permit that might affect an endangered species.

Sedated — chemically quieted.

Septicemia — the presence of pathogenic microorganisms or toxins in the blood.

Serosa — refers to the outside layer of an organ, such as the serosal surface of the intestine, or the lining of a body cavity.

Serosal surface — the external surface of an organ or a tissue within the body.

Serotype — a taxonomic subdivision of a microorganism, based on characteristic antigens or proteins.

Serovar — a taxonomic subdivision of a microorganism similar to serotype (above) but usually more specific.

Shorebirds — birds that feed at the edge of shallow water, along mudflats, and in shallow wetlands. Typically, these birds feed on invertebrates and include such species as American avocet, black-necked stilt, curlews, plovers, phalaropes, sandpipers, yellowlegs, and sanderling.

Signs — observable evidence of disease in animals (similar to symptoms in humans).

Sloughing — shedding of dead cells or dead tissue from living structures or tissues.

Slugs — terrestrial, snail-like mollusks that have a long, fleshy body and only a rudimentary shell.

Small mammals — mice to rabbits, racoons etc.; a general term used in wildlife management to group species of small to moderate size.

Small rodents — see rodents; rodents of small size, such as rats and mice.

Songbirds — small perching and singing birds, typically of the order Passeriformes, including sparrows, finches, and cardinals.

Sowbugs — see isopods.

Splenomegaly — enlargement of the spleen.

Spore — refers to a resistant stage, usually of bacteria or fungi, by which some microorganisms survive unfavorable environmental conditions and then develop into active life forms during favorable environmental conditions.

Sporogony — sporulation that involves multiple fission of a sporont (schizogony), resulting in the production of a sporocysts and sporozoites.

Sporont — a zygote of coccidian protozoa.

Sporozoite — the elongate nucleated infective stage of coccidian protozoan parasites.

Sporulation — the formation or libertion of spores.

Squab — a nestling pigeon that has not fledged.

Sternum — the breastbone.

Subcutaneous — under the skin.

Systemic — affecting the entire body.

Tapeworms — segmented parasitic flatworms; also referred to as cestodes.

Teal — small, swift-flying waterfowl of the genus Anas.

Tegument — the covering of an organ or the body.

Tenosynovitis — inflammation of the tendon sheath.

Teratogenic — causing embryonic deformities due to abnormal differentiation and development of cells.

Thermoregulation — regulation of the internal temperature of the body by various physiological processes.

Thorax — the part of the body between the neck and the respiratory diaphragm (in mammals), encased by the ribs.

Thorny-headed worms — acanthocephalan parasites.

Thymus gland — a lymph-gland-like organ involved in cellular immunity, located in the neck or upper thoracic cavity.

Torticollis — twisting or rotation of the neck causing an unnatural position of the head.

Toxic — poisonous.

Toxicosis — the condition of being poisoned.

Trematodes — flat, unsegmented parasitic worms; flatworms, flukes.

Trichomonids — protozoan parasites of the genus Trichomonas.

Trophic level — refers to an animal's position in the food chain. Species at higher trophic levels are, to a greater or lesser extent, dependent upon species in preceding trophic levels as sources of energy.

Tumor (neoplasm) — growths within organs and tissues of the body that result from the abnormal progressive multiplication of cells in a manner uncontrolled by the body.

Ubiquitous — found everywhere.

Ulceration — crater-like lesions in the skin and other tissues.

Ungulates — hoofed mammals.

Unthrifty appearance — an expression used in animal husbandry to describe an animal that is unkempt and dirty. Usually hair or feathers are soiled by excrement.

Upland gamebirds — game birds found in terrestrial habitats. Includes species such as ring-necked pheasant, quails, grouse, wild turkey, etc.

Upper digestive tract — the portion of the gastrointestinal tract that extends from the anterior opening of the esophagus in the region of the mouth to the stomach, but not including the intestines.

Ureter — the tubular structure that transports urates from the kidneys to the cloaca of birds.

Vascular system — blood circulation system.

Vector — an insect or other living organism that carries and transmits a disease agent from one animal to another.

Vegetative form — in bacteria, an active, growing, multiplying stage of development as opposed to a "spore," or a resistant resting stage.

Velogenic — refers to highly virulent strains of Newcastle disease virus that are capable of producing severe disease in the host.

Ventriculus — the stomach of a bird.

Verminous peritonitis — inflammation of the peritoneal cavity caused by parasites, usually nematodes.

Vertebrates — animals with backbones.

Viremia — the presence of virus in the blood.

Virulence — the disease-producing ability of a microorganism, generally indicated by the severity of the infection in the host and the ability of the agent to invade or cause damage or both to the host's tissues.

Virulent — the degree to which an infectious agent produces adverse effects on the host; a highly virulent organism may produce severe disease, including death.

Virus shedding — discharge of virus from body openings by way of exudate, excrement, or other body wastes or discharges.

Viscera — the internal organs, particularly of the thoracic and abdominal cavities.

Viscerotropic — possessing an affinity for visceral organs; a disease that acts primarily on the soft internal tissues of the body such as the heart, lungs, liver, and digestive tract.

Voluntary muscle — muscle normally under control of the individual.

Voucher specimen — specimens deposited in scientific collections that are representative of a species or a subgrouping of a species.

Wading birds — long-necked, long-legged birds that feed by wading in wetlands and catching prey with their bills. Includes egrets, herons, ibises, roseate spoonbills, flamingos, and bitterns.

Waterbirds — birds that require aquatic habitat.

Waterfowl — species of the Family Anatidae; ducks, geese, and swans. Does not include American coot.

Whistling ducks — the fulvous whistling duck or the tropical black-bellied tree duck.

Yeasts — single-celled, usually rounded fungi that produce by budding.

Zooplankton — minute animal organisms that in combination with counterparts from the plant kingdom constitute the plankton (minute free-floating organisms) of natural waters.

Zygote — a cell resulting from the union of a male and a female gamete, until it divides; the fertilized ovum.

Index

A

Acanthocephaliasis 189, 241–243
Acanths, see Acanthocephaliasis
Acaricides 285
Acid precipitation 339
Aflatoxicosis, see Aflatoxin poisoning
Aflatoxin poisoning 129, 260, 267–269
Agricultural waste 131
Air, gasping for 105
Air sacs, cheesy plaques in 131–132
Alachlor 285
Albatross
    plastic debris in stomach of 363
    tick paralysis in 258
Alcids, chlamydiosis in 111
Aldicarb 285, 287, 292, 351
Aldrin 295–297, 303
Algacides 285
Algal bloom 261, 263–264
Algal toxins 261, 263–266
Alkali poisoning, see Avian botulism
Allergic disease 128
All-terrain vehicle 21, 27
    disinfection of 33
    transport of live animals in 64–65
American avocet, lead poisoning in 320
American black duck
    duck plague in 143
    sarcocystis in 220
American coot
    algal toxin poisoning in 264
    aspergillosis in 130
    gizzard worms in 236
    lead poisoning in 320
    salmonellosis in 103
    trematodes in 251, 253
American egret, chlamydiosis in 112
American goldeneye, avian pox in 163
American goldfinch, mycoplasmosis in 115
American kestrel
    chlorinated hydrocarbon poisoning in 296
    staphylococcosis in 123
American wigeon
    algal toxin poisoning in 264
    nasal leeches in 245
    sarcocystis in 220
American woodcock, woodcock reovirus

disease in 185–186
Amidostomiasis, see Gizzard worms
Amidostomum, see Gizzard worms
Amphibians, Eustrongylides in 223–224
Analytical laboratories 371–372
Anatoxin 263, 265
Anemia 195, 207–208, 231, 257–258, 309–310
Anesthesia
    as chemical restraint 58
    euthanasia with inhalant anesthetics 50
Animal Care and Use Committees, institutional 54
Animal containers 65–66
Animal lists 19
Animal marking 44, 61–63
    criteria for 61–62
    professional and ethical considerations in 62–63
Animal population data 19, 22
Animal release guidelines 70
Animal relocation 32–34, 69
Animal Welfare Act (1985) 53–54, 71
Anoxia 344
Antimycin 285
Apapane, hemosporidiosis in 194, 196–197
Appearance, unthrifty 95, 105, 208, 257
Appetite, loss of 118, 154, 195, 223, 268
Arbovirus 171
Aroclor® 303, 305
Arsenic 285
Arthritis 107, 185
Ascites 172
Aspergillosis 96, 127–133
Aspergillus, see Aspergillosis
Aspergillus flavus, see Aflatoxin poisoning
Aspergillus niger 137
Aspergillus parasiticus, see Aflatoxin poisoning
Asper mycosis, see Aspergillosis
Asphyxiation 129
Ataxia 172, 177, 290
Atrazine 285
Auk
    avian cholera in 79
    oil toxicosis in 311
Avian botulism 4, 55, 74, 260, 271–281
Avian cholera 74–92
Avian diphtheria, see Avian pox
Avian hemorrhagic septicemia, see Avian cholera
Avian influenza 140, 181–184

Avian malaria, see Hemosporidiosis
Avian pasteurellosis, see Avian cholera
Avian pox 140, 163–169
Avian trichomoniasis, see Trichomoniasis
Avicides 285
Avitrol® 285
Azinphos-methyl 287

B

Bacillary white diarrhea, see Salmonellosis
Bacterial diseases 74
Bald eagle
    avian pox in 163, 165, 168
    barbiturate poisoning in 349–350
    chlorinated hydrocarbon poisoning in 296, 298
    cyanide poisoning in 342
    electrocution of 359
    eustrongylidiosis in 225
    herpesvirus disease in 159
    lead poisoning in 317, 319–321
    mercury poisoning in 338–339
    oil toxicosis in 314
    pesticide poisoning in 292
    scoliosis in 355, 367
    staphylococcosis in 123
Ballast waters 264
Band-tailed pigeon, trichomoniasis in 201, 203
Barbiturate poisoning 349–350, 353
Barn owl, chlorinated hydrocarbon poisoning in 296
Barred owl, avian pox in 165
Bay ducks
    lead poisoning in 318
    oil toxicosis in 311
Bayluscide® 285
Bean goose, heartworm of swans and geese in 234
Beaver, tularemia in 124
Behavior, altered 290
Benzene hexachloride 303
Beuthanasia-D Special® 51
Bewick's swan, heartworm of swans and geese in 234
Big game, Pasteurella multocida in 77
Bill
    blood-stained 146
    cheesy lesions around 203
    discharge from 122, 319
    rapid opening and closing of 131
    scratching at 245
Bioaccumulation 295
Biomagnification 295
Biotoxins 260–261
Bird banding activity 109, 123
Bird feeding stations 100, 103–105, 107–109, 116, 119, 130–131, 164–165, 202–203, 206
Birds of prey
    acanthocephaliasis in 242
    avian cholera in 75
    eustrongylidiosis in 223
    mycoplasmosis in 115–116
    trichomoniasis in 203, 206
    tuberculosis in 93–94
Bird tick 257
Biting fly 193
Biting louse 233–234
Biting midge 193, 195
Bittern, oil toxicosis in 311
Blackbird
    aspergillosis in 130
    erysipelas in 121
    pesticide poisoning in 288
    salmonellosis in 100, 103
Black-crowned night heron, mercury poisoning in 338
Black duck
    aflatoxin poisoning in 267
    chlorinated hydrocarbon poisoning in 301
    duck plague in 143–144
    lead poisoning in 321
    mercury poisoning in 340
    necrotic enteritis in 122
    trematodes in 249
Black fly 193–195
Black-footed albatross, chlorinated hydrocarbon poisoning in 295
Black-footed penguin, herpesvirus disease in 159
Blackhead, see Histomoniasis
Black-necked stilt, lead poisoning in 320
Blindness 105, 118, 247, 267, 290
Blood, thin and watery 195, 197
Blood lead level 323, 332
Blood sample 8
    collection from live animals 57–59
    shipment of blood tubes 14–15
Blood smear, in hemosporidiosis 198
Bluebird, pesticide poisoning in 288
Blue-green bloom 261
Blue grouse, tularemia in 124
Blue jay
    chlorinated hydrocarbon poisoning in 296
    mycoplasmosis in 119
Blue-winged teal
    aflatoxin poisoning in 267
    avian pox in 164
    chlorinated hydrocarbon poisoning in 296
    duck plague in 141
    trematodes in 249
Bobwhite quail
    avian pox in 164, 169
    chlorinated hydrocarbon poisoning in 297, 301
    gizzard worms in 236

herpesvirus disease in 159
histomoniasis in 257
lead poisoning in 320, 333
polychlorinated biphenyl poisoning in 305
tularemia in 124
Body temperature, reduced ability to regulate 290
Bomyl® 285
Booted eagle, herpesvirus disease in 159
Botulinum toxin 271–281
Botulism, avian, see Avian botulism
Brachial vein, blood collection from live animals 57
Brain
cholinesterase levels in 375–376
reddening of surface of 348
Brandt's cormorant, algal toxin poisoning in 264
Brant's goose, lead poisoning in 318
Breathing
impaired 165, 168, 195, 223, 247, 290
rapid 105, 290
Brevetoxin 263–265
Broadifacoum 285
Brolga crane, inclusion body disease of cranes in 153
Brooder pneumonia, see Aspergillosis
Brown-headed cowbird
polychlorinated biphenyl poisoning in 305
salmonellosis in 104
Brown pelican
algal toxin poisoning in 264–265
erysipelas in 121
lead poisoning in 317, 321
mercury poisoning in 338
oil toxicosis in 311
Budgerigar, candidiasis in 135
Bufflehead, nasal leeches in 245
Buildings, collisions with 365
Bullfinch, avian pox in 169
Bumblefoot 123
Burial of carcasses 21, 32
Burn marks 359–360
Bursa of Fabricus, atrophy of 305–306
Buteo, electrocution of 357
Buzzard, tuberculosis in 93–94

C

Cadmium 285
California gull, lead poisoning in 320
California herring gull, Newcastle disease in 175
California quail, chlorinated hydrocarbon poisoning in 297
Call, change in tone of 319
Canada goose
aflatoxin poisoning in 267–268
aspergillosis in 130
avian cholera in 81, 91
avian pox in 163–164
caught in plastic debris 363
chlorinated hydrocarbon poisoning in 296–297
duck plague in 141, 143, 146
gizzard worms in 236, 238
heartworm of swans and geese in 234
histomoniasis in 257
ingestion of dry soybeans 356
lead poisoning in 317–319, 321, 324, 327, 331, 333
necrotic enteritis in 122
oil toxicosis in 313–314
transport of live animals 66
tumor in 364
weather-related losses of 366
Canary
erysipelas in 121
herpesvirus disease in 159
salmonellosis in 100
Candida albicans, see Candidiasis
Candidiasis 128, 135–136
Canker, see Trichomoniasis
Cannon net 35
Canvasback
gizzard worms in 236
nasal leeches in 245
oil toxicosis in 311
Cape cormorant, avian cholera in 81
Captan 285
Captive-reared animals, release of 70
Capture devices 33, 35–36
Capture myopathy 58, 60, 364
Carbamate pesticides 4, 287–293, 351
Carbaryl 287
Carbofuran 285, 287, 289, 292, 351
Carbon dioxide, euthanasia with 50–51
Carbon dioxide projected dart 61
Carbon-in-pulp mill process 342, 345
Carbon monoxide, euthanasia with 50–51
Carcass, see also Specimen
disposal of 21–32, 43
dogs to aid in locating 21
euthanized, barbiturate poisoning in scavengers 349–350
fish 280
labeling of 12
lines of dead birds 276, 278
removal of 21, 41, 43
plastic bags/containers 21, 25, 43
specimen to submit to laboratory 7
transport of 21, 25
vertebrate 272, 279–280
Carcass-maggot cycle, of avian botulism 272–273, 279
Carnivores
Pasteurella multocida in 77
Sarcocystis in 219–222

Cassin's finch, salmonellosis in 104
Castor beans 261, 356
Cat(s), Mycobacterium avium in 93
Catalogued collections 56–58
Cattle, see Livestock
Cecal cores 257
Cedar waxwing
  avian cholera in 80
  salmonellosis in 104
Ceratopagonid fly 193–195
Cervical dislocation, as euthanasia method 49–50
Cestodes 189, 254
Cheilospirura spinosa, see Gizzard worms
Chemical euthanasia 50–51
Chemical ice pack 14
Chemical names 382–384
Chemical restraint 58, 61
Chemical toxins 284–285
  miscellaneous 351–353
  sources of assistance for investigating mortality events
    370–372
Chestnut-backed chickadee, salmonellosis in 104
Chicken
  aspergillosis in 130
  avian botulism in 275
  avian cholera in 82
  avian influenza in 181, 183
  avian pox in 165
  candidiasis in 135
  erysipelas in 121
  herpesvirus disease in 157–158
  histomoniasis in 257
  inclusion body disease of cranes in 153–154
  intestinal coccidiosis in 210
  mycoplasmosis in 115–116
  nematodes in 255
  Newcastle disease in 175–179
  new duck disease in 122
  salmonellosis in 100
  ticks on 257
  tracheal worms in 229
  trichomoniasis in 203
  tuberculosis in 94
  ulcerative enteritis in 123
Chipping sparrow, salmonellosis in 104
Chlamydia psittaci, see Chlamydiosis
Chlamydiosis 74, 111–114
Chlordane 296–297, 301, 303
Chloride poisoning 353
Chlorinated hydrocarbon poisoning 295–302, 352
Chlorine 285
Chlorine bleach 33–34
Chlorophacinone 285
Chlorpyrifos 289, 292, 352

Choke cherry 261
Cholera, avian, see Avian cholera
Cholinesterase, brain levels, normal values 375–376
Cholinesterase inhibitors 287–293
Chromium 285
Chronic respiratory disease, see Mycoplasmosis
Chukar partridge
  chlorinated hydrocarbon poisoning in 296–297
  eastern equine encephalomyelitis in 171
  mercury poisoning in 338
  mycoplasmosis in 117–118
  tuberculosis in 96
  ulcerative enteritis in 122
Ciguatoxin 264
Circular movements, involuntary 172
Cloaca, cheesy plaques in 151
Cloacotaenia 254
Clophens® 305
Clostridial enterotoxemia 74
Clostridium botulinum, see Avian botulism
Clostridium colinum, see Ulcerative enteritis
Clostridium perfringens, see Necrotic enteritis
Closure of area 33–34
Clothing, protective, see Protective clothing
Coccidian 189
Coccidiasis, see Intestinal coccidiosis
Coccidiosis
  intestinal, see Intestinal coccidiosis
  renal, see Renal coccidiosis
Cockatiel, chlamydiosis in 111
Cold stress, reduced tolerance to 290
Colibacillosis 125–126
Colonial nesting birds
  avian pox in 164
  salmonellosis in 99, 103–105
Comatose bird 349
Common buzzard, herpesvirus disease in 159
Common crane, inclusion body disease of cranes in
    153–154
Common egret
  chlamydiosis in 114
  eustrongylidiosis in 225
  mercury poisoning in 338
Common eider
  acanthocephaliasis in 242
  intestinal coccidiosis in 211
  lead poisoning in 317
Common goldeneye, avian pox in 164
Common loon
  lead poisoning in 317, 320–321
  mercury poisoning in 338
Common merganser
  erysipelas in 121

mercury poisoning in 338
Common moorhen, trematodes in 252
Common murre
    algal toxin poisoning in 264
    avian pox in 164, 169
Common names of wildlife 377–381
Common scoter, avian pox in 164
Common tern
    algal toxin poisoning in 264
    avian influenza in 183
    mercury poisoning in 338–339
Composting of carcasses 21, 32–33
Conjunctivitis, herpesvirus 158
Contagious epithelioma, see Avian pox
Conversion table, of units of measurement 385–386
Convulsions 83, 85, 105, 107, 146, 208, 265, 290, 298, 300, 306
Cooper's hawk, chlorinated hydrocarbon poisoning in 296
Coot
    avian cholera in 78–81, 91
    chlorinated hydrocarbon poisoning in 296
    gizzard worms in 236–237
    inclusion body disease of cranes in 153
    lead poisoning in 319
    mycoplasmosis in 117
    pesticide poisoning in 288
    salmonellosis in 100, 103
    selenium poisoning in 335
    trematodes in 249, 252
Copper 285
Copper sulfate 285
Cormorant
    avian cholera in 79
    chlorinated hydrocarbon poisoning in 298
    erysipelas in 121
    eustrongylidiosis in 225
    herpesvirus disease in 158–159
    mercury poisoning in 338
    Newcastle disease in 175–176
    oil toxicosis in 311
    polychlorinated biphenyl poisoning in 303
    renal coccidiosis in 216
    salmonellosis in 100, 103
Corn
    aflatoxin in 267–268
    fusariotoxin in 269
Cornea, cloudy 247
Coturnix quail, polychlorinated biphenyl poisoning in 303
Coughing 231
Coumaphos 352
Counter® 285
Cowbird
    aspergillosis in 130

salmonellosis in 103
Cracker shell 32, 34
Crane
    avian cholera in 79–80
    avian influenza in 181
    capture myopathy in 364
    chemical poisoning in 353
    chlamydiosis in 112
    disseminated visceral coccidiosis of cranes in 207, 212
    eastern equine encephalomyelitis in 171–173
    erysipelas in 121
    herpesvirus disease in 157–159
    inclusion body disease of cranes in 140, 153–156
    intestinal coccidiosis in 207, 210–211
    lead poisoning in 319
    oil toxicosis in 311
    organophosphate poisoning in 352
    salmonellosis in 100, 103
    tuberculosis in 93–94
Crane herpes, see Inclusion body disease of cranes
Crop, sour, see Candidiasis
Crow
    aspergillosis in 130
    avian cholera in 78–80, 85, 91
    candidiasis in 135
    carbamate poisoning in 351
    chemical poisoning in 353
    erysipelas in 121
    organophosphate poisoning in 352
    pesticide poisoning in 288
    rodenticide and repellent poisoning in 353
    salmonellosis in 103
    tuberculosis in 93–94
Crustaceans, acanthocephalans in 242
Cryptosporidium, see Intestinal coccidiosis
Curlew, salmonellosis in 100
Cyanide poisoning 261, 341–345, 353
Cyanobacterial bloom 263, 265
Cyanobacterial toxins 264
Cyanosis 257
Cyathocotyle buchiensis 249–251
Cyathostoma bronchialis, see Tracheal worms
Cylindrospermopsin 264

D

Dabbling ducks, sarcocystis in 220
Dactylaria gallopova 137
Daphnia, as nematode host 255
DDD 295–296
DDE 295, 298, 300
DDT 287, 296–298, 301
Decapitation 50
Deer, Mycobacterium avium in 93
Definitive host 188

Deformity 306, 335, 367
Dehydration 207–208, 250
Demeton 285
Demoiselle crane, inclusion body disease of cranes in 153
Deoxynivalenol 269
Depression 105, 118, 122, 157, 172, 257, 267, 300
Diacetoxyscirpenol 269
Diagnostic assistance, sources of 370–372
Diarrhea 95, 105, 122, 172, 290
    bright green 319
    greenish 122
    rust-colored 112
    watery 208
Diazinon 285, 289, 292, 351–352
Dicrotophos 292
Dieldrin 287, 295–297, 300–301, 303, 352
Die-off, description of location of 5–6
Digestive tract, see Gastrointestinal tract; Intestine
Dimethoate 287, 292, 352
Dinoflagellate bloom 259, 263
Diphacinone 285
Diphtheria, avian, see Avian pox
Diquat® 285
Direct life cycle 188–190
Disease control operations
    analyses 42
    disease outbreak summary 48
    field communications systems 44
    during field research 67–69
    planning phase of
        biological data records 19, 22–23, 41
        identification of needs 19–20, 41
    response activities
        animal relocation 32–34
        carcass disposal 21–32, 43
        carcass removal 21, 41, 43
        disinfection 33–34, 38–39, 43
        establishing control of area 41
        personnel for 34, 40
        problem identification 21, 41
        response modifications 40
    for specific diseases
        for acanthocephaliasis 243
        for aflatoxin poisoning 268
        for algal toxins 266
        for aspergillosis 131
        for avian botulism 279–280
        for avian cholera 88–91
        for avian influenza 184
        for avian pox 169
        for barbiturate poisoning 350
        for chlamydiosis 113
        for chlorinated hydrocarbon poisoning 302
        for cyanide poisoning 345
        for duck plague 39, 151
        for eastern equine encephalomyelitis 174
        in electrocutions 360
        for eustrongylidiosis 225
        for fusariotoxin poisoning 270
        for gizzard worms 239
        for hemosporidiosis 199
        for herpesvirus disease 160
        for inclusion body disease of cranes 154–156
        for intestinal coccidiosis 211–213
        for lead poisoning 332–333
        for mercury poisoning 340
        for mycoplasmosis 119
        for nasal leeches 248
        for Newcastle disease 178–179
        for oil toxicosis 313–314
        for pesticide poisoning 292–293
        for polychlorinated biphenyl poisoning 306–307
        for renal coccidiosis 218
        for salmonellosis 107–109
        for salt poisoning 348
        for sarcocystis 221–222
        for selenium poisoning 336
        for tracheal worms 231
        for trichomoniasis 206
        for tuberculosis 95–96
        for woodcock reovirus disease 186
    station disease contingency plan 45–47
    surveillance activities 42, 44
    wildlife population and habitat management 44
    wildlife sampling and monitoring 44
Disease introduction 67–69
Disease onset, estimation of 4
Disinfectant 33–34
Disinfection 33–34, 38–39, 43
Dissection of bird 10–11
Disseminated visceral coccidiosis of cranes 207, 212
Disulfoton 292, 352
Diving birds
    mercury poisoning in 338
    oil toxicosis in 311
Diving ducks
    aspergillosis in 130
    gizzard worms in 237
    lead poisoning in 318, 332
    nasal leeches in 245
    pesticide poisoning in 288
    sarcocystis in 221
    trematodes in 249
Dog
    location of carcasses by 21
    Mycobacterium avium in 93
Domestic animal disease programs 370
Domoic acid 263–265

Double-crested cormorant
    entangled in fishing line 362
    lead poisoning in 320–321
    mercury poisoning in 338
    Newcastle disease in 175–178
    polychlorinated biphenyl poisoning in 303
    renal coccidiosis in 217
    salmonellosis in 103
Dove
    avian cholera in 79
    carbamate poisoning in 351
    chemical poisoning in 353
    chlamydiosis in 111
    erysipelas in 121
    hemosporidiosis in 194
    herpesvirus disease in 158
    intestinal coccidiosis in 210
    lead poisoning in 319, 332–333
    Newcastle disease in 176
    organophosphate poisoning in 352
    pesticide poisoning in 288
    salmonellosis in 103
    trichomoniasis in 201–203, 206
Drive trap 36
Droopiness 105, 122, 146
Drop net 33
Droppings
    blood-stained 83
    fawn-colored or yellow 83
Drowning 148, 277–279, 309–310, 356, 367
Drowsiness 83, 172, 349
Dry ice
    euthanasia with 51
    for specimen shipment 14, 17
Duck
    acanthocephaliasis in 242
    aflatoxin poisoning in 267
    algal toxin poisoning in 264
    aspergillosis in 130
    avian botulism in 277
    avian cholera in 78–79
    candidiasis in 135
    chlamydiosis in 111, 114
    chlorinated hydrocarbon poisoning in 296
    duck hepatitis in 361
    duck plague in 141–151
    erysipelas in 121
    gizzard worms in 236–237
    hemosporidiosis in 194–195
    herpesvirus disease in 157
    intestinal coccidiosis in 210
    lead poisoning in 318
    mycoplasmosis in 115, 117
    nasal leeches in 248
    necrotic enteritis in 122
    nematodes in 255
    Newcastle disease in 175
    new duck disease in 122
    oil toxicosis in 311
    renal coccidiosis in 216–217
    rictin poisoning in 356
    salmonellosis in 100, 103, 107
    salt encrustation of 347
    staphylococcosis in 123
    tracheal worms in 229
    tularemia in 124
Duck disease, see Avian botulism
Duck hepatitis 361–362
Duck leeches, see Nasal leeches
Duck plague 39, 140–151
Duck virus enteritis, see Duck plague
Dung piles 101
Dusky song sparrow, salmonellosis in 104
DVE, see Duck plague

E

Eagle
    avian cholera in 78–79, 91
    carbamate poisoning in 351
    chemical poisoning in 353
    chlorinated hydrocarbon poisoning in 352
    cyanide poisoning in 342
    electrocution of 357
    erysipelas in 121
    herpesvirus disease in 158
    lead poisoning in 317, 322
    organophosphate poisoning in 352
    pesticide poisoning in 288
    rodenticide and repellent poisoning in 353
    salmonellosis in 100
    staphylococcosis in 123
    strychnine poisoning in 351
    tularemia in 124
Eagle owl, herpesvirus disease in 159
Eared grebe
    algal toxin poisoning in 264
    erysipelas in 121
    salmonellosis in 103
Earthworms
    chlorinated hydrocarbons in 295
    histomonads in 257
    tracheal worm carriage 229–230
East African crowned crane, inclusion body disease of cranes in 153
Eastern equine encephalomyelitis 140, 171–174
Eastern screech owl, avian pox in 165
Eastern sleeping sickness of horses, see Eastern equine encephalomyelitis

Echinuria uncinata 255–256
Ectoparasites 188, 249, 257–258
Edema 305–306, 344
EEE, see Eastern equine encephalomyelitis
Egg production, decreased 183
Eggshell, thin 298–300, 303, 306
Egret
    avian cholera in 79
    chemical poisoning in 353
    chlamydiosis in 112
    chlorinated hydrocarbon poisoning in 298
    eustrongylidiosis in 225
    lead poisoning in 321
    oil toxicosis in 311
    salmonellosis in 99–100, 103
Eider
    aspergillosis in 131
    avian cholera in 80, 82, 91
    oil toxicosis in 311
    renal coccidiosis in 217
Eimeria, see Intestinal coccidiosis; Renal coccidiosis
Electrocution 356–360
Emaciation 95, 131, 165, 185–186, 195, 208, 217, 223, 237, 242, 256, 298, 301, 305, 312, 319, 339
Embryo mortality 306, 335
Embryonic deformity 335
Emerging infectious disease 140
Emu, eastern equine encephalomyelitis in 171
Encephalomyelitis 158
Endoparasites 188, 249
Endosulfan 297
Endrin 295–297, 301
Enflurane 50
English sparrow, salmonellosis in 104, 106
Enteritis 211, 249
    duck virus, see Duck plague
    necrotic, see Necrotic enteritis
    ulcerative, see Ulcerative enteritis
    woodcock reovirus 185
Environmental factors, records of 4
Epithelioma, contagious, see Avian pox
Epizootiology 3
Epomidiostomum, see Gizzard worms
Erysipelas 121
Erysipilothrix rhusopathiae, see Erysipelas
Escherichia coli, see Colibacillosis
Esophagus
    cheesy, raised plaques along 146, 150–151
    hard, cheesy lesions of 205
    impaction of 320, 328–329
    light-colored granulomas in 211–212
    yellow, cheesy nodules in 106
Ethion 287
Ethylene glycol poisoning 285, 353

European blackbird, erysipelas in 121
European bullfinch, salmonellosis in 100
European jay, aspergillosis in 130
European starling
    mycoplasmosis in 116, 118
    polychlorinated biphenyl poisoning in 305
    salmonellosis in 103
European thrush, acanthocephaliasis in 242
Eustrongylides, see Eustrongylidiosis
Eustrongylidiosis 223–228
Euthanasia 7, 32, 49–51
    chemical 50–51
    guidelines for field research 54, 67, 70
    physical 49–50
    selection of method of 49
Euthanasia agents 50–51, 285
    barbiturate poisoning of scavengers 349–350
Eutrophication 225, 261
Evening grosbeak, salmonellosis in 104, 106
Exertional myopathy, see Capture myopathy
Exhaustion 309–310
Exsanguination 50
Eye
    cheesy plaques in 131
    discharge from 112, 122
    drainage from 118
    erosions on surface of 348
    inflamed 118
    leeches in 245, 247
    protrusion of 290
    puffy 118
    swollen 118
Eyelids
    crusty 118
    drooping of 290, 339
    paralysis of 277–278
    pasted 105
    swollen 118

F

Face, puffy 203–204
Falcon
    avian cholera in 78–79
    electrocution of 357
    herpesvirus disease in 157–159
    intestinal coccidiosis in 210
    pesticide poisoning in 288
    salmonellosis in 100, 103
    trichomoniasis in 203
    tuberculosis in 93
Famphur 289, 292, 351–352
Favus, fowl, see Ringworm
Feathers
    burn marks on 359

loss of 137, 258
oil contamination of 309–310, 312
ruffled 105, 112, 122, 135, 183, 257, 339
Feces
bile-stained 319
green-colored 319, 325
Feeding difficulties 112, 165
Fenamiphos 292
Fences, collisions with 365
Fenclors® 305
Fensulfothion 292
Fenthion 285, 289, 292, 352
Ferruginous hawk
avian pox in 165
staphylococcosis in 123
"Field hospital," for waterfowl 55
Field research
animal disposal at completion of study 70
animal marking 61–63
blood and tissue collections 57–59
conditions for confined wildlife 54–55, 63–64
disease considerations 67–69
euthanasia 67, 70
housing and maintenance of field sites 54–55, 63–64
investigator disturbance and impacts 56
professional society guidelines for 71
restraint and handling of wildlife 58–60
safety of personnel 70
surgical and medical procedures 65–67
transport of live animals 64–65
wildlife in 53–71
wildlife observations and collections 56–58
Wildlife Society guidelines for 53–71
Filarial heartworm, see Heartworm of swans and geese
Finch
avian cholera in 79
avian pox in 164
erysipelas in 121
herpesvirus disease in 159
Fish
acanthocephalans in 242
Eustrongylides in 223–225
mercury in 339
Pfiesteria mortality in 265
Fish-eating birds
chlorinated hydrocarbon poisoning in 295
eustrongylidiosis in 225
polychlorinated biphenyl poisoning in 303
renal coccidiosis in 216
Fishing line, discarded 362
Fishing sinkers 333
Flamingo
capture myopathy in 364
lead poisoning in 320

Flea 258
Flight
awkward 265
erratic 83, 319
inability to fly 146, 148, 267, 277–278, 309
reluctance to fly 318
Flukes, see Trematodes
Fluorine poisoning 353
Fly 258
Fly larvae, tracheal worm carriage 229–230
Fonofos 292, 352
Food chain 295, 303, 337
Foot webs, congestion in 265
Formalin 8
Fowl cholera, see Avian cholera
Fowl favus, see Ringworm
Fowl pest, see Avian influenza
Fowl plague, see Avian influenza
Fowl pox, see Avian pox
Fowl tick 257
Fowl typhoid, see Salmonellosis
Francisella tularensis, see Tularemia
Franklin's gull, algal toxin poisoning in 264
Frogs, acanthocephalans in 242
Frounce, see Trichomoniasis
Fulmar, mercury poisoning in 338
Fulvous whistling duck, chlorinated hydrocarbon poisoning in 296–297
Fungal disease 128
cutaneous 137
miscellaneous 137
subcutaneous 137
systemic 137
Fungicides 285
Funnel trap 35–36
Fusariomycotoxicosis, see Fusariotoxin poisoning
Fusariotoxin poisoning 269–270
Fusarium, see Fusariotoxin poisoning

G
Gadwall
aflatoxin poisoning in 267
duck plague in 141
nasal leeches in 245
sarcocystis in 220
Gait
abnormal 112
unsteady 319
Gallbladder, distended and bile-filled 320, 330
Gallinaceous birds
chlorinated hydrocarbon poisoning in 296, 298
mercury poisoning in 338
Gannet
mercury poisoning in 338

Newcastle disease in 177
   salmonellosis in 103
Gapes, see Tracheal worms
Gape worm, see Tracheal worms
Gaping 131, 229, 247
Gastiotaenia 254
Gastrointestinal tract, see also Intestine
   dye in 291
   hemorrhage in 146, 149, 151, 269, 309
   inflammation and ulceration of 269–270
   inflammation of 310
   moist, necrotic lesions of 165
Geographic distribution
   of aflatoxin poisoning 267–269
   of aspergillosis 130
   of avian botulism 275–277
   of avian cholera 76, 80–84
   of avian influenza 182–183
   of avian pox 164–166
   of candidiasis 135
   of chlamydiosis 111
   of chlorinated hydrocarbon poisoning 295
   of cyanide poisoning 342–344
   of duck plague 141–145
   of eastern equine encephalomyelitis 171
   of electrocutions 358
   of eustrongylidiosis 223, 226–227
   of fusariotoxin poisoning 269
   of heartworm of swans and geese 234
   of hemosporidiosis 195
   of herpesvirus disease 157, 159
   of inclusion body disease of cranes 153–154
   of intestinal coccidiosis 208, 210
   of lead poisoning 322–323, 333
   of mercury poisoning 339
   of mycoplasmosis 117
   of nasal leeches 245–246
   of Newcastle disease 176–178
   of oil toxicosis 311–312
   of pesticide poisoning 289
   of polychlorinated biphenyl poisoning 303–305
   of renal coccidiosis 217
   of salmonellosis 100, 104
   of salt poisoning 347
   of sarcocystis 220
   of selenium poisoning 335
   of tracheal worms 229
   of trichomoniasis 201, 203–204
   of tuberculosis 94
   of woodcock reovirus disease 185
Gizzard
   green-stained 320, 330
   hemorrhages on surface of 83
   lead pellets in 320, 331

Gizzard pads, inflammation of 254
Gizzard worms 235–239
Glaucous-winged gull, lead poisoning in 320
Globe, collapse of 247
Glossy ibis, eastern equine encephalomyelitis in 171
Gloves 7, 21
   disposable and reusable 21, 24
Goats, see Livestock
Golden eagle
   avian pox in 165
   barbiturate poisoning in 349–350
   electrocution of 357–358
   herpesvirus disease in 159
   lead poisoning in 317, 319–320
   mercury poisoning in 338–339
   staphylococcosis in 123
Goldfinch, salmonellosis in 100, 104
Gold mining 341–345
Goose
   acanthocephaliasis in 242
   algal toxin poisoning in 264
   avian cholera in 79
   candidiasis in 135
   chlamydiosis in 111
   duck plague in 141–151
   erysipelas in 121
   fusariotoxin poisoning in 269
   gizzard worms in 236–237
   heartworm of swans and geese in 233–234
   hemosporidiosis in 194–195
   intestinal coccidiosis in 210
   lead poisoning in 318
   mycoplasmosis in 115
   nasal leeches in 245
   nematodes in 254–255
   oil toxicosis in 311
   pesticide poisoning in 288
   renal coccidiosis in 210–217
   salmonellosis in 100, 103
   sarcocystis in 221
   tracheal worms in 229
   tularemia in 124
Goshawk
   chlorinated hydrocarbon poisoning in 296
   mercury poisoning in 339
Gout 367
   visceral 172–173, 348
Grackle
   aspergillosis in 130
   avian cholera in 79–80
   salmonellosis in 103
Grain
   aflatoxin in 267
   drugged, to capture birds 36

    fusariotoxin in 269–270
    mercury-treated 339
    moldy 129, 131
Grain storage facility 206
Granuloma 131, 225
Grasshopper, gizzard worm carriage 236, 239
Gray partridge, chlorinated hydrocarbon poisoning in 297
Great blue heron
    eustrongylidiosis in 223, 226
    lead poisoning in 320–321
    mercury poisoning in 338
Great cormorant, algal toxin poisoning in 264
Great egret
    eustrongylidiosis in 225
    lead poisoning in 320
Greater sandhill crane, lead poisoning in 320
Great horned owl
    chlorinated hydrocarbon poisoning in 296
    electrocution of 357
    herpesvirus disease in 159–160
Grebe
    avian cholera in 79
    chemical poisoning in 353
    chlorinated hydrocarbon poisoning in 296
    erysipelas in 121
    gizzard worms in 236
    nasal leeches in 245
    oil toxicosis in 311
Green-winged teal
    aflatoxin poisoning in 267
    avian botulism in 281
    avian pox in 163–164
    erysipelas in 121
    trematodes in 249
Grinding pads, degeneration of 237–238
Grouse
    candidiasis in 135
    gizzard worms in 235
    histomoniasis in 257
    lead poisoning in 317, 319, 332
    mercury poisoning in 338
    nematodes in 254–255
    salmonellosis in 100, 103
    tularemia in 124
    ulcerative enteritis in 123
Growth retardation 135
Guillemot
    oil toxicosis in 311
    salmonellosis in 103
Guinea fowl
    candidiasis in 135
    duck hepatitis in 361
    erysipelas in 121
    herpesvirus disease in 157
    mycoplasmosis in 115–116
    nematodes in 255
    salmonellosis in 100
Gull
    algal toxin poisoning in 264
    aspergillosis in 130
    avian botulism in 274
    avian cholera in 78–80, 91
    avian influenza in 181, 183
    candidiasis in 135
    capture myopathy in 364
    carbamate poisoning in 351
    chemical poisoning in 353
    chlamydiosis in 111–112
    chlorinated hydrocarbon poisoning in 298
    eastern equine encephalomyelitis in 171–172
    erysipelas in 121
    gizzard worms in 237
    intestinal coccidiosis in 210
    lead poisoning in 319, 321
    mercury poisoning in 338
    oil toxicosis in 311
    pesticide poisoning in 288
    renal coccidiosis in 216
    salmonellosis in 99–100, 103–104, 107
    sarcocystis in 221
    tuberculosis in 93–94
    tularemia in 124
Gunshot, as euthanasia method 50, 56
Gyrfalcon, herpesvirus disease in 159

H

Haemoproteus, see Hemosporidiosis
Halothane 50
Handling of live animals 58–60
    disease due to 364
Harlequin duck, avian pox in 164
Hatchet-breast appearance 319, 325
Hatchlings, disease in 361
Hawaiian forest birds
    avian pox in 166
    hemosporidiosis in 195
Hawaiian goose, avian pox in 163
Hawk
    avian cholera in 79
    carbamate poisoning in 351
    chemical poisoning in 353
    erysipelas in 121
    intestinal coccidiosis in 210
    lead poisoning in 319
    organophosphate poisoning in 352
    pesticide poisoning in 288
    rodenticide and repellent poisoning in 353
    salmonellosis in 100, 103

trichomoniasis in 203
tuberculosis in 94
tularemia in 124
Hawk-headed parrot, salmonellosis in 100
Hazardous Materials Regulations 17
Hazing 32
Head
arched back 290
fluid beneath skin of 269–270
inability to hold erect 277
puffy or swollen 319, 327
pulled back and twisted to side 148
resting on back 83, 85
shaking of 157, 223, 231, 245
side-to-side movement of 265
twisting of 177–179, 265
unnatural positioning of 131
Head droop 265, 269
Head restraint 59
Heap leach system 341–345
Heart
enlarged 234
excess fluids around 305
fibrinous covering on surface of 122
hemorrhage on surface of 83, 86–87, 146, 150
light patches on 211, 213
pale foci or spots on 234
Heartworm 188
Heartworm of swans and geese 233–234
Hemorrhage
in gastrointestinal tract 146, 149, 151, 269, 309
on gizzard surface 83
on heart surface 83, 86–87, 146, 150
in intestine 154, 249–251, 268, 291, 293
in liver 265, 268, 361
in thymus gland 154
trauma related 366
Hemorrhagic septicemia, avian, see Avian cholera
Hemosporidiosis 187–189, 193–199
Hemozoin pigment 195, 199
Hepatitis
duck, see Duck hepatitis
herpesvirus 157–158
woodcock reovirus 185
Hepatomegaly 112–113, 172
Heptachlor 296–297, 301, 352
Herbicides 285
Heron
aspergillosis in 130
avian botulism in 274
avian cholera in 79
chlamydiosis in 111–112
chlorinated hydrocarbon poisoning in 298

eustrongylidiosis in 225
oil toxicosis in 311
salmonellosis in 99–100, 103
Herpesvirus
duck plague 141–151
inclusion body disease of cranes 153–156
Herpesvirus disease 157–161
Herring gull
algal toxin poisoning in 264
aspergillosis in 131
avian influenza in 183
erysipelas in 121
lead poisoning in 320
mercury poisoning in 338
oil toxicosis in 314
salmonellosis in 99
staphylococcosis in 123
Heterakis gallinarum, as vector for Histomonas 257
Hexaconazole 285
High voltage transmission lines 357–360
Histomonas meleagridis, see Histomoniasis
Histomoniasis 256–257
History, see Specimen history
Holding devices 58–59
Hooded crane, inclusion body disease of cranes in 153
Horned lark, chlorinated hydrocarbon poisoning in 297
Horse, see Livestock
House finch
avian pox in 169
mycoplasmosis in 115–119
salmonellosis in 104
House finch conjunctivitis, see Mycoplasmosis
House sparrow
chlorinated hydrocarbon poisoning in 297
eastern equine encephalomyelitis in 171
salmonellosis in 104, 107
Housing, of confined wildlife 54–55, 63
Huddling together 105
Human health risks
of aflatoxin 269
of algal toxins 266
of aspergillosis 133
of avian botulism 280
of avian cholera 91–92
of avian influenza 184
of avian pox 169
of bacterial diseases 74
of candidiasis 136
of chlamydiosis 113–114
of chlorinated hydrocarbons 302
of cyanide 345
of eastern equine encephalomyelitis 174
of eustrongylidiosis 227
of fusariotoxin 270

of lead 333–334
of mercury 340
of Newcastle disease 179
of oil pollution 314
of pesticides 293
of polychlorinated biphenyls 307
of salmonellosis 100, 109
of selenium 336
of staphylococcosis 123–124
of tuberculosis 98
Hungarian partridge
　histomoniasis in 157
　lead poisoning in 320
　Newcastle disease in 175
　transport of live animals 66
Hydrocyanic acid poisoning, see Cyanide poisoning
Hydrogen sulfide poisoning 353
Hydropericardium 185
Hypothermia 309–310, 312, 350
Hypovolemic shock 249

I

Ibis
　chlamydiosis in 112
　eustrongylidiosis in 225
Ice-related problems 365–367
Inactivity 208
Incineration of carcasses 21, 28–31
　above-ground methods for 21, 29–30
　in-trench methods for 21, 29
　layering of carcasses for 32
Incinerator, portable 21, 28
Inclusion body, intranuclear 160
Inclusion body disease of cranes 140, 153–156
Inclusion body disease of falcons, see Herpesvirus disease
Incoordination 105, 172, 265, 290, 298, 300, 305–306, 339–340
Indian blue peafowl, mycoplasmosis in 118
Indirect life cycle 188–191
Infectious laryngotracheitis 158
Infectious sinusitis, see Mycoplasmosis
Influenza, avian, see Avian influenza
Inhalant anesthetics, euthanasia with 50
Insecticide poisoning 295–302
Intermediate host 188
Intestinal coccidiosis 207–213
Intestine
　fibrous-to-caseous core of necrotic debris in 252–253
　hemorrhage in 154, 249–251, 268, 291, 293
　lesions of enteritis 122–123
　light-colored areas in 211
　necrosis of 157
　necrotic, crumbly cores in 106–107
　nodules in 95

　raised tunnels on 223, 225, 227
　reddened 291, 312
　thick, yellowish fluid in 88
　white nodules on 242
Isazophos 285
Isoflurane 50
Iso-neosolaniol 269
Isospora, see Intestinal coccidiosis

J

Jackdaw
　candidiasis in 135
　salmonellosis in 100
　tracheal worms in 229
Japanese quail, polychlorinated biphenyl poisoning in 305
Java finch, chlamydiosis in 112
Jay
　avian cholera in 79
　pesticide poisoning in 288
Jugular vein, blood collection from live animals 57

K

Kanechlors® 305
Keel, prominent 217
Kesterson Reservoir 335
Kestrel
　chlorinated hydrocarbon poisoning in 298
　herpesvirus disease in 159
　tuberculosis in 93
Kidney
　congested 300
　enlarged 217, 305
Kill trap 56
Kingfisher, intestinal coccidiosis in 210
King parrot, chlamydiosis in 112
Kite, candidiasis in 135
Kittiwake, avian influenza in 183

L

Lameness 95, 195, 208, 220
Landfill 101, 107, 303, 349–350
Laryngotracheitis, infectious 158
Lasso 285
Laughing gull, lead poisoning in 320
Laysan albatross
　avian pox in 164–165, 167–169
　lead poisoning in 317, 320–321
Laysan duck, nematodes in 255–256
Lead poisoning 285, 317–334
Leaf litter 295
Leeches, nasal, see Nasal leeches
Leech fixation procedure 248
Leg, extended to rear 265
Lesser sandhill crane, lead poisoning in 320

Lesser scaup
    aflatoxin poisoning in 267
    algal toxin poisoning in 264–265
    intestinal coccidiosis in 207
    nasal leeches in 245
    trematodes in 251
Lethal injection 51
Lethargy 83, 85, 95, 118, 154, 242, 249, 267, 290, 298, 300, 305–306
Leucocytozoon, see Hemosporidiosis
Leyogonimus polyoon 252–253
Lice 258
Life cycle
    of acanthocephalans 241–242
    of coccidia 207–209, 215–216
    direct 188–190
    of Echinuria uncinata 255
    of Eustrongylides 223–224
    of gizzard worms 235–236
    of hemosporidia 193–194
    of Histomonas meleagridis 256
    indirect 188–191
    of Sarcocystis 219–220
    of Sarconema 233–234
    of tracheal worms 229–230
    of trematodes 249–252
    of Trichostrongylus tenuis 254
Light sensitivity 146
Limberneck, see Avian botulism
Lindane 297
Listlessness 122, 135, 195, 203
Little blue heron, chlamydiosis in 113
Liver
    amyloid deposits in 96
    bile stained 125–126
    congested 123, 300
    enlarged 195–196, 305
    enlarged, swollen pale 268
    fibrinous covering on surface of 122
    gray spots within 123
    hemorrhage in 265, 268, 361
    large, pale grey areas of necrosis in 257
    light patches on 211, 213
    necrotic lesions of 150, 160
    nodules in 95
    paratyphoid nodules in 106
    swollen 125–126, 154
    white-to-yellow spots on 83, 87, 124, 154–155
    yellow areas of tissue death in 123
Livestock
    avian influenza virus in 184
    botulism in 274
    eastern equine encephalomyelitis virus in 172–173
    euthanized carcasses 349–350
    Mycobacterium avium in 93–94
    Pasteurella multocida in 77
    Salmonella in 107–108
Live trap 56–57
Living conditions, of confined wildlife 54–55, 63
Long-billed dowitcher, lead poisoning in 320
Long-eared owl, herpesvirus disease in 159
Loon
    aspergillosis in 130
    avian botulism in 274
    avian cholera in 79
    eustrongylidiosis in 225
    mercury poisoning in 339
    nasal leeches in 245
    oil toxicosis in 311
    renal coccidiosis in 216
    ringworm in 137
Lorikeet, candidiasis in 135
Lory, candidiasis in 135
Louisiana pneumonitis, see Chlamydiosis
Louse fly 193–194
Lovebird, candidiasis in 135
Lung
    cheesy plaques in 131–132
    congested 113, 300, 344
    dark red, granular 131, 133
    fluid accumulation in 348
    nodules in 95
    purple areas of firmness in 183

M

M-44 sodium cyanide ejector 341–345
Macaw
    chlamydiosis in 111
    tuberculosis in 93
Maggots, carcass-maggot cycle of avian botulism 272–273, 279
Magpie
    avian cholera in 79
    mercury poisoning in 339
    pesticide poisoning in 292
    salmonellosis in 100, 103
Malaria, avian, see Hemosporidiosis
Mallard
    aflatoxin poisoning in 267
    aspergillosis in 129, 131
    avian botulism in 281
    avian cholera in 85
    avian influenza in 181, 183
    avian pox in 163–164
    chlorinated hydrocarbon poisoning in 296–297, 299
    duck hepatitis in 361–362
    duck plague in 141, 143–144, 146, 151

lead poisoning in 317, 321, 324, 328, 330
mercury poisoning in 340
mycoplasmosis in 117
nasal leeches in 247
necrotic enteritis in 122
Newcastle disease in 175, 177, 179
pesticide poisoning in 287, 292
polychlorinated biphenyl poisoning in 303, 305
sarcocystis in 220
selenium poisoning in 335
staphylococcosis in 123
trematodes in 249
Malnutrition 367
Manchurian crane, inclusion body disease of cranes in 153
Mange 258
Man-made structures, collisions with 365–366
Manure, contaminated 97, 107
Maps 19, 23
Marbled godwit, lead poisoning in 320
Marek's disease 158, 364
Marine birds
aspergillosis in 130
avian influenza in 181, 183
avian pox in 163, 166, 169
eastern equine encephalomyelitis in 171
herpesvirus disease in 158
oil toxicosis in 311
renal coccidiosis in 216
tick paralysis in 258
Marking, see Wildlife marking
Martin, avian cholera in 79
Measurement, units of, conversion table 385–386
Medial-metatarsal vein, blood collection from live animals 57
Medical procedures, wildlife 65–67
major procedures 67
minor procedures 67
Meningoencephalitis 131, 183
Mercury poisoning 337–340
Merganser
eustrongylidiosis in 225
lead poisoning in 318
mercury poisoning in 338
nasal leeches in 245
oil toxicosis in 311
sarcocystis in 221
Metals 285
Methamidiphos 292
Methiocarb 285, 287, 292, 351
Methoxyflurane 50
Mexacarbate 287
Microcystin 263
Microsporum gallinae 137
Minimata disease, see Mercury poisoning

Mink
botulism in 274
Mycobacterium avium in 93
polychlorinated biphenyl poisoning in 303
Minnows, Eustrongylides in 223–224
Mirex 297
Mite 258
Mollusc(s), as trematode hosts 249
Molluscicides 285
Moniliasis, see Candidiasis
Monocrotophos 289, 292, 352
Mosquito 163–165, 171–174, 193–195
Motor vehicle
disinfection of 33, 38
transport of live animals in 64–65
Mottle duck, sarcocystis in 220
Mourning dove
avian pox in 164, 169
lead poisoning in 320, 333–334
salmonellosis in 103
trichomoniasis in 201, 203–205
Mouth
hard, cheesy lesions of 205
inflammation of 205
inflammation and ulceration of 269–270
lesions of 205
moist, necrotic lesions of 165
mucous discharge from 83
Murre
avian influenza in 183
oil toxicosis in 311
Muscle
cyst-like bodies in 195, 197
necrosis of 364
rice-grain cysts in 220–221
wasting of 95
Muscovy duck
algal toxin poisoning in 265
duck plague in 141, 151
inclusion body disease of cranes in 153–154
Museum collections 56–57
Mute swan
avian pox in 163–164, 169
heartworm of swans and geese in 234
lead poisoning in 318, 321
salmonellosis in 103
staphylococcosis in 123
trematodes in 249
Mycobacteriosis, see Tuberculosis
Mycobacterium avium, see Tuberculosis
Mycoplasma gallisepticum, see Mycoplasmosis
Mycoplasma meleagridis, see Mycoplasmosis
Mycoplasma synovia, see Mycoplasmosis
Mycoplasmosis 74, 115–119

Mycosis, see Fungal disease
Mycotic pneumonia, see Aspergillosis
Mycotoxicosis 128
Mycotoxins 137, 260–261, 267–270
Myiasis 258
Mynah, chlamydiosis in 111
Myocarditis 185
Myopathy, capture, see Capture myopathy

N

Nares, bleeding from 290
Nasal discharge 265
    thick, mucous-like, roppy 83
Nasal leeches 188, 245–248
National Response Center 313
National Wildlife Health Center, shipment of specimens to 14–16
ND, see Newcastle disease
Neck
    abnormal position of 319
    fluid beneath skin of 269–270
    paralysis of muscles of 277–278
    twisting of 157, 177–179
Necropsy kit 8
Necrotic enteritis 74, 122
Nematocidices 285
Nematodes 189, 254–256
    eustrongylidiosis 223–228
    gizzard worms 235–239
    tracheal worms 229–231
Nene goose, tuberculosis in 93
Neosaxitoxin 264
Neosolaniol 269
Nervous system disorder 122
Nesting, on power poles 357–360
Neurologic symptoms 172–173, 265, 269, 290
Newcastle disease 140, 175–179
    asymptomatic lentogenic 176
    lentogenic 175–176
    mesogenic 176
    neurotrophic velogenic 175–176
    velogenic 175
    viscerotrophic velogenic 176
New duck disease 122
Nictitating membrane
    paralysis of 277
    stringy, cheeselike material beneath 247
Nitrapyrin 285
Nitrous oxide 50
Nodularin 263
Northern cardinal, salmonellosis in 104
Northern fulmar, algal toxin poisoning in 264
Northern harrier
    avian cholera in 78
    lead poisoning in 320
Northern mockingbird, salmonellosis in 104
Northern pintail
    avian cholera in 85
    lead poisoning in 317, 325
    nasal leeches in 245
    sarcocystis in 220
Northern shoveler
    erysipelas in 121
    nasal leeches in 245
    sarcocystis in 220, 222
No-see-ums 194–195
Nuthatch, avian cholera in 79
Nutritional disease 356, 367
Nutritional needs, of confined wildlife 63–64
NVND, see Newcastle disease

O

Observation of wildlife, guidelines for field research 56–58
OCs, see Chlorinated hydrocarbon poisoning
Ocular discharge 265
Oil Pollution Act (1990) 313–314
Oil spill 130, 164, 283, 309–315
Oil toxicosis 285, 309–315
Okadaic acid 264
Oldsquaw
    avian cholera in 81
    oil toxicosis in 311
Oligochaetes, Eustrongylides in 223–225
Opisthotonos 361
Opportunistic infection 128, 137
Oral discharge 265
Organochlorines, see Chlorinated hydrocarbon poisoning
Organophosphorus pesticides 4, 287–293, 352
Oriole, avian cholera in 79
Ornithosis, see Chlamydiosis
Osmoregulatory dysfunction 309–310
Osprey
    chlorinated hydrocarbon poisoning in 298
    mercury poisoning in 338
Osteoporosis 185
Ostrich
    avian influenza in 181
    eastern equine encephalomyelitis in 171
    intestinal coccidiosis in 210
    salmonellosis in 100
Over-straining disease, see Capture myopathy
Owl
    avian cholera in 79
    chemical poisoning in 353
    chlorinated hydrocarbon poisoning in 301, 352
    electrocution of 357
    herpesvirus disease in 157–159
    intestinal coccidiosis in 210

lead poisoning in 319
mercury poisoning in 338–339
organophosphate poisoning in 352
pesticide poisoning in 288
renal coccidiosis in 216
salmonellosis in 100, 103
trichomoniasis in 203
tuberculosis in 94
Owl herpesvirus, see Herpesvirus disease
Oxamyl 292

P

Pacific loon, algal toxin poisoning in 264
Parakeet
candidiasis in 135
chlamydiosis in 111, 113
erysipelas in 121
gizzard worms in 235
Paralysis 157, 172, 177, 179, 220, 265, 269, 277, 290
Paralytic shellfish poisoning 261, 263–264
Paramyxovirus-1, see Newcastle disease
Parasitic disease 188–189
Paratenic host 188, 192
Parathion 289, 292, 352
Paratyphoid, see Salmonellosis
Parrot
candidiasis in 135
chlamydiosis in 111–113
erysipelas in 121
intestinal coccidiosis in 210
mycoplasmosis in 115
salmonellosis in 100
tuberculosis in 93
Parrot disease, see Chlamydiosis
Parrot-fever, see Chlamydiosis
Partridge
candidiasis in 135
gizzard worms in 235
histomoniasis in 257
lead poisoning in 317
mercury poisoning in 339
mycoplasmosis in 115–116
salmonellosis in 100, 103
tuberculosis in 94
ulcerative enteritis in 123
Passerines
chlorinated hydrocarbon poisoning in 296
mercury poisoning in 339
pesticide poisoning in 287–288
Pasteurella anatipestifer, see New duck disease
Pasteurella multocida, see Avian cholera
Pasteurellosis, avian, see Avian cholera
PCBs, see Polychlorinated biphenyls
Peafowl
candidiasis in 135
erysipelas in 121
herpesvirus disease in 157
mycoplasmosis in 115, 117
salmonellosis in 100
Peanuts
aflatoxin in 267
fusariotoxin in 269–270
Pectoral sandpiper, lead poisoning in 320
Pelican
avian cholera in 79
chemical poisoning in 353
chlorinated hydrocarbon poisoning in 298
erysipelas in 121
eustrongylidiosis in 225
gizzard worms in 237
intestinal coccidiosis in 210
mercury poisoning in 338
oil toxicosis in 311
sarcocystis in 221
Penguin
aspergillosis in 130
avian cholera in 79
erysipelas in 121
hemosporidiosis in 195
oil toxicosis in 311
salmonellosis in 103–104
Penis, prolapsed 146–147
Pentobarbital poisoning, see Barbiturate poisoning
Peregrine falcon
avian pox in 165
capture myopathy in 364
chlorinated hydrocarbon poisoning in 296, 301
herpesvirus disease in 159–160
lead poisoning in 320
mercury poisoning in 339
mycoplasmosis in 115
Pericarditis 113, 125, 185
Perihepatitis 125
Peritonitis 225, 227
verminous, see Eustrongylidiosis
Permethrin 285
Pesticides 4, 284–285, 287–293, 351
Petrel, tick paralysis in 258
Petroleum, see Oil toxicosis
Pfiesteria, in fish 265
Pfiesteria toxin 264
Pheasant
avian botulism in 274–275
avian cholera in 79
avian influenza in 181, 183
candidiasis in 135
chlorinated hydrocarbon poisoning in 297, 299, 301
duck hepatitis in 361

Pintail
  eastern equine encephalomyelitis in 171–173
  erysipelas in 121
  gizzard worms in 235
  herpesvirus disease in 157–158
  histomoniasis in 257
  intestinal coccidiosis in 210
  lead poisoning in 317, 319, 332
  mercury poisoning in 339–340
  mycoplasmosis in 115, 117–118
  Newcastle disease in 175
  new duck disease in 122
  salmonellosis in 100, 103
  salt poisoning in 347
  tracheal worms in 229, 231
  trichomoniasis in 203
  tuberculosis in 93–94, 96
  tularemia in 124
  ulcerative enteritis in 123
Phenoclors® 305
Phenol 285
Phorate 287, 289, 292, 352
Phosmet 285
Phosphamidon 292
Photographic record 3–4, 8
Phycotoxins, see Algal toxins
Physical euthanasia 49–50
Physical restraint 58–59
Pigeon
  avian cholera in 79
  candidiasis in 135
  chlamydiosis in 111–112
  chlorinated hydrocarbon poisoning in 301
  eastern equine encephalomyelitis in 171
  erysipelas in 121
  gizzard worms in 236
  hemosporidiosis in 194
  herpesvirus disease in 157–159
  intestinal coccidiosis in 210
  lead poisoning in 332
  mercury poisoning in 339
  mycoplasmosis in 115–116
  nematodes in 255
  Newcastle disease in 176
  salmonellosis in 103, 107
  trichomoniasis in 201–203, 205–206
  tuberculosis in 94
  ulcerative enteritis in 123
Pigeon herpes encephalomyelitis virus, see Herpesvirus
    disease
Pigs, see Livestock
Piloerection 290
Pine sisken, salmonellosis in 104
Pinnipeds, Pasteurella multocida in 77

  avian botulism in 281
  duck plague in 141
  lead poisoning in 326
Piscicides 285
Pitfall, as live trap 57
Plague, fowl, see Avian influenza
Plant toxins 261
Plasmodium, see Hemosporidiosis
Plastic debris 361–363
Plover, oil toxicosis in 311
Plumbism, see Lead poisoning
Pneumonia
  brooder, see Aspergillosis
  mycotic, see Aspergillosis
Pneumonitis, Louisiana, see Chlamydiosis
Pododermatitis, see Bumblefoot
Poison Control Centers 371
Poisonous organism 260
Polychlorinated biphenyls 284, 303–307
Population movement, documentation of 5
Posture, aberrant 299
Potassium bromide 285
Pouch scratching 265
Poultry
  avian botulism in 274
  avian cholera in 77–78, 80, 82, 89
  avian influenza in 183–184
  candidiasis in 135
  erysipelas in 121
  fusariotoxin poisoning in 269
  herpesvirus disease in 158
  intestinal coccidiosis in 210
  mycoplasmosis in 115, 117–118
  nematodes in 254
  Newcastle disease in 175–179
  salmonellosis in 99–100, 103–105, 107, 109
  subcutaneous mycosis in 137
  tracheal worms in 229
  trichomoniasis in 203, 206
  tuberculosis in 94
  tumor in 364
Poultry slaughterhouse 114
Power lines 280, 357–360
Power poles 357–360
Pox, avian, see Avian pox
Prairie falcon
  herpesvirus disease in 159
  lead poisoning in 320
Problem area
  closure of 33–34
  description of location of 5–6
Propane exploder 32, 34
Protective clothing 7, 21, 24
  disinfection of 33, 38

Protein imbalance 367
Protozoan parasites 189, 256–257
Proventriculus
   cheesy, raised plaques along 146, 150
   impaction of 320, 328–329
   nematodes in 255–256
   raised tunnels on 223, 225
Prussic acid poisoning, see Cyanide poisoning
Pseudotuberculosis, see Aspergillosis
Psittacine herpesvirus, see Herpesvirus disease
Psittacosis, see Chlamydiosis
Ptarmigan, tularemia in 124
Puddle ducks
   avian cholera in 81
   gizzard worms in 237
   lead poisoning in 318, 332
   nasal leeches in 245
   sarcocystis in 221
Puffin
   avian influenza in 183
   oil toxicosis in 311
   renal coccidiosis in 216
Pullorum disease, see Salmonellosis
Punkies 194
Pupils, contraction of 290
Purple finch, salmonellosis in 104
Pyralenes® 305

Q

Quail
   avian cholera in 80
   avian influenza in 181
   avian pox in 164
   candidiasis in 135
   chemical poisoning in 353
   chlorinated hydrocarbon poisoning in 296
   gizzard worms in 235
   herpesvirus disease in 158
   intestinal coccidiosis in 210
   lead poisoning in 319
   mercury poisoning in 338
   mycoplasmosis in 115
   nematodes in 255
   salmonellosis in 100, 103
   trichomoniasis in 203
   tuberculosis in 93–94
   ulcerative enteritis in 122–123
Quail disease, see Ulcerative enteritis

R

Rabbit
   Mycobacterium avium in 93
   Pasteurella multocida in 77
Raccoon, Pasteurella multocida in 78

Rail
   avian cholera in 79
   intestinal coccidiosis in 210
   lead poisoning in 319
Raptors
   aspergillosis in 130
   avian botulism in 274
   avian cholera in 80, 89
   avian influenza in 181
   avian pox in 163, 166
   chlamydiosis in 112
   chlorinated hydrocarbon poisoning in 295, 298
   eastern equine encephalomyelitis in 171
   electrocution of 357
   gizzard worms in 237
   herpesvirus disease in 158–159
   lead poisoning in 317, 319, 332
   mercury poisoning in 338–339
   mycoplasmosis in 119
   pesticide poisoning in 287–288, 291
   sarcocystis in 221
   traumatic injuries in 365
   trichomoniasis in 202–203
   tuberculosis in 94, 97
   tularemia in 124
Ratites, avian influenza in 181
Raven
   aspergillosis in 130
   erysipelas in 121
Razorbill, salmonellosis in 103
Red blood cells, in hemosporidiosis 198–199
Red-breasted merganser
   eustrongylidiosis in 226
   lead poisoning in 321
   mercury poisoning in 338
Red-crowned crane, inclusion body disease of cranes in 154
Red grouse, nematodes in 255
Redhead
   avian cholera in 81
   avian pox in 164
   duck plague in 141
   lead poisoning in 317, 321
   nasal leeches in 245–246
   staphylococcosis in 123
Red-headed falcon, herpesvirus disease in 159
Red-tailed hawk
   avian pox in 165
   lead poisoning in 320
   mercury poisoning in 340
   staphylococcosis in 123
Red-tailed tropicbird, avian pox in 164–165
Red-throated loon, mercury poisoning in 338
Red tide 261, 265

Red tide toxins, see Algal toxins
Red-winged blackbird
    pesticide poisoning in 287
    polychlorinated biphenyl poisoning in 305
Regurgitation of food 223
Rehabilitation, of oiled birds 314–315
Renal coccidiosis 215–216
Rendering of carcasses 21, 32
Reovirus, woodcock 140, 185–186
Repellents 353
Reproductive impairment 290, 309–310, 335
Reptiles
    Eustrongylides in 223–224
    Salmonella in 99
Research, see Field research
Respiratory distress 118, 131, 157, 183, 229
Respiratory syndromes 125, 185
Respiratory tract, moist, necrotic lesions of 165
Restraint 58–60
    for blood and tissue collection from live animals 57
    chemical 58, 61
    physical 58–59
Rhea, avian influenza in 181
Rice breast disease, see Sarcocystis
Rictin poisoning 356
Righting reflex, loss of 265
Ring-billed gull
    chlamydiosis in 113
    Newcastle disease in 175
    staphylococcosis in 123
Ring-dove
    polychlorinated biphenyl poisoning in 303
    salmonellosis in 100
Ringed turtle dove, herpesvirus disease in 159
Ring-necked duck, nasal leeches in 245
Ring-necked pheasant
    eastern equine encephalomyelitis in 172
    lead poisoning in 320
    mercury poisoning in 338
    mycoplasmosis in 118
    pesticide poisoning in 287
    polychlorinated biphenyl poisoning in 305
    tracheal worms in 231
Ringworm 137
Road kill 365
Road salt 347–348
Robin
    chlorinated hydrocarbon poisoning in 296
    gizzard worms in 235
    salmonellosis in 104
    ulcerative enteritis in 123
Rock dove
    lead poisoning in 323
    salmonellosis in 100
    salt poisoning in 347
Rocket net 33, 35
Rodent(s)
    eastern equine encephalomyelitis virus in 172
    Pasteurella multocida in 77–78
    Salmonella in 99, 107, 109
Rodenticides 285, 353
Rook
    salmonellosis in 100, 103
    tracheal worms in 229
Rosella, candidiasis in 135
Ross' goose, lead poisoning in 318
Rotenone 285
Rough-legged hawk
    avian pox in 165
    lead poisoning in 320
Roundworm, see Nematodes
Royal tern, lead poisoning in 320–321
Ruddy duck, nasal leeches in 245
Ruddy turnstone, avian influenza in 183
Ruffed grouse
    avian pox in 169
    gizzard worms in 235–236
    tularemia in 124
    tumor in 365
Rufous-sided towhee, salmonellosis in 104

S

Sage grouse, tularemia in 124
Saker falcon, mycoplasmosis in 117–118
Salmonellosis 74, 99–109
Salt encrustation 347–348
Salt gland 347
    disruption of 300
    impaired functioning of 310
    swollen 312–313
Salt poisoning 284, 347–348, 353
Sandfly 193
Sandhill crane
    aflatoxin poisoning in 267
    chlorinated hydrocarbon poisoning in 297
    fusariotoxin poisoning in 269–270
    inclusion body disease of cranes in 153–154
    intestinal coccidiosis in 207–208
    lead poisoning in 321
    tracheal worms in 229
    tumor in 365
Sandpiper, oil toxicosis in 311
Sarcocystis 188, 219–222
Sarcocystis rileyi, see Sarcocystis
Sarcocystosis, see Sarcocystis
Sarconema, see Heartworm of swans and geese
Sarcosporidiosis, see Sarcocystis
Sarus crane, inclusion body disease of cranes in 153

Saxitoxin, see also Paralytic shellfish poisoning 264–265
Scaled quail, lead poisoning in 320, 333
Scare devices 32, 34
Scaup
   aspergillosis in 130
   lead poisoning in 317
   oil toxicosis in 311
Scavenger species
   avian cholera in 78, 89
   barbiturate poisoning in 349–350
   cyanide poisoning in 342
   lead poisoning in 332
   Pasteurella multocida in 75
   tuberculosis in 94
   tularemia in 124
Scientific names of wildlife 377–381
Scoliosis 355, 367
Scoter
   avian cholera in 81
   avian pox in 163
   oil toxicosis in 311
Seabirds
   mercury poisoning in 338–339
   oil toxicosis in 311
Sea ducks
   avian cholera in 82
   gizzard worms in 237
   lead poisoning in 318
   nasal leeches in 245
   oil toxicosis in 311
   sarcocystis in 221
   tuberculosis in 96
Seasonality
   of acanthocephaliasis 242
   of aflatoxin poisoning 267–269
   of aspergillosis 130
   of avian botulism 271, 276
   of avian cholera 82
   of avian influenza 183
   of avian pox 164–166
   of barbiturate poisoning 349
   of chlamydiosis 112
   of chlorinated hydrocarbon poisoning 295–298
   of cyanide poisoning 342
   of duck plague 144–145
   of eastern equine encephalomyelitis 172
   of electrocutions 358
   of eustrongylidiosis 223
   of fusariotoxin poisoning 269
   of heartworm of swans and geese 234
   of hemosporidiosis 195
   of herpesvirus disease 157
   of intestinal coccidiosis 208
   of lead poisoning 318

   of mercury poisoning 339
   of mycoplasmosis 118
   of nasal leeches 245–246
   of Newcastle disease 177
   of oil toxicosis 311
   of pesticide poisoning 290
   of renal coccidiosis 217
   of salmonellosis 104–105
   of salt poisoning 347
   of sarcocystis 220
   of tracheal worms 229
   of tuberculosis 94–95
   of woodcock reovirus disease 185
Sedation 58, 349
Seed-eating birds, pesticide poisoning in 288
Seed treatment 295, 337, 339
Seizures, see Convulsions
Selenium poisoning 284–285, 335–336, 353, 367
Selenosis, see Selenium poisoning
Serovars, Salmonella 99
Sewage effluent 107
Sewage lagoon 101
Sewage sludge 99
Sewage treatment plant 108
Sewage wastewater 107
Shag
   algal toxin poisoning in 264
   Newcastle disease in 177
Sharp-tailed grouse
   chlorinated hydrocarbon poisoning in 297, 301
   tularemia in 124
Shearwater
   mercury poisoning in 338
   renal coccidiosis in 216
Sheep, see Livestock
Shipment of live animals 64–66
Shipment of specimen 13–17
   basic supplies for 13, 373–374
   blood tubes 14–15
   by commercial carrier 17
   containers for 13–16
   cooling and refrigeration 14–16
   federal regulations for 17
   labeling requirements for 17
   to National Wildlife Health Center 14–16
   preventing breakage and leakage 13
Shivering 105
Shorebirds
   aspergillosis in 130
   avian botulism in 274
   avian cholera in 79–80
   avian influenza in 181, 183–184
   avian pox in 163
   candidiasis in 135

    chemical poisoning in 353
    chlamydiosis in 111–112
    eastern equine encephalomyelitis in 171–172
    gizzard worms in 237
    intestinal coccidiosis in 210
    lead poisoning in 319, 332
    necrotic enteritis in 122
    organophosphate poisoning in 352
    pesticide poisoning in 288
    sarcocystis in 221
Short-eared owl, avian cholera in 78
Shoveler
    avian botulism in 281
    chlorinated hydrocarbon poisoning in 296
    lead poisoning in 318
    mycoplasmosis in 117
Shrike, tularemia in 124
Silver mining 341–345
Sinusitis, infectious, see Mycoplasmosis
SI units, conversion table 385–386
Six-pack ring 363
Skin
    burn marks on 359–360
    nodular tubercular lesions on 95–96
    wart-like nodules on featherless areas 165, 167
Skua
    avian cholera in 79
    salmonellosis in 104
Skull, fluid accumulation in 335
Skunk 219
Sleepaway® 51
Slug, tracheal worm carriage 229–230
Slurry runoff 99
Snails
    tracheal worm carriage 229–230
    as trematode hosts 249–250, 252
Snake, acanthocephalans in 242
Sneezing 231, 245
Snow goose
    aflatoxin poisoning in 267
    avian cholera in 75, 77, 79, 81, 91
    chlamydiosis in 114
    chlorinated hydrocarbon poisoning in 296
    gizzard worms in 236, 239
    heartworm of swans and geese in 234
    lead poisoning in 317–318, 324, 328
    necrotic enteritis in 122
    new duck disease in 122
Snowy egret
    chlamydiosis in 112, 114
    eustrongylidiosis in 226–227
    lead poisoning in 320
Snowy owl, herpesvirus disease in 159
Sodium pentobarbital poisoning, see Barbiturate poisoning

Songbirds
    algal toxin poisoning in 264
    aspergillosis in 130
    avian botulism in 274–275
    avian cholera in 80
    avian influenza in 181
    avian pox in 163, 166, 169
    candidiasis in 135
    carbamate poisoning in 351
    chemical poisoning in 353
    chlamydiosis in 111–112
    chlorinated hydrocarbon poisoning in 298, 352
    eastern equine encephalomyelitis in 171–173
    erysipelas in 121
    herpesvirus disease in 158
    intestinal coccidiosis in 210
    mycoplasmosis in 115–116
    organophosphate poisoning in 352
    rodenticide and repellent poisoning in 353
    salmonellosis in 103–107
    tick paralysis in 257
    tracheal worms in 229
    trichomoniasis in 203
    tuberculosis in 94
Sooty shearwater, algal toxin poisoning in 264
Sora rail, lead poisoning in 320
Sour crop, see Candidiasis
Sparrow
    avian cholera in 79–80
    candidiasis in 135
    salmonellosis in 100, 107
    tuberculosis in 93
Sparrowhawk, mercury poisoning in 339
Specimen
    choice of 7–9
    collection of 7–12
    basic supplies and equipment for 8, 373–374
    killed specimens for catalogued collections/museums 56–58
    dissection of bird 10–11
    labeling of 8, 12
    preservation of 7–12, 373–374
    shipment of, see Shipment of specimen
Specimen history 3–6
Specimen history form 369
Spectacled eider, lead poisoning in 317
Sphaeridiotrema globulus 249, 251–252
Spleen
    amyloid deposits in 96
    atrophy of 305–306
    congested 123
    necrotic lesions of 160
    nodules in 95
    swollen 154

yellow-white spots on 154–155
Splenomegaly 112–113, 172, 195–196
Spoonbill, eustrongylidiosis in 225
Spore, Clostridium botulinum 271
Squeeze chute 59
Staggering 105
Stanley crane, inclusion body disease of cranes in 153
Staphylococcosis 123–124
    bumblefoot 123
    septicemic 123–124
Starlicide® 285
Starling
    avian cholera in 79–80
    candidiasis in 135
    chlorinated hydrocarbon poisoning in 296
    erysipelas in 121
    mercury poisoning in 340
    salmonellosis in 103
    tracheal worms in 229
    tuberculosis in 93
Starvation 309
Station brochures 19
Stellar's jay, aspergillosis in 130
Sternum, light-colored granulomas of 211–212
Stilt, avian cholera in 79
Stomach worm, see Gizzard worms
Stork
    erysipelas in 121
    herpesvirus disease in 158–159
Strychnine 351
Stunning and exsanguination 50
Styrofoam coolers, for shipment of specimens 13–15
Subcutaneous fat, loss of 319–320, 326
Suffocation 205
Surf scoter, nasal leeches in 245
Surgical procedures, wildlife 65–67
Swallow
    chemical poisoning in 353
    pesticide poisoning in 288
Swallowing difficulty 205, 223
Swan
    acanthocephaliasis in 242
    avian cholera in 79
    duck plague in 141, 151
    erysipelas in 121
    gizzard worms in 236–237
    heartworm of swans and geese in 233–234
    hemosporidiosis in 194
    lead poisoning in 317–318, 333
    nasal leeches in 245, 248
    oil toxicosis in 311
    renal coccidiosis in 216
    salmonellosis in 100, 103
    sarcocystis in 221

    tracheal worms in 229
    trematodes in 249
Swift, avian cholera in 79
Swim-in trap 33
Swimming, circular 83, 85, 148
Syngamiasis, see Tracheal worms
Syngamus trachea, see Tracheal worms
Systox® 285

T

T-2 toxicosis, see Fusariotoxin poisoning
Tailings ponds 342
Tail vein, blood collection from live animals 59
Talon (claw), vegetation clenched in 290–291
Talon® (pesticide) 285
Tapeworms, see Cestodes
Tarsal vein, blood collection from live animals 59
TB, see Tuberculosis
Teal
    lead poisoning in 318
    nasal leeches in 245
    sarcocystis in 220
Tears, excessive 290
Temephos 287
Temik 285
Tenosynovitis 185
Terbufos 285, 292, 352
Tern
    avian cholera in 91
    carbamate poisoning in 351
    chemical poisoning in 353
    chlamydiosis in 111–112
    erysipelas in 121
    gizzard worms in 237
    mercury poisoning in 338
    salmonellosis in 99–100, 103–104
    sarcocystis in 221
    tularemia in 124
Theromyzon, see Nasal leeches
Thiram 285
Thirst 146, 290
Thorny-headed worms, see Acanthocephaliasis
Thrasher, avian cholera in 79
Throat, inflammation of 205
Thrush (bird)
    avian cholera in 79
    pesticide poisoning in 288
Thrush (disease), see Candidiasis
Thymus gland
    atrophy of 185
    hemorrhage in 154
Tick 69, 124, 257–258
Tick paralysis 257–258
Tissue sample

collection of 7–9
collection from live animals 57–58
containers for 8–9
fixation of 8
labeling of 12
Toes, clenched 177, 265
Tongue, ulcerative "cold sore" under 146–148
Toxaphene 296–297
Toxic gas, euthanasia with 50–51
Tracheal worms 229–231
Tranquilizer gun 58
Transformers 357–360
Transmission lines 357–360
Transport host 188, 192
Transport myopathy, see Capture myopathy
Transport of live animals 64–65
confinement during shipping 65–66
Trapping of wildlife 33, 35
Traumatic injury 290, 356, 365–366
Tree sparrow, salmonellosis in 104
Trematodes 189, 249–253
Tremors 105, 172, 177, 208, 265, 267, 290, 298, 300, 305–306, 339
Trichomonas gallinae, see Trichomoniasis
Trichomoniasis 188–189, 201–206
Trichophyton gallinae, see Ringworm
Trichostrongylidosis 254–255
Trichostrongylus tenuis, see Trichostrongylidosis
Trichothecene 260, 269
Trichothecene mycotoxicosis, see Fusariotoxin poisoning
Triumph® 285
Trumpeter swan
acanthocephaliasis in 242
avian pox in 163–164
heartworm of swans and geese in 234
lead poisoning in 321
nasal leeches in 245, 247
Tuberculosis 93–98
Tularemia 124
Tumbling 107
Tumors 356, 364–365
herpesvirus-related 157
Tundra swan
aspergillosis in 130
avian pox in 163–164
heartworm of swans and geese in 234
lead poisoning in 317–318, 321, 324, 328
nasal leeches in 245
Turkey
avian cholera in 82
avian influenza in 181, 183
avian pox in 165
candidiasis in 135
chlamydiosis in 111

erysipelas in 121
hemosporidiosis in 197–198
herpesvirus disease in 158
histomoniasis in 257
intestinal coccidiosis in 210
lead poisoning in 319
mercury poisoning in 339
mycoplasmosis in 115–116
nematodes in 255
Newcastle disease in 176
new duck disease in 122
salmonellosis in 99–100
ticks on 257
tracheal worms in 229
trichomoniasis in 203
tuberculosis in 93–94
wild, see Wild turkey
Turkey vulture
lead poisoning in 320
mercury poisoning in 339
Turtles, Salmonella in 99
Typhoid, fowl, see Salmonellosis

U
Ulcerative enteritis 74, 122–123
Upland gamebirds
aspergillosis in 130
avian botulism in 274
avian cholera in 80
avian influenza in 181, 183
avian pox in 163, 166, 169
chlamydiosis in 111–112
eastern equine encephalomyelitis in 171–172
lead poisoning in 319, 333
mycoplasmosis in 115, 119
pesticide poisoning in 288
ulcerative enteritis in 123
Uric acid salts 217, 348

V
Venipuncture 57–58
Venomous organism 260
Vent
bloodstained 249
bloody discharge from 146
bright green staining of 327
pasted 105
Ventricular nematodiasis, see Gizzard worms
Ventriculus, raised tunnels on 223, 225
Verminous peritonitis, see Eustrongylidiosis
Viral disease 140
Visceral gout 172–173
Vision, impaired 118
Vitamin deficiency 367

Voice changes 172
Vomiting 265, 269, 290
Vomitoxicosis, see Fusariotoxin poisoning
Vomitoxin 269
Vulture
   avian botulism in 274
   avian cholera in 78
   pesticide poisoning in 288
   tuberculosis in 94
VVND, see Newcastle disease

W

Wading birds
   avian cholera in 79–80
   avian pox in 163
   eustrongylidiosis in 223, 225, 227
   gizzard worms in 237
   mercury poisoning in 338
   necrotic enteritis in 122
   oil toxicosis in 311
   pesticide poisoning in 288
   sarcocystis in 220–221
Walk-in trap 33
Warfarin 285
Wariness, loss of 146
Warty nodules, on featherless areas 165, 167
Wastewater discharge 97, 101, 107, 225, 280, 309, 335
Wastewater pits, oil-field 312
Wastewater sites 94
Watercraft
   for carcass collection 21, 26
   disinfection of 33
   transport of live animals in 64–65
Water deprivation, see Salt poisoning
Water drawdown 279
Waterfowl
   aspergillosis in 130
   avian botulism in 274–275
   avian cholera in 75, 78, 80, 91
   avian influenza in 181, 183–184
   avian pox in 163, 169
   capture myopathy in 364
   carbamate poisoning in 351
   chemical poisoning in 353
   chlamydiosis in 111–112
   chlorinated hydrocarbon poisoning in 296, 298, 301
   eastern equine encephalomyelitis in 171–172
   fusariotoxin poisoning in 269
   gizzard worms in 235
   hemosporidiosis in 194
   herpesvirus disease in 158–159
   intestinal coccidiosis in 208
   lead poisoning in 317, 321–322, 332
   mycoplasmosis in 116
   nasal leeches in 245
   necrotic enteritis in 122
   new duck disease in 122
   organophosphate poisoning in 352
   pesticide poisoning in 287–288, 293, 351
   renal coccidiosis in 216
   rodenticide and repellent poisoning in 353
   sarcocystis in 220
   tracheal worms in 229
   trematodes in 249
   trichomoniasis in 203
   tuberculosis in 94
Water quality 4
Waxwing, avian cholera in 79
Weakness 112, 195, 220, 237, 269, 290, 339
Weather 356, 365–367
Weaver, herpesvirus disease in 159
Weight loss 112, 118, 195, 203, 208, 231, 250, 257, 268, 290, 306
Wenyonella, see Intestinal coccidiosis
Western duck sickness, see Avian botulism
Western grebe, chlorinated hydrocarbon poisoning in 296
Western sandpiper, lead poisoning in 320
Wet pox 165, 168–169
Whirl-Pak bags 8–9
Whistling duck, lead poisoning in 318
White-faced ibis, chlorinated hydrocarbon poisoning in 299
White-fronted goose
   avian cholera in 80
   heartworm of swans and geese in 234
   necrotic enteritis in 122
White ibis, lead poisoning in 320–321
White-naped crane, inclusion body disease of cranes in 153
White Pekin duck
   algal toxin poisoning in 265
   duck plague in 141, 143–144, 151
   eastern equine encephalomyelitis in 171
   inclusion body disease of cranes in 153–154
   trematodes in 250
White pelican
   entangled in fishing line 362
   lead poisoning in 320–321
   mercury poisoning in 338
   Newcastle disease in 175
   polychlorinated biphenyl poisoning in 303
White-tailed eagle, mercury poisoning in 339
White-tailed tropicbird, avian pox in 169
White-winged dove, chlamydiosis in 114
White-winged scoter, nasal leeches in 245
Whooper swan, lead poisoning in 321
Whooping crane
   avian cholera in 91
   capture myopathy in 60

eastern equine encephalomyelitis in 171–173
  intestinal coccidiosis in 207
  traumatic injuries in 366
  tuberculosis in 97
Wildlife, in field research 53–71
Wildlife disease programs 370
Wildlife marking 44, 61–63
Wildlife Society, guidelines for use of wildlife in field research 53–71
Wild turkey
  avian pox in 164–165
  capture myopathy in 364
  gizzard worms in 235
  hemosporidiosis in 194
  herpesvirus disease in 159
  histomoniasis in 257
  lead poisoning in 320
  mycoplasmosis in 115
  tracheal worms in 229
Wing(s)
  flapping of 267
  roof-shaped 319, 324
  violent beating of 300
Wing droop 105, 131, 177–179, 257, 269, 319, 321, 324
Woodcock
  lead poisoning in 319
  renal coccidiosis in 216
  woodcock reovirus in 185–186
Woodcock reovirus 140, 185–186
Wood duck
  aspergillosis in 131
  avian pox in 164
  duck plague in 141
  lead poisoning in 318, 321
Woodpecker
  avian cholera in 79
  intestinal coccidiosis in 210
Wood pigeon
  candidiasis in 135
  erysipelas in 121
  salmonellosis in 100
  tuberculosis in 95
Written history 3

Y
Yellowleg, avian cholera in 79
Yellow-naped Amazon parrot, mycoplasmosis in 115
Young birds, disease in 361

Z
Zearalenone 269
Zearalenone toxicosis, see Fusariotoxin poisoning
Zectran® 285
Zinc 285
Ziram 285
Zooplankton bloom 255–256

REPORT DOCUMENTATION PAGE		*Form Approved* *OMB No. 0704-0188*

Public reporting burden for this collection of information is estimated to average 1 hour per response, including the time for reviewing instructions, searching existing data sources, gathering and maintaining the data needed and completing and reviewing the collection of information. Send comments regarding this burden estimate or any other asses of this collection of information, including suggestions for reducing this burden, to Washington Headquarters Services, Directorate for Information Operations and Reports, 1215 Jefferson Davis Highway, Suite 1204, Arlington, VA 22202-4302, and to the Office of Management and Budget, Paperwork Reduction Project (0704-0188), Washington, DC 20503.

1. AGENCY USE ONLY (Leave Blank)	2. REPORT DATE 1999	3. REPORT TYPE AND DATES COVERED Final.
4. TITLE AND SUBTITLE Field Manual of Wildlife Diseases, General Field Procedures and Diseases of Birds		**5. FUNDING NUMBERS** 99/00-3204-40A40
6. AUTHOR(S) Milton Friend and J. Christian Franson, Technical Editors		
7. PERFORMING ORGANIZATION NAME(S) AND ADDRESS(ES) U.S. Geological Survey, Biological Resources Division, National Wildlife Health Center, 6006 Schroeder Rd., Madison, WI 53711		**8. PERFORMING ORGANIZATION REPORT NUMBER** ITR 1999-001
9. SPONSORING/MONITORING AGENCY NAMES(S) AND ADDRESS(ES) U.S. Geological Survey, 12201 Sunrise Valley Dr., Reston, VA 20192		**10. SPONSORING/MONITORING AGENCY REPORT NUMBER** ITR 1999-001

11. SUPPLEMENTARY NOTES
Supersedes U.S. Fish and Wildlife Service Resource Publication 167 (1987).

12a. DISTRIBUTION/AVAILABILITY STATEMENT	12b. DISTRIBUTION CODE

13. ABSTRACT (Maximum 200 words).
The "Field Manual of Wildlife Diseases, General Field Procedures and Diseases of Birds" presents practical, current information and insights about wild bird illnesses and the procedures to follow when ill birds are found or epidemics occur. Section 1 of the Manual provides information about general field procedures. Sections 2 through 5 describe various bird diseases. Sections 6 and 7 provide information about toxins that affect birds, and Section 8 describes miscellaneous diseases and hazards that affect birds. Manual lists institutions and laboratories that offer diagnostic services; sources of supplies for collecting, preserving, and shipping specimens; and it contains color illustrations for dissection.

14. SUBJECT TERMS Bird diseases; wildlife diseases; bacterial diseases; fungal diseases; viral diseases; parasitic diseases; biotoxins; chemical toxins; general field procedures.			15. NUMBER OF PAGES 440
			16. PRICE CODE
17. SECURITY CLASSIFICATION OF REPORT Unclassified	18. SECURITY CLASSIFICATION OF THIS PAGE	19. SECURITY CLASSIFICATION OF ABSTRACT	20. LIMITATION OF ABSTRACT

www.ingramcontent.com/pod-product-compliance
Lightning Source LLC
Chambersburg PA
CBHW081233180526
45171CB00005B/409